THE FLUID MECHANICS OF ASTROPHYSICS AND GEOPHYSICS

Mathematical Aspects of Natural Dynamos

SELECTED BY GRENOBLE SCIENCES

edited by

Emmanuel Dormy
& Andrew M. Soward

GRENOBLE SCIENCES
UNIVERSITE JOSEPH FOURIER

CRC Press
Taylor & Francis Group
Boca Raton London New York

CRC Press is an imprint of the
Taylor & Francis Group, an informa business

The Fluid Mechanics of Astrophysics and Geophysics

A series edited by
Andrew Soward
University of Exeter UK

Michael Ghil
University of California, Los Angeles, USA

Founding Editor: Paul Roberts
University of California, Los Angeles, USA

Volume 1
Solar Flare Magnetohydrodynamics
Edited by E. R. Priest

Volume 2
Stellar and Planetary Magnetism
Edited by A. M. Soward

Volume 3
Magnetic Fields in Astrophysics
Ya. B. Zeldovitch, A. A. Ruzmaikin and D. D. Sokoloff

Volume 4
Mantle Convection
Plate tectonics and global dynamics
Edited by W. R. Peltier

Volume 5
Differential Rotation and Stellar Convection
Sun and solar-type stars
G. Rüdiger

Volume 6
Turbulence, Current Sheets and Shocks in Cosmic Plasma
Edited by S. I. Vainshtein, A. M. Bykov and I. N. Toptygin

Volume 7
Earth's Deep Interior
The Doornbos memorial volume
Edited by D. J. Crossley

Volume 8
Geophysical and Astrophysical Convection
Edited by P. A. Fox and R. M. Kerr

Volume 9
Advances in Nonlinear Dynamos
Edited by A. Ferriz-Mas and M. Núñez

Volume 10
Magnetohydrodynamics and The Earth's Core
Selected works of Paul Roberts
Edited by A. M. Soward

Volume 11
Earth's Core and Lower Mantle
Edited by C. A. Jones, A. M. Soward and K. Zhang

Volume 12
Fluid Dynamics and Dynamos in Astrophysics and Geophysics
Edited by A. M. Soward, C. A. Jones, D. W. Hughes and N. O. Weiss

Volume 13
Mathematical Aspects of Natural Dynamos
Edited by E. Dormy and A. M. Soward

This edition published by Taylor & Francis under license.

Chapman & Hall/CRC
Taylor & Francis Group
6000 Broken Sound Parkway NW, Suite 300
Boca Raton, FL 33487-2742

First issued in paperback 2019

ISBN-13: 978-1-58488-954-0 (hbk)
ISBN-13: 978-0-367-37546-1 (pbk)

Library of Congress Cataloging-in-Publication Data

Dormy, Emmanuel.
Mathematical aspects of natural dynamos / Emmanuel Dormy and Andrew M. Soward.
p. cm. -- (The fluid mechanics of astrophysics and geophysics ; 13)
Includes bibliographical references and index.
ISBN-13: 978-1-58488-954-0 (alk. paper)
ISBN-10: 1-58488-954-3 (alk. paper)
1. Dynamo theory (Cosmic physics)--Mathematical models. 2. Fluid dynamics--Mathematical models. 3. Geodynamics--Mathematical models. I. Soward, Andrew M. II. Title. III. Series.

QC809.M25D67 2007
523.01'887--dc22 2007013027

Visit the Taylor & Francis Web site at
http://www.taylorandfrancis.com

and the CRC Press Web site at
http://www.crcpress.com

Grenoble Sciences

Grenoble Sciences pursues a triple aim:

- to publish works responding to a clearly defined project, with no curriculum or vogue constraints,
- to guarantee the selected titles' scientific and pedagogical qualities,
- to propose books at an affordable price to the widest scope of readers.

Each project is selected with the help of anonymous referees, followed by an average one-year interaction between the authors and a Readership Committee. Publication is then confided to the most adequate publishing company by Grenoble Sciences.

(Contact: Tel.: (33) 4 76 51 46 95 – Fax: (33) 4 76 51 45 79
E-mail: Grenoble.Sciences@ujf-grenoble.fr)

Scientific Director of Grenoble Sciences: Jean BORNAREL,
Professor at the Joseph Fourier University, France

Grenoble Sciences is supported by the French Ministry of Education and Research and the "Région Rhône-Alpes".

Front cover illustration composed by Alice GIRAUD *with following pictures*
Galaxies NGC 891 and M 51:
courtesy of the Canada-France-Hawaii Telescope/J.C.Cuillandre/Coelum;
Earth and Jupiter: NASA (NSSDC & JSC) (Apollo 17, HST and Voyager);
Sun: SOHO (ESA & NASA)

CONTENTS

CONTRIBUTORS

Daniel Brito
Lab. Geophysique Interne et Tectonophysique
Batiment IRIGM / U. Joseph Fourier, BP 53
F–38041 Grenoble Cedex 9, France

Friedrich H. Busse
Institute of Physics
University of Bayreuth
D–95440 Bayreuth, Germany

Philippe Cardin
Lab. Geophysique Interne et Tectonophysique
Batiment IRIGM / U. Joseph Fourier, BP 53
F–38041 Grenoble Cedex 9, France

Benoît Desjardins
C.E.A./D.I.F.
B.P. 12
F–91680 Bruyères le Châtel, France

Emmanuel Dormy
MAG (IPGP–ENS)
Département de Physique, LRA
Ecole Normale Supérieure
24 rue Lhomond
F–75231 Paris Cedex 05, France

Stephan Fauve
Laboratoire de Physique Statistique
Département de Physique
Ecole Normale Supérieure
24 rue Lhomond
F–75231 Paris Cedex 05, France

David R. Fearn
Department of Mathematics
University of Glasgow
University Gardens
Glasgow G12 8QW, U.K.

Andrew D. Gilbert
Centre for GAFD
Mathematics Research Institute
University of Exeter
Harrison Building, North Park Road
Exeter, EX4 4QE, U.K.

Raymond Hide
Department of Mathematics
Imperial College London
London SWT 2BZ, U.K.

David W. Hughes
Department of Applied Mathematics
University of Leeds
Leeds LS2 9JT, U.K.

Christopher A. Jones
Department of Applied Mathematics
University of Leeds
Leeds, LS2 9JT, U.K.

Irene Moroz
Mathematical Institute
University of Oxford
24–29 St. Giles'
Oxford OX1 3LB, U.K.

François Pétrélis
Laboratoire de Physique Statistique
Ecole Normale Supérieure
24 rue Lhomond
F–75231 Paris Cedex 05, France

Michael R.E. Proctor
Department of Applied Mathematics
and Theoretical Physics
University of Cambridge
Wilberforce Road,
Cambridge CB3 0WA, U.K.

Paul H. Roberts
Department of Mathematics
University of California Los Angeles
Los Angeles, CA 90095–1567, U.S.A.

Anvar Shukurov
Department of Mathematics
University of Newcastle
Newcastle, NE1 7RU, U.K.

Radostin D. Simitev
Institute of Physics
University of Bayreuth
D–95440 Bayreuth, Germany

Andrew M. Soward
Centre for GAFD
Mathematics Research Institute
University of Exeter
Harrison Building, North Park Road
Exeter, EX4 4QE, U.K.

Steven M. Tobias
Department of Applied Mathematics
University of Leeds
Leeds, LS2 9JT, U.K.

Nigel O. Weiss
Department of Applied Mathematics
and Theoretical Physics
University of Cambridge
Wilberforce Road
Cambridge CB3 0WA, U.K.

List of Tables

List of Notations

We use the following notations as a standard. When deviations from this convention are needed, introducing a risk of confusion, this is clearly specified in the text.

PREFACE

The idea behind this book originated at a meeting held in Caramulo in Portugal in September 2003. The participants agreed that, though the field of natural dynamos (planetary, stellar and galactic) was rapidly evolving and attracting the interest of researchers in other branches of fluid mechanics, there was no comprehensive introductory book for researchers or graduate students entering this research area. The organisers therefore decided to take advantage of the broad-based knowledge of the invited lecturers at the conference to assemble a "multi-authored monograph". Despite the obvious contradiction in this phrase, it does reflect the spirit in which this book was prepared. While each section of the book was written by specialists in different aspects of this subject, a concerted effort has been made to provide a unified presentation, which develops concepts in a coherent order and, where feasible, uses consistent notation.

The first part of the book is devoted to the theoretical background that forms the foundation of dynamo theory and which is necessary to describe and understand natural dynamos. The first chapter introduces the governing equations and outlines kinematic dynamo theory. Although linear equations are often considered as simple, the reader will see how even the kinematic theory of dynamo action can raise subtle issues. The second chapter turns to nonlinear effects. These include amplitude saturation, but also intricate dynamics such as polarity reversals. Because of angular momentum conservation, most natural objects are rapidly rotating. This induces very specific effects in their relevant fluid dynamics. These are discussed in the third and last chapter of this part.

In the second part of the book, we turn our attention to natural dynamos and their modelling. Amongst natural dynamos, the one which we know best is without doubt our own planet, the Earth. We therefore begin the fourth chapter with a description of the Earth's magnetic field and our present understanding of its characteristics. In the following chapter, we turn to the other planets of our solar system. In the sixth chapter we study the magnetic field of stars, including our own star, the Sun. The seventh chapter addresses dynamo action on an even larger scale, that of galaxies.

Finally, we describe experiments conducted over the years which try to model natural dynamos. We conclude with some speculations about future research directions in this rapidly evolving subject.

The understanding of the origin of magnetic fields in astrophysics and geophysics provides a considerable challenge. The authors in this book convey their interest and enthusiasm for their individual field of research. We hope that the reader, whatever his background or research experience, will find that this book has reached our desired objective. That is to bridge the gap between mathematicians, physicists, geophysicists and astrophysicists, each working on natural dynamos using their own specific approaches. We hope that the reader will come to enjoy the complexities of this fascinating area of research as much as the many authors of this book do.

Emmanuel Dormy and Andrew Soward

ACKNOWLEDGEMENTS

The editors are grateful to Dr. Lara Silvers and Dr. Pierre Lesaffre for their help in identifying many typos in a preliminary version of this book. They also wish to thank Dr. Frank Lowes, Dr. Frank Stefani, Dr. Ulrich Müller and Dr. Robert Stieglitz for providing photographs of dynamo experiments to illustrate Chapter 8 (respectively figures 8.1, 8.2, and 8.3) and Sylvie Bordage for her continual help with the preparation of the figures.

ERRATUM

If typos are found in this book, an erratum will be made available at the following website: `http://www.phys.ens.fr/~dormy/MAND`

Part I

Foundations of dynamo theory

CHAPTER 1

INTRODUCTION TO SELF-EXCITED DYNAMO ACTION

Benoît Desjardins, Emmanuel Dormy
Andrew Gilbert & Michael Proctor

The theory of self-excited dynamo action discussed throughout this volume was first suggested by Sir Joseph Larmor in 1919 to account for the magnetic field of sunspots. This concept was later formalised mathematically by Walter Elsasser (1946). The objective of this first chapter is to introduce the subject and provide the necessary background for the later developments. As such, we derive the relevant equations and discuss the usual approximations in Section 1.1, before introducing the concept of a homogeneous self-excited dynamo in Section 1.2. Having dispensed with these preliminaries, the existing theoretical results and necessary conditions for dynamo action are then presented in Section 1.3 and the essential distinction between steady and time-dependent velocities then follows in Section 1.4. In Section 1.5, we then introduce the concept of mean field electromagnetism, which will be a reoccuring topic throughout the book. Finally, in Section 1.6, we discuss the difficult large magnetic Reynolds number limit, which is relevant for astrophysical problems.

1.1. GOVERNING EQUATIONS

1.1.1. MAGNETIC INDUCTION

The common aspect among all natural objects described in this volume is their ability to maintain their own magnetic field. While the magnetic field is maintained, it does vary in both space and time and the equation that governs its evolution is known as the "induction equation", which we will derive below. However, before we continue, it is helpful to note here that we will discuss throughout the book a variety of conducting fluids ranging from molten iron in the Earth's core to ionised gas in stars and galaxies, Despite this variety, the evolution of the magnetic field in all these fluids can be accurately captured by the induction equation.

THE INDUCTION EQUATION

To begin our derivation we need to start from Maxwell's equations which are

$$\boldsymbol{\nabla}\times\mathbf{E} = -\partial_t\mathbf{B}\,, \qquad \boldsymbol{\nabla}\times\mathbf{B} = \mu\,\mathbf{j} + \varepsilon\,\mu\,\partial_t\mathbf{E}\,, \tag{1.1a,b}$$

$$\boldsymbol{\nabla}\cdot\mathbf{B} = 0\,, \qquad \boldsymbol{\nabla}\cdot\mathbf{E} = \rho_c/\varepsilon\,, \tag{1.1c,d}$$

where the following notation $\partial_t\,\cdot \equiv \partial\cdot/\partial t$ has been used. In these equations $\mathbf{B}$ is the magnetic induction (often referred to as the magnetic field), $\mathbf{E}$ is the electric field, $\mathbf{j}$ is the electric current density, ρ_c is the charge density, μ is the magnetic permeability, and ε the dielectric constant. To continue with our derivation, we will assume the free-space value for the magnetic permeability, i.e. $\mu \simeq \mu_0 = 4\pi \times 10^{-7}$ and also $\varepsilon \simeq \varepsilon_0 = (\mu_0 c^2)^{-1}$, then equation (1.1b) can be rewritten as

$$\boldsymbol{\nabla}\times\mathbf{B} = \mu\,\mathbf{j} + c^{-2}\,\partial_t\mathbf{E}\,. \tag{1.2}$$

Now the last term can be neglected provided the typical velocity of the phenomena (i.e. the ratio of a typical length to a typical timescale) is small compared to the speed of light c. This is the case for all the objects dicussed in this book. Therefore we will neglect this term in the rest of the derivation and so

$$\boldsymbol{\nabla}\times\mathbf{B} = \mu\,\mathbf{j}\,. \tag{1.3}$$

An additional constitutive relation is now required, and this is known as Ohm's law, which relates electric currents to the electric field via the electrical conductivity, σ, by the relation

$$\mathbf{j} = \sigma\mathbf{E}\,. \tag{1.4}$$

These equations are valid in a reference frame at rest. However, in general the fluids that we will consider are not at rest, and so it is necessary to introduce some modifications for the case of a moving medium at velocity u. Following standard electromagnetic theory, we write

$$\begin{cases} \mathbf{E}' = (1-\gamma_u)\,\dfrac{\mathbf{u}\cdot\mathbf{E}}{|\mathbf{u}|^2}\,\mathbf{u} + \gamma_u\,(\mathbf{E}+\mathbf{u}\times\mathbf{B})\,, \\ \mathbf{B}' = (1-\gamma_u)\,\dfrac{\mathbf{u}\cdot\mathbf{B}}{|\mathbf{u}|^2}\,\mathbf{u} + \gamma_u\left(\mathbf{B}-\dfrac{\mathbf{u}\times\mathbf{E}}{c^2}\right)\,, \end{cases} \tag{1.5a,b}$$

where $\gamma_u = (1-|\mathbf{u}|^2/c^2)^{-1/2}$ is known as the Lorentz factor.

Following our assumption that $|\mathbf{u}| \ll c$, the Lorentz factor can be set equal to unity. Then, from equation (1.1a), it follows that $|\mathbf{E}| \sim |\mathbf{u}|\,|\mathbf{B}|$, where $|\mathbf{u}|$ is the ratio of a typical length to a typical time. So that the only modification associated with the displacement of the reference frame is

$$\mathbf{E}' = \mathbf{E}+\mathbf{u}\times\mathbf{B}\,, \tag{1.6}$$

and so Ohm's law becomes

$$\mathbf{j} = \sigma\,(\mathbf{E}+\mathbf{u}\times\mathbf{B})\,. \tag{1.7}$$

For simplicity, let us now assume that the medium is in a neutral state, by which we mean that we take

$$\rho_c \equiv Z_i\,n_i - e\,n_e = 0\,, \tag{1.8}$$

where ρ_c is the charge density, Z_i is the average charge of ions in the medium, and n_e and n_i are the number densities respectively of free electrons and ions. It should be noted that, as stressed by Roberts (1967), this assumption cannot be rigorously valid in a conducting fluid, since the divergence of equation (1.7) together with equations (1.1d) and (1.3) implies that

$$\boldsymbol{\nabla}\cdot(\mathbf{u}\times\mathbf{B}) = -\rho_c/\varepsilon\,. \tag{1.9}$$

Because $\boldsymbol{\nabla}\cdot(\mathbf{u}\times\mathbf{B}) \neq 0$ the charge density cannot be exactly vanishing. However, by assuming that ε is negigibly small, we can neglect ρ_c in the sequel.

To continue with our derivation, we note that electric currents are present in the medium provided $\mathbf{u}_e \neq \mathbf{u}_i$, and so

$$\mathbf{j} = Z_i\,n_i\,\mathbf{u}_i - e\,n_e\,\mathbf{u}_e\,, \tag{1.10}$$

then using equation (1.8) we obtain

$$\mathbf{j} = -e\,n_e\,\mathbf{u}'_e\,, \quad \text{where} \quad \mathbf{u}'_e = \mathbf{u}_e - \mathbf{u}_i\,. \tag{1.11}$$

Formally, three equations of motion should now be established, namely one for each: neutrals, ions and electrons. However, much of the work to understand the maintenance of magnetic fields is carried out in what is referred to as single-fluid magnetohydrodynamics (MHD), where the key assumption is that collisions occur often enough to mechanically couple all three components. This assumption requires special care concerning ions and neutrals. It is important to point out that while the single-fluid MHD approximation is clearly valid for the Earth's core or for solar dynamics, in some weakly ionised plasmas relevant to the interstellar medium (ISM), the drift of charged particles with respect to the neutrals can become significant. This effect is referred to as ambipolar drift, or ambipolar diffusion.[1] This effect will not be considered at this stage.

The relative velocity of electrons to ions, $\mathbf{u}'_e$, can be estimated from the amplitude necessary to produce electric currents compatible with the observed magnetic fields for the geophysical and astrophysical applications addressed in this book.

From equation (1.3) and equation (1.11) we write

$$|\mathbf{u}'_e| \simeq \frac{|\mathbf{B}|}{\mu\, L\, |e|\, n_e}\,, \tag{1.12}$$

which can be used to obtain the following rough estimates of $\mathbf{u}'_e$.

For the case of the Earth's core

$$|\mathbf{u}'_e| \simeq \frac{10^{-4}}{4\pi \times 10^{-7} \times 10^{6} \times 2 \times 10^{-19} \times 10^{29}} \simeq 10^{-20}\mathrm{m\,s^{-1}}\,. \tag{1.13a}$$

For the case of the Sun's interior

$$|\mathbf{u}'_e| \simeq \frac{10^{-1}}{4\pi \times 10^{-7} \times 2 \times 10^{8} \times 2 \times 10^{-19} \times 10^{29}} \simeq 10^{-14}\mathrm{m\,s^{-1}}\,. \tag{1.13b}$$

In galaxies

$$|\mathbf{u}'_e| \simeq \frac{5 \times 10^{-10}}{4\pi \times 10^{-7} \times 10^{20} \times 2 \times 10^{-19} \times 10^{3}} \simeq 10^{-8}\mathrm{m\,s^{-1}}\,. \tag{1.13c}$$

In all these cases the velocity of the flow $|\mathbf{u}|$ (i.e. for both the ions and neutrals) is much larger than $|\mathbf{u}'_e|$. The typical velocity $|\mathbf{u}|$ in the slow moving liquid iron that resides in the Earth's core is of the order of $10^{-4}\mathrm{m\,s^{-1}}$ and it is much larger in the Sun and in galaxies. These are thus extremely small deviations from the mean

1 The term "ambipolar diffusion" can be slightly misleading, since this effect is not strictly equivalent to resistive diffusion. In particular, it preserves magnetic topology (as will be discussed later for ideal MHD). Still, this effect acts to damp fluctuations on small scales.

velocity. This demonstrates that, in these contexts, the single-fluid MHD approximation is valid. Therefore, from here on we will assume $\mathbf{u}_n \simeq \mathbf{u}_i \simeq \mathbf{u}_e$, but retain the small difference $\mathbf{u}_e' = \mathbf{u}_e - \mathbf{u}_i$ only as a source of magnetic induction.

The curl of equation (1.7), together with relation (A.21) and equation (1.1c), now yields

$$\partial_t \mathbf{B} = \boldsymbol{\nabla} \times (\mathbf{u} \times \mathbf{B}) + \eta \Delta \mathbf{B}, \qquad (\text{where } \Delta \equiv \boldsymbol{\nabla}^2), \tag{1.14}$$

the coefficient $\eta = 1/(\sigma\,\mu)$ is referred to as the *magnetic diffusivity*, assumed here to be constant. One must not forget that there is an additional constraint provided by equation (1.1c), namely

$$\boldsymbol{\nabla} \cdot \mathbf{B} = 0, \tag{1.15}$$

and note that, provided this constraint is satified at a given time, equation (1.14) will ensure it remains satisfied for all time.

Using relation (A.27), equation (1.14) can be rewritten as

$$\partial_t \mathbf{B} + \mathbf{u} \cdot \boldsymbol{\nabla} \mathbf{B} = \mathbf{B} \cdot \boldsymbol{\nabla} \mathbf{u} + \eta \Delta \mathbf{B}. \tag{1.16}$$

At times, when considering the evolution of the magnetic field, it is helpful to work with the vector potential, $\mathbf{A}$, and not $\mathbf{B}$ itself. Equation (1.15) can conveniently be used to rewrite the magnetic field in terms of $\mathbf{A}$, i.e.

$$\mathbf{B} = \boldsymbol{\nabla} \times \mathbf{A}. \tag{1.17}$$

INFLUENCE ON MATTER

We have seen above how the motion of a conducting fluid can affect the magnetic induction, but this is just part of the story. The induction equation derived above is just a convenient starting point in dynamo theory. With this equation alone, and a prescribed flow, the magnetic field is governed by a linear equation (this is referred to as the “kinematic dynamo” problem, and will be discussed later in this chapter). The magnetic energy can, in this case, grow exponentially and reach unrealistic values. This does not happen in reality because the magnetic field can change the flow via a force, the Lorentz force, in such a way as to avoid this scenario.

The Lorentz force density is given by

$$\begin{aligned} \mathbf{F}_L &= n_i Z_i (\mathbf{E} + \mathbf{u}_i \times \mathbf{B}) - n_e e (\mathbf{E} + \mathbf{u}_e \times \mathbf{B}) \\ &= \mathbf{j} \times \mathbf{B} = {\mu_0}^{-1} (\boldsymbol{\nabla} \times \mathbf{B}) \times \mathbf{B}, \end{aligned} \tag{1.18}$$

where equations (1.8) and (1.10) have been used. This force density applies to the single-fluid described above. It can be expanded as

$${\mu_0}^{-1} (\boldsymbol{\nabla} \times \mathbf{B}) \times \mathbf{B} = {\mu_0}^{-1} \left[(\mathbf{B} \cdot \boldsymbol{\nabla}) \mathbf{B} - \tfrac{1}{2} \boldsymbol{\nabla} |\mathbf{B}|^2 \right], \tag{1.19}$$

where the first term is known as the "magnetic tension", and the second as the "magnetic pressure".

1.1.2. THERMODYNAMIC EQUATIONS

In the case of planets and stars, it is expected that convection is the main way by which motions are generated in the fluid. We will assume, for simplicity, a single driving mechanism for convection in this section. However, we note here that this assumption is not fully valid for example when investigating the Earth's core, for which compositional as well as thermal driving need to be considered. A similar set of equations can however be recovered in this case by introducing a codensity variable accounting for both of these effects. For a rigorous derivation of the equations in this more complicated case (and including turbulence modelling), the reader should refer to Braginsky and Roberts (1995, 2003).

We use standard notations, and so P, ρ and T, respectively denote the pressure, density and temperature. We assume that the equation of state of the fluid is given by the following three thermodynamic coefficients, namely the expansion coefficient at constant pressure, α_P , the specific heat at constant pressure, c_P , and the polytropic coefficient, γ , which can be expressed as

$$\alpha_P = -\frac{T}{\rho}\frac{\partial \rho}{\partial T}\bigg|_P , \qquad c_P = \frac{\partial H}{\partial T}\bigg|_P , \qquad \gamma = \frac{\rho}{P}\frac{\partial P}{\partial \rho}\bigg|_S , \tag{1.20a,b,c}$$

where H denotes the specific enthalpy, and S denotes the specific entropy of the system.

From the second principle of thermodynamics, we deduce that

$$\mathrm{d}E = T\mathrm{d}S + \frac{P}{\rho^2}\mathrm{d}\rho \, , \tag{1.21}$$

where $E = H - P/\rho$ denotes the specific internal energy, and so

$$c_P = T\frac{\partial S}{\partial T}\bigg|_P . \tag{1.22}$$

Finally, we introduce for convenience

$$\alpha_S = -\frac{1}{\rho}\frac{\partial \rho}{\partial S}\bigg|_P = \frac{\alpha_P}{c_P} . \tag{1.23}$$

All thermodynamic relations are deduced from equation (1.21) and the preceding three coefficients. Indeed, from equation (1.21) we can write

$$\frac{\mathrm{d}P}{P}\left(\frac{1}{\gamma} + \frac{P\alpha_P^2}{\rho T c_P}\right) = \frac{\mathrm{d}\rho}{\rho} + \frac{\alpha_P \mathrm{d}T}{T} \, , \tag{1.24a}$$

$$\frac{\mathrm{d}S}{c_P} = \frac{\mathrm{d}T}{T} - \frac{P\alpha_P}{\rho T c_P}\frac{\mathrm{d}P}{P}\,, \tag{1.24b}$$

and
$$\alpha_S \mathrm{d}S = \frac{1}{\gamma}\frac{\mathrm{d}P}{P} - \frac{\mathrm{d}\rho}{\rho}\,. \tag{1.24c}$$

Introducing the heat production, δQ, the heat flux, $\mathbf{q}$, and the rate of internal dissipation per unit volume, $\mathcal{E}$ (including viscous and ohmic dissipation), we can rewrite the second principle of Thermodynamics, using the fact that $T\mathrm{d}S = \delta Q = -\boldsymbol{\nabla}\cdot\mathbf{q}$, as

$$\rho\,\mathrm{D}_t E - \frac{P}{\rho}\mathrm{D}_t\rho = -\boldsymbol{\nabla}\cdot\mathbf{q} + \mathcal{E}\,, \quad \text{where} \quad \mathrm{D}_t \equiv \partial_t + \mathbf{u}\cdot\boldsymbol{\nabla} \tag{1.25a,b}$$

denotes the lagrangian derivative.

This expression, together with Fourier's law of heat conduction for the temperature T (introducing the thermal conduction coefficient k)

$$\mathbf{q} = -k\boldsymbol{\nabla}T\,, \qquad \text{yields} \qquad \rho T\,\mathrm{D}_t S = \boldsymbol{\nabla}\cdot(k\boldsymbol{\nabla}T) + \mathcal{E}\,. \tag{1.26a,b}$$

It is also useful to define the thermal diffusivity $\kappa \equiv k/(\rho\, c_P)$.

1.1.3. NAVIER-STOKES EQUATION

The compressible Navier-Stokes equations include the continuity equation,

$$\partial_t\rho + \boldsymbol{\nabla}\cdot(\rho\,\mathbf{u}) = 0\,, \qquad \text{or} \qquad \mathrm{D}_t\rho = -\rho\,\boldsymbol{\nabla}\cdot\mathbf{u}\,, \tag{1.27a,b}$$

and the momentum equation, written in a rotating reference frame

$$\rho \mathrm{D}_t\mathbf{u} + 2\,\rho\,\boldsymbol{\Omega}\times\mathbf{u} = -\boldsymbol{\nabla}P - \rho\,\boldsymbol{\nabla}\Phi - \boldsymbol{\nabla}\cdot\boldsymbol{\tau} + \boldsymbol{\mathcal{F}}\,, \tag{1.27c}$$

where $\boldsymbol{\Omega}$ is the rotation vector (in the direction of the rotation axis and with the rotation rate as magnitude), $\boldsymbol{\mathcal{F}}$ represents all remaining body forces (including the Lorentz force), and $\boldsymbol{\tau}$ is the viscous stress tensor, with components

$$\tau_{ij} = -2\rho\nu S_{ij}\,, \qquad S_{ij} = \varepsilon_{ij} - \tfrac{1}{3}\left(\boldsymbol{\nabla}\cdot\mathbf{u}\right)\delta_{ij}\,, \tag{1.27d,e}$$

ν being the kinematic viscosity, S_{ij} the strain rate tensor

$$S_{ij} = \tfrac{1}{2}\left(\partial_i u_j + \partial_j u_i\right)\,, \tag{1.27f}$$

and Φ includes the gravity potential Φ_g as well as the centrifugal potential Φ_Ω.

The apparent gravity field is then provided by $\mathbf{g} = -\boldsymbol{\nabla}\Phi$, and here there are two contributions to Φ. In a non rotating problem, the gravity potential is simply obtained from

$$\Delta\Phi_g = 4\pi G\rho\,. \tag{1.28}$$

In a rotating fluid, this potential is complemented by the effect of the centrifugal potential

$$\Phi_\Omega = -\tfrac{1}{2}\,\Omega^2 s^2\,, \qquad \Delta\Phi_\Omega = -2\Omega^2\,. \tag{1.29a,b}$$

For a galactic disc, density is low and so centrifugal effects are essential, because they balance the radial component of gravity. As a result, the apparent gravity is oriented along the axis of rotation.

For much denser objects, such as the Earth or the Sun, the role of the centrifugal effect is much smaller. It essentially acts to flatten equipotential surfaces. This effect is minute for these objects, which are almost spherical bodies. One can then assume for simplicity that gravity potential varies only with radius. On a given sphere of radius r, and outward normal $\mathbf{n}$:

$$\int_{S(r)} \boldsymbol{\nabla}\Phi\cdot\mathbf{n}\,\mathrm{d}S = 4\pi\,G\int_{V(r)} \rho\,\mathrm{d}V\,, \tag{1.30}$$

masses at larger radii cancel their contributions. So for a sphere of uniform density, gravity is proportional to radius, we have

$$\mathbf{g} = -\tfrac{4}{3}\pi G\rho r\,\mathbf{e}_r. \tag{1.31}$$

Note that using equation (1.27a,b), the energy equation (1.25a) can be rewritten as

$$\rho\,\mathrm{D}_t E + P\,\boldsymbol{\nabla}\cdot\mathbf{u} = \boldsymbol{\nabla}\cdot(k\boldsymbol{\nabla}T) + \mathcal{E}\,. \tag{1.32}$$

This set of equations needs to be completed by an equation of state relating P, E, ρ, and T as described in the previous section.

The equations described above are appropriate to describe the dynamics of galaxies. Simpler models can however be derived for convection in planets and stars, and this is the purpose of the next section.

THE ANELASTIC APPROXIMATION

The anelastic approximation relies on two simplifying assumptions (see Ogura & Phillips, 1962; Gough, 1969). The first one consists in filtering out acoustic waves, while the second implies linearising fluctuating variables around the reference state. Both can be achieved by an appropriate asymptotic expansion.

We begin by rewriting the above set of equations in dimensionless form. As such we introduce a typical velocity U_*, length L_*, density ρ_*, and temperature T_*. Equation (1.24b) provides $P_* = \rho_* T_* c_{P*}/\alpha_{p_*} = \rho_* T_*/\alpha_{s*}$. Having defined four units, and since nine parameters define our problem $(L, T_*, \rho_*, c_{P*}, \alpha_{p_*}, \nu, k, G, \Omega)$, five independent non-dimensional combinations can be constructed. We define the Reynolds number Re, the Rossby number Ro (measuring the ratio between the rotation and the hydrodynamic timescale), the Froude number Fr (measuring inertia versus gravity forces), the ratio $\mathcal{X}$ of gravitational to pressure forces, and finally the Prandtl number Pr (measuring the ratio of the thermal diffusion timescale to the viscous timescale),

$$\mathrm{Re} = \frac{U_* L_*}{\nu}, \qquad \mathrm{Ro} = \frac{U_*}{\Omega L_*}, \qquad \mathrm{Fr} = \frac{U_*^2}{L_* g_*} = \frac{U_*^2}{4\pi L_*^2 G \rho_*}, \tag{1.33a,b,c}$$

$$\mathcal{X} = \frac{\rho_* g_* L_*}{P_*} = \frac{\alpha_{s*} \rho_* 4\pi L_*^2 G}{T_*}, \qquad \mathrm{Pr} = \frac{\nu \rho_* c_P}{k} = \frac{\nu}{\kappa}. \tag{1.33d,e}$$

The equation of mass conservation retains its form, whereas the momentum and energy equation can be rewritten (note that ρ is now dimensionless) as

$$\rho \mathrm{D}_t \mathbf{u} + \frac{2}{\mathrm{Ro}} \rho \mathbf{k} \times \mathbf{u} = -\frac{1}{\mathcal{X}\mathrm{Fr}} \boldsymbol{\nabla} P + \frac{1}{\mathrm{Fr}} \rho \mathbf{e_g} + \frac{2}{\mathrm{Re}} \boldsymbol{\nabla} \cdot (\rho \mathbf{S}), \tag{1.34}$$

where $\mathbf{k}$ denotes the unit vector along the rotation axis, $\Omega \mathbf{k} = \boldsymbol{\Omega}$.

To simplify the following development, we have dropped here the forcing term $\mathcal{F}$. However, the Lorentz force will be re-introduced later in the resulting equations. This simplification although convenient, is not necessary. For a full treatment, including the Lorentz force, the reader is refered to Lantz & Fan (1999).

We can also assume at this stage that we are dealing with a perfect gas, and so c_P can be regarded as constant. The final equation, the entropy equation, then becomes

$$\rho T \, \mathrm{D}_t S = \frac{1}{\mathrm{PrRe}} \Delta T - \frac{2\mathcal{X}\mathrm{Fr}}{\alpha_{S*}\,\mathrm{Re}} \rho \mathbf{S} : \boldsymbol{\varepsilon}. \tag{1.35}$$

Let us stress again that, although we use the same symbols as previously, all quantities are now dimensionless. In addition, note that we used the notation ":" for the double contraction of two tensors, i.e.

$$\mathbf{S} : \boldsymbol{\varepsilon} = Tr(\mathbf{S} \cdot \boldsymbol{\varepsilon}) = S_{ij}\, \varepsilon_{ji}. \tag{1.36}$$

Heat transfer is particularly important for stars and planets as it will induce convective motion directly related to dynamo action. Since we are dealing with convection, it is helpful to define a reference state. The best reference state is the neutrally stable one, with constant S (this is not the diffusive state). The governing equations can then be rewritten in terms of deviations from this reference state. This leads to the "anelastic approximation". These deviations are assumed to be small compared to the reference state. This assumption is well justified in a strongly convective state and away from boundary layers.

The reference state mentioned above is assumed to be fully decoupled from possible nonlinear correlations of the perturbed state, so that the dynamics of ρ_a, $\mathbf{u}_a$ and S_a are given assuming an isentropic equilibrium ($\boldsymbol{\nabla} S_a = \mathbf{0}$). Finally, let us recall that in the limit of no thermal or radiative conduction, entropy S_a is uniform, and the corresponding temperature profile is the *adiabatic* profile T_a.

All quantities are expanded as

$$\rho = \rho_0 + \varepsilon_\rho \rho_1 \,, \qquad P = P_0 + \varepsilon_P P_1 \,, \tag{1.37a,b}$$

$$T = T_0 + \varepsilon_T T_1 \,, \qquad S = S_0 + \varepsilon_S S_1 \,. \tag{1.37c,d}$$

Linearization of the equation of state around the reference state provides

$$\varepsilon_\rho = \varepsilon_P = \varepsilon_T = \varepsilon_S = \varepsilon \,.$$

We assume here that all quantities in this expansion $(\rho_0, \rho_1, P_0, P_1, T_0, T_1, S_0, S_1)$ are order one. As such, the mass conservation equation becomes

$$\partial_t(\rho_0 + \varepsilon\rho_1) + \boldsymbol{\nabla} \cdot [(\rho_0 + \varepsilon\rho_1)\mathbf{u}] = 0 \,, \tag{1.38}$$

and to leading order (because ρ_0 is not a function of time) this is

$$\boldsymbol{\nabla} \cdot (\rho_0 \mathbf{u}) = 0 \,. \tag{1.39}$$

Neglecting higher order terms ensures the filtering of elastic waves out of the resulting model, hence the name "anelastic".

The conservation of momentum equation can be expressed in a similar manner, i.e.

$$\begin{aligned}(\rho_0 + \varepsilon\rho_1)\,[\partial_t \mathbf{u} + (\mathbf{u}\cdot\boldsymbol{\nabla})\mathbf{u}] + \frac{1}{\mathcal{X}\mathrm{Fr}}\boldsymbol{\nabla}(P_0 + \varepsilon P_1) + \frac{2}{\mathrm{Ro}}(\rho_0 + \varepsilon\rho_1)\,\mathbf{k}\times\mathbf{u} \\ = \frac{1}{\mathrm{Fr}}(\rho_0 + \varepsilon\rho_1)\boldsymbol{\nabla}(\Phi_0 + \varepsilon\Phi_1) - \frac{2}{\mathrm{Re}}\boldsymbol{\nabla}\cdot[(\rho_0 + \varepsilon\rho_1)\mathbf{S}] \,.\end{aligned} \tag{1.40}$$

The coupling between energy and Navier-Stokes equations in this limiting process, necessary for convection to occur, requires $\varepsilon \sim Fr$. This scaling reveals at leading order $(1/\varepsilon)$

$$\frac{1}{\mathcal{X}}\boldsymbol{\nabla} P_0 = -\rho_0 \boldsymbol{\nabla}\Phi_0 \,. \tag{1.41}$$

Equation (1.41), together with equation (1.24c), provides the balance relevant for the reference state.

At the next order (ε^0), we get

$$\rho_0\,[\partial_t \mathbf{u} + (\mathbf{u}\cdot\boldsymbol{\nabla})\mathbf{u}] + \frac{1}{\mathcal{X}}\boldsymbol{\nabla} P_1 + \frac{2}{\mathrm{Ro}}\rho_0 \mathbf{k}\times\mathbf{u} = -\rho_0\boldsymbol{\nabla}\Phi_1 - \rho_1\boldsymbol{\nabla}\Phi_0 - \frac{2}{\mathrm{Re}}\boldsymbol{\nabla}\cdot(\rho_0\,\mathbf{S}) \,. \tag{1.42}$$

It can be useful to manipulate this expression, following Braginsky and Roberts (1995, 2003), by making use of thermodynamic relations. From equation (1.24c)

$$\mathbf{0} = \frac{1}{\gamma}\frac{\boldsymbol{\nabla} P_0}{P_0} - \frac{\boldsymbol{\nabla}\rho_0}{\rho_0}\,, \tag{1.43}$$

while from equations (1.24b) and (1.24c), we obtain

$$S_1 = \frac{1}{T_0}T_1 - \frac{1}{\rho_0 T_0}P_1\,, \qquad \text{and} \qquad \alpha_S S_1 = \frac{1}{\gamma}\frac{P_1}{P_0} - \frac{\rho_1}{\rho_0}\,. \tag{1.44a,b}$$

Hence it follows that

$$\begin{aligned}
-\frac{1}{\mathcal{X}}\boldsymbol{\nabla} P_1 - \rho_0\boldsymbol{\nabla}\Phi_1 - \rho_1\boldsymbol{\nabla}\Phi_0 &= -\rho_0\boldsymbol{\nabla}\left(\frac{P_1}{\mathcal{X}\rho_0} + \Phi_1\right) - \frac{P_1}{\mathcal{X}\rho_0}\boldsymbol{\nabla}\rho_0 + \rho_1\boldsymbol{\nabla}\Phi_0 \\
&= -\rho_0\boldsymbol{\nabla}\left(\frac{P_1}{\mathcal{X}\rho_0} + \Phi_1\right) - \frac{P_1}{\mathcal{X}\rho_0}\boldsymbol{\nabla}\rho_0 - \left(\frac{1}{\gamma}\frac{P_1}{P_0} - \alpha_S S_1\right)\rho_0\boldsymbol{\nabla}\Phi_0 \quad \text{from (1.44b)} \\
&= -\rho_0\boldsymbol{\nabla}\left(\frac{P_1}{\mathcal{X}\rho_0} + \Phi_1\right) - \rho_0\alpha_S S_1\mathbf{g}_0 \quad \text{from (1.43) and (1.41).}
\end{aligned}$$

Thus equation (1.42) can be rewritten in the more compact form

$$\partial_t\mathbf{u}+\mathbf{u}\cdot\boldsymbol{\nabla}\mathbf{u}+\frac{2}{\mathrm{Ro}}\mathbf{k}\times\mathbf{u} = -\boldsymbol{\nabla}\left(\frac{P_1}{\mathcal{X}\rho_0} + \Phi_1\right) - \alpha_S S_1\mathbf{g}_0 - \frac{2}{\rho_0\,\mathrm{Re}}\boldsymbol{\nabla}\cdot(\rho_0\,\mathbf{S})\,. \tag{1.45}$$

In the more general case, when more than one driving mechanism is considered (e.g. thermal and chemical in the Earth's core), it can be convenient to introduce a unique variable in the momentum equation. This can be achieved by introducing a co-density variable C (see Braginsky & Roberts, 2003), which reduces in our simpler case to $C = -\alpha_S S_1$.

Turning to the entropy equation we have

$$\begin{aligned}
(\rho_0 + \varepsilon\rho_1)(T_0 + \varepsilon T_1)\left(\partial_t(S_0 + \varepsilon S_1) + \mathbf{u}\cdot\boldsymbol{\nabla}(S_0 + \varepsilon S_1)\right) &= \frac{1}{\mathrm{Re}\,\mathrm{Pr}}\Delta(T_0 + \varepsilon T_1) \\
&\quad - \frac{2\mathcal{X}\mathrm{Fr}}{\alpha_{S*}\,\mathrm{Re}}(\rho_0 + \varepsilon\rho_1)\,\mathbf{S} : \boldsymbol{\varepsilon}\,.
\end{aligned} \tag{1.46}$$

At order ε

$$\rho_0 T_0\left(\partial_t S_1 + \mathbf{u}\cdot\boldsymbol{\nabla} S_1\right) = \frac{1}{\mathrm{Re}\,\mathrm{Pr}}\Delta T_1 - \frac{2\mathcal{X}}{\alpha_{S*}\,\mathrm{Re}}\rho_0\,\mathbf{S} : \boldsymbol{\varepsilon}\,. \tag{1.47}$$

Equations (1.39), (1.45), and (1.47) constitute the anelastic system.

The system governing the slow evolution of the reference state is

$$\boldsymbol{\nabla} S_0 = 0\,, \qquad \rho_0 \boldsymbol{\nabla} P_0 = \gamma P_0 \boldsymbol{\nabla} \rho_0\,, \tag{1.48a,b}$$

$$\frac{1}{\mathcal{X}} \boldsymbol{\nabla} P_0 = -\rho_0 \boldsymbol{\nabla} \Phi_0\,, \qquad \Delta \Phi_0 = \rho_0\,. \tag{1.48c,d}$$

This yields the adiabatic temperature profile, $\boldsymbol{\nabla} T_0 = -\mathcal{X} \boldsymbol{\nabla} \Phi_0$, and is completed by the equation of state relating P_0, ρ_0, T_0.

Convection above this reference state is then governed by

$$\partial_t \mathbf{u} + (\mathbf{u} \cdot \boldsymbol{\nabla})\mathbf{u} + \frac{2}{\mathrm{Ro}} \mathbf{k} \times \mathbf{u} = -\boldsymbol{\nabla} \left[\frac{P_1}{\mathcal{X} \rho_0} + \Phi_1 \right] - \alpha_S S_1 \mathbf{g}_0 - \frac{2}{\mathrm{Re}\,\rho_0} \boldsymbol{\nabla} \cdot (\rho_0\,\mathbf{S})\,, \tag{1.49a}$$

$$\boldsymbol{\nabla} \cdot (\rho_0 \mathbf{u}) = 0\,, \qquad \rho_0 T_0 \left(\partial_t S_1 + \mathbf{u} \cdot \boldsymbol{\nabla} S_1\right) = \frac{1}{\mathrm{Re}\,\mathrm{Pr}} \Delta T_1 - \frac{2\mathcal{X}}{\alpha_{S*}\,\mathrm{Re}} \rho_0\, \mathbf{S} : \boldsymbol{\varepsilon}\,. \tag{1.49b,c}$$

No separate equation is needed for the quantity $\boldsymbol{\nabla} \left[P_1/\mathcal{X}\rho_0 + \Phi_1 \right]$ since it acts as a Lagrange multiplier to satisfy $\boldsymbol{\nabla} \cdot (\rho_0 \mathbf{u}) = 0$.

Let us conclude this section by stressing that under the approximation discussed above $\Phi_\Omega \ll \Phi_g$, the reference state only depends on the radial coordinate. The anelastic system can then be introduced as a decomposition of each variable f into a spherically averaged reference state, $\overline{f}$, and a perturbation, f', i.e.

$$f(r, \theta, \phi, t) = \overline{f}(r, t) + \varepsilon\, f'(r, \theta, \phi, t)$$

(e.g. Gough, 1969; Latour *et al.*, 1976). This formulation allows one to introduce a slow evolution of the reference state (not necessarily compatible with the above expansion).

As discussed at the start of the section, we have derived, under this approximation, a set of equations in their simplest form. Other important effects, such as turbulent transport coefficients (expected to be dominant in the solar convection zone) can be introduced to add further complexity to the models. The effects of compositional convection –a major ingredient to the Earth's core dynamics– can also be envisaged. For a detailed treatment, including both thermal and compositional convection (and also including the effect of turbulent motions), the reader is refered to Braginsky and Roberts (1995, 2000).

THE BOUSSINESQ APPROXIMATION

When the region of fluid is thin enough (in a sense to be clarified later) a more drastic approximation can be introduced, which is refered to as the Boussinesq approximation. It is often used for thin layers of fluid in the laboratory, but has been used as a starting point in understanding natural dynamos (see Chapters 4, 5 and 6). In such thin layers, pressure effects are negligible and the adiabatic temperature profile, T_0, can be assumed to be constant. This allows significant simplifications in the equations, which we will now discuss.

If $\mathcal{X} \ll \mathcal{O}(1)$, i.e. if the size of the system is small compared to the typical depth of an adiabatic gas ($P_*/\rho_* g_*$), compressibility of the fluid under its own weight can safely be neglected.

For all quantities x expanded above in $x_1 + \varepsilon x_1$ [see equations (1.37a–d)], following Malkus (1964), we now introduce a second expansion in terms of $\mathcal{X}$,

$$x_0 = x_{00} + \mathcal{X} x_{01}\,, \qquad x_1 = x_{10} + \mathcal{X} x_{11}\,. \tag{1.50a,b}$$

System (1.48) at order $\mathcal{X}^{-1}$ reveals

$$\boldsymbol{\nabla} P_{00} = \mathbf{0}\,, \tag{1.51}$$

and it follows that the temperature and density of the reference profile are constant.

System (1.49) at order $\mathcal{X}^{-1}$ gives

$$\boldsymbol{\nabla} P_{10} = \mathbf{0}\,, \tag{1.52}$$

while at order $\mathcal{X}^0$ it yields

$$\partial_t \mathbf{u} + (\mathbf{u}\cdot\boldsymbol{\nabla})\mathbf{u} + \frac{2}{\mathrm{Ro}}\mathbf{k}\times\mathbf{u} = -\boldsymbol{\nabla}\Pi - \alpha_S S_{10}\mathbf{g}_{00} + \frac{2}{\mathrm{Re}\,\rho_0}\Delta\mathbf{u}\,, \tag{1.53a}$$

$$\boldsymbol{\nabla}\cdot\mathbf{u} = 0\,, \qquad \rho_{00}T_{00}\left(\frac{\partial S_{10}}{\partial t} + \mathbf{u}\cdot\boldsymbol{\nabla} S_{10}\right) = \frac{1}{\mathrm{Re\,Pr}}\Delta T_{10}\,. \tag{1.53b,c}$$

All gradient terms have conveniently been written as $\boldsymbol{\nabla}\Pi$ in equation (1.53a), this term being a Lagrange multiplier to satisfy equation (1.53b). It follows from equation (1.52) that

$$S_{10} = \frac{T_{10}}{T_{00}}\,. \tag{1.54}$$

It is usual in the Boussinesq formalism to introduce the coefficient of "thermal expansion" α. It is defined, using dimensional variables, by

$$\frac{\delta\rho}{\rho} = \alpha\,\delta T\,, \tag{1.55}$$

and relates to the previously introduced α_S and α_P through

$$\alpha = -\frac{1}{\rho}\left(\frac{\partial\rho}{\partial T}\right)\bigg|_P = \frac{\alpha_S c_P}{T_*} = \frac{\alpha_P}{T_*}\,. \tag{1.56}$$

In dimensionless from (for clarity, we introduce here a different symbol) it yields

$$\alpha' = \alpha_S c_P = \alpha_P \,. \tag{1.57}$$

System of equations (1.49) then becomes,

$$\partial_t \mathbf{u} + (\mathbf{u}\cdot\boldsymbol{\nabla})\mathbf{u} + \frac{2}{\mathrm{Ro}}\mathbf{k}\times\mathbf{u} = -\boldsymbol{\nabla}\Pi - \alpha' T_{10}\mathbf{g}_0 + \frac{1}{\mathrm{Re}}\Delta\mathbf{u}\,, \tag{1.58a}$$

$$\boldsymbol{\nabla}\cdot\mathbf{u} = 0\,, \qquad \partial_t T_{10} + \mathbf{u}\cdot\boldsymbol{\nabla}T_{10} = \frac{1}{\mathrm{Re}\,\mathrm{Pr}}\Delta T_{10}\,. \tag{1.58b,c}$$

In the Boussinesq approximation the entropy and the temperature are equivalent up to a scaling factor given by equation (1.54). To recover a more classical dimensionless formalism, let us assume that a super adiabatic entropy gradient is maintained accross the system. This gradient provides the natural unit for temperature, while the velocity scale U_* can be set to κ/L_0.

One can then introduce the Rayleigh number, Ra, and the Ekman number, E, which are formed as

$$\mathrm{Ra} = \frac{\alpha\,\Delta T\, g_*\, L_*^3}{\nu\,\kappa}\,, \qquad \text{and} \qquad \mathrm{E} = \frac{\nu}{\Omega\, L_*^2}\,. \tag{1.59a,b}$$

Using equation (1.37), one recovers a familiar, and much used, system

$$\partial_t \mathbf{u} + (\mathbf{u}\cdot\boldsymbol{\nabla})\mathbf{u} + \frac{2\,\mathrm{Pr}}{\mathrm{E}}\mathbf{k}\times\mathbf{u} = -\boldsymbol{\nabla}\Pi - \mathrm{Ra}\,\mathrm{Pr}\,T\,\mathbf{g}_0 + \mathrm{Pr}\;\Delta\mathbf{u}\,, \tag{1.60a}$$

$$\boldsymbol{\nabla}\cdot\mathbf{u} = 0\,, \qquad \partial_t T + \mathbf{u}\cdot\boldsymbol{\nabla}T = \Delta T\,. \tag{1.60b,c}$$

Finally, it is often useful to decompose the temperature field in two contributions, which are a steady contribution satisfying the boundary conditions (and balancing an internal source term, if there is one), and a perturbation with homogeneous boundary conditions (and governed by a homogeneous equation)

$$T = T_s + \Theta\,.$$

Provided $\boldsymbol{\nabla}T_s \times \boldsymbol{\nabla}\Phi = \mathbf{0}$ (as will be the case under for a perfectly spherical problem) the resulting system is

$$\partial_t \mathbf{u} + (\mathbf{u}\cdot\boldsymbol{\nabla})\mathbf{u} + \frac{2\,\mathrm{Pr}}{\mathrm{E}}\mathbf{k}\times\mathbf{u} = -\boldsymbol{\nabla}\Pi - \mathrm{Ra}\,\mathrm{Pr}\,\Theta\,\mathbf{g}_0 + \mathrm{Pr}\,\Delta\mathbf{u}\,, \tag{1.61a}$$

$$\boldsymbol{\nabla}\cdot\mathbf{u} = 0\,, \qquad \partial_t\Theta + \mathbf{u}\cdot\boldsymbol{\nabla}T_s + \mathbf{u}\cdot\boldsymbol{\nabla}\Theta = \Delta\Theta\,. \tag{1.61b,c}$$

To bring this section to an end, let us stress that since all gradient terms are included in the $\boldsymbol{\nabla}\Pi$ term, when the magnetic field is included and the Lorentz force is non-zero, the magnetic pressure term will therefore not enter the dynamical balance.

This term can produce buoyant effects in regions of localised intense field. Such magnetic buoyancy is believed to be of particular importance for solar dynamics. It is possible to construct approximations which retain this dynamical effect while considering simple incompressible fluids. This is done in a similar manner to how the thermal buoyancy has been retained here. We refer the reader to Spiegel & Weiss for such a derivation, which is achieved at the cost of relaxing (1.1c).

1.1.4. BOUNDARY CONDITIONS

When investigating a planet, a star, or a galaxy, it is convenient to consider a bounded finite volume of space, $\mathcal{D}$, in which the relevant physics are to be investigated. While the fluid can often be assumed to remain within this volume, the magnetic field on the other hand cannot easily be artificially confined. The first, and most natural assumption is to assume that the outside world (i.e. the complementary domain to the finite volume of interest, ${}^c\mathcal{D}$) consists of vacuum and is insulating. No electric current can therefore escape the volume, $\mathcal{D}$, and the resulting $\boldsymbol{\nabla} \times \mathbf{B} = \mathbf{0}$ in ${}^c\mathcal{D}$, together with $\boldsymbol{\nabla} \cdot \mathbf{B} = 0$ imply that the field in ${}^c\mathcal{D}$ derives from a potential

$$\mathbf{B} = -\boldsymbol{\nabla}\Phi\,, \qquad \text{and} \qquad \Delta\Phi = 0\,. \tag{1.62a,b}$$

The above relation on the field in the complementary domain provides the necessary conditions to compute the field evolution in $\mathcal{D}$ once continuous quantities across $\partial\mathcal{D}$ are identified. Equation (1.1c) implies that $\mathbf{n} \cdot \mathbf{B}$ is continuous across the boundary, while equation (1.1a) implies the continuity of $\mathbf{n} \times \mathbf{E}$ ($\mathbf{n}$ is the normal to the boundary). These can be used to reduce the induction problem to a closed integro–differential formulation on $\mathcal{D}$ (e.g. Iskakov & Dormy, 2004, 2005).

It is important to note that, while this choice of boundary condition is a very natural one and will in fact be the only one used in this book, some astrophysical bodies (like the Sun) are bounded by a conducting corona. For such corona, nothing can balance the Lorentz force in the momentum equation. As a result, the field has to relax to a state for which the Lorentz force vanishes. Such state is known as a "force-free" state. Interestingly, the field is then prescribed from the momentum equation rather than the induction equation. From $(\boldsymbol{\nabla} \times \mathbf{B}) \times \mathbf{B} = \mathbf{0}$, one deduces that

$$\boldsymbol{\nabla} \times \mathbf{B} = \alpha\,\mathbf{B} \qquad \text{with} \qquad (\mathbf{B} \cdot \boldsymbol{\nabla})\,\alpha = 0\,, \tag{1.63a,b}$$

where α is real, and must not be confused with the notation α in mean field theory (although the α in mean field theory, that is discussed in Section 1.5, relates $\boldsymbol{\nabla} \times \mathbf{B}$ to $\mathbf{B}$ as well, it derives from a different physical reasoning). Unless α is artificially assumed to be uniform in ${}^c\mathcal{D}$, the resulting problem is nonlinear and very difficult to address (even determining the necessary conditions on $\partial\mathcal{D}$ to determine $\mathbf{B}$ in ${}^c\mathcal{D}$

is not a trivial issue. So far, to the authors knowledge, no dynamo model has been produced with this type of bounding domain (even in the linear approximation). Solar models presently rely on the matching to a potential field as expressed by equation (1.62) (see Chapter 6).

Boundary conditions on thermodynamic quantities are, depending on the problem are usually of either the Dirichlet type (fixed value) or of the Neumann typed (fixed flux). Boundary conditions on the fluid flow usually require non-penetration of the fluid at the boundary which translates to

$$\mathbf{n} \cdot \mathbf{u} = 0\,. \tag{1.64}$$

While this condition is sufficient when viscosity is omitted, additional conditions are needed if viscosity is retained. These usually amount for the configurations investigated in this book to either "no-slip" (1.65a) or "stress-free" (1.65b) conditions, namely

$$\mathbf{n} \times \mathbf{u} = \mathbf{0}\,, \qquad \text{or} \qquad \mathbf{n} \cdot \boldsymbol{\nabla}\,(\mathbf{n} \times \mathbf{u}) = \mathbf{0}\,. \tag{1.65a,b}$$

1.2. HOMOGENEOUS DYNAMOS

1.2.1. DISC DYNAMO

Here we introduce the dynamo instability on an apparently for a very simple device: the "homopolar dynamo" or "disc dynamo". Consider a conducting disc of radius r, free to rotate on its axis [see Figure 1.1(a)]. If one places a permanent magnet under the disc and rotates the disc at the angular velocity Ω, then an electromotive force will be driven between the axis and the rim of the disc. If a conducting wire connects the rim of the disc to the axis then an electric current will be driven through this wire. This setup was originally introduced by Faraday in 1831; it is a dynamo (it converts kinetic energy to magnetic energy), but it is not a "self-excited dynamo", since it relies on a permanent magnet. Introducing the magnetic flux through the disc $\Phi = B\pi r^2$, we can quantify this electromotive force, $\mathcal{E}$, by integrating $\mathbf{u} \times \mathbf{B}$ across the disc. Assuming for simplicity a uniform and vertical field $\mathbf{B} = B\mathbf{e}_z$ one finds

$$\mathcal{E} = \frac{\Omega B r^2}{2} = \frac{\Phi\Omega}{2\pi}\,. \tag{1.66a}$$

If the permanent magnet is now replaced with a solenoid of self-inductance L [see Figure 1.1(b)], an instability problem is presented. If the rotation rate is small enough, the electrical resistance will damp any initial magnetic perturbation. If the rotation rate is sufficient (in a way we will immediately quantify), then the system undergoes a "bifurcation" and an initial perturbation of electric currents, I, can be amplified exponentially by self-excited dynamo action.

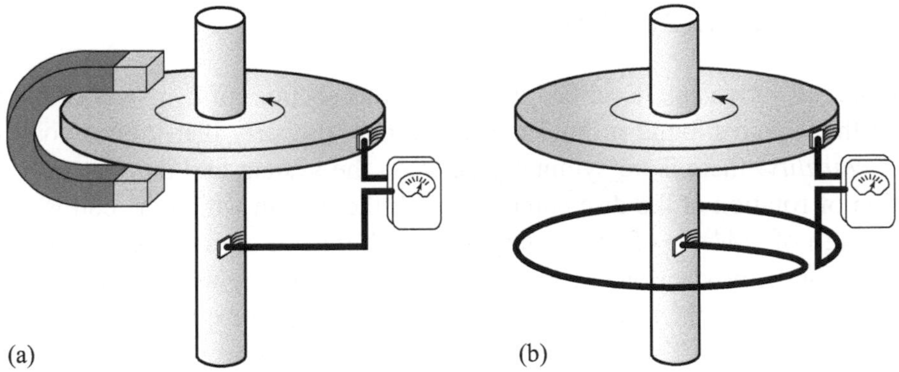

(a) (b)

Figure 1.1 - (a) The original Faraday disc dynamo. (b) The homopolar self excited dynamo.

Let us introduce the mutual inductance, M, between the solenoid and the disc, which allows us, using $\Phi = MI$, to write

$$\mathcal{E} = \frac{M\Omega I}{2\pi}. \tag{1.66b}$$

Now if R is the electrical resistance of the complete circuit, the governing equation for the electric currents in the system is

$$L\frac{\mathrm{d}I}{\mathrm{d}t} + RI = \frac{M\Omega I}{2\pi}. \tag{1.67}$$

It follows that the system is unstable if

$$\Omega > \Omega_c = \frac{2\pi R}{M}. \tag{1.68}$$

In reality, the value of Ω_c for an experimental setup is too high to be achieved. Therefore, while this setup offers a simple description of a self-excited dynamo, it cannot be constructed as such in practice (e.g. Rädler & Rheinhardt, 2002).

It is worth stressing here that this mathematical description of the physical setup is oversimplified and further developments and refinements will be dicussed later in the book. Furthermore, we only consider here a linear problem. The currents here appear to grow indefinitely. This is because the Lorentz force acting on the disc to slow it down has been neglected. This force is the essence of a third setup that can also be constructed using such a disc configuration, which is the Barlow wheel. In this setup, no torque is externally applied to the disc. Instead a battery replaces the current-meter of Figure 1.1(a), and the interaction between this current and the externally applied magnetic field causes the disc to rotate.

1.2.2. CHIRALITY AND GEOMETRY

The simple disc dynamo described in the previous section, of course does not possess all the features found in fluid dynamos. One property that it does possess is that of *chirality*; there is no symmetry between the system and its reflection. The direction of rotation of the disc compared with the way in which the coil is wound (i.e. the sign of ΩM), is of crucial importance. It will be seen that chirality is very important for the production of large scale magnetic fields by fluid dynamos, though it is not essential for the production of local small scale fields; this is accomplished by stretching instead. The disc dynamo has no stretching properties, which on the face of things would suggest that magnetic energy could not be increased. However the disc dynamo is not a fluid, and current is constrained to flow in the wires and through the disc. This corresponds to a highly anisotropic electrical conductivity, while in a homogeneous fluid dynamo one expects the conductivity to be isotropic, at least to a first approximation. The key to a successful dynamo is to get the currents to flow in such a way that the resulting fields reinforce those previously existing - not a trivial task for homogeneous fluid bodies! In general currents will wish to take the shortest paths and unless the flow fields are sufficiently complicated they will simply not be able to produce the correct topology for sustained growth.

In fact it is notable that astrophysical bodies such as the Earth and Sun in which large scale fields are generated *do* in fact possess symmetry under reflection and exchange by rotation of North and South poles. So while local properties of motion in these bodies are chiral, the net lack of chirality distinguishes them from the disc problem.

1.2.3. BASIC MECHANISMS OF DYNAMO ACTION

The dynamo process is in essence a way of turning mechanical energy into magnetic energy. To see this we can take the scalar product of the induction equation (1.14) with **B**, integrate over some suitable domain and obtain, after some integration by parts and ignoring all boundary terms:

$$\frac{1}{2}\frac{\mathrm{d}}{\mathrm{d}t}\int |\mathbf{B}|^2 \mathrm{d}\mathbf{x} = \int \mathbf{B}\cdot(\mathbf{B}\cdot\boldsymbol{\nabla}\mathbf{u})\mathrm{d}\mathbf{x} - \eta\int |\boldsymbol{\nabla}\mathbf{B}|^2 \mathrm{d}\mathbf{x}\,, \qquad (1.69)$$

The second term here is negative and represents the conversion of energy into heat due to Ohmic losses. The first term (due to induction) can be rewritten (in the case $\boldsymbol{\nabla}\cdot\mathbf{u} = 0$) as $-\int \mathbf{u}\cdot[(\mathbf{B}\cdot\boldsymbol{\nabla})\mathbf{B}]\,\mathrm{d}\mathbf{x}$ and this is just the negative of the work done by the velocity field against the Lorentz force. Clearly there can be no growth of magnetic energy, let alone total magnetic flux, unless the induction term is effective. We can see how induction can act to increase magnetic energy by ignoring the effects of diffusion entirely. We are left with the reduced system

$$\partial_t \mathbf{B} = \boldsymbol{\nabla} \times (\mathbf{u} \times \mathbf{B}) . \tag{1.70}$$

This is formally identical to the vorticity equation for $\boldsymbol{\omega} = \boldsymbol{\nabla} \times \mathbf{u}$ in an inviscid fluid, and we can therefore take over many results about the kinematics of vorticity (but not, note, of the dynamic aspects, since in MHD we do not have $\mathbf{B} = \boldsymbol{\nabla} \times \mathbf{u}$!). In particular, Faraday's law that the total flux threading a material element is conserved, is completely equivalent to Kelvin's circulation theorem, i.e. $\oint_{\mathcal{C}} \mathbf{u} \cdot \mathrm{d}\mathbf{x} = \int_{\mathcal{S}} \boldsymbol{\omega} \cdot \mathrm{d}\mathbf{S}$ is constant for material curves $\mathcal{C}$ spanned by material surfaces $\mathcal{S}$. This has the corollary that "vortex lines move with the fluid" (Kelvin). For magnetic fields the analogous "freezing-in" result is called Alfvén's Theorem. Consider then vortex stretching. In an extensional flow involving contraction in two directions and expansion in the third, a material tube of vortex lines aligned with the expanding direction has constant total vorticity at every cross section. Since the cross sectional area is diminishing, the local vorticity must increase, and so since the volume is fixed the integral of $|\boldsymbol{\omega}|^2$ also increases. Exactly the same argument can be applied to magnetic fields, with the result that such stretching flows can increase magnetic energy. Note, however that the total magnetic flux is not increased, so this mechanism as it stands is not able to account for any increases in, for example, dipole moments in conducting spheres. In addition, in a finite domain stretching must be accompanied by folding, as in kneading dough, and this second action will in general bring oppositely directed fields together, where they will cancel due to Ohmic dissipation. This does not always happen though, as can be seen from the Vainshtein-Zeldovich dynamo (the *Stretch–Twist–Fold, or STF mechanism*) leads to the effective doubling of the energy of a loop of flux, as shown in Figure 1.2. This is the most dramatic example of a number of transformations of the space that can lead to net stretching. More explicit examples of the consequences of folding and stretching are given in Section 1.6. There are outstanding questions as to whether such folding can exist throughout a homogeneous fluid; in general some cancellation will occur. In particular, when fields and flows are two–dimensional there is always too much folding, cancellation always dominates stretching and fields will decay. A simple example is provided by the non-dimensional flow field $\mathbf{u} = (-x, 0, z)$ (where the timescale has been based on a typical velocity U_0 and a typical length $\mathcal{L}$), with $\mathbf{B} = (0, 0, B(x, t))$. From (1.14) we can see, introducing the *magnetic Reynolds number*, $\mathrm{Rm} = U_0\mathcal{L}/\eta$, that B obeys

$$\partial_t B - x\,\partial_x B = B + \mathrm{Rm}^{-1}\,\partial_{xx} B . \tag{1.71}$$

If $B(x, 0) = \mathrm{Re}\left\{\beta_0\, \mathrm{e}^{\mathrm{i}k_0 x}\right\}$ then

$$B(x, t) = \mathrm{Re}\left\{\beta_0 \exp\left[t - k_0^2(\mathrm{e}^{2t} - 1)/2\mathrm{Rm}\right] \exp\left(\mathrm{i}k_0\, \mathrm{e}^t x\right)\right\} , \tag{1.72}$$

so that $|B|$ eventually decays superexponentially. This is due to diffusion acting on the exponentially increasing gradients caused by folding. In spite of this, however,

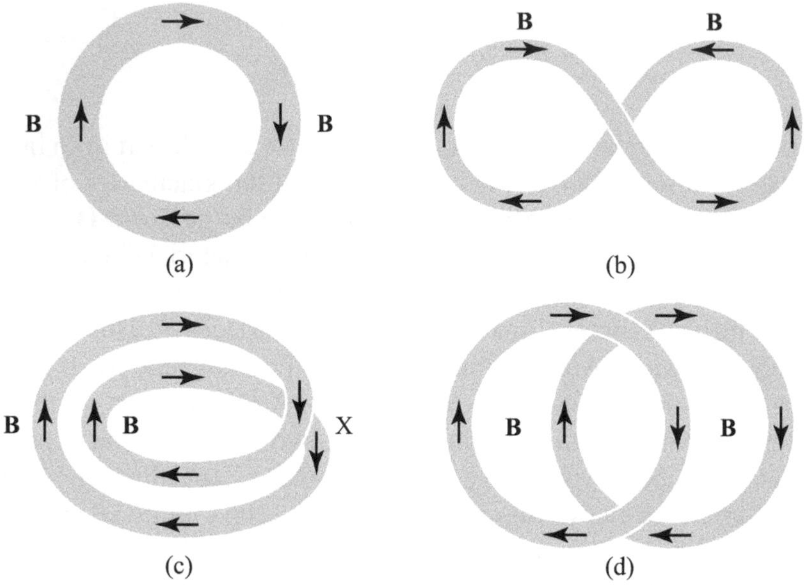

Figure 1.2 - Sketch of the stretch–twist–fold (STF) mechanism. The final magnetic flux is doubled.

we can have transient growth of magnetic energy for long times $\sim \ln(\mathrm{Rm}/k_0^2)$. As $\mathrm{Rm} \to \infty$ energy can increase indefinitely. This example is instructive in that it points up the singular nature of the infinite Rm limit; the limits of large times and large conductivity cannot be interchanged.

1.2.4. FAST AND SLOW DYNAMOS

An important application of dynamo theory is to astrophysical applications, in which we need to understand the behaviour of dynamo growth rates when Rm is very large. When Rm is of order unity, the two intrinsic timescales, associated with the turnover time and the Ohmic diffusion rate are comparable, but at large Rm the turnover/advective timescale is much shorter, while the Ohmic time is longer than any recognisable magnetic process. Thus we ask; can magnetic energy (or magnetic flux or dipole moment) grow at a rate independent of η as $\eta \to 0$? This leads to the distinction between fast and slow dynamos. The subject is treated in much greater detail in Section 1.6: here we give only a brief outline, concentrating on the problem of growth of flux at large Rm.

For a *slow dynamo* growth rates (on the advective timescale) $\to 0$ as $\mathrm{Rm} \to \infty$, while for a *fast dynamo* growth rates (or at least the $\limsup$ if there are many modes) do not tend to zero at large Rm. In this case the field appears on all scales as

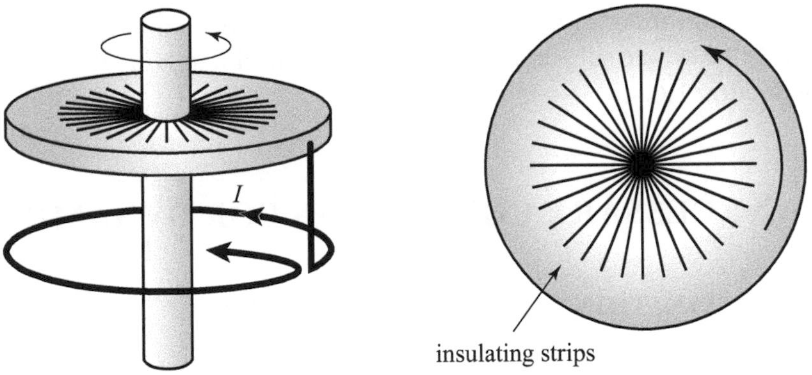

Figure 1.3 - The segmented Faraday dynamo (Moffatt, 1979). The insulating strips in the inner part of the disc ensure that the current is radial there.

$\mathrm{Rm} \to \infty$, and diffusion can never be neglected. This important point was first made by Moffatt and Proctor (1985). While as we have seen it is easy to produce an increase in magnetic energy if diffusion is entirely neglected, an increase of magnetic *flux* of dipole moment can only occur due to the presence of diffusion (as shown by Faraday's Law). This is necessary to get round flux conservation as diffusion becomes negligible. The Faraday disc dynamo has been discussed in Section 1.2.1. Here we examine a modification introduced by Moffatt (1979), which illustrates the role of diffusion in preventing fast dynamo action. This is the segmented Faraday dynamo (see also the brief discussion in Section 2.8). It is best understood by reference to Figure 1.3; the difference from the usual single disc dynamo geometry, as shown in Figure 1.3 is that currents are constrained to move radially on the disc except near the outer edge.

We can write down simple equations relating current in the wire I, current round the disc J, the angular velocity Ω and the fluxes through the wire and disc, respectively, Φ_I, Φ_J. We obtain

$$\Phi_I = LI + MJ, \quad \Phi_J = MI + L'J, \quad RI = \Omega\Phi_J - \frac{\mathrm{d}\Phi_I}{\mathrm{d}t}, \quad R'J = -\frac{\mathrm{d}\Phi_J}{\mathrm{d}t}. \quad (1.73)$$

We seek solutions $\propto \mathrm{e}^{pt}$. As for the usual dynamo, we find growth if $\Omega M > R$. The growth rate is

$$p_+ = \frac{\sqrt{(RL' + R'L)^2 + 4R'(\Omega M - R)(LL' - M^2)} - (RL' + R'L)}{2(LL' - M^2)}. \quad (1.74)$$

We can see that $p_+ > 0$ for all $\Omega > R/M$ but $p_+ \sim \sqrt{\Omega R'}$ as $\Omega \to \infty$. Thus the growth rate is controlled by diffusion and not exclusively by advection, and in particular the growth rate tends to zero on the advective timescale Ω^{-1}.

We shall discuss further aspects of fast and slow dynamo action in realistic flows in Section 1.3; the whole subject of the fast dynamo problem is treated in much more detail in Section 1.6.

1.3. NECESSARY CONDITIONS FOR DYNAMO ACTION

1.3.1. DEFINITIONS OF DYNAMO ACTION

In this section, we describe various rigorous results concerning dynamo action. It is helpful first to give a precise definition of what is meant by dynamo action: the definition depends on the geometry considered. We can consider either a bounded conductor surrounded by insulator, or magnetic fields and flows defined in a periodic box. Many generalisations are possible (for example, one could consider the effects of an external stationary conductor, as was done by Proctor, 1977a), but the details complicate the analysis.

Case 1: *Finite conductor.*

Suppose $\mathbf{B}$ is defined in a finite volume $\mathcal{D}$, surrounded (in ${}^c\mathcal{D}$) by an insulator. In ${}^c\mathcal{D}$ we have $\boldsymbol{\nabla} \times \mathbf{B} = \mathbf{0}$, with all components of $\mathbf{B}$ continuous at $\partial\mathcal{D}$, because there are no surface currents. We suppose no currents at infinity, so that $|\mathbf{B}| \sim \mathcal{O}(|\mathbf{x}|^{-3})$ as $|\mathbf{x}| \to \infty$.

Case 2: *Periodic dynamo.*

$\mathbf{B}$ is defined in a periodic domain $\mathcal{D} \in \mathbb{R}^3$, with $\int_{\mathcal{D}} \mathbf{B}\, d\mathbf{x} = \mathbf{0}$.

In each case $\mathbf{u}$ satisfies $\boldsymbol{\nabla}\cdot\mathbf{u} = 0$, and has time-bounded norm (for Case 2, we choose a frame so that the mean value of $\mathbf{u}$ vanishes. Several different norms can be defined, for example $\mathsf{U} \equiv \max_{\mathcal{D}}(|\mathbf{u}|)$, $\mathsf{S} \equiv \max_{\mathcal{D},i,j}(|\partial_j u_i|)$, $\mathsf{E}^{1/2} \equiv \left(\int_{\mathcal{D}} |\boldsymbol{\nabla}\mathbf{u}|^2 d\mathbf{x}\right)^{1/2}, \ldots$. In Case 1, we suppose that $\mathbf{u} = \mathbf{0}$ on $\partial\mathcal{D}$ (this is not strictly necessary for some of the bounds but aids the analysis). Then we can define the magnetic energy $\mathcal{M} = \frac{1}{2}\int |\mathbf{B}|^2\, \mathrm{d}\mathbf{x}$ where the integral is over $\mathbb{R}^3$ in Case 1, or over $\mathcal{D}$ in Case 2. The usual requirement for dynamo action is that $\mathcal{M}$ does not tend to zero as $t \to \infty$.

1.3.2. NON-NORMALITY OF THE INDUCTION EQUATION

In the next subsections we give several criteria which, if violated, rule out dynamo action. These are *necessary conditions*. It is notable that there are no general sufficient conditions known for dynamo action; working dynamos can only be found by explicit integration of particular flows. This is because the induction equation, considered as a parabolic linear operator, is *non-normal*; when $\mathbf{u}$ is independent of time, the eigenvectors found by looking for solutions $\propto \exp(pt)$ are not orthogonal, and so even when all eigenvectors p have negative real part, i.e. when we have a non-dynamo, the magnetic energy can still increase for some time. The condition that the energy decays is much stronger than that the spectrum is in the left hand half plane. The situation is analogous to that of the stability of shear flows, for which the energy stability result of Orr gives a bound on the Reynolds number that is far below observed stability thresholds. A simple example of this effect is provided by the interaction of a purely zonal flow with a meridional field in a sphere. For large Rm the zonal field increases more rapidly than the meridional field decays, leading to transient growth of the magnetic energy, but the meridional field eventually decays and the whole system runs down.

1.3.3. FLOW VELOCITY BOUNDS

If we nonetheless try to find conditions for the decay of the magnetic energy, we focus on (1.69), which gives us in Case 1

$$\frac{\mathrm{d}\mathcal{M}}{\mathrm{d}t} = \mathcal{P} - \eta\mathcal{J}\,, \tag{1.75}$$

where $\mathcal{P}$ and $\mathcal{J}$ can take the alternative forms:

$$\mathcal{P} = \int_{\mathcal{D}} \mathbf{B}\cdot(\mathbf{B}\cdot\boldsymbol{\nabla}\mathbf{u})\,\mathrm{d}\mathbf{x} = \int_{\mathcal{D}} (\mathbf{u}\times\mathbf{B})\cdot(\boldsymbol{\nabla}\times\mathbf{B})\,\mathrm{d}\mathbf{x}\,, \tag{1.76}$$

$$\mathcal{J} = \int_{\mathcal{D}} |\boldsymbol{\nabla}\times\mathbf{B}|^2\,\mathrm{d}\mathbf{x} = \int_{\mathbb{R}^3} |\boldsymbol{\nabla}\mathbf{B}|^2\,\mathrm{d}\mathbf{x}\,, \tag{1.77}$$

(for Case 2, we have the same results, but all integrals are taken over $\mathcal{D}$).

In order to construct the proofs we shall need a *Poincaré inequality*. Defining $\mathcal{F} = \frac{1}{2}\mathcal{J}/\mathcal{M}$, we have $\mathcal{F} \geq c^{-2}$; $c \propto (\int_{\mathcal{D}} \mathrm{d}\mathbf{x})^{1/3}$. For a sphere of radius a, $c = a/\pi$, while for a periodic cube of side a, $c = a/2\pi$. The proof of this result can either be done by the standard methods of variational calculus, or by expressing the magnetic field in terms of spherical harmonics.

Using the above inequality together with (1.75) in the case that $\mathcal{P} = 0$ (stationary conductor) we have the result that $\mathrm{d}(\ln\mathcal{M})/\mathrm{d}t \leq -2\eta c^{-2}$, so that the magnetic energy decays exponentially at a finite rate. It is not surprising, then that a finite velocity is needed for dynamo action to be possible. We can find bounds on each of the three norms defined above. We have the following bounds on $\mathcal{P}$:

$$(a) \quad \mathcal{P} \leq \mathsf{U} \int_{\mathcal{D}} |\mathbf{B} \cdot \nabla \mathbf{B}| d\mathbf{x} \leq \mathsf{U}(2\mathcal{M})^{1/2} \mathcal{J}^{1/2}, \quad \text{Childress (1969)},$$
$$(b) \quad \mathcal{P} \leq \mathsf{S}(2\mathcal{M}), \quad \text{Backus (1958)},$$
$$(c) \quad \mathcal{P} \leq \mathsf{E}^{1/2} \left(\int_{\mathcal{D}} |\mathbf{B}|^4 d\mathbf{x} \right)^{1/2} \leq \mathsf{E}^{1/2} c_1 (2\mathcal{M})^{1/4} \mathcal{J}^{3/4}, \quad \text{Proctor (1979)},$$

where c_1 is a dimensionless constant (Proctor, 1979, gives the value 4). Using these results, we can get three bounds on the exponential growth rate $\sigma = \mathrm{d}(\ln \mathcal{M})/\mathrm{d}t$:

$$(a) \quad \tfrac{1}{2}\sigma \leq \mathcal{F}^{1/2}(\mathsf{U} - \eta c^{-1}),$$
$$(b) \quad \tfrac{1}{2}\sigma \leq \mathsf{S} - \eta c^{-2},$$
$$(c) \quad \tfrac{1}{2}\sigma \leq \mathcal{F}^{3/4}(c_1 \mathsf{E}^{1/2} - \eta c^{-1/2}).$$

So if $\mathcal{M}$ is not to tend to zero we must have $\mathsf{U} > \eta/c$, $\mathsf{S} > \eta/c^2$, $\mathsf{E} > \eta^2/cc_1^2$. (The first result can be proved under the less restrictive assumption $\mathbf{u} \cdot \mathbf{n} = 0$ on $\partial\mathcal{D}$.) Because $\mathcal{F}$ has a minimum value we can get upper bounds on σ in cases (a) and (c) that do not involve $\mathcal{F}$:

$$(a) \quad \tfrac{1}{2}\sigma \leq \max \left[(\mathsf{U}/c - \eta c^{-2}), \frac{\mathsf{U}^2}{4\eta} \right],$$
$$(c) \quad \tfrac{1}{2}\sigma \leq \max \left[(c_1 \mathsf{E}^{1/2} c^{-3/2} - \eta c^{-2}), \frac{27\, c_1^4\, \mathsf{E}^2}{256\, \eta^3} \right].$$

It is notable that none of these bounds involves the kinetic *energy* $\mathcal{K} = \frac{1}{2}\int_{\mathcal{D}} |\mathbf{u}|^2 d\mathbf{x}$ of the velocity field. In fact a working dynamo can be found with arbitrarily small energy. Consider a velocity field $\mathbf{u}$ in a sphere of radius R surrounded by stationary conductor. For a steady dynamo the induction equation is invariant under $\mathbf{x} \to \mathbf{x}/R$, $\mathbf{u} \to R\mathbf{u}$, $\mathcal{K} \to R\mathcal{K}$. Thus as $R \to 0$ the necessary energy $\to 0$. The argument can be extended to the case where the conductor is replaced outside some large radius by an external insulator.

1.3.4. GEOMETRICAL CONSTRAINTS

These conditions are of two kinds; restrictions on the nature of *flows* that can give growing field, and constraints on the types of *field* that can be sustained by dynamo action. In the first category, until recently the best result was the *toroidal theorem* of Elsasser (1946), Bullard & Gellman (1954) (see also Moffatt, 1978). For Case 1, if we multiply equation (1.14) by $\mathbf{r} \equiv r\,\mathbf{e}_r$ and integrate then we obtain (defining $P = \mathbf{B} \cdot \mathbf{r}$, $Q = \mathbf{u} \cdot \mathbf{r}$),

$$\partial_t P + \mathbf{u} \cdot \nabla P = \mathbf{B} \cdot \nabla Q + \eta \Delta P \ \text{ in } \mathcal{D}, \tag{1.78}$$

with $\Delta P = 0$ in ${}^c\mathcal{D}$, and $P, \partial P/\partial r$ continuous on $\partial\mathcal{D}$.

If $\nabla \cdot \mathbf{u} = 0$, we can separate $\mathbf{u}$ into toroidal and poloidal parts $\mathbf{u}_T$, $\mathbf{u}_P$ (see Appendix B), where

$$\mathbf{u} = \mathbf{u}_T + \mathbf{u}_P \equiv \nabla \times (\phi \mathbf{r}) + \nabla \times \nabla \times (\psi \mathbf{r}). \tag{1.79}$$

It follows that $Q = L_2 \psi$, where L_2 is the angular momentum operator, defined as

$$L_2 = (\mathbf{r} \cdot \nabla)^2 - r^2 \Delta. \tag{1.80}$$

A similar decomposition can be made for $\mathbf{B}$, with

$$\mathbf{B}_T = \nabla \times (T\,\mathbf{r}), \ \mathbf{B}_P = \nabla \times \nabla \times (S\,\mathbf{r}), \ \text{with } P = L_2 S. \tag{1.81a,b,c}$$

If therefore the velocity field is toroidal, $\psi = 0$ and so Q also vanishes. Then equation (1.78) reduces to a sourceless diffusion-type equation for P, so that

$$\frac{1}{2}\frac{\mathrm{d}}{\mathrm{d}t}\int_{\mathcal{D}} P^2 \mathrm{d}\mathbf{x} = -\eta \int_{\mathbb{R}^3} |\nabla P|^2 \mathrm{d}\mathbf{x} < -\eta\, c^{-2} \int_{\mathcal{D}} P^2 \mathrm{d}\mathbf{x} \ \Rightarrow |P| \to 0. \tag{1.82}$$

Once $|P|$ and so $|\mathbf{B}_P|$ becomes negligible the equation for the toroidal part of the induction equation can also be simplified. Now $\mathbf{u}$, $\mathbf{B}$ are both toroidal, and

$$\nabla \times (\mathbf{u} \times \mathbf{B}_T) = \nabla \times [-\mathbf{r}(\mathbf{u} \cdot \nabla T)]. \tag{1.83}$$

After "uncurling" (integrating and setting the arbitrary function of r that arises to zero without loss of generality), we obtain

$$\partial_t T + \mathbf{u} \cdot \nabla T = \eta \Delta T, \ \text{with } T = 0 \ \text{ on } \partial\mathcal{D}. \tag{1.84}$$

Apart from the boundary conditions this is the same equation as satisfied by P, and we can show by similar means that $\int_{\mathcal{D}} T^2 d\mathbf{x} \to 0$ (exponentially) also. While this result does not rule out a transient increase in the magnetic energy of $\mathbf{B}_T$, which depends upon mean square gradients of T, it can be shown that if the magnetic energy does *not* tend to zero then $\mathcal{F}$ must increase without bound, and so eventually Childress' result above will be violated, giving a contradiction. Thus a dynamo is impossible.

A similar result holds in cartesian coordinates (Case 2), when $\mathbf{u} \cdot \mathbf{z} = 0$, then

$$\partial_t B_z + \mathbf{u} \cdot \nabla B_z = \mathbf{B} \cdot \nabla u_z + \eta \Delta B_z, \tag{1.85}$$

and we can apply exactly analogous reasoning (Zeldovich, 1957).

Busse (1975a) used equation (1.78) when $Q \neq 0$ to obtain a bound on the ratio of toroidal and poloidal field energies. We have

$$\begin{aligned}\frac{1}{2}\frac{\mathrm{d}}{\mathrm{d}t}\int_{\mathcal{D}} P^2 \mathrm{d}\mathbf{x} &= -\int_{\mathcal{D}} Q\,\mathbf{B}\cdot\boldsymbol{\nabla}P \mathrm{d}\mathbf{x} - \eta\int_{\mathbb{R}^3}|\boldsymbol{\nabla}P|^2 \mathrm{d}\mathbf{x}\\ &\leq \max_{\mathcal{D}} Q\left(2\mathcal{M}\cdot 2\int_{\mathbb{R}^3}|\mathbf{B}_P|^2\,\mathrm{d}\mathbf{x}\right)^{1/2} - 2\eta\int_{\mathbb{R}^3}|\mathbf{B}_P|^2\,\mathrm{d}\mathbf{x}\end{aligned}$$

where the inequality $\int_{\mathbb{R}^3}|\mathbf{B}_P|^2\mathrm{d}\mathbf{x} \leq \frac{1}{2}\int_{\mathbb{R}^3}|\boldsymbol{\nabla}P|^2\mathrm{d}\mathbf{x}$ has been used (see, for example, Proctor, 2004). Then we have the result that

$$\max_{\mathcal{D}} Q \geq \eta\left(\frac{1}{\mathcal{M}}\int_{\mathbb{R}^3}|\mathbf{B}_P|^2\,d\mathbf{x}\right)^{1/2}. \tag{1.86}$$

Though this result may be useful in interpreting geomagnetic data, it is not of course an anti–dynamo theorem. Nonetheless it turns out that (as might be expected) dynamo action can be ruled out if the poloidal flow is sufficiently weak for any given toroidal flow. In fact it is possible to find inequalities for time derivatives of P^2 and T^2, namely (choosing some constant $\mu > 0$)

$$\begin{aligned}\frac{1}{2}\frac{\mathrm{d}}{\mathrm{d}t}\left(\int_{\mathcal{D}}(P^2+\mu T^2)\mathrm{d}\mathbf{x}\right) &\leq \left(\frac{aU_P}{\sqrt{2}}-\eta\right)(\mathcal{P}^2+\mu\mathcal{T}^2)\\ &\quad+\left[a^2U_P+\frac{\mu}{2}(U_T+U_P)\right]\mathcal{P}\mathcal{T}, \end{aligned}\tag{1.87}$$

where $\mathcal{P}^2 = \int_{\mathbb{R}^3}|\boldsymbol{\nabla}P|^2\mathrm{d}\mathbf{x}$, $\mathcal{T}^2 = \int_{\mathcal{D}}|\boldsymbol{\nabla}T|^2\mathrm{d}\mathbf{x}$, and U_P, U_T are the maxima of $|\mathbf{u}_P|$, $|\mathbf{u}_T|$ respectively in $\mathcal{D}$. For an appropriate choice of μ, we can show that the best possible condition under which the left hand side is negative definite is

$$a^2U_P\,(U_T+U_P) - 2\left(\eta-\frac{aU_P}{\sqrt{2}}\right)^2 < 0 \quad \text{or} \quad a^2U_PU_T + 2\sqrt{2}\eta aU_P < 2\eta^2 \tag{1.88a,b}$$

(Proctor, 2004). Poincaré inequalities may be used to show that the integrals of both P^2 and T^2 decay exponentially, and this implies eventual decay of the magnetic energy as argued above. The result, expressions (1.88a,b), does not rule out dynamo action when the velocity field $\mathbf{u}$ is purely poloidal; and indeed there are examples in the literature of dynamos with purely poloidal velocity fields. A classic example is provided by the twin-torus dynamo of Gailitis (Gailitis, 1970), see Figure 1.4.

As regards constraints on the field, the main result is *Cowling's Theorem* (Cowling, 1934): An axisymmetric magnetic field cannot be maintained by dynamo action. It should be noted that if **B** is axisymmetric then so is **u** but the converse is not true,

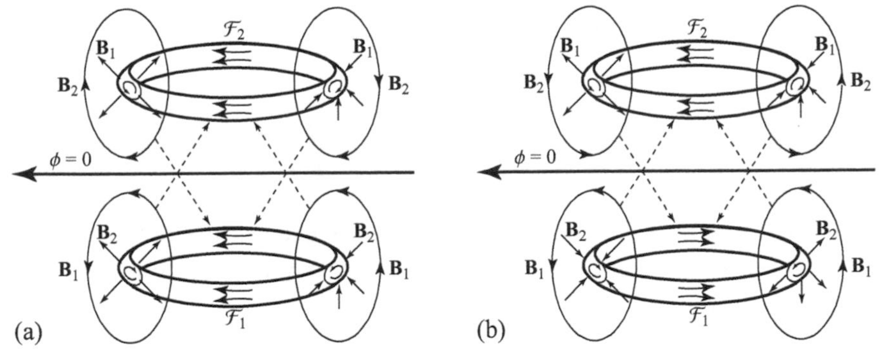

Figure 1.4 - The poloidal dynamo of Gailitis (after Gailitis 1970). The flow is axisymmetric, while the magnetic field is proportional to $e^{i\phi}$. Two different parities of solution are shown. Suffix 1 refers to fields generated by the lower ring, suffix 2 those due to the upper ring. For more details see e.g. Fearn *et al.* (1988)

and the dynamo of Gailitis (1970) above provides an example of an axisymmetric flow field which acts as a dynamo for non-axisymmetric fields. There are several proofs of this in various cases. We first follow the proof of Braginsky (1964a). We again assume $\boldsymbol{\nabla}\cdot\mathbf{u} = 0$, and that the conducting region $\mathcal{D}$ is spherical. Since $\mathbf{B}, \mathbf{u}$ are axisymmetric, we can separate the zonal and meridional parts of (1.14) by writing [in polar coordinates (s, ϕ, z)];

$$\mathbf{B} = B\mathbf{e}_\phi + \boldsymbol{\nabla}\times(A\mathbf{e}_\phi) = B\,\mathbf{e}_\phi + \mathbf{B}_P\,, \qquad \mathbf{u} = \mathbf{u}_P + U\,\mathbf{e}_\phi\,. \tag{1.89a,b}$$

Since there are no imposed zonal currents, we get

$$\frac{\partial}{\partial t}A + \frac{1}{s}\,\mathbf{u}_P\cdot\boldsymbol{\nabla}(sA) = \frac{1}{\mathrm{Rm}}\left(\Delta - \frac{1}{s^2}\right)A\,, \tag{1.90a}$$

$$\frac{\partial}{\partial t}B + s\,\mathbf{u}_P\cdot\boldsymbol{\nabla}\left(\frac{B}{s}\right) = s\,\mathbf{B}_P\cdot\boldsymbol{\nabla}\left(\frac{U}{s}\right) + \frac{1}{\mathrm{Rm}}\left(\Delta - \frac{1}{s^2}\right)B\,. \tag{1.90b}$$

Further simplification ensues if we write $A = \chi/s$, $B = \psi s$, $U = \Omega s$. Then we obtain the alternative system

$$\partial_t\chi + \mathbf{u}_P\cdot\boldsymbol{\nabla}\chi = \eta\left(\Delta - \frac{2}{s}\frac{\partial}{\partial s}\right)\chi\,, \tag{1.91a}$$

$$\partial_t\psi + \mathbf{u}_P\cdot\boldsymbol{\nabla}\psi = \mathbf{B}_P\cdot\boldsymbol{\nabla}\Omega + \eta\left(\Delta + \frac{2}{s}\frac{\partial}{\partial s}\right)\psi\,, \tag{1.91b}$$

with $(\Delta - (2/s)\partial_s)\chi = \psi = 0$ in ${}^c\mathcal{D}$ and $\chi \sim \mathcal{O}(|\mathbf{x}|^{-1})$ as $|\mathbf{x}| \to \infty$. It is notable that the toroidal field does not appear in the equation for χ. The analysis now proceeds in a similar manner to that for the toroidal theorem. We form the poloidal "energy equation"

$$\frac{1}{2}\frac{\mathrm{d}}{\mathrm{d}t}\int_{\mathcal{D}} \chi^2 \mathrm{d}\mathbf{x} = \eta \int_{\mathcal{D}} \chi \left(\Delta - \frac{2}{s}\frac{\partial}{\partial s}\right) \chi \mathrm{d}\mathbf{x} = -\eta \int_{\mathbb{R}^3} |\nabla \chi|^2 \mathrm{d}\mathbf{x} \leq -\eta\, c_3^2 \int_{\mathcal{D}} \chi^2 \mathrm{d}\mathbf{x}\,. \tag{1.92}$$

It is then clear that $\chi^2 \rightarrow 0$, and so by arguments used in the previous subsection, eventually the poloidal field will decay also. When χ is negligible, we can form a similar relation for ψ and show similarly that

$$\begin{aligned}\frac{1}{2}\frac{\mathrm{d}}{\mathrm{d}t}\int_{\mathcal{D}} \psi^2 \mathrm{d}\mathbf{x} &= \eta \int_{\mathcal{D}} \psi \left(\Delta + \frac{2}{s}\frac{\partial}{\partial s}\right) \psi \,\mathrm{d}\mathbf{x} \\ &= -\eta \left(\int_{\mathcal{D}} |\nabla \psi|^2 \mathrm{d}\mathbf{x} + 2\pi \int_{-a}^{a} \psi(0,z)^2 \,\mathrm{d}z \right),\end{aligned} \tag{1.93}$$

and so $\psi^2 \rightarrow 0$ also. We can prove very similar results for fields (and so flows) that are independent of z.

There are other types of proofs of Cowling's theorem, which allow us to generalise the problem to permit η to depend on position. They show the impossibility of the maintenance of a steady magnetic field against Ohmic decay when there is a neutral curve on which the meridional field vanishes at an O-type neutral point. Suppose that this is at $\mathbf{X}$, and consider a small meridional circle $S_\mathcal{E}$ centred at $\mathbf{X}$, boundary $C_\mathcal{E}$, radius ε, with $B_\mathcal{E} \equiv (2\pi\varepsilon)^{-1} \oint_{C_\mathcal{E}} |\mathbf{B}_P| \mathrm{d}\mathbf{x}$,

$$(\max_{\mathcal{D}} |\mathbf{u}|) B_\mathcal{E} S_\mathcal{E} \geq \int_{S_\mathcal{E}} (\mathbf{u}_P \times \mathbf{B}_P) \cdot \mathrm{d}\mathbf{x} = \int_{S_\mathcal{E}} \eta(\mathbf{x}) \nabla \times \mathbf{B}_P \cdot \mathrm{d}\mathbf{x} \sim 2\pi\varepsilon B_\mathcal{E} \eta(\mathbf{X})\,. \tag{1.94}$$

This leads to a contradiction as $S_\varepsilon \sim \varepsilon^2$. The neutral ring argument, while in some sense more general than the Braginsky proof in that the field does not have to be exactly axisymmetric, is more limited in other ways, since the result of the proof is to rule out steady fields (for steady flows) and so has nothing to say about exponential decay. Fuller details are given in Moffatt (1978) and Fearn *et al.* (1988).

When the flow is not incompressible useful results are harder to find. The equation for χ is still correct. Since $\chi(0,z) = 0$ and $\chi \rightarrow 0$ as $|\mathbf{x}| \rightarrow \infty$, there must exist a positive maximum of χ, at $\mathbf{X}(t)$ where $\nabla\chi = \mathbf{0}$, $\Delta\chi \leq 0$. This rules out a growing dynamo with a poloidal field. Hide & Palmer (1982) have argued that if $\Delta\chi(\mathbf{X}) = 0$ for all time then χ becomes non-differentiable near $\mathbf{X}$ and so $\mathbf{B}_0 \rightarrow \mathbf{0}$. The arguments used are appealing but are hard to formulate rigorously. They have been criticised by Ivers & James (1984). These authors have used maximum principles to show that both poloidal and toroidal fields decay exponentially, but the bounds for the decay rates so far found are not useful, in that the associated decay times are much longer than that of any astrophysical body. The question of how far a properly selected compressible flow in a sphere can reduce the Ohmic decay rate for an axisymmetric field remains partially open.

1.4. STEADY AND TIME-DEPENDENT VELOCITIES

In this short section, we discuss the differences between the dynamo properties of steady and time-dependent flow fields. This is necessary because so much of our intuition on the efficacy of dynamo action is based on thinking about steady flows, and these can be misleading in the general case.

1.4.1. TWO SIMPLE EXAMPLES

Smooth, steady flows $\mathbf{u}$ are not usually efficient as dynamos at large Rm, because there is not enough stretching. In particular, smooth axisymmetric or 2D flows cannot be fast dynamos if they are steady, since there is then no exponential stretching of material lines (the relation between stretching properties of the flow and growth rates at large Rm has been discussed earlier, and will be treated in much more detail in the following). On the other hand time-dependent flows can be very efficient as dynamos, even if they have a very simple Eulerian form. As an example consider two related flows, the so-called [G.O.] Roberts (Roberts, 1970) and Galloway–Proctor (GP) (Galloway & Proctor, 1992) flows

$$\text{Roberts flow:}\quad \mathbf{u}(x,y) \propto \boldsymbol{\nabla}\times[\psi(x,y)\,\mathbf{e}_z] + \gamma\,\psi(x,y)\,\mathbf{e}_z\,,\qquad \psi = \sin x \sin y\,; \tag{1.95}$$

$$\text{GP-flow:}\quad \mathbf{u}(x,y,t) \propto \boldsymbol{\nabla}\times[\psi(x,y,t)\,\mathbf{e}_z] + \gamma\,\psi(x,y,t)\,\mathbf{e}_z\,,\qquad \psi = \sin(y+\varepsilon\sin\omega t) + \cos(x+\varepsilon\cos\omega t)\,. \tag{1.96}$$

The Roberts flow has three components, but depends only on x and y. It has a fixed cellular pattern; there is no stretching except at the cell corners. The GP-flow has a very similar cellular structure in the Eulerian flow, but the cellular pattern rotates. The consequences for the stretching properties are profound; there is stretching (positive Liapunov exponent) almost everywhere (see Figure 1.5). We can find dynamo action for both these flows by looking for fields of the form $\mathbf{B} = \mathrm{Re}\left\{\widetilde{\mathbf{B}}(x,y,t)\mathrm{e}^{\mathrm{i}kx}\right\}$. Then the growth rate (for the GP-flow the average growth rate over one time period of the flow) depends on Rm and k.

For the Roberts flow the optimum growth rate occurs at large wavenumber[2] k for

2 The scale k^{-1}, though small compared to the cell size, is long compared to the thin boundary layer scale $\mathrm{Rm}^{-1/2}$ for field near stagnation points.

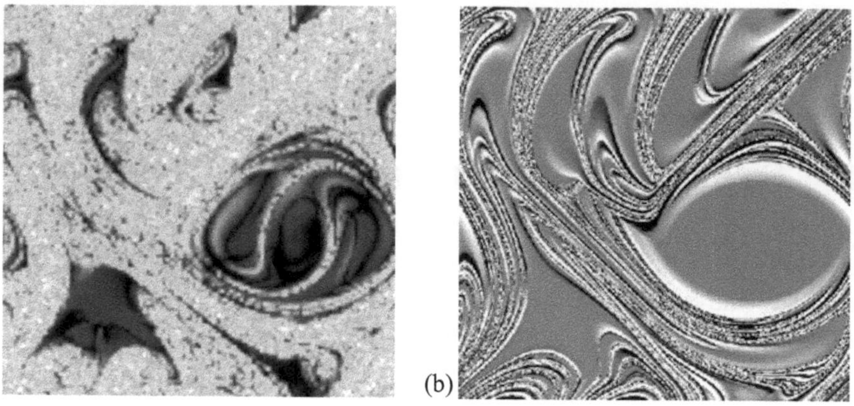

(a) (b)

Figure 1.5 - Chaos in the GP-flow at time $t \approx 20$. (a) Finite-time Liapunov exponents (courtesy of D.H. Hughes) for $\omega = 1$, $\varepsilon = 1$, showing there is exponential stretching almost everywhere. (b) Normal field B_z (courtesy of F. Cattaneo). Note the large regions of multiply folded field. (**See colour insert**.)

$\mathrm{Rm} \gg 1$, and in fact $k \sim (\mathrm{Rm}^{1/2}/\ln \mathrm{Rm})$. As $\mathrm{Rm} \to \infty$ the optimum growth rate is $\sim \mathcal{O}(\ln(\ln \mathrm{Rm})/\ln \mathrm{Rm})$, see Figure 1.6. So this flow is not (quite) a fast dynamo.

The GP-flow is completely different. The growth rate is $\mathcal{O}(1)$ for large Rm, and the optimum wavenumber also $\mathcal{O}(1)$. Here the flow is chaotic, and though there are thin flux structures, chaotic regions near the stagnation points do not scale with Rm. The choice of k for optimum growth is presumably related to the widths of these structures. Time dependent flows of this type have proved a fertile ground for extensive numerical simulation of fast dynamo properties.

1.4.2. PULSED FLOWS

Another important aspect of time-dependent flows is that many restrictions that would prevent dynamo action for the instantaneous flow field do not apply when the flow is time dependent. This is associated with the non-normality of the induction equation, as discussed above. As a particular example we show how the Toroidal and Zeldovich theorems can be got round for time-dependent flows. Consider the *pulsed Beltrami flow* (Soward, 1993).

$$\mathbf{u} = \begin{cases} (0, \sin x, \cos x) & (0 \le t \le \tau) \\ (\sin y, 0, \cos y) & (\tau \le t \le 2\tau), \quad \text{etc.} \end{cases} \tag{1.97}$$

This is a planar flow at all but isolated discrete times, but during each interval τ we can have transient growth, and this can lead to dynamo action. The development is

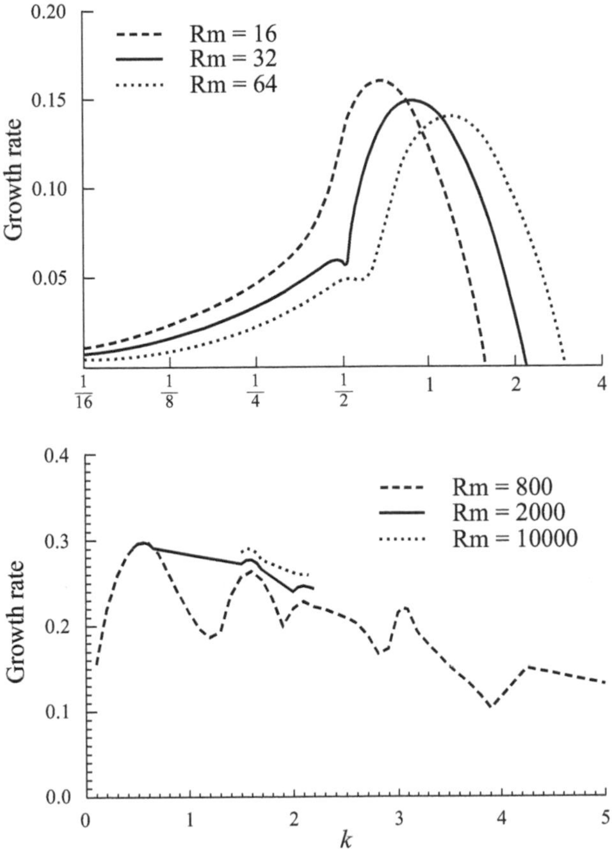

Figure 1.6 - Growth rates for the Roberts and GP flows, as functions of Rm and k. Top figure shows the Roberts flow, with peak growth rates decreasing at large Rm, and the critical k increasing. Bottom figure shows the same data for the GP flow with $\varepsilon = \omega = 1$. Note the convergence of the growth rate and critical wavenumber for large Rm.

most easily seen when we set $\eta = 0$ (for small η the results are almost the same as long as τ is not too large). In the interval $(0 \le t \le \tau)$ consider the horizontally averaged field $\overline{\mathbf{B}} \equiv \mathrm{Re}\left\{\widetilde{\mathbf{B}}_H \exp(\mathrm{i}kz)\right\}$, then

$$\overline{\mathbf{B}}_H(\tau) = J_0(k\tau)\overline{\mathbf{B}}_H(0) - \mathrm{i}\tau J_1(k\tau)\overline{B_x}(0)\,\mathrm{e}_y\,, \tag{1.98}$$

which can be large for large τ. If we add (small) diffusion, we still get growth, provided τ is much less than the diffusion time. Then the second pulse can refold and stretch the field and give further enhancement. A more complicated version of this kind of flow is one that arises in thermal convection, where there is a homoclinic connection between two different planar flows. In this case the flow is not a dynamo, because the interval between switching of the flows tends to infinity. The addition of noise to the system, however, will render the switching time finite and can induce instability. For further details see Gog *et al.* (1999).

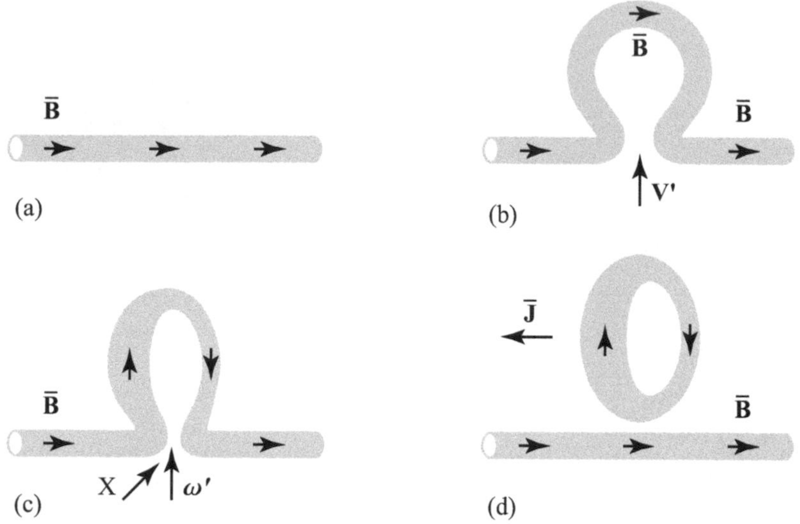

Figure 1.7 - The cyclonic event mechanism as envisaged by Parker (after Roberts, 1994). The uniform field in (a) is pulled up in (b), twisted in (c), and then reconnects to form a field loop with a normal component (and so EMF) anti-parallel to the original field (d).

1.5. TWO–SCALE DYNAMOS

1.5.1. THE TWO–SCALE CONCEPT AND PARKER'S MODEL

The dynamo flows we have already met: Roberts, GP and pulsed flows and extensions to 3D flows such as the ABC model (see Section 1.6, and Childress & Gilbert 1995) are small scale dynamos. The magnetic field has scales comparable to that of $\mathbf{u}$. But if B exists on two distinct scales then dynamo action can be easily verified. Perhaps the simplest model is that of Parker (1955). Suppose that small scale "cyclonic events" act on a uniform field. If the velocity of these small-scale motions has non-zero *helicity*, i.e. $\mathbf{u} \cdot \boldsymbol{\nabla} \times \mathbf{u} \neq 0$, then the field is twisted by the motion as in Figure 1.7. By Ampère's Law (1.3) there is generated an electromotive force (EMF) parallel to the original field. The sign of this EMF is opposite to the helicity for short-lived events. However for longer lived events there is not in general any such clear correlation. If these helical motions are distributed isotropically then any EMF perpendicular to the field will cancel out when an average is taken over all events. When this new EMF is incorporated, we get an extra term $\boldsymbol{\nabla} \times \alpha\mathbf{B}$ on the r.h.s. of equation (1.14); this new term is called the α–effect. An extended discussion including nonlinear effects is given in Section 2.7 and Chapter 6.

Parker's model of the solar magnetic field supposes that the large scale field is axisymmetric. The crucial role of the α–effect is to sustain poloidal from toroidal field. The same mechanism is also capable of sustaining toroidal from poloidal field, but is ignored in his model in favour of the much more effective role of zonal shears. We then obtain the model system

$$\partial_t A + \frac{1}{s}\,\mathbf{u}_P \cdot \boldsymbol{\nabla}(sA) = \alpha B + \frac{1}{\mathrm{Rm}}\left(\Delta - \frac{1}{s^2}\right) A\,, \tag{1.99a}$$

$$\begin{aligned} \partial_t B + s\,\mathbf{u}_P \cdot \boldsymbol{\nabla}\left(\frac{B}{s}\right) &= [\boldsymbol{\nabla} \times (\alpha \boldsymbol{\nabla} \times \mathbf{B}_P)] \\ &\quad + s\,\mathbf{B}_P \cdot \boldsymbol{\nabla}\left(\frac{U}{s}\right) + \frac{1}{\mathrm{Rm}}\left(\Delta - \frac{1}{s^2}\right) B\,. \end{aligned} \tag{1.99b}$$

We discuss solutions of this equation below when we have looked at a more systematic derivation.

1.5.2. MEAN FIELD ELECTRODYNAMICS

We now suppose formally that the magnetic and velocity fields exist on a small scale ℓ and a large scale L, and/or on short and long timescales. We may then define some average over the short scales (denoted by $\overline{\cdots}$) and write $\mathbf{B} = \overline{\mathbf{B}} + \mathbf{B}'$, $\mathbf{u} = \overline{\mathbf{u}} + \mathbf{u}'$, etc. Then, taking the average,

$$\partial_t \overline{\mathbf{B}} = \boldsymbol{\nabla} \times \boldsymbol{\mathcal{E}} + \boldsymbol{\nabla} \times (\overline{\mathbf{u}} \times \overline{\mathbf{B}}) - \boldsymbol{\nabla} \times (\eta \boldsymbol{\nabla} \times \overline{\mathbf{B}})\,, \tag{1.100}$$

where $\boldsymbol{\mathcal{E}} \equiv \overline{\mathbf{u}' \times \mathbf{B}'}$.

In order to calculate $\boldsymbol{\mathcal{E}}$ we need to find $\mathbf{B}'$, whose equation is

$$\begin{aligned} \partial_t \mathbf{B}' &= \boldsymbol{\nabla} \times (\overline{\mathbf{u}} \times \mathbf{B}') + \boldsymbol{\nabla} \times (\mathbf{u}' \times \overline{\mathbf{B}}) \\ &\quad + \boldsymbol{\nabla} \times (\mathbf{u}' \times \mathbf{B}' - \overline{\mathbf{u}' \times \mathbf{B}'}) - \boldsymbol{\nabla} \times (\eta \boldsymbol{\nabla} \times \mathbf{B}')\,. \end{aligned} \tag{1.101}$$

This equation can only be solved in special cases, but we can make some general remarks about the nature of $\boldsymbol{\mathcal{E}}$. Clearly, for fixed $\mathbf{u}'$, $\mathbf{B}'$ depends linearly on $\overline{\mathbf{B}}$ and so $\boldsymbol{\mathcal{E}}$ is a linear functional of $\overline{\mathbf{B}}$. Assuming the simplest possible local relation, we obtain the expression

$$\mathcal{E}_i = \alpha_{ij}\overline{B}_j - \beta_{ijk}\,\partial_j \overline{B}_k + \ldots \tag{1.102}$$

α_{ij} is a pseudo-tensor; the symmetric part is non-zero only if the statistics of $\mathbf{u}$ lack mirror-symmetry. The anti-symmetric part, on the other hand, acts like a velocity and because of this it can only be non-zero if the statistics lack homogeneity, or if there is anisotropy combined with broken reflection symmetry. If we suppose that the statistics of $\mathbf{u}'$ are isotropic but not mirror-symmetric, then $\alpha_{ij} = \alpha\delta_{ij}$.

We can relate the pseudo-scalar α to the helicity of the small-scale flow. Both arise from broken mirror-symmetry, and we can give explicit relations in limiting cases.

We can similarly simplify the second term in the expansion for $\boldsymbol{\mathcal{E}}$, for in the isotropic case $\beta_{ijk} = \beta\varepsilon_{ijk}$, which can be identified as a "turbulent magnetic diffusivity".

All the foregoing assumes that $\mathbf{B}'$ owes its existence entirely to $\overline{\mathbf{B}}$. In this case, in particular, the value of α can be determined simply by making $\overline{\mathbf{B}}$ uniform, in which case $\mathcal{E}_i$ is exactly $\alpha_{ij}\overline{B}_j$. However, as we have already seen, when Rm is large enough there is a possibility, indeed a likelihood that a small-scale field can exist even when $\overline{\mathbf{B}} = \mathbf{0}$. It is hard to see how to interpret the α–effect in this situation since any "mean-field" effect has to exist on top of an already equilibrated small-scale field. The problem is then intrinsically nonlinear and so beyond the scope of this section, though it will be considered in the next chapter.

Supposing that indeed $\boldsymbol{\mathcal{E}}$ owes its existence to $\overline{\mathbf{B}}$, we can see that the α–effect can lead to dynamo action. Consider (writing $\eta + \beta = \eta'$)

$$\partial_t \overline{\mathbf{B}} = \boldsymbol{\nabla} \times (\alpha \overline{\mathbf{B}}) - \boldsymbol{\nabla} \times (\eta' \boldsymbol{\nabla} \times \overline{\mathbf{B}}) \,. \tag{1.103}$$

If α, β are uniform, we get solutions of form $\mathrm{Re}\left\{\widehat{\mathbf{B}}\exp(\mathrm{i}\mathbf{k}\cdot\mathbf{x} + pt)\right\}$, with $(p + \eta' k^2)^2 = \alpha^2 k^2$, so $p_+ > 0$ for all sufficiently small k. It can thus be seen that mean-field dynamo action is inevitable on all sufficiently large scales, provided only that $\alpha \neq 0$.

The α tensor will take more general forms with lower symmetry of flow statistics. In a sphere, when there are two preferred directions, namely the rotation $\boldsymbol{\Omega}$ and the radial vector $\mathbf{r}$, we will get the more general form

$$\boldsymbol{\mathcal{E}} = \alpha_1(\boldsymbol{\Omega}\cdot\mathbf{r})\overline{\mathbf{B}} + \alpha_2\mathbf{r}(\boldsymbol{\Omega}\cdot\overline{\mathbf{B}}) + \alpha_3\boldsymbol{\Omega}(\mathbf{r}\cdot\overline{\mathbf{B}}) + \ldots \tag{1.104}$$

Note that both rotation and a preferred direction would seem necessary for an α–effect.

A detailed discussion of possible forms of $\boldsymbol{\mathcal{E}}$ in various cases is given by Krause & Rädler (1980).

As explained above it is hard to calculate α in the general case. There are two special cases in which analytical progress can be made:

(a) If Rm, based on the small length scale ℓ, is very small, then there is no small-scale dynamo. We calculate α by approximating the equation for $\mathbf{B}'$ by

$$\mathbf{0} = \overline{\mathbf{B}}\cdot\boldsymbol{\nabla}\mathbf{u}' + \eta\Delta\mathbf{B}' \,, \tag{1.105}$$

with $\overline{\mathbf{B}}$ uniform. If we consider, as an example, $\mathbf{u}'$ in the simple Fourier form $\propto \mathrm{Re}\left\{\mathrm{e}^{\mathrm{i}\mathbf{k}\cdot\mathbf{x}}\right\}$ then we have $B'_i = \mathrm{i}\overline{B}_j k_j u'_i/\eta k^2$ so

$$\mathcal{E}_i = \alpha_{ij}\,\overline{B}_j = \mathrm{i}\,\varepsilon_{ipq}\,k_j\,\overline{u'^*_p u'_q}\;\overline{B}_j/\eta k^2 \,. \tag{1.106}$$

If we choose coordinates in which $\mathbf{k} = (0, 0, k)$ then $\mathcal{E}_i = \alpha_{ij}\overline{B}_j$ where $\alpha_{ij} = \alpha\delta_{i3}\delta_{j3}$ and $\alpha\eta k^2 = -\varepsilon_{ijk}k_j\overline{u_i'^* u_k'}$. The latter quantity is just the helicity, and so as predicted from Parker's ansatz we see that α has the opposite sign to the helicity. Adding together many modes of this type, we can reproduce α due to any velocity field.

(b) The "short-sudden" approximation. This is used when the small-scale Rm is *large*, and thus is harder to justify. In general the fluctuating field $\mathbf{B}'$ will be much larger than the mean field, and so extra assumptions have to be made to simplify the equations. We suppose that the fluctuating velocity field, and so the fluctuating magnetic field, becomes decorrelated on a time τ_c short enough that the correlated part of $\mathbf{B}'$ is again small compared to $\overline{\mathbf{B}}$. We ignore diffusion. Then $\partial_t\mathbf{B}' \approx \overline{\mathbf{B}}\cdot\boldsymbol{\nabla}\mathbf{u}'$. This can be solved to give $B_i' \approx \tau_c\overline{\mathbf{B}}\cdot\boldsymbol{\nabla}u_i'$, so in the isotropic case

$$\alpha = -\frac{\tau_c}{3}\,\overline{\mathbf{u}'\cdot\boldsymbol{\nabla}\times\mathbf{u}'}\,. \tag{1.107}$$

Again we see that α is anticorrelated with helicity.

The approximations involved in both these limits essentially ignore the self-interaction of $\mathbf{u}'$ and $\mathbf{B}'$ in the $\mathbf{B}'$ equation. The equation becomes intractable when these terms are not ignored, and so apart from these extreme cases it is hard to give useful results. However there is one result available without approximation in Gruzinov & Diamond (1994). If we suppose the fields and flow statistically steady with uniform imposed field $\overline{\mathbf{B}}$ (and periodic boundary conditions for simplicity), and, using the vector potential introduced in (1.17), write $\mathbf{B}' = \boldsymbol{\nabla}\times\mathbf{A}'$, we then have

$$\begin{aligned}\partial_t\mathbf{A}' = &-\boldsymbol{\nabla}\Phi + \overline{\mathbf{u}}\times\mathbf{B}' + \mathbf{u}'\times\overline{\mathbf{B}}\\ &+\mathbf{u}'\times\mathbf{B}' - \overline{\mathbf{u}'\times\mathbf{B}'} - \eta\boldsymbol{\nabla}\times\mathbf{B}'\,,\end{aligned} \tag{1.108}$$

so (ignoring boundary terms that arise from integration by parts)

$$0 = \tfrac{1}{2}\left(\overline{\mathbf{B}'\cdot\partial_t\mathbf{A}' + \mathbf{A}'\cdot\partial_t\mathbf{B}'}\right) = -\overline{\mathbf{B}}\cdot\boldsymbol{\mathcal{E}} - \eta\overline{\mathbf{B}'\cdot\boldsymbol{\nabla}\times\mathbf{B}'}\,. \tag{1.109}$$

This holds without approximation if boundary terms are ignored. Thus in the isotropic case

$$\alpha|\overline{\mathbf{B}}|^2 = -\frac{\eta}{3}\,\overline{\mathbf{B}'\cdot\boldsymbol{\nabla}\times\mathbf{B}'}\,. \tag{1.110}$$

This result gives some guidance about the behaviour of α as the small-scale Rm increases. In particular, it shows that diffusion must be included in any proper model of α. If α is independent of η at large Rm, leading to a fast mean field dynamo, and we posit that $|\mathbf{B}'| \sim \eta^a|\overline{\mathbf{B}}|$, $|\boldsymbol{\nabla}\times\mathbf{B}'| \sim \eta^{-1/2}|\mathbf{B}'|$, and is intermittent with a filling factor $\sim \eta^b$, then $2a + b = -1/2$. Possible solutions include $b = 1/2$, $a = -1/2$ giving sheet-like fields, while if the fields are primarily tubes rather than sheets we might expect $a = -1$, so $b = 3/2$.

1.5.3. MEAN FIELD MODELS

If the α–effect is accepted as a model of the effects of small-scale flows on the large scale field, then Cowling's theorem does not apply, since now toroidal field can sustain poloidal field, and so we can investigate axisymmetric models. Physical considerations (the role of the Coriolis force in inducing helicity) suggest that in a rotating body such as the Earth or the Sun α is odd about the equator. Similar considerations suggest that the zonal flow U should be even, so we can get two types of field structure: (i) Dipole: where B is odd about the equator, and A is even. (ii) Quadrupole; A is odd, B is even. Examples of fields of the two types are shown in Figure 1.9.

Most models are one of two types: (i) "α^2", with U neglected. This has been used to model stationary e.g. planetary dynamos; (ii) "$\alpha\omega$" in which the α term in (1.99b) is neglected, as in the Parker model. α^2–models typically give steady dynamos (real growth rates) while $\alpha\omega$–models usually give cyclic dynamos (complex growth rates). We can understand the latter in terms of dynamo waves. We use cartesian geometry; let

$$A = A(x,t)\,, \qquad B = B(x,t)\,, \qquad \mathbf{B}_p \cdot \boldsymbol{\nabla} U \sim \omega \partial_x A\,, \tag{1.111a,b,c}$$

where x is a variable corresponding to latitude [the term (1.111c) is referred to as the ω–effect]. Substituting into (1.99), and modelling radial derivatives with a constant damping term, we obtain the simplified system [compare to Equation (6.1a,b) in Chapter 6].

$$\partial_t A = \alpha B + \eta\left(\partial_{xx} A - K^2 A\right)\,, \quad \partial_t B = \omega\,\partial_x A + \eta\left(\partial_{xx} B - K^2 B\right)\,. \tag{1.112a,b}$$

This has travelling wave solutions with $A, B \propto \exp\left[ik(x - ct)\right]$ when

$$\alpha\omega = \pm 2\eta^2 (k^2 + K^2)^2/k\,, \qquad c = -\alpha\omega/[2\eta(k^2 + K^2)]\,. \tag{1.113}$$

Note that the modulus of the *dynamo number*, $\mathrm{D} = \alpha\omega/\eta^2 K^3$, takes a minimum value $16/3\sqrt{3}$ when $k = K/\sqrt{3}$, also note the definite sign of the wave speed c which depends on the sign of D.

In a spherical geometry, $\alpha\omega$–models can be used to give models of the solar cycle (butterfly diagram) by identifying large B with regions of sunspot eruption. Forms of α, U and any meridional velocity are prescribed, and the equations solved numerically as an eigenvalue problem to obtain marginal (periodic solutions). A particularly comprehensive study was carried out by Roberts (1972). While these kinematic studies are now overshadowed by the dynamical studies reported on later, it is interesting to note that travelling waves of activity, similar to the Parker waves, can be seen propagating latitudinally. The direction of propagation depends on the sign

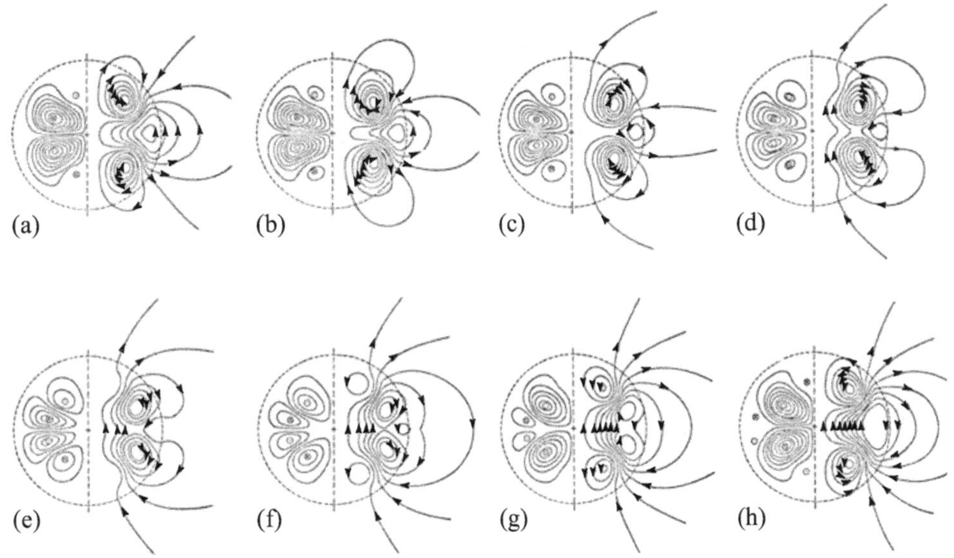

Figure 1.8 - Oscillatory Dipolar solutions for an $\alpha\omega$–dynamo (from Roberts, 1972).

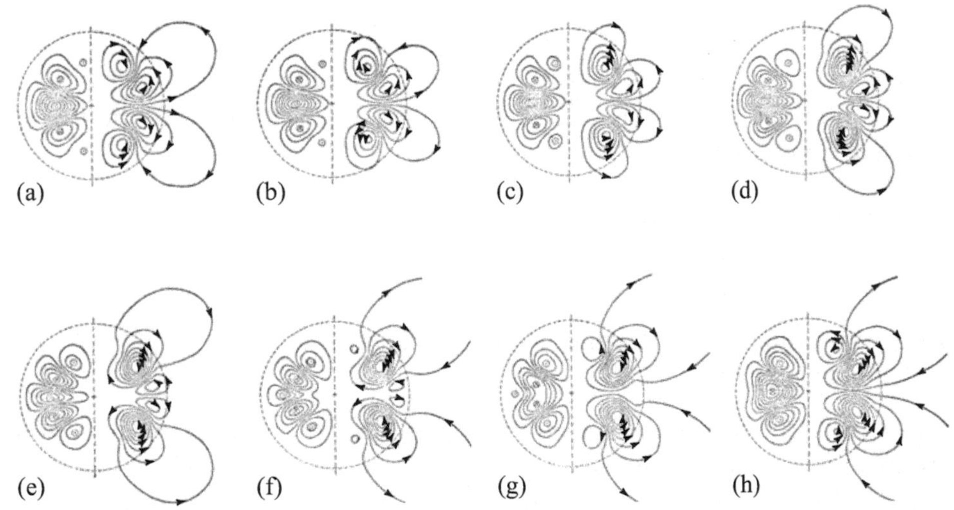

Figure 1.9 - As above, but quadrupolar solutions.

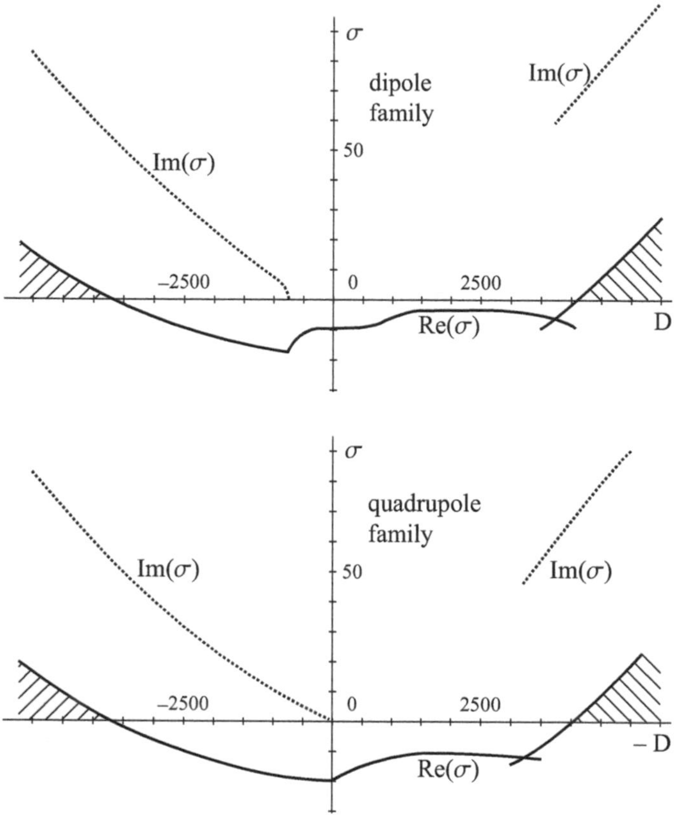

Figure 1.10 - Real (solid lines) and imaginary (dotted lines) parts of the growth rate for typical $\alpha\omega$–dynamos (from Roberts, 1972). Unstable regions are shaded. The x–axis is reversed in the lower figure. Note the similarity between the figures, as suggested in Proctor (1977b).

of the dynamo number, and since on the Sun the waves move towards the equator, we can make some deductions about the dynamics leading to α. The associated frequency of oscillation also emerges from the calculation and is comparable with the turbulent diffusion time.

There is an interesting near symmetry, associated with the adjoint dynamo problem, between dipole (quadrupole) modes with $\alpha, \mathbf{u}$, and quadrupole (dipole) modes with $\alpha, -\mathbf{u}$ (Proctor, 1977b). This is illustrated in Figure 1.10, which shows growth rates for a particular dynamo model. The figures for the different parities are very similar, though the x–axis, measuring the dynamo number is reversed in the right-hand figure.

We can relate these results to the well known butterfly diagram of the solar cycle (shown in Figure 6.3, page 284 and discussed in Section 6.1). If we identify the sites of sunspot activity with maxima of B, (since we believe that sunspots are manifestations of large toroidal fields through the magnetic buoyancy instability), then the equatorward propagation of the disturbances will lead to a picture like the observations.

These global models of dynamo action have been superseded by models in which the shear is concentrated just below the convective zone of the sun, and so the α–effect is separated spatially from the shear. This "interface model" (Parker, 1993), which also leads to dynamo waves, will be discussed in detail along with its dynamical consequences in Chapter 6.

1.6. LARGE MAGNETIC REYNOLDS NUMBERS

Let us now turn to the evolution of magnetic fields under the induction equation at large magnetic Reynolds number, as explained in Section 1.2.4. We will begin by giving a formal definition, before discussing the motivation for such studies and presenting various examples. For further information and more references than can easily be provided here see the reviews Childress (1992), Bayly (1994), Soward (1994), Childress & Gilbert (1995) and Gilbert (2003).

Suppose we have a given incompressible flow $\mathbf{u}$ with a typical length scale $\mathcal{L}$ and velocity scale U, and the magnetic diffusivity is η. Then after non-dimensionalisation using these scales, the induction equation (1.16) becomes

$$\partial_t \mathbf{B} + \mathbf{u} \cdot \boldsymbol{\nabla} \mathbf{B} = \mathbf{B} \cdot \boldsymbol{\nabla} \mathbf{u} + \varepsilon \Delta \mathbf{B} \,, \tag{1.114a}$$

where $\varepsilon^{-1} \equiv \mathrm{Rm} = U\mathcal{L}/\eta$ is the magnetic Reynolds number, and

$$\boldsymbol{\nabla} \cdot \mathbf{B} = 0 \,, \qquad \boldsymbol{\nabla} \cdot \mathbf{u} = 0 \,. \tag{1.114b,c}$$

For a given flow $\mathbf{u}$ and an $\varepsilon > 0$, dynamo action may take place, the fastest growing magnetic field mode having an exponential growth rate $\gamma(\varepsilon)$; for example for a steady flow

$$\mathbf{B}(\mathbf{x}, t) \propto \mathbf{b}(\mathbf{x}) \mathrm{e}^{\sigma t}, \qquad \gamma = \mathrm{Re}\,\{\sigma\} \,. \tag{1.115}$$

The flow $\mathbf{u}$ is a *fast dynamo* if the *fast dynamo exponent*

$$\gamma_0 \equiv \lim_{\varepsilon \to 0} \gamma(\varepsilon) \tag{1.116}$$

is positive; otherwise it is a *slow dynamo*. For a fast dynamo, magnetic field growth occurs on the turnover time-scale of the underlying flow $\mathbf{u}$ (on which we first non-dimensionalised), independently of molecular diffusion. A slow dynamo operates on a slower, diffusion-limited time-scale, as we shall see in some examples below.

Why study fast dynamos? Before answering this question, it is best to widen the scope of our enquiry: our interest is in dynamo mechanisms (fast and slow) at large Rm, the structure of magnetic fields, and the saturation of dynamo instabilities (in which case (1.114a) must be supplemented by an equation for $\mathbf{u}$). Mathematically, the limit $\mathrm{Rm} \to \infty$ or $\varepsilon \to 0$ in equation (1.114a) is a singular limit as ε multiplies the highest derivative, and so this requires careful treatment by numerical codes, or by asymptotic means. Taking this limit allows a clear subdivision of dynamos and unstable magnetic modes into different families, as we shall see. This classification can be useful even if Rm is not particularly large in an application; however in many astrophysical applications Rm is very large, and dynamo processes do appear to operate on fast time-scales; for example in the Sun Rm is of the order of 10^8 and the magnetic field oscillates on the fast, 11-year solar cycle.

Finally, developing mathematical tools to cope with fast dynamos is a considerable challenge with wider application, for example to vorticity and passive scalar transport in complex flows (e.g. Reyl, Antonsen & Ott, 1998; Fereday *et al.* 2002). The induction equation (1.114a) is challenging because the behaviours as $\varepsilon \to 0$ and for $\varepsilon = 0$ are markedly different at large times. If one simply sets $\varepsilon = 0$, then the induction equation corresponds to advecting a vector field $\mathbf{B}$ in the given flow $\mathbf{u}$, field lines being frozen in the fluid. The field will gain finer and finer scales, and the magnetic energy will grow because of field stretching. Because of this reduction of scale, there are no well-behaved eigenfunctions for a general flow in the case $\varepsilon = 0$ (Moffatt & Proctor, 1985). Now suppose diffusion is introduced: this can have very dramatic effects because of the fine scales in the field. For example for a typical planar flow $\mathbf{u}(x, y, t) = (u_1, u_2, 0)$, the magnetic energy grows indefinitely for $\varepsilon = 0$, but for any $\varepsilon > 0$ it eventually decays, in keeping with the anti–dynamo theorem for planar flows discussed in Section 1.3.4.

In this short review we will consider examples of slow and fast dynamos in flows and mappings, but only make passing reference to issues of dynamo saturation; these will be taken up in Chapter 2.

1.6.1. SLOW DYNAMOS IN FLOWS

Perhaps the simplest example of a slow dynamo is the Ponomarenko dynamo (e.g. Ponomarenko, 1973; Gilbert, 1988; Ruzmaikin, Sokoloff & Shukurov, 1988). In cylindrical polar coordinates (r, θ, z),

$$\mathbf{u} = r\Omega(r)\,\mathbf{e}_\theta + U(r)\,\mathbf{e}_z\,; \tag{1.117}$$

this is a swirling helical flow, depending only on radius r. Here we focus on the case of a smooth flow, although Ponomarenko's original paper had piecewise constant U and Ω. Related flows were studied by Lortz (1968).

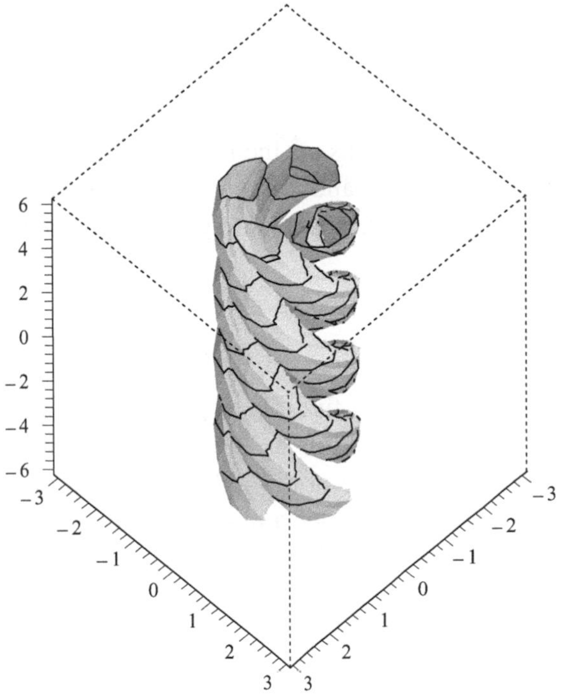

Figure 1.11 - Magnetic field in the Ponomarenko dynamo at large magnetic Reynolds number $\mathrm{Rm} = \varepsilon^{-1}$ forms spiralling tubes of field localised near the resonant stream surface (from Gilbert, 2003).

We may consider a magnetic mode $\mathbf{B} = \mathbf{b}(r)\,\exp\left[\mathrm{i}m\theta + \mathrm{i}kz + \sigma t\right]$, classified by wavenumbers m and k. In this case the induction equation (1.114a) for b_r and b_θ becomes

$$[\sigma + \mathrm{i}\, m\,\Omega(r) + \mathrm{i}\, k\, U(r)]\; b_r = \varepsilon\,\left[(\Delta_m - r^{-2})\, b_r - 2\,\mathrm{i}\, m\, r^{-2} b_\theta\right] , \tag{1.118a}$$

$$[\sigma + \mathrm{i}\, m\,\Omega(r) + \mathrm{i}\, k\, U(r)]\; b_\theta = r\,\Omega'(r)\, b_r + \varepsilon\,\left[(\Delta_m - r^{-2})\, b_\theta + 2\,\mathrm{i}\, m\, r^{-2} b_r\right] . \tag{1.118b}$$

We can drop the b_z equation as b_z can be reconstructed from the condition $\boldsymbol{\nabla}\cdot\mathbf{B} = 0$. The basic mechanism can be seen in these two equations, and can be described as of $\alpha\omega$–type. The stretching of radial field by the gradient of angular velocity $\Omega'(r)$ generates b_θ field in equation (1.118b) (an ω–effect), while diffusion of b_θ field in curved geometry can generate radial field by the last term in (1.118a) (broadly speaking, an α–effect).

To obtain formulae for growth rates at small ε, we rescale, so as to capture the fastest growing modes, setting

$$m = \varepsilon^{-1/3} M\,, \quad k = \varepsilon^{-1/3} K\,, \quad r = a + \varepsilon^{1/3} s\,. \tag{1.119a,b,c}$$

Here we are seeking a mode localised at a radius a (whose significance we will discover shortly) in the interior of the fluid. We scale the growth rate as

$$\sigma = \varepsilon^{-1/3}\sigma_0 + \sigma_1 + \varepsilon^{1/3}\sigma_2 + \cdots\,, \tag{1.120}$$

and for the field, set

$$b_r = \varepsilon^{1/3} b_{r0}(s) + \cdots\,, \qquad b_\theta = b_{\theta 0}(s) + \cdots\,, \tag{1.121a,b}$$

These expansions are then to be substituted into (1.118a,b) and the flow field (1.117) also Taylor-expanded about $r = a$ in powers of s. When this is done, corresponding powers of ε are equated, to give at the leading two orders:

$$\sigma_0 + \mathrm{i}\, M\, \Omega(a) + \mathrm{i}\, K\, U(a) = 0\,, \tag{1.122a}$$

$$\mathrm{i}\, M\, \Omega'(a) + \mathrm{i}\, K\, U'(a) = 0\,, \quad \sigma_1 = 0\,. \tag{1.122b,c}$$

The first simply fixes σ_0 as purely imaginary, advection of the magnetic field mode by the flow at radius a. The second condition implies that a mode with given (m, k) tends to localise at the radius a where the shear of the flow is aligned with field lines, assuming such a radius exists; if it does not, then we may expect the mode to localise at a boundary.

At the next order we obtain from (1.118) coupled parabolic cylinder equations, which may be written in the form

$$(c_0 + \mathrm{i}\, c_2\, s^2 - \partial_s^2)\, b_{r0} = -2\,\mathrm{i}\, M\, a^{-2}\, b_{\theta 0}\,, \tag{1.123a}$$

$$(c_0 + \mathrm{i}\, c_2\, s^2 - \partial_s^2)\, b_{\theta 0} = a\, \Omega'(a)\, b_{r0}\,, \tag{1.123b}$$

where $c_0 = \sigma_2 + M^2/a^2 + K^2$ and $2c_2 = M\Omega''(a) + KU''(a)$. These coupled differential equations can be rewritten as

$$\mathcal{P}_+ \mathcal{P}_- b_{r0} = 0\,, \tag{1.124a}$$

with $$\mathcal{P}_\pm \equiv (c_0 \pm d + \mathrm{i}\, c_2\, s^2 - \partial_s^2), \quad d \equiv (-2\,\mathrm{i}\, M\, \Omega'(a)/a)^{1/2}\,. \tag{1.124b,c}$$

The parabolic cylinder operators $\mathcal{P}_+$ and $\mathcal{P}_-$ commute and so the solution for b_{r0} is a linear combination of solutions to the two equations $\mathcal{P}_\pm b_{r0} = 0$. Putting these into canonical form gives

$$\left[\partial_\sigma^2 - (\tfrac{1}{4}\sigma^2 + c_\pm)\right]\, b_{r0} = 0\,, \tag{1.125}$$

with $$\sigma = s(4\,\mathrm{i}\, c_2)^{1/4}\,, \qquad c_\pm = (c_0 \pm d)/(4\,\mathrm{i}\, c_2)^{1/2}\,, \tag{1.126a,b}$$

and solutions that decay for $s \to \pm\infty$ exist only if $c_\pm = -j - \frac{1}{2}$ for $j = 0, 1, 2, \ldots$. This gives eigenvalues of the original dynamo problem.

Finally returning to the original variables gives leading order growth rates,

$$\gamma \equiv \operatorname{Re}\sigma \simeq \mp\sqrt{\varepsilon |m\Omega'(a)|/a} - (j + \tfrac{1}{2})\sqrt{\varepsilon |m\Omega''(a) + kU''(a)} - \varepsilon(m^2/a^2 + k^2)\,. \tag{1.127}$$

This formula was derived for $m, k = \mathcal{O}(\varepsilon^{-1/3})$, but is in fact valid for all m, k. The fastest growing modes have scales $m, k = \mathcal{O}(\varepsilon^{-1/3})$ and $\gamma = \mathcal{O}(\varepsilon^{1/3})$, and so this provides a slow dynamo. The resulting magnetic fields have spiralling tubes along which the field is approximately directed; for example, an $m = 2$ mode is illustrated schematically in Figure 1.11.

An important feature of the formula (1.127) is that the first two terms scale in precisely the same way with m (and k) and ε, while the last term can always be made subdominant at small ε by taking m (and k) small enough. Taking the upper sign, and $j = 0$, for a dynamo to occur at large Rm for some mode (m, k) it follows that the first, positive term must dominate the second, negative term, and this only occurs at the given resonant surface $r = a$ provided the purely geometrical condition, obtained with the help of (1.122b),

$$r \left| \frac{\Omega''(r)}{\Omega'(r)} - \frac{U''(r)}{U'(r)} \right| < 4\,, \tag{1.128}$$

is met there. One can write down flows for which this is not satisfied, and so which would not be dynamos at large Rm, even though they appear well-endowed with helical streamlines.

This example can be generalised away from strictly circular geometry to allow more general stream surfaces (Gilbert & Ponty, 2000). As an example of an application, the resulting theory gives excellent predictions of the instability threshold for these Ponomarenko modes in a study (Plunian, Marty & Alémany, 1999) of dynamo instabilities in model nuclear reactor flows, even at moderate Rm. Such modes can also occur in convective cellular flows (e.g. Ponty, Gilbert & Soward, 2001). A smooth flow of the form (1.117) can give slow dynamo action, but if $\Omega(r)$ and $U(r)$ have discontinuities at some radius $r = a$, then fast dynamo action can occur, with growth rates $\gamma = \mathcal{O}(1)$ for modes with $m, k = \mathcal{O}(\varepsilon^{-1/2})$ (Gilbert, 1988); we will not discuss this further here. Some aspects of the saturation of smooth Ponomarenko dynamos are studied in Bassom & Gilbert (1997) for $\mathrm{Re} \gg \mathrm{Rm} \gg 1$: the flow adopts a layered structure, with solid body rotation in a broad region surrounding the radius a and where the α–effect and field are concentrated. Outside are thin layers where the shear and ω–effect are significant.

These Ponomarenko modes, with spiralling tubes of field alternating in direction, are rather localised; for example a mode would sit in one cell of a convective flow. They are far from the mean field dynamos which are traditionally studied by means of an α–effect and discussed in Section 1.5. The best laminar flow to study which allows such large-scale field generation is the Roberts (1970) flow, which was introduced in Section 1.4,

$$\mathbf{u} = (\sin x \cos y, -\cos x \sin y, K \sin x \sin y)\,, \qquad K = \sqrt{2}\,. \tag{1.129a,b}$$

This is a Beltrami flow, with vorticity $\boldsymbol{\nabla} \times \mathbf{u} = K\mathbf{u}$ proportional to the flow itself. It thus provides a steady solution to the Euler equation, and is a member of the ABC family of flows; the general ABC flow is given by

$$\mathbf{u} = (C \sin z + B \cos y, A \sin x + C \cos z, B \sin y + A \cos x)\,, \qquad (1.130)$$

where A, B and C are parameters (and (1.129a,b) is obtained by setting $A = B = 2^{-1/2}$, $C = 0$, rescaling and rotating axes through $\pi/4$). At low Rm the Roberts flow provides an α–effect dynamo, destabilising large-scale magnetic field modes (e.g. Moffatt, 1978). The field is dominated by diffusion; the flow is a small perturbation to the field on the scales of the flow, but one which has a large-scale destabilising effect. A nonlinear study within this low Rm model reveals an inverse cascade of magnetic energy to large scales (Gilbert & Sulem, 1990).

At large Rm, however, the field tends to localise on stream surfaces. The flow is independent of z; there is an array of square helical cells, in which the flow is spiralling, where dynamos can exist. However the key new feature is the network of hyperbolic stagnation points $(x, y) = (n\pi, m\pi)$, joined by straight-line separatrices: new magnetic modes appear, localised on this network. A mode $\mathbf{B} \propto \exp(\mathrm{i}\,k\,z + \sigma\,t)$ with wavenumber k in z has growth rate

$$\gamma \equiv \sigma = \alpha\,k - \varepsilon\,k^2\,, \qquad \alpha = -\tfrac{1}{2}\,k\,\varepsilon^{1/2}\,G\,, \qquad G \simeq 1.0655\,. \qquad (1.131\text{a,b,c})$$

(Childress, 1979; Soward, 1987). This is valid for $k = \mathcal{O}(1)$, but the growth rate increases with k, and the above equation is suggestive of a maximum growth rate $\gamma = \mathcal{O}(1)$ for $k = \mathcal{O}(\varepsilon^{-1/2})$, that is, a fast dynamo. A delicate analysis (Soward, 1987) shows that the maximum growth rate is in fact given by

$$\gamma = \mathcal{O}((\log\log\varepsilon^{-1})/\log\varepsilon^{-1})\,, \qquad k = \mathcal{O}(\varepsilon^{-1/2}/\sqrt{\log\varepsilon^{-1}})\,. \qquad (1.132\text{a,b})$$

Is this a fast dynamo? Not technically, as the growth rate still goes to zero as $\varepsilon \to 0$ and so the dynamo is slow. However the decay is only logarithmic in ε, and what is a logarithm between friends? In view of our opening remarks in this chapter, this is therefore still an interesting and important slow dynamo mechanism; for example, similar Roberts modes are found in the study of Plunian, Marty & Alémany (1999). One important feature to note is that the fastest growing magnetic field modes have a very small lengthscale in z. They are extended in x and y (unlike the Ponomarenko modes), but the magnetic energy is entirely at the diffusive scales, $k \simeq \mathcal{O}(\varepsilon^{-1/2})$. In the Roberts dynamo diffusion is still playing a crucial role in the amplification process, and the field has to adopt diffusive scales to benefit. This should be contrasted with the fast dynamos below, where the magnetic fields have typically a power-law spread of energy over a range of scales, from the full scale of the flow down to diffusive scales.

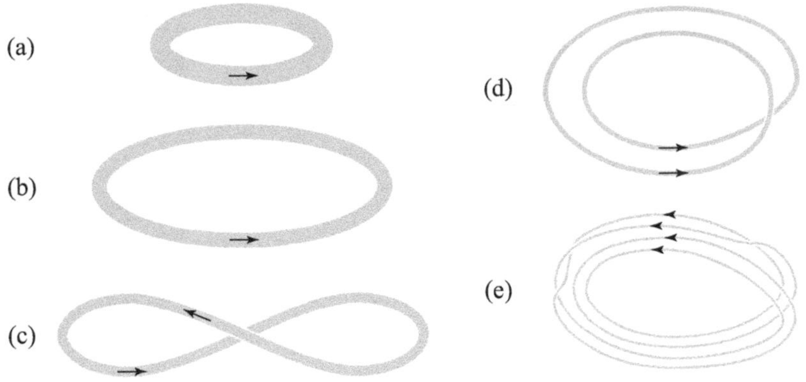

Figure 1.12 - The stretch–twist–fold dynamo: an initial flux tube (a), is stretched (b), twisted (c) and folded (d), to obtain a doubled flux tube. (e) a folded flux tube after two STF cycles.

1.6.2. THE STRETCH–TWIST–FOLD PICTURE

In so far as finding fast dynamos, the problem with the flows so far discussed is that diffusion is crucially involved in the amplification process. In fact, in rough terms, these steady flows have dynamos of an $\alpha\omega$–type at large Rm. Field perpendicular to stream surfaces is stretched out along stream surfaces by the flow, giving strong field parallel to stream surfaces (an ω–effect); in curved geometry weak diffusion acts on this parallel field to generate perpendicular field (an α–effect). This $\alpha\omega$–cycle allows the field to grow and the dynamo to operate. To avoid the dynamo process being limited by diffusion as in these examples, it is necessary for advection by the fluid flow to do all the amplification itself without relying on diffusion. The simplest picture of how this may be achieved is in the stretch–twist–fold (STF) dynamo (see Section 1.2.3; Vainshtein & Zeldovich, 1972), depicted in Figure 1.12 (see also Figure 1.2).

In this figure the flow is not given explicitly. Instead the action of the flow is shown on a tube of field frozen into the fluid; we may think of the perfectly conducting case $\varepsilon = 0$ for the moment. The initial tube (a) is stretched to twice its length, its cross section being halved, giving (b). This doubles the field strength and so multiplies the energy by four. The field is then twisted into a figure-of-eight (c) and folded (d), to give a tube of similar structure to the original in (a). If this process is repeated, with a time period $T = 1$, then the energy at time $t = n$ will be $E_n = 2^{2n} E_0$, corresponding to a growth rate $\gamma = \log 2$. Now let us reintroduce weak diffusion; this will begin to play a role when the field scale becomes of order $\varepsilon^{1/2}$, and will begin to smooth and reconnect the field (Moffatt & Proctor, 1985). Because the action of the STF moves has been to bring tubes of field largely into alignment, one would expect diffusion not to lead to a wholesale destruction of field, but simply to

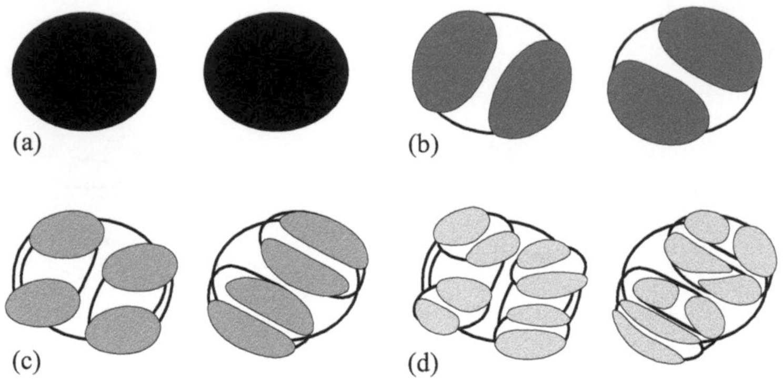

Figure 1.13 - Cross sections of the torus (a) initially, and (b) after one STF iteration, (c) two iterations and (d) three iterations. Shading indicates regions containing field, and white is field-free fluid.

smooth the fine structure in the field, giving $\gamma \simeq \log 2$ for $0 < \varepsilon \ll 1$ and so a fast dynamo with $\gamma_0 \simeq \log 2$.

There are a number of problems with realising the STF picture in practice. The first is that it is not easy to specify a fluid flow to apply the STF moves (Moffatt & Proctor, 1985). But even in such a flow (or iterated mapping), the field rapidly becomes unmanageable (Vainshtein *et al.*, 1996), for the reasons indicated schematically in Figure 1.13. Starting with a magnetic field (black) in a torus, whose cross section is shown in (a), the doubled up field in (b) will entrain field-free fluid (white) and so some field will lie outside the original torus. As the stretch, twist, fold operations are repeated (c,d) the bundle of field lines and entrained fluid will increase in volume until the whole fluid volume contains strands of field, and it is necessary to understand the global nature of the fluid flow and folding of field, a problem that has not been addressed. The field lines also become tangled up in a complicated fashion (see Gilbert, 2002a), with poorly-understood implications for diffusion of field.

Nonetheless, the STF moves provide a useful picture of how a fast dynamo with growth rate $\gamma_0 \simeq \log 2$ might operate. This is only a *picture*, hard to realise in practice (for example in a convective fluid flow!), but informative nonetheless. The key points to bear in mind are: first, the flow has chaotic particle trajectories, as the length of the field lines in the tube doubles with each period. In fact Lagrangian chaos in a smooth fluid flow is a necessary ingredient for fast dynamo action; technically the topological entropy h of the flow must be positive (Finn & Ott, 1988; Klapper & Young, 1995), as we discuss further below. Such chaotic flows are easy to realise; but the second key ingredient in a fast dynamo is constructive alignment of magnetic field vectors. The STF moves tend to bring fields close with similar orientation, which minimises the possible destruction of field through magnetic diffusion.

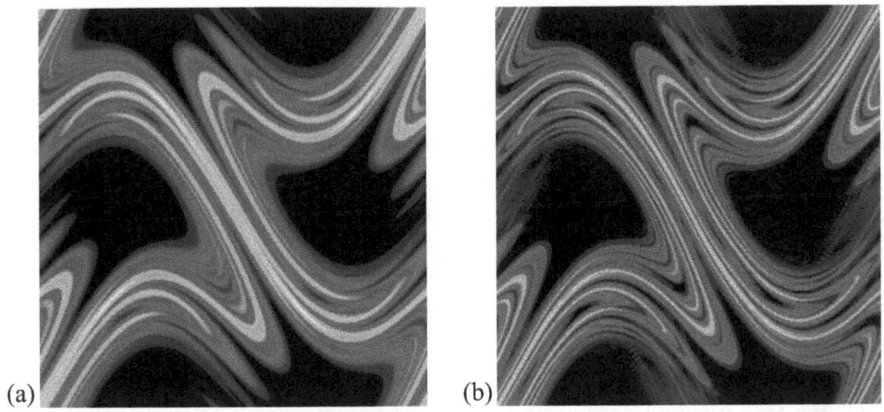

Figure 1.14 - Eigenfunctions of Otani's flow for $k = 0.8$ and (a) $\varepsilon = 5 \times 10^{-4}$ and (b) $\varepsilon = 5 \times 10^{-5}$. The magnitude of the magnetic field is shown, with black indicating zero field. (**See colour insert**.)

1.6.3. FAST DYNAMOS IN SMOOTH FLOWS

The numerical study of dynamo action in chaotic flows began with investigation of steady ABC flows defined by equation (1.130) (Galloway & Frisch, 1986). However these are generally three-dimensional, having complex stream line topology, and solving the induction equation is computationally intensive. It is easier to deal with two-dimensional flows $\mathbf{u}(x, y, t)$ (independent of z), and the best-studied examples are essentially variants of (1.129), for which time dependence is introduced and results in a breaking up of the separatrices joining hyperbolic stagnation points, to give chaotic layers. One example is the flow of Otani (1993),

$$\mathbf{u}(x, y, t) = 2\cos^2 t\,(0, \sin x, \cos x) + 2\sin^2 t\,(\sin y, 0, -\cos y)\,, \qquad (1.133)$$

which is similar to an example studied by Galloway & Proctor (1992) and discussed in Section 1.4.1. Growing magnetic fields take a Floquet form

$$\mathbf{B}(x, y, z, t) = \mathrm{e}^{\mathrm{i}k\,z+\sigma t}\,\mathbf{b}(x, y, t)\,, \qquad (1.134)$$

in which $\mathbf{b}$ is periodic in time, period 2π. The z–wave number k is a parameter and for each diffusivity ε, the mode with maximum growth rate may be found. Numerical study (Otani, 1993) shows good evidence for fast dynamo action with

$$\gamma_0 \simeq 0.39\,, \qquad k \simeq 0.8\,. \qquad (1.135\text{a,b})$$

Note that the value of k at which growth rates are maximised does not depend on ε; the magnetic field has a large-scale component, unlike in the slow Ponomarenko and Roberts dynamos discussed above.

However while numerical studies show that the convergence of $\gamma(\varepsilon)$ to γ_0 is rapid as $\varepsilon \to 0$, the magnetic eigenfunctions become more and more complicated, as indicated in Figure 1.14. This shows a snapshot of magnetic energy (averaged over z) plotted as a function of (x, y). In the centre are bands of field, resulting from chaotic stretching and folding in the flow in the (x, y)–plane. In the large black, field-free regions the flow has islands of KAM surfaces with insignificant stretching.[3]

The action of the flow is to fold field in the plane, giving the belts of field dominating the centre of the picture. This would not give any kind of constructive alignment of field vectors, however, without the shearing motion in z, which advects field up and down, giving changes of sign of field by virtue of the $\mathrm{e}^{\mathrm{i}kz}$ dependence on z; see (1.134). By this means bands in the centre of the picture have fields that are largely aligned. This "stretch–fold–shear" mechanism amplifies a large-scale field component, while creating a cascade of fluctuations to small scales; these fluctuations are smoothed out by diffusion, which plays a relatively passive role.

While the above flow of Otani (1993) has been written down without any obvious link to real astrophysical fluid flows, the above mechanism of stretching and folding in the (x, y) plane and shearing in z is very natural and can occur, for example, in convection. Two-dimensional time-dependent convective eddies can give chaotic folding in the plane containing the roll axes, while the influence of rotation (natural in an astrophysical body) can drive flows along their axes (Kim, Hughes & Soward, 1999; Ponty, Gilbert & Soward, 2001).

The flows of Otani (1993) and Galloway & Proctor (1992) have also been studied in dynamical regimes, where the given fluid flow is now driven by a prescribed body force until the field grows and becomes dynamically involved through the Lorentz force. Studies indicate that the field saturates through suppression of the Lagrangian chaos and alpha effect in the flow, although there is also some evidence that an inverse cascade of magnetic energy to large scales may occur in spatially extended systems, on long time-scales; see for example, Maksymczuk & Gilbert (1998) and Cattaneo *et al.* (2002).

1.6.4. FAST DYNAMOS IN MAPPINGS

Studying fast dynamo action in flows such as Otani's above, or an ABC flow, is extremely difficult. The problem is that it is not just the individual Lagrangian trajectories that are important, but how ensembles of trajectories lead to folding of magnetic field. Most progress in understanding has been obtained by studying dynamo action in models for which the fluid flow is replaced by a mapping.

3 KAM=Kolmogorov-Arnold-Moser: this refers to regions where trajectories are not chaotic and lie on surfaces.

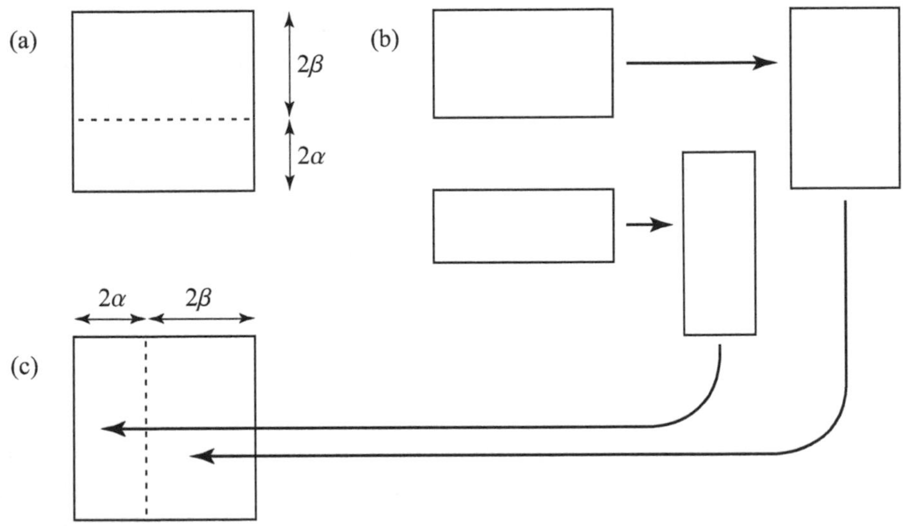

Figure 1.15 - A baker's map with uneven stretching, as described in the text (from Gilbert 2006).

Perhaps the simplest mapping that can be considered is the stacked Baker's map with uneven stretching (Finn & Ott, 1988): this discontinuous map of a square, say $[-1, 1]^2$, to itself is depicted in Figure 1.15. The map M is defined by a parameter α with $0 < \alpha < 1$ and we set $\beta = 1 - \alpha$. The unit square is cut at a horizontal level $y = -1 + 2\alpha$ into two pieces. The first is stretched by a factor α^{-1}, changing its dimensions in (x, y) coordinates from $2 \times 2\alpha$ to $2\alpha \times 2$; see (b). The second piece is stretched by a factor β^{-1}, going from $2 \times 2\beta$ to $2\beta \times 2$. Finally the two squares are reassembled in (c), stacked together, and this completes the mapping process. This mapping can be thought of as a simplified model for the STF picture, giving the doubling up of the tubes of flux in the presence of uneven stretching (Finn & Ott, 1988). The map M may be written as

$$M(x,y) = \begin{cases} (\alpha(x+1)-1, \alpha^{-1}(y+1)-1) & \text{for } y < \Upsilon\,; \\ (\beta(x-1)+1, \beta^{-1}(y-1)+1) & \text{for } y \geq \Upsilon\,, \end{cases} \tag{1.136}$$

with $\Upsilon = -1 + 2\alpha \equiv 1 - 2\beta$.

We imagine starting with a field $\mathbf{B}(x) = b(x)\,\mathbf{e}_y$ and using the Cauchy solution, it may be checked that the action of M is to replace $b(x)$ with the field Tb, where

$$Tb(x) = \begin{cases} \alpha^{-1}b(\alpha^{-1}(x+1)-1) & \text{for } x < \Upsilon\,; \\ \beta^{-1}b(\beta^{-1}(x-1)+1) & \text{for } x \geq \Upsilon\,. \end{cases} \tag{1.137}$$

T is called the dynamo operator (without diffusion). Ignoring diffusion for the present, we may imagine iterating this operator on an initial unit magnetic field

$b_0(x) = 1$, possessing flux $\Phi_0 = 2$ through any horizontal line $y = \text{constant}$. Applying the map once yields two rectangles of field, one of width 2α and strength α^{-1}, and one of width 2β, strength β^{-1}: the flux $\Phi_1 = 4$ has been doubled. Iterating the map we see that $\Phi_n = 2^{n+1}$. If we can ignore the effects of diffusion we have a dynamo with growth rate $\gamma_0 = \log 2$ as in the STF picture, if we agree that each iteration of the mapping takes unit time. We would expect the effect of weak diffusion to be unimportant, as the fields that emerge through repeated application of M are all pointing in the same direction (Finn & Ott, 1990).

The key feature that the stacked Baker's map highlights is that the rate of growth of flux can be different from the Liapunov exponent, a popular measure of how chaotic a system is. To measure this quantity we imagine how a y–directed vector attached to a typical point (x, y) is stretched as the map M is iterated. Since on average a proportion α of the iterates $M^n(x, y)$ will lie in $y < \Upsilon$, where the vector will be stretched by a factor α^{-1}, and a proportion β in $y > \Upsilon$, with stretching by β^{-1}, the Liapunov exponent will be

$$\lambda_{\text{Liap}} = \alpha \log \alpha^{-1} + \beta \log \beta^{-1} . \tag{1.138}$$

This is *less* than the fast dynamo growth rate $\gamma_0 = \log 2$, except in the special case $\alpha = \beta = 1/2$, of even stretching. This at first seems surprising, as magnetic field is composed of vectors, and surely both γ_0 and λ_{Liap} measure the stretching rate of vectors! In fact there is a difference in the averaging processes involved. In the case of magnetic field, in computing a flux, we are weighting more heavily the more stretched vectors, by integrating $b(x)\,dx$, whereas a Lipaunov exponent involves a *typical* point, with weighting dx in the sense of a measure. Equivalently, stronger magnetic fields tend to concentrate in the regions of higher stretching, and so give a different weight in the average.

A more useful quantity to measure as a diagnostic in a fast dynamo is the rate of stretching h_{line} of material lines (which could be thought of as field lines in the absence of diffusion). If the reader experiments with placing a line, say $x = y$ in the square $[-1, 1]^2$ (see Figure 1.15), and then iterating the map M on all the points constituting the line, he or she will soon find that the line length approximately doubles with each iteration, giving an asymptotic value $h_{\text{line}} = \log 2$, which is the same as the fast dynamo growth rate γ_0. Like magnetic field, material lines tend to concentrate in the regions of high stretching (with the consequent inequality $\lambda_{\text{Liap}} \leq h_{\text{line}}$).

This then suggests the general result that the fast dynamo exponent γ_0 should not exceed h_{line}. In fact in two dimensions the exponent h_{line} may be identified with the topological entropy h, and so the result one might expect is

$$\gamma_0 \leq h ; \tag{1.139}$$

this was argued by Finn & Ott (1988) and proved rigorously (under some natural smoothness conditions) by Klapper & Young (1995). The fact that γ_0 can be less

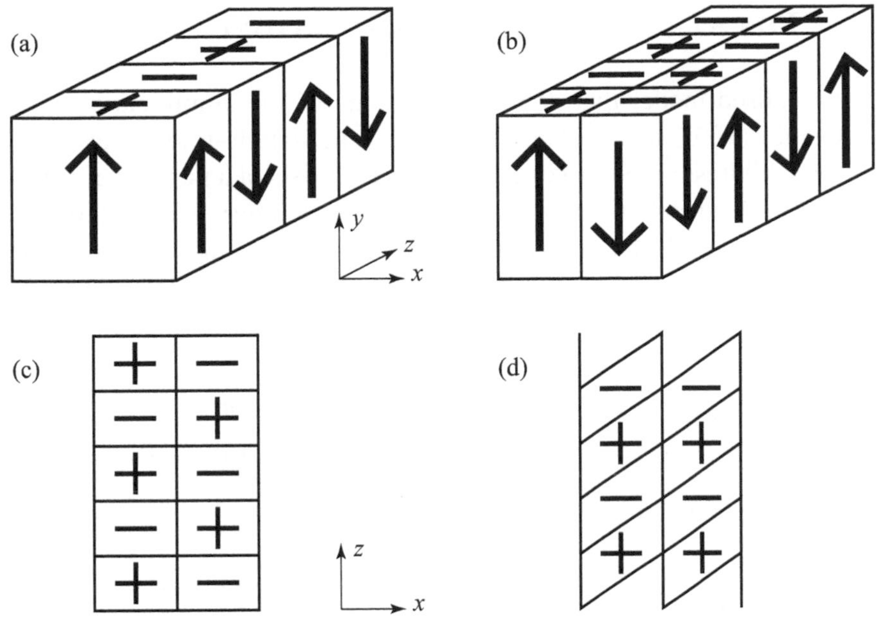

Figure 1.16 - The stretch–fold–shear map. (a) Magnetic field depending on z is stretched and folded with a Baker's map in the (x, y)–plane to give (b). In (c) the field orientation is shown in the (x, z)–plane, which after the shear operation gives (d) (from Gilbert, 2002b).

than h is easily understood: the Baker's map in (1.136) above gives perfect alignment of field in the vertical: all vectors point in the $+y$ direction with our given initial condition. If instead there is folding of field in a more realistic scenario, which can be modelled using a Baker's map with several cuts and rotating one or more rectangles of field at each iteration, the flux Φ_n, a signed quantity, will tend to grow less quickly than the rate of stretching of material lines. This aspect can also be characterised by a *cancellation exponent* (Du *et al.*, 1994). If there are no sign changes in the field, the cancellation exponent would be zero, and we would have $\gamma_0 = h$.

While the uneven Baker's map model is an interesting and useful way to explore these considerations of uneven stretching and cancellations, it suffers from the fact that it is derived from the STF picture, which has the shortcomings and problems mentioned above. Note that for $\alpha = \beta = 1/2$ the Baker's map is trivial (doubling all field vectors and no cancellations), and so probably too simple to model what occurs in a typical fluid flow!

1.6.5. THE STRETCH–FOLD–SHEAR MODEL

Another idealised model that does capture some of the amplification mechanism seen in Otani's (1993) flow and similar flows, is the stretch–fold–shear (SFS) model of Bayly & Childress (1988, 1989). This model consists of a number of components. The first is a folded Baker's map (with uniform stretching), which maps the square $-1 \le x, y \le 1$ to itself, by stretching and folding. This is like the process depicted in Figure 1.15 with $\alpha = \beta = 1/2$, except that the second rectangle is rotated through π before reassembly, representing the folding of a sheet of field. The map is defined by

$$M_1(x,y,z) = \begin{cases} (\frac{1}{2}(x-1), 1+2y, z) & \text{for } y < 0\,; \\ (\frac{1}{2}(1-x), 1-2y, z) & \text{for } y \ge 0\,. \end{cases} \tag{1.140}$$

The action of this on a magnetic field

$$\mathbf{B}(x,y,z) = \mathrm{e}^{\mathrm{i}\,k\,z} b(x)\,\mathbf{e}_y + \text{complex conjugate} \tag{1.141}$$

is shown in Figure 1.16(a,b), giving one fold of field in the (x, y)–plane. If this map were now simply repeated, the effect would be to obtain ever finer alternating bands of magnetic field in this plane, vulnerable to diffusion. There is plenty of stretching, but no constructive folding. The flux through a line $x =$ constant would become zero after one iteration and remain so thereafter. In this case, we would have γ_0 negative, but $h_{\text{line}} = \log 2$.

Thus a second ingredient is required, a shear in the z–direction, shown in a top-down view going from (c) to (d). The action of the shear is to bring upward pointing field (+) approximately into alignment with other upward fields, and similarly downward pointing field (−). This corresponds to the mapping

$$M_2(x,y,z) = (x, y, z + \alpha x)\,, \tag{1.142}$$

where α is a shear parameter (not related to the previous α, and not intended to imply an α–effect!). The alignment is only approximate, but intended to capture the basic mechanism observed in flows such as Otani's, in which belts of field are drawn out and folded in the (x, y)–plane, and then sheared in the z–direction (Bayly & Childress, 1988).

In this way we obtain the SFS dynamo model: the field is first stretched and folded (by M_1) and then sheared (by M_2). Acting on the complex field $b(x)$ in (1.141) above gives a field Tb, with

$$Tb(x) = \begin{cases} 2\mathrm{e}^{-\mathrm{i}\,\alpha\,k\,x}\, b(1+2x) & \text{for } x < 0\,; \\ -2\mathrm{e}^{-\mathrm{i}\,\alpha\,k\,x}\, b(1-2x) & \text{for } x \ge 0\,. \end{cases} \tag{1.143}$$

T is again the dynamo operator without diffusion. For diffusion we employ suitable boundary conditions and allow the field $b(x)$ to diffuse for unit time according to $\partial_t b = \varepsilon \partial_{xx} b$. Possible boundary conditions (only employed at $x = -1, 1$) include insulating (I), perfectly conducting (C) and periodic (P),

$$b(1) = b(-1) = 0 \quad (\mathrm{I}), \quad \partial_x b(1) = \partial_x b(-1) = 0 \quad (\mathrm{C}), \quad b(x) \text{ periodic} \quad (\mathrm{P}). \tag{1.144a,b,c}$$

The diffusion step may be written as mapping b to $H_\varepsilon b$, where H_ε is another operator involving heat kernels (for further details see Gilbert, 2002b, 2005).

Finally the SFS dynamo operator with diffusion is written $T_\varepsilon = H_\varepsilon T$. The magnetic field is most easily discretised using Fourier series, and eigenvalues λ for $T_\varepsilon b = \lambda b$ sought numerically using matrix eigenvalue solvers. If the mapping and diffusion are assumed to take a time unity, then the corresponding magnetic growth rate is

$$\sigma = \log \lambda \,, \tag{1.145}$$

and we refer to λ as the growth factor. An eigenvalue λ then corresponds to a growing magnetic mode provided that $|\lambda| > 1$. Our aim is to understand the properties of eigenvalues of T_ε in the limit as $\varepsilon \to 0$. If eigenvalues remain bounded above $|\lambda| = 1$ in the limit, then the SFS model is a fast dynamo.

Figure 1.17(a,b,c) shows the modulus $|\lambda|$ of the leading eigenvalues λ as a function of α (with $k = 1$ set without loss of generality) for the (I), (C) and (P) boundary conditions given above, at $\varepsilon = 10^{-5}$. We see that it is necessary in all cases to increase the shear parameter α above about $\pi/2$ to obtain growing modes. There has to be sufficient constructive alignment for the dynamo to operate.

We also see that the modes with the larger values of $|\lambda|$, certainly $|\lambda| > 1$, are robust to the kinds of boundary condition employed, though the picture is rather different for marginal and decaying modes with $|\lambda| \leq 1$. Further numerical study (not set out here) indicates that the more robust eigenvalues, with $|\lambda| > 1$, appear to show convergence to positive values as $\varepsilon \to 0$, although individual magnetic modes $b(x)$ show increasingly fine structure in this limit. Thus there is good evidence for fast dynamo action in the SFS model (Bayly & Childress, 1988, 1989).

This leaves open the mathematical question: how can we prove fast dynamo action, and obtain some information about these growth rates for small positive diffusivity, $0 < \varepsilon \ll 1$? We need to set out a sensible problem for zero diffusion, and then treat diffusion as a perturbation. The key idea of Bayly & Childress (1989) is to note that while the (diffusionless) operator T tends to reduce the scales of a magnetic field and generally has no eigenfunctions, its adjoint T^* (in L^2), given by

$$T^* c(x) = \mathrm{e}^{-\mathrm{i}\,\alpha\,\frac{1}{2}(x-1)} c\left(\tfrac{1}{2}(x-1)\right) - \mathrm{e}^{-\mathrm{i}\,\alpha\,\frac{1}{2}(1-x)} c\left(\tfrac{1}{2}(1-x)\right) , \tag{1.146}$$

instead tends to expand scale and average. T^* can possess smooth eigenfunctions even at zero diffusion, unlike T.

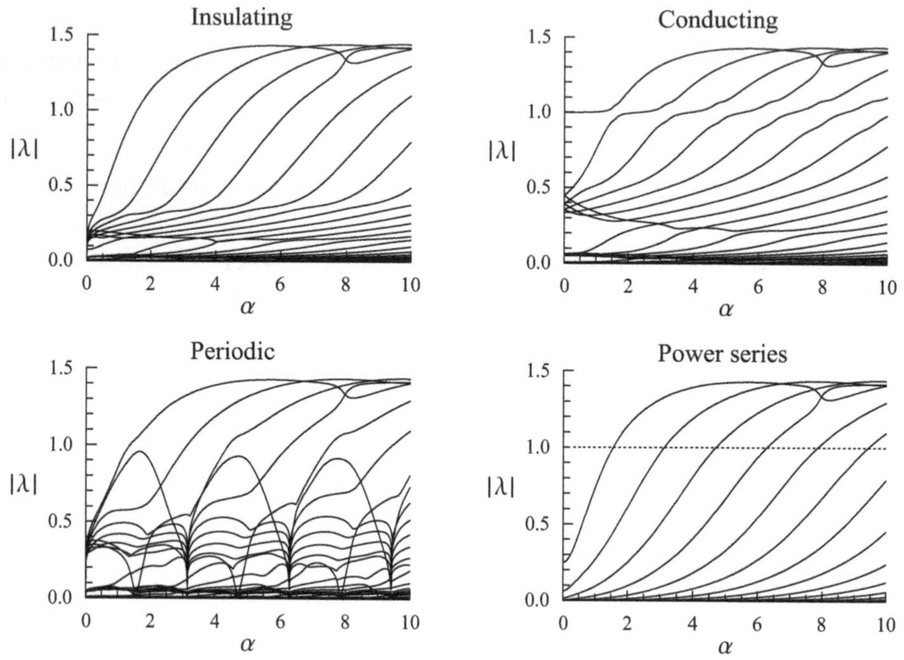

Figure 1.17 - Moduli of eigenvalues $|\lambda|$ plotted against α for the SFS model. The boundary conditions are (a) insulating, (b) perfectly conducting and (c) periodic, with $\varepsilon = 10^{-5}$. In (d) eigenvalues are obtained using a power series and $\varepsilon = 0$ (from Gilbert, 2005).

If we seek an eigenfunction $T^*c = \lambda c$, and expand the function $c(x)$ in a basis x^n, that is as a power series about the origin, we obtain a matrix for T^* whose eigenvalues may be found numerically. Figure 1.17(d) gives growth factors obtained in this way for zero diffusion. The results are very close to those obtained in Figure 1.17(a,b,c) in the presence of weak diffusion, particularly for larger values of $|\lambda|$. Plainly most branches in Figure 1.17(d) are relatively robust to diffusion, though this depends on boundary conditions and the size of $|\lambda|$.

One branch that shows particular sensitivity to diffusion and boundary conditions is the horizontal branch in 1.17(d), for which the adjoint eigenfunction of (1.146) is given analytically by

$$c(x) = \mathrm{e}^{\mathrm{i}\,\alpha\,(x-1)} - \mathrm{e}^{\mathrm{i}\,\alpha\,(1-x)} = 2\,\mathrm{i}\,\sin\alpha(x-1)\,, \quad \lambda = \mathrm{e}^{\mathrm{i}\,\alpha}\,. \tag{1.147a,b}$$

This branch only survives for conducting boundary conditions. Current research (Gilbert, 2005) is aimed at understanding the effects of diffusion and boundary conditions in the SFS model. The aim is to be able to use perturbation theory to write, for a given branch and value of α,

$$\lambda(\varepsilon) \simeq \lambda(0) + C\varepsilon^q\,, \tag{1.148}$$

where $\lambda(0)$ is the complex growth factor obtained by means of a power series for zero diffusion, as shown in Figure 1.17(d). The term $C\varepsilon^q$ is the diffusive correction, which is dependent on the structure of the mode and boundary conditions, and $q > 0$ is the condition on the exponent q for the branch to survive the effects of diffusion.

CHAPTER 2

NONLINEARITIES AND SATURATION

Stephan Fauve, Raymond Hide,
David Hughes, Irene Moroz,
François Pétrélis

In this chapter, we focus on the effects of nonlinearity. After general considerations in Section 2.1, we investigate the role of nonlinearity in saturating the growth of the magnetic field for a dynamo with a spatially periodic flow in Section 2.2. We then turn to the more general situation of saturation in the vicinity of the dynamo threshold; first in the low Re limit in Section 2.3 and then in the high Re limit in Section 2.4. We then address the issue of dynamo saturation in flows strongly affected by rotation (with planetary applications in mind) in Section 2.5, and we present some conjectures for the magnetic field generated by a turbulent flow when Re and Rm are both large in Section 2.6. Mean field dynamo saturation (with astrophysical applications in mind) is discussed in Section 2.7. Finally, we return in Section 2.8 to the apparently simple disc dynamo model used as an introductory example in Chapter 1 to show how rich the nonlinear dynamical behaviour can be.

2.1. GENERAL CONSIDERATIONS

The study of dynamo action is motivated both by laboratory experiments and by observations of astrophysical or geophysical magnetic fields. Recently, the first homogeneous fluid dynamos have been successfully demonstrated: in Karlsruhe (Stieglitz & Müller, 2001) using a flow in an array of pipes set up in order to mimic a spatially periodic flow proposed by G.O. Roberts (1972), and in Riga (Gailitis *et al.*,

2001) using a Ponomarenko-type flow (Ponomarenko, 1973). Although there were no doubts about self-generation of magnetic fields by Roberts or Ponomarenko-type laminar flows, these experiments have displayed several interesting features:

- The observed thresholds are in rather good agreement with theoretical predictions (Busse *et al.*, 1996; Rädler *et al.*, 1998; Gailitis *et al.*, 2002a) made by considering only the laminar mean flow and neglecting the small-scale turbulent fluctuations that are present in both experiments.

- The nature of the dynamo bifurcation, stationary for the Karlsruhe experiment or oscillatory (Hopf) in the Riga experiment, is also in agreement with laminar models.

- On the contrary, the saturation level of the magnetic field, due to the Lorentz force back reaction on the flow, cannot be predicted with a laminar flow model. It has been shown indeed that different scaling laws exist in the supercritical dynamo regime depending on the magnitude of the Reynolds number (Pétrélis & Fauve, 2001).

These observations raise several questions: we do not discuss here the effect of turbulence on the dynamo threshold (see Fauve & Pétrélis, 2003) or the characteristics of magnetic field fluctuations (see Bourgoin *et al.*, 2002) but rather try to understand the scaling law for the mean magnetic field amplitude above the dynamo threshold. To this end, we take into account the back reaction of the magnetic field on the velocity field. We thus try to solve the dynamic dynamo problem, or in other words, to find a nonlinear equation for the amplitude of the linearly unstable mode at the bifurcation. Solving this equation determines the subcritical or supercritical nature of the bifurcation and in the later case, the amplitude of the magnetic field as a function of the distance to the dynamo threshold.

Both the Karlsruhe and Riga experiments operate in the vicinity of dynamo threshold (typically 10% above threshold) and it is unlikely that a laboratory experiment could reach high Rm values (say 10 times critical) because the power needed to drive a turbulent flow increases like the cubic power of its mean velocity. It is also possible that the Geodynamo does not operate too far from threshold, but it is not the case of other astrophysical objects for which huge values of Rm can be reached. Weakly nonlinear theory is of little use in these situations as well as in the case of a strongly subcritical bifurcation that may be the case of the Geodynamo.

Magnetic fields exist on a wide range of scales in astrophysics. Their orders of magnitude as well as some associated relevant parameters for planets, stars and our galaxy are given in Table I. It is perhaps meaningless to try to compare these data because these astrophysical objects have strongly different physical properties.

Table I - Approximate parameters and magnetic field strength of some astrophysical objects (Zeldovich *et al.*, 1983). The size L is based on the typical length (radius for spheres, thickness for discs) of the conducting region (not of the full object). The magnetic field strength is an averaged one (in the case of the Sun, the field can locally be 10^3 times stronger). The resistivity η is usually based on molecular estimates, except in the Galaxy for which it represents "ambipolar diffusion".

	the Galaxy	Sun	Jupiter	Earth core	White dwarfs	Neutron stars
$\lvert\mathbf{B}\rvert$ (T)	10^{-10}	10^{-4}	4×10^{-4}	10^{-4}	10^2 - 10^4	10^6 - 10^9
ρ (kg m^{-3})	10^{-21}	1	10^3	10^4	10^5–10^{12}	10^{13}–10^{18}
L (m)	10^{19}	2×10^8	5×10^7	3×10^6	10^6	10^4–10^6
η (m^2 s^{-1})	10^{17}	10^3	10	2		
$\lvert\mathbf{B}\rvert^2 L^3/2\mu_0$	10^{43}	4×10^{22}	10^{22}	2×10^{17}		
$\lvert\mathbf{B}\rvert^2 L\eta/2\mu_0$	10^{22}	10^9	4×10^7	2×10^5		

However, we may observe that the strength of the magnetic field, $\mathbf{B}$, is not strongly related to the size of the object L, but seems to increase with its density ρ. If instead of looking at the intensity of the magnetic field, we consider the typical magnetic energy of the object $\langle|\mathbf{B}|^2\rangle L^3/2\mu_0$ (μ_0 is the magnetic permeability of vacuum), we find the expected ordering from the galaxy to the Earth. We may also consider the typical value of the Joule dissipation, which is obtained by dividing the magnetic energy by the characteristic magnetic diffusion time L^2/η . As such, we get an idea of the minimum amount of power which is necessary to maintain the magnetic field against Joule dissipation. Again, we observe the expected ordering from the galaxy to the Earth. Note that these values have certainly been underestimated for two principal reasons. First, they are estimated from the visible (poloidal) part of the magnetic field, and are thus strongly underestimated if the azimuthal field inside the body is large compared to the poloidal component. Second, we have assumed that the length scale of the gradients of the magnetic field is the size L of the conducting medium. Magnetic energy at smaller scales will lead to a shorter diffusion timescale and thus to a higher dissipated power.

Let us first recall the induction equation (1.14) and the Navier-Stokes equation (1.27) and restrict considerations to incompressible flows ($\boldsymbol{\nabla}\cdot\mathbf{u}=0$). Further, we neglect the Coriolis force, but include the Lorentz force (1.18). Therefore,

$$\frac{\partial \mathbf{B}}{\partial t} = \boldsymbol{\nabla}\times(\mathbf{u}\times\mathbf{B}) + \eta\Delta\mathbf{B}\,, \tag{2.1a}$$

$$\frac{\partial \mathbf{u}}{\partial t} + (\mathbf{u}\cdot\boldsymbol{\nabla})\mathbf{u} = -\boldsymbol{\nabla}\left(\frac{p}{\rho} + \frac{|\mathbf{B}|^2}{2\mu_0}\right) + \nu\Delta\mathbf{u} + \frac{1}{\mu_0\rho}(\mathbf{B}\cdot\boldsymbol{\nabla})\mathbf{B}\,. \tag{2.1b}$$

The flow is created, either by moving solid boundaries or by a body force added to the Navier-Stokes equation. We have to develop equations (2.1a,b) close to the dynamo threshold in order to derive an amplitude equation for the growing magnetic field. If the dynamo bifurcation is found to be supercritical, this allows us to calculate the saturated magnetic field.

Thus, even in the simplest configuration, the problem involves three dimensionless parameters. One may choose the Reynolds number, Re, the magnetic Reynolds number, Rm, and the *Lundquist number*, $\mathrm{S} = \langle|\mathbf{B}|^2\rangle\mu_0(\sigma L)^2/\rho$, leading in general to the following form of law

$$\frac{\langle|\mathbf{B}|^2\rangle\mu_0(\sigma L)^2}{\rho} = f(\mathrm{Rm}, \mathrm{Re})\,. \tag{2.2}$$

Another possible choice is obtained by replacing Re by the *magnetic Prandtl number*, $\mathrm{Pm} \equiv \mathrm{Rm}/\mathrm{Re} = \mu_0\sigma\nu = \nu/\eta$. For most fluids, $\mathrm{Pm} \ll 1$ i.e. $\mathrm{Re} \gg \mathrm{Rm}$.

In general, the analytic determination of f using weakly nonlinear perturbation theory in the vicinity of the dynamo threshold is tractable only in the unrealistic case $\mathrm{Pm} \gg 1$ such that the dynamo bifurcates from a laminar flow ($\mathrm{Re} \ll 1$). For $\mathrm{Pm} \ll 1$, many hydrodynamic bifurcations occur first and the flow becomes turbulent before the dynamo threshold.

We first present the structure of the perturbation analysis of weakly nonlinear theory in the vicinity of the dynamo threshold in the tractable case $\mathrm{Re} \ll 1$. We then discuss the realistic situation ($\mathrm{Re} \gg 1$) and, using dimensional or phenomenological arguments, show that the expression of the generated magnetic field as a function of the fluid parameters strongly differs from the case $\mathrm{Re} \ll 1$.

Astrophysical or geophysical dynamos involve many more parameters due to the nature of the driving of the flow. A particularly important one is the global rotation rate. We shortly review how this may affect the saturation of the magnetic field. Finally, we discuss some conjectures in the limit of Re and Rm both large for a turbulent flow without global rotation.

2.2. SATURATION OF A DYNAMO GENERATED BY A PERIODIC FLOW

It has been shown by G.O. Roberts (1970, 1972) that many spatially periodic flows generate a magnetic field at a large scale compared to their spatial periodicity. In that case the weakly nonlinear problem above the bifurcation threshold is also more easily tractable (Gilbert & Sulem, 1990). We recall some of these linear and nonlinear results obtained for periodic flows and that have been recently used to discuss the results of the Karlsruhe experiment (Tilgner & Busse, 2001; Rädler *et al.*, 2002a, 2002b).

2.2.1. SCALE SEPARATION

We consider a spatially periodic velocity field with wavelength ℓ and zero mean value, and we assume that a magnetic field $\mathbf{B}_0$ is generated on a spatial scale L. A magnetic field with spatial periodicity ℓ is generated by the interaction of $\mathbf{B}_0$ with the flow. We thus write

$$\mathbf{B} = \mathbf{B}_0 + \mathbf{b}\,, \tag{2.3}$$

with $\langle \mathbf{b} \rangle = 0$, where $\langle \cdot \rangle$ stands for the spatial average over one wavelength ℓ. Inserting (2.3) in the induction equation, and averaging over space, we get the evolution equation for the mean field $\mathbf{B}_0$

$$\partial_t \mathbf{B}_0 = \boldsymbol{\nabla} \times \langle \mathbf{u} \times \mathbf{B} \rangle + \eta \Delta \mathbf{B}_0\,. \tag{2.4}$$

Subtracting (2.4) from the induction equation, we get the evolution equation for the fluctuating field $\mathbf{b}$

$$\partial_t \mathbf{b} = \boldsymbol{\nabla} \times (\mathbf{u} \times \mathbf{B}_0 + \mathbf{u} \times \mathbf{b} - \langle \mathbf{u} \times \mathbf{b} \rangle) + \eta \Delta \mathbf{b}\,. \tag{2.5}$$

We have to find $\mathbf{b}$ as a function of $\mathbf{B}_0$ using equation (2.5) in order to get a closed equation for the mean field from (2.4). Equation (2.5) may be solved easily if $b = |\mathbf{b}|$ is small compared to $B_0 = |\mathbf{B}_0|$; we then have at leading order a diffusion equation for $\mathbf{b}$ with a source term depending on $\mathbf{B}_0$ and the velocity field. Then, we get

$$\eta \frac{b}{\ell^2} \sim \frac{u B_0}{\ell}, \quad \text{thus} \quad b \sim \frac{u\ell}{\eta} B_0\,. \tag{2.6}$$

Using this expression for b in order to estimate $\langle \mathbf{u} \times \mathbf{b} \rangle$, which does not depend any more on ℓ after being averaged, we get from (2.4) the following condition for dynamo onset on $u_c = |\mathbf{u}|$:

$$\frac{u_c^2 \ell}{\eta} \frac{B_0}{L} \sim \frac{\eta B_0}{L^2}, \quad \text{thus} \quad u_c \sim \frac{\eta}{\sqrt{L\ell}}\,. \tag{2.7}$$

We first observe that $b \sim \sqrt{\ell / L} B_0 \ll B_0$ provided that $\ell \ll L$. In this limit, the magnetic Reynolds number defined on each eddy of size ℓ is thus very small whereas the one defined on L is large. We observe that the relevant definition here for the magnetic Reynolds number would be

$$\mathrm{Rm}_2 \equiv \frac{|\mathbf{u}|\sqrt{L\ell}}{\eta}\,, \tag{2.8}$$

the critical value of which for dynamo onset is of order one. Consequently, even if the above mechanism works, we cannot reach the dynamo onset just by increasing scale separation. For η and $|\mathbf{u}|$ fixed, it does not help to decrease ℓ. Scale separation makes it possible to keep the magnetic Reynolds number small at dynamo onset if it is defined on the scale of each eddy, ℓ. In this limit, the field $\mathbf{B}_0$ is not strongly distorted by the fluid motion. This allows easier analytical calculations.

2.2.2. The G.O. Roberts dynamo

We consider the spatially periodic flow (see Section 1.4.1) with velocity field

$$\mathbf{u}(x,y,z) = (U\sin ky,\, U\cos kx,\, V(\sin kx + \cos ky))\ . \tag{2.9}$$

We have $\langle \mathbf{u}\rangle = 0$ and the mean helicity is $\mathcal{H} = \langle \mathbf{u}\cdot\boldsymbol{\nabla}\times\mathbf{u}\rangle = -2kUV$. Assuming that b is small compared to $B_0 = |\mathbf{B}_0|$, we get from equation (2.5)

$$\mathbf{b} \approx \frac{1}{\eta\, k}\,(U\,B_2\cos ky,\, -U\,B_1\sin kx,\, V\,B_1\cos kx - V\,B_2\sin ky)\ , \tag{2.10}$$

where $\mathbf{B}_0 = (B_1, B_2, B_3)$. We thus reach

$$\langle \mathbf{u}\times\mathbf{b}\rangle \approx \frac{UV}{\eta k}\begin{pmatrix} 1 & 0 & 0\\ 0 & 1 & 0\\ 0 & 0 & 0\end{pmatrix}\mathbf{B}_0\,. \tag{2.11}$$

We observe that if a large scale field exists along the x or y–axis, the cooperative effect of small scale periodic fluctuations is to drive a current parallel to the large scale field. This has been understood by Parker (1955) and is due to the helical nature of the flow. Any field B_1 along the x–axis is distorted in the vertical (x,z)–plane by the z–component of the flow of amplitude V. The field is twisted out of the (x,z)–plane by the toroidal component of the flow of amplitude U. This drives field loops in the (y,z)–plane, i.e. a current parallel to x, which generates a magnetic field with a non-zero component along the y–axis, B_2. B_2 can then regenerate B_1 through the same process. The mean electromotive force $\langle \mathbf{u}\times\mathbf{b}\rangle$ in the mean field equation (2.4) was described by Steenbeck & Krause (1966) as the "α–effect" (see for instance Krause & Rädler, 1980). In this terminology, the G.O. Roberts' dynamo is an α^2–dynamo. Defining $\alpha = UV/\eta k$, we have for a mean field of the form $\mathbf{B}_0(z,t) = (B_1, B_2, 0)$, where B_1 and B_2 satisfy

$$\frac{\partial B_1}{\partial t} = -\alpha\frac{\partial B_2}{\partial z} + \eta\frac{\partial^2 B_1}{\partial z^2}\,, \qquad \frac{\partial B_2}{\partial t} = \alpha\frac{\partial B_1}{\partial z} + \eta\frac{\partial^2 B_2}{\partial z^2}\,. \tag{2.12a,b}$$

Defining $A = B_1 + \mathrm{i}B_2$, we get

$$\frac{\partial A}{\partial t} = -\mathrm{i}\alpha\frac{\partial A}{\partial z} + \eta\frac{\partial^2 A}{\partial z^2}\,. \tag{2.13}$$

The linear stability analysis of the solution $A = 0$ (i.e. $\mathbf{B}_0 = 0$) is straightforward. We consider normal modes of the form $A \propto \exp(\eta t \pm \mathrm{i} K z)$ and get from equation (2.13) the dispersion relation

$$\eta = \pm|\alpha K| - \eta K^2, \tag{2.14}$$

which shows that there exists a branch of unstable modes at long enough wavelength ($K < |\alpha|/\eta$).

We observe that dynamo action vanishes if $U \to 0$ or $V \to 0$ in agreement with antidynamo theorems. It is interesting to consider the behaviour of α when the magnetic Reynolds number becomes larger. The calculation of b should be performed at higher orders in equation (2.5). Solving perturbatively this equation for b as an expansion in powers of $U/\eta k$, one gets

$$\alpha = \frac{UV}{\eta\, k}\left(1 - \frac{U^2}{2\,\eta^2\, k^2} + ...\right). \tag{2.15}$$

α increases linearly with V, but its behaviour as a function of U is more complex. It first increases but reaches a maximum and then decreases as U is increased. This behaviour is due to the expulsion of the transverse field by the rotating eddies, as already shown in Rädler *et al.* (1998) by numerically solving (2.5). It has been found that α decreases toward zero at large Rm. Note however that the large Rm limit should be considered carefully. As stated above, the great simplification of scale separation results from the fact that the magnetic Reynolds number evaluated on the small scale of the flow is small whereas the one evaluated on the large scale of the mean field is large. This is clearly apparent in our second order result (2.15). Truncating the expansion in $|\mathbf{u}|/\eta\, k$ is not accurate if Rm is too large such that the magnetic Reynolds number related to the azimuthal motion of the eddies becomes of order 1.

The α–effect has been demonstrated experimentally by directly measuring the mean electromotive force generated by a helical flow of liquid sodium in the presence of an external magnetic field (Steenbeck *et al.*, 1968). Self-generation of a magnetic field by the α–effect has been achieved recently, using a periodic arrangement of counter-rotating and counter-current helical vortices that mimic G.O. Roberts' flow. Axial and azimuthal sodium flows are driven by pumps in an array of helical ducts immersed in a cylinder (Karlsruhe experiment, Stieglitz & Müller, 2001).

2.2.3. SATURATION OF DYNAMOS DRIVEN BY THE α–EFFECT

Saturation of an α–dynamo may involve the generation of a large scale flow generated by the large scale magnetic field (Malkus & Proctor, 1975). If this large scale flow is not forbidden by the geometrical configuration, it is likely to exist without a magnetic field and to play a role already at the level of the kinematic dynamo problem. On the contrary, if any large scale flow is forbidden, as in the Karlsruhe experiment, the saturation is due to the modification of the small scale velocity field which reduces the elecromotive force related to the α–effect. In that case, the perturbation method based on scale separation can be easily extended to the study of the dynamic dynamo problem, as shown in the case of the G.O. Roberts' flow (Gilbert & Sulem, 1990). The mean field equation (2.4) is unchanged, but the mean electromotive force $\langle \mathbf{u} \times \mathbf{b} \rangle$ should be calculated using both equation (2.5) and the Navier-Stokes equation (2.1b). The simplest way to generate the G.O. Roberts' flow is to add a body force $\mathbf{f} = -\nu \Delta \mathbf{u}^0$ to (2.1b) where $\mathbf{u}^0$ is given by (2.9). In the presence of a magnetic field, we have to leading order

$$\eta \Delta \mathbf{b} \approx -(\mathbf{B}_0 \cdot \boldsymbol{\nabla})\mathbf{u}\,, \quad \nu \Delta \mathbf{u} + \frac{(\mathbf{B}_0 \cdot \boldsymbol{\nabla})\mathbf{b}}{\rho \mu_0} + \mathbf{f} \approx \mathbf{0}\,. \tag{2.16a,b}$$

The first equation is formally unchanged compared to the kinematic calculation although $\mathbf{u}$ is no longer prescribed, but should be obtained by solving the linear system (2.16a,b). The velocity field $\mathbf{u}^0$ in the absence of magnetic field is modified by the Lorentz force. Note that $(\mathbf{B} \cdot \boldsymbol{\nabla})\mathbf{B} \approx (\mathbf{B}_0 \cdot \boldsymbol{\nabla})\mathbf{b}$ up to terms of order $\sqrt{\ell / L} \ll 1$ from the assumption of scale separation. Solving (2.16a,b), we get for the electromotive force

$$\langle \mathbf{u} \times \mathbf{b} \rangle \approx \frac{UV}{\eta k} \left(1 + \frac{\sigma B_2^2}{\rho \nu k^2} \right)^{-2} \begin{pmatrix} 1 & 0 & 0 \\ 0 & 1 & 0 \\ 0 & 0 & 0 \end{pmatrix} \mathbf{B}_0\,. \tag{2.17}$$

We thus find that the α–effect saturates when the magnetic field amplitude increases because of the action of the Lorentz force on the velocity field. This saturation should not be confused with that observed for large $|\mathbf{u}|$ in (2.15) which is a linear effect due to flux expulsion. Defining

$$\widetilde{B_i}^2 = \frac{\sigma B_i^2}{\rho \nu k^2}\,, \tag{2.18}$$

we obtain from the mean field equation (2.4)

$$\frac{\partial \widetilde{B}_1}{\partial t} = -\alpha \frac{\partial}{\partial z} \left[\frac{\widetilde{B}_2}{(1 + \widetilde{B}_2^2)^2} \right] + \eta \frac{\partial^2 \widetilde{B}_1}{\partial z^2}\,, \tag{2.19a}$$

$$\frac{\partial \widetilde{B}_2}{\partial t} = \alpha \frac{\partial}{\partial z} \left[\frac{\widetilde{B}_1}{(1 + \widetilde{B}_1^2)^2} \right] + \eta \frac{\partial^2 \widetilde{B}_2}{\partial z^2}\,. \tag{2.19b}$$

Numerical simulation of these equations shows that the magnetic field cascades to large spatial scales during the saturation process (Gilbert & Sulem, 1990).

We thus observe that the saturated mean magnetic field obeys the following scaling law

$$\langle B_0^2 \rangle \propto \frac{\rho\nu}{\sigma\ell^2} \left(\mathrm{Rm} - \mathrm{Rm}_c\right), \tag{2.20}$$

with $\mathrm{Rm} = \sqrt{UV}\sqrt{Ll}/\eta$.

In the case of an isotropic flow, a nonlinear evolution equation for the mean field can be easily obtained by symmetry considerations. We get

$$\partial_T \mathbf{B}_0 = \alpha \boldsymbol{\nabla} \times \left(1 - \gamma \mathbf{B}_0^2\right) \mathbf{B}_0 + \eta\, \Delta \mathbf{B}_0 \,. \tag{2.21}$$

In the absence of a large scale flow, we expect similar nonlinearities in the case of α^2–dynamos generated by small scale turbulent fluctuations. Phenomenological descriptions leading to equations of the form (2.21) have been proposed (Kraichnan, 1979; Meneguzzi *et al.*, 1981; Gruzinov & Diamond, 1994). We do not expect however that γ corresponds to the laminar scaling when the Reynolds number of the flow is large (see below). Different scaling laws have been also proposed in relation to the helicity injection rate and dynamics.

2.3. SATURATION IN THE LOW Re LIMIT IN THE VICINITY OF THE DYNAMO THRESHOLD

2.3.1. A PONOMARENKO TYPE DYNAMO AS A TRACTABLE PROBLEM WITHOUT SCALE SEPARATION

In the absence of scale separation, it is much more difficult to derive an amplitude equation for the magnetic field in the vicinity of the bifurcation threshold. We have performed such a calculation using the following trick. We slightly modified Ponomarenko's original configuration (a cylinder in solid body rotation and translation along its axis, embedded in an infinite static medium of the same conductivity with which it is in perfect electrical contact) by considering that the rotating cylinder is hollow and filled with a liquid metal of the same conductivity. This gives a very simple flow, i.e. solid body rotation and translation, which is the simplest way to avoid turbulence at dynamo onset. The kinematic dynamo problem is thus the same as that studied by Ponomarenko. However, above the dynamo threshold, the flow is modified by the Lorentz force and is expected to saturate the growth of the magnetic field.

We will not present here the calculation of the amplitude equation (see Nuñez *et al.*, 2001) but simply show the structure of the perturbation analysis.

2.3.2. STRUCTURE OF THE PERTURBATION ANALYSIS

The structure of the weakly nonlinear analysis above threshold is as follows: the forcing generates a velocity field $\mathbf{u}_f$ and the dynamo bifurcates for $\mathbf{u}_f = \mathbf{u}_c$, i.e. $\mathrm{Rm} = \mathrm{Rm}_c$. We write (2.1a) in the form

$$\mathcal{L}\left(\mathbf{B}^{(0)}\right) = 0\,, \tag{2.22}$$

where $\mathbf{B}^{(0)}$ is the neutral mode at threshold and $\mathcal{L}$ is a linear operator that depends on the bifurcation structure (stationary or Hopf bifurcation). In the case of the Ponomarenko dynamo, we have a Hopf bifurcation with neutral modes of the form (Ponomarenko, 1973)

$$\mathbf{B}^{(0)} = A(T)\mathbf{B}_p + c.c. = A(T)\mathbf{b}_p(r)\exp i(m\theta + kz + \omega_0 t) + c.c., \tag{2.23}$$

where (r, θ, z) are cylindrical coordinates and c.c. stands for the complex conjugate of the previous expression.

The flow is forced slightly above threshold, $\mathbf{u}_f = \mathbf{u}_c + \varepsilon\mathbf{u}_d + \cdots$, with $\varepsilon = (\mathrm{Rm} - \mathrm{Rm}_c)/\mathrm{Rm}_c \ll 1$. In addition, the leading order flow distortion by the Lorentz force, $\varepsilon\mathbf{u}^{(1)}$, yields

$$\mathbf{u} = \mathbf{u}_f + \varepsilon\mathbf{u}^{(1)} + \cdots\,. \tag{2.24}$$

For $\mathbf{B}$ we have

$$\mathbf{B} = \sqrt{\varepsilon}\left(\mathbf{B}^{(0)} + \varepsilon\mathbf{B}^{(1)} + \cdots\right)\,. \tag{2.25}$$

We first compute $\mathbf{u}^{(1)}$ from equation (2.1b) at order ε:

$$\partial_t\mathbf{u}^{(1)} + (\mathbf{u}_c\cdot\boldsymbol{\nabla})\,\mathbf{u}^{(1)} + \left(\mathbf{u}^{(1)}\cdot\boldsymbol{\nabla}\right)\mathbf{u}_c = -\frac{1}{\rho}\boldsymbol{\nabla}\left(p_1 + \frac{|\mathbf{B}^{(0)}|^2}{2\mu_0}\right) + \nu\Delta\mathbf{u}^{(1)} + \frac{1}{\mu_0\rho}\left(\mathbf{B}^{(0)}\cdot\boldsymbol{\nabla}\right)\mathbf{B}^{(0)}\,. \tag{2.26}$$

If $\mathrm{Pm} \gg 1$, the flow is laminar at the dynamo threshold, and the Lorentz force is mostly balanced by the modification of the viscous force, thus

$$|\mathbf{u}^{(1)}| \propto \frac{|\mathbf{B}^{(0)}|^2 L}{\mu_0\rho\nu}\,. \tag{2.27}$$

We get from equation (2.1a) at order ε:

$$\mathcal{L}\left(\mathbf{B}^{(1)}\right) = \partial_T\mathbf{B}^{(0)} - \boldsymbol{\nabla}\times\left(\mathbf{u}_d\times\mathbf{B}^{(0)}\right) - \boldsymbol{\nabla}\times\left(\mathbf{u}^{(1)}\times\mathbf{B}^{(0)}\right), \tag{2.28}$$

where $T = \varepsilon t$ is the slow timescale of $\mathbf{B}^{(0)}$ slightly above threshold. The amplitude equation for $\mathbf{B}^{(0)}$ that governs the saturation of the magnetic field is obtained by applying the solvability condition to equation (2.28):

$$\langle\mathbf{C}|\mathcal{L}(\mathbf{B}^{(0)})\rangle = \langle\mathbf{C}|\partial_T\mathbf{B}^{(0)}\rangle - \langle\mathbf{C}|\boldsymbol{\nabla}\times(\mathbf{u}_d\times\mathbf{B}^{(0)})\rangle - \langle\mathbf{C}|\boldsymbol{\nabla}\times(\mathbf{u}^{(1)}\times\mathbf{B}^{(0)})\rangle = 0\,, \tag{2.29}$$

where $\langle \mathbf{a}|\mathbf{b}\rangle = \int \mathbf{a}\cdot\mathbf{b}\,\mathrm{d}\mathbf{x}$. It follows that

$$\left\langle \mathbf{C}|\partial_T \mathbf{B}^{(0)}\right\rangle = \left\langle \mathbf{C}|\boldsymbol{\nabla}\times(\mathbf{u}_d\times\mathbf{B}^{(0)})\right\rangle + \left\langle \mathbf{C}|\boldsymbol{\nabla}\times(\mathbf{u}^{(1)}\times\mathbf{B}^{(0)})\right\rangle , \tag{2.30}$$

where $\mathbf{C}$ is an eigenvector of the adjoint problem.

The first term on the right hand side of equation (2.30) corresponds to the linear growth rate of the magnetic field whereas the second describes the nonlinear saturation due to the modified velocity field $\mathbf{u}^{(1)}$. For nonlinearly saturated solutions, we thus get $u_d \propto u^{(1)}$. In the vicinity of threshold, $\mu_0\sigma L(u_f - u_c) \propto \mathrm{Rm} - \mathrm{Rm}_c$, and we obtain

$$\langle|\mathbf{B}|^2\rangle \propto \frac{\rho\nu}{\sigma L^2}\,(\mathrm{Rm}-\mathrm{Rm}_c). \tag{2.31}$$

2.3.3. THE LAMINAR SCALING

We call (2.31) the "laminar scaling", obtained for $\mathrm{Re} \ll 1$ and characterised by the fact that $B \to 0$ if $\nu \to 0$ with all the other parameters fixed.

For a Ponomarenko type flow, we obtained a supercritical bifurcation (Nuñez *et al.*, 2001). The leading order nonlinear effects tend to saturate the growing magnetic field because the Lorentz force slows down the motion and hence diminishes the induction. For the magnetic field at saturation $\mathbf{B}_{\mathrm{sat}}$, we obtained

$$\mathbf{B}_{\mathrm{sat}} = 2.82\sqrt{\frac{\rho\nu}{\sigma R^2}}\,\sqrt{\mathrm{Rm}-\mathrm{Rm}_c}\;\mathrm{Re}\left\{\mathbf{B}_p\right\} , \tag{2.32}$$

where $\mathbf{B}_p$ is the neutral mode of the Ponomarenko dynamo.

The magnetic energy has the form of equation (2.31), what we called the laminar scaling because the Lorentz force is balanced by the perturbation in velocity through a viscous term. Close to onset, there is obviously no equipartition of energy because the magnetic energy tends to zero with $\mathrm{Rm} - \mathrm{Rm}_c$ while the kinetic energy is finite. Neither is there any simple balance between viscous dissipation and Joule dissipation. For Joule dissipation, we have $P_j \propto \int |\mathbf{j}|^2 \mathrm{d}V \propto \int |\boldsymbol{\nabla}\times\mathbf{B}|^2\mathrm{d}V \propto (\mathrm{Rm}-\mathrm{Rm}_c)$. Concerning viscous dissipation P_ν, it is proportional to the square of the stress tensor. This tensor is linear in the total velocity and is thus proportional to $\mathbf{u}^{(1)}$ because the stress tensor of $\mathbf{u}_f$ is zero (solid body rotation and translation). Hence $P_\nu \propto |\mathbf{u}^{(1)}|^2 \propto (\mathrm{Rm}-\mathrm{Rm}_c)^2$. In this particular case, with no viscous dissipation at onset, we observe that most of the input power is dissipated by Joule effect close to the dynamo onset. In more complex laminar flows, Joule dissipation is of course negligible compared to viscous dissipation just above the dynamo threshold.

More realistic helical flow geometries have been considered (Bassom & Gilbert, 1997), but the saturating magnetic field has only been computed in the limit $\mathrm{Re} \gg$

$\mathrm{Rm} \gg 1$ for which it is difficult to have controlled approximations. However, the result also shares the main property of the laminar scaling, $B \to 0$ if $\nu \to 0$ with all the other parameters fixed.

2.4. SATURATION IN THE HIGH Re LIMIT IN THE VICINITY OF THE DYNAMO THRESHOLD

2.4.1. DIMENSIONAL ARGUMENTS

We show now that we can take advantage of the characteristics of experimental dynamos to find the correct scaling of the magnetic field above the dynamo threshold (Pétrélis & Fauve, 2001). We have already mentioned that $\mathrm{Pm} \ll 1$ for most fluids. More precisely, $\mathrm{Pm} < 10^{-5}$ for all liquid metals. Thus, the Reynolds number is larger than several millions at the dynamo threshold (Rm_c is in the range $10 - 100$). In addition, the power needed to generate this turbulent flow increases like the cubic power of the driving velocity. Consequently, most experimental dynamos should:

- (i) bifurcate from a strongly turbulent flow regime,

- (ii) operate in the vicinity of their bifurcation threshold.

Although (i) makes almost impossible any realistic analytical calculation or direct numerical simulation, the above two characteristics allow an estimation of the non-linearly saturated magnetic field above Rm_c using dimensional analysis. Our goal is thus to find the expression of f in equation (2.2) in the limits (i) $\mathrm{Re} \to \infty$ and (ii) $\mathrm{Rm} - \mathrm{Rm}_c \to 0$: (i) implies that the momentum is mostly transported by turbulent fluctuations. Consequently, using the basic assumption of fully developed turbulence, we can neglect the kinematic viscosity, thus Re. (ii) implies that the dependence of $\langle B^2 \rangle$ on Rm is proportional to $\mathrm{Rm} - \mathrm{Rm}_c$, as expected for a supercritical bifurcation close to threshold. In other words, $U = |\mathbf{u}|$ is no longer a free parameter, but should take approximately the value corresponding to the dynamo threshold. Thus, (i) and (ii) reduce the number of parameters from 6 to 4, and the saturated value of the magnetic field can be obtained using dimensional analysis, to give

$$\langle |\mathbf{B}|^2 \rangle \propto \frac{\rho}{\mu_0 (\sigma L)^2} \left(\mathrm{Rm} - \mathrm{Rm}_c\right). \tag{2.33}$$

There is no paradox in the fact that the saturated magnetic field is inversely proportional to the square of the electric conductivity and to the square of the typical lengthscale of the flow. This does not mean that one should have σ and L small in order to observe large values of $|\mathbf{B}|$ since $\mathrm{Rm} = \mathrm{Rm}_c$ will be then achieved for a larger flow velocity. Using the typical velocity $U_c = |\mathbf{u}|$ at dynamo threshold, we can write (2.33) in the form, $\langle|\mathbf{B}|^2\rangle/\mu_0\rho U_c^2 \propto (\mathrm{Rm} - \mathrm{Rm}_c)/\mathrm{Rm}_c^2$, which shows that the system is very far from equipartition of energy in the vicinity of the dynamo threshold. We emphasise also that the *interaction parameter*, $\mathrm{N} = \sigma L\langle|\mathbf{B}|^2\rangle/\rho|\mathbf{u}|$, is much smaller than one. It is such that

$$\mathrm{N} \propto \mathrm{Rm} - \mathrm{Rm}_c\,. \tag{2.34}$$

2.4.2. HIGH Re DYNAMOS CLOSE TO THE BIFURCATION THRESHOLD

For $\mathrm{Pm} \ll 1$ or $\mathrm{Re} \gg 1$, we can recover the "turbulent scaling" (2.33) using the structure of the perturbation analysis presented for laminar dynamos. The only difference is that if $\mathrm{Re} \gg 1$, we have to balance the Lorentz force with the inertial instead of the viscous terms in (2.26). We thus get $|\mathbf{B}_{\mathrm{laminar}}| \propto |\mathbf{B}_{\mathrm{turbulent}}|\,\mathrm{Pm}^{1/2}$; consequently the two scalings strongly differ for experiments using liquid metals ($\mathrm{Pm} < 10^{-5}$).

It may be instructive to replace ν by the turbulent viscosity, $\nu_{\mathrm{T}} \propto |\mathbf{u}|\,L$, in the laminar scaling (2.31). Using $|\mathbf{u}| \approx \mathrm{Rm}_c/\mu_0\sigma L$, we have

$$\langle|\mathbf{B}|^2\rangle \propto \frac{\rho\nu_{\mathrm{T}}}{\sigma L^2}\,(\mathrm{Rm} - \mathrm{Rm}_c) \propto \frac{\rho}{\mu_0(\sigma L)^2}\,(\mathrm{Rm} - \mathrm{Rm}_c)\,. \tag{2.35}$$

We thus recover the turbulent scaling. However, dimensional arguments of the previous section do not require any assumption about the turbulent viscosity expression and are thus clearer.

The Karlsruhe (Stieglitz & Müller, 2001) and Riga (Gailitis *et al.*, 2001) experiments have recently reported values of the saturated mean magnetic field of order $10\,\mathrm{mT}$, roughly 10% above threshold. Both experiments used liquid sodium ($\mu_0\sigma \approx 10\,\mathrm{m}^{-2}\,\mathrm{s}$, $\rho \approx 10^3\,\mathrm{kg\ m}^{-3}$). The inner diameter of the Riga experiment is $L = 0.25\,\mathrm{m}$. The spatial periodicity of the flow used in the Karlsruhe experiment is of the same order of magnitude, within a cylinder of radius $0.85\,\mathrm{m}$ and height $0.7\,\mathrm{m}$. The presence of two length scales in the Karlsruhe experiment makes the comparison with our analysis more difficult, but we can easily compare the results of the Riga experiment with our "turbulent" scaling in (2.33) and "laminar" scaling in (2.31), that predict a saturated field of order $10\,\mathrm{mT}$ (respectively $10\,\mu\mathrm{T}$). Taking into account the qualitative nature of our analysis, we conclude that the "turbulent scaling" is in agreement with the experimental observations whereas the "laminar scaling" predicts a field that is orders of magnitude too small. The "turbulent scaling" also gives a correct order of magnitude for the Karlsruhe experiment if its spatial period is taken as the relevant lengthscale in (2.33). We thus note that the above experiments display a very interesting feature: turbulent fluctuations can be neglected when computing the dynamo threshold; indeed, the observed thresholds are in rather good agreement with those predicted by solving the kinematic dynamo problem for the mean flow alone. However, the high value of Re has a very strong effect on the value of the saturated magnetic field above the dynamo threshold.

We emphasise that the correct identification of the dominant transport mechanism of momentum is essential to estimate the order of magnitude of the saturated magnetic field above dynamo threshold. The reason is that it determines the flow distortion by the Lorentz force and thus the saturation mechanism of the field.

A laminar model of the flow thus generally leads to a wrong estimate of the magnetic field amplitude although it sometimes correctly predicts the dynamo threshold. This does not seem to have been fully understood in the early literature on dynamical dynamo models. It is of course possible to recover correct orders of magnitude for the field by using ad hoc turbulent transport coefficients. However, this is not very useful and may even hide the simplicity of the result.

We have shown that a simple scaling law given by (2.33) for the mean magnetic field generated by laboratory dynamos can be found because they bifurcate from a high Reynolds number flow and operate close to the dynamo onset (Pétrélis & Fauve, 2001). It would be interesting to test the validity of this scaling law in existing laboratory experiments. This has not been done yet, but may be achieved both in Karlsruhe and Riga experiments by varying the temperature of liquid sodium and thus its conductivity, σ.

2.5. EFFECT OF ROTATION

2.5.1. WEAK AND STRONG FIELD REGIMES OF THE GEODYNAMO

We first recall some general features displayed by several geodynamo models (for a detailed review, see Chapter 4 and Roberts, 1988). Rotation imposes a strong constraint on the flow that tends to become nearly two-dimensional. The length scale ℓ of the flow in a direction perpendicular to $\mathbf{\Omega}$ is thus much smaller than that along $\mathbf{\Omega}$, $\ell \ll L$. When convection is generated in a rotating sphere, the flow concentrates in columns of diameter $\ell \propto L\,\mathrm{E}^{1/3}$, where $\mathrm{E} = \nu/\Omega L^2$ is the Ekman number (see Chapter 3, and Roberts, 1968, Busse, 1970). This type of flow can generate a large scale magnetic field on length scale L via an α–effect (Busse, 1975b). Plane layer models (Childress & Soward, 1972; Soward, 1974) display most of the important features of spherical geometries: increasing the rotation rate too much delays the linear instability onset of self-generation because more and more power is necessary to overcome dissipation at small scale ℓ. However, for finite amplitude magnetic fields, the Lorentz force suppresses the rotational constraint and allows large scale motions, leading to much smaller viscous and ohmic dissipation. A subcritical "strong field" branch thus exists below the linear stability onset (St. Pierre, 1993) in addition to the "weak field" branch that bifurcates continuously at the linear dynamo threshold.

Only the weak field branch has been computed analytically, with different models (Childress & Soward, 1972; Soward, 1974; Busse, 1975b, 1976). These computations assume the flow to be laminar with a simple geometry. Consequently, the saturated magnetic field is governed by the low Reynolds number scaling (2.31), thus $\langle|\mathbf{B}|^2\rangle_{\text{weak}} \propto \rho\nu/\sigma L^2$. The weak field regime may be stable above the linear threshold (depending on the model) but it becomes unstable for an order one Chandrasekhar number ($\mathrm{Q} = \mathrm{N\,Re}$). The system then jumps to the strong field regime. It is belived that its scaling corresponds to a balance between the Coriolis and the Lorentz forces (known as the magnetostrophic balance), thus

$$\langle|\mathbf{B}|^2\rangle_{\text{strong}} \propto \rho\Omega/\sigma\,. \tag{2.36}$$

For the Earth, taking $\rho \approx 10^4$ kg m^{-3}, $\sigma \approx 3\,10^5$ S m^{-1} and $L \approx 3\,10^6$ m, gives $B_{\text{weak}} \approx 5 \times 10^{-2}\,\text{nT}$ (0.5 μG). This is orders of magnitude too small, whereas the strong field scaling, $\sqrt{\rho\Omega/\sigma} \approx 1\,\text{mT}$ (10 G), looks better.

A very interesting feature of dynamos generated by rapidly rotating flows is thus the subcritical nature of the bifurcation. Consequently, the questions related to the effect of rotation on the linear dynamo threshold are of secondary importance. The mean magnetic energy of finite amplitude dynamo solutions deserves more attention and is strongly affected by rotation.

2.5.2. FURTHER COMMENTS ON WEAK AND STRONG FIELD REGIMES

We first note that the form of $|\mathbf{B}_{\text{weak}}|$ given above is oversimplified. The important aspect is that $|\mathbf{B}_{\text{weak}}| \to 0$ if $\nu \to 0$ with all other parameters fixed. However, the length scale in the expression of $|\mathbf{B}_{\text{weak}}|$ is likely to involve both L and $\ell \propto L\mathrm{E}^{1/3}$, and thus to be a function of the rotation rate, Ω. But, note that even if we replace L by ℓ, we obtain

$$\langle|\mathbf{B}|^2\rangle_{\text{weak}} \propto \frac{\rho\nu}{\sigma L^2}\,\mathrm{E}^{-2/3}\,, \tag{2.37}$$

thus changing the field by a factor 10^5. This gives $5\,\mu$T (50 mG), which is still too small for the mean field value in the core of the Earth.

As a second step, we may try to incorporate the effect of turbulence since we have already emphasised that it strongly affects the mean magnetic energy. This can be done phenomenologically, starting from the laminar scaling with length scale $\ell \propto L\mathrm{E}^{1/3}$, and then replacing ν by the turbulent viscosity $\nu_\mathrm{T} \propto |\mathbf{u}_\mathrm{T}|\ell_\mathrm{T}$, where $|\mathbf{u}_\mathrm{T}|$ is the typical velocity scale on length ℓ_T. In the vicinity of the dynamo threshold, we have $\mathrm{Rm}_c \approx \mu_0\sigma|\mathbf{u}_\mathrm{T}|\ell_\mathrm{T}$, and we get the turbulent scaling for the magnetic energy with a length scale

$$\ell_\mathrm{T} \approx \ell\left(\frac{\mathrm{Rm}_c}{\mathrm{Pm}}\right)^{1/3} = \left(\frac{L}{\mu_0\sigma\Omega}\,\mathrm{Rm}_c\right)^{1/3}. \tag{2.38}$$

This gives a more realistic length scale for the diameter of the columns than the laminar one (a few tenth of kilometers rather than a few tenth of meters). We thus obtain a third possible scaling of the magnetic energy

$$\langle|\mathbf{B}|^2\rangle_\mathrm{turb} \propto \frac{\rho\nu}{\sigma L^2}\left(\frac{\mathrm{Rm}_c}{\mathrm{Pm}\,\mathrm{E}^2}\right)^{1/3} = \rho\left(\frac{\Omega^2}{\mu_0\sigma^4 L^2}\,\mathrm{Rm}_c\right)^{1/3}, \tag{2.39}$$

giving a more realistic value of the order of a Gauss for the mean field.

We finally note that we obtain the strong field scaling from the weak one by replacing ν by ΩL^2. This only means that, instead of the Stokes term, we have to balance the additional Coriolis term, $2\boldsymbol{\Omega}\times\mathbf{u}^{(1)}$, in (2.26) with the Lorentz force. However, such a scaling does not seem to require a subcritical bifurcation. If the Coriolis term is the dominant one, weakly nonlinear perturbation theory will lead to

$$|\mathbf{u}^{(1)}| \propto \frac{|\mathbf{B}^{(0)}|^2}{\mu_0\rho\Omega L}. \tag{2.40}$$

This gives

$$\langle|\mathbf{B}|^2\rangle \propto \frac{\rho\Omega}{\sigma}\,(\mathrm{Rm} - \mathrm{Rm}_c). \tag{2.41}$$

We obtain the strong field scaling, but without assuming that there is a balance between the Lorentz force and the total Coriolis force. Only the Coriolis force related to the velocity perturbation balances the Lorentz force; this gives the additional term $\mathrm{Rm} - \mathrm{Rm}_c$ in the expression of the mean magnetic energy. Although this looks fine, it is not obvious that a perturbative analysis can be worked out that way and we should be cautious in the absence of an explicit analytical example leading to the strong field scaling.

2.5.3. SCALINGS OF MAGNETIC ENERGY USING DIMENSIONAL CONSIDERATIONS

The weak field scaling given by expression (2.37) yields too small field values, but the turbulent expression (2.39) and the strong field expression (2.36) only differ by roughly an order of magnitude in the case of the Earth. Although their expressions are different, both give possible values of the field for the Earth if we take into account the qualitative nature of our analysis.

It may be interesting to understand the strong field scaling as follows: we already noticed that altough magnetic fields exist in a wide range of scales in astrophysics, their values do not seem to be primarily determined by the size L of astrophysical objects. As a very rough approximation, assume that $\langle|\mathbf{B}|^2\rangle$ does not depend on L and also neglect ν since the flow is turbulent (at least at small enough scales). We are then left with 6 parameters, $B = |\mathbf{B}|$, ρ, μ_0, σ, $U = |\mathbf{u}|$, Ω, from which we can construct two dimensionless numbers, for instance $B^2/\mu_0\rho U^2$ and $\mathrm{Rm}\,\mathrm{Ro} = \mu_0\sigma U^2/\Omega$. We thus get

$$B^2 = \mu_0\rho U^2\, g(\mathrm{Rm}\,\mathrm{Ro})\,. \tag{2.42}$$

Close to the dynamo threshold, g bifurcates from zero and behaves like $\mathrm{Rm}\,\mathrm{Ro} - (\mathrm{Rm}\,\mathrm{Ro})_c$, with $(\mathrm{Rm}\,\mathrm{Ro})_c = \mu_0\sigma U_c^2/\Omega$. Consequently we obtain

$$B^2 \propto \frac{\rho\Omega}{\sigma}\left[\mathrm{Rm}\,\mathrm{Ro} - (\mathrm{Rm}\,\mathrm{Ro})_c\right]\,, \tag{2.43}$$

and we recover the strong field scaling. Note that we expect it to be valid if $\mathrm{Ro} \ll 1$ (dominant rotation) but for $\mathrm{Rm}\,\mathrm{Ro}$ large enough to generate the dynamo. We do not expect L to be the relevant length scale for the strong field regime, but the smaller scale U_c/Ω. We observe that the flow is turbulent on this scale at dynamo onset ($U_c^2/\nu\Omega \gg 1$ since $\mathrm{Re}\,\mathrm{Ro} \gg \mathrm{Rm}\,\mathrm{Ro}$).

2.6. SCALING LAWS IN THE LIMIT OF LARGE Rm AND Re

Finally, we will consider the case of astrophysical flows where both Rm and Re are very large. Neither laboratory experiments, nor direct numerical simulations are possible in this range of Rm and Re. The only way is to try to guess scaling laws for the magnetic field using some simple hypothesis. We thus consider again the minimum set of parameters, U, L, ρ, ν, μ_0, σ. We note that discarding global rotation is certainly invalid for most astrophysical objects. However, even in the simplest case of a homogeneous isotropic turbulent flow, with an integral velocity U in a domain of size L, no clear-cut result exists neither for the dynamo threshold, nor for the scaling of the magnetic energy. We will shortly review the problem of the dynamo threshold of a turbulent flow and then discuss possible scalings for the magnetic energy.

2.6.1. EFFECT OF TURBULENCE ON THE DYNAMO THRESHOLD

Taking into account the minimum set of parameters, U, L, ρ, ν, μ_0, σ, dimensional analysis gives for the dynamo threshold Rm_c

$$\mathrm{Rm}_c = F(\mathrm{Re}). \tag{2.44}$$

For a given geometry and a large scale flow, the unknown function, F, represents how Rm_c depends on the fluid properties. Finding the behaviour of F in the limit of large Re will show how turbulent fluctuations affect the dynamo threshold. This is still an open problem, even in the case of a homogeneous isotropic turbulent flow with zero mean and without helicity. Recent direct numerical simulations show that Rm_c keeps increasing with Re at the highest possible resolution without any indication of a possible saturation (Schekochihin *et al.*, 2004). However, if one assumes that the magnetic field is a large scale quantity, i.e. is not affected by the value of viscosity in the limit of large Re according to the usual phenomenology of turbulence, we immediately get that, if dynamo action is possible in the limit of large Re, its threshold is given by $\mathrm{Rm}_c \to \mathrm{constant}$ in this limit.

A lot of work has been performed on the determination of Rm_c as a function of Re for turbulent dynamos in the limit of large Re (or small Pm). Eddy–damped quasi–normal Markovian approximation (EDQNM) closures (see page 90) have predicted $\mathrm{Rm}_c \approx 30$ for non helical flows (Léorat *et al.*, 1981). The agreement with the above simple argument is not really surprising since these closures keep only the large scales. A lot of analytical studies have been also performed, mostly following Kazantsev's model (Kazantsev, 1968). Kazantsev considered a random homogeneous and isotropic velocity field, δ–correlated in time and with a wave number spectrum of the form k^{-p}. He showed that for p large enough, generation of a homogeneous isotropic magnetic field with zero mean value, takes place. This is a nice model, but its validity is limited to large Pm for which the magnetic field has a much larger time scale than the velocity field. In this case, assuming that the velocity field is δ–correlated in time is probably a reasonable approximation. However, Kazantsev's model has been also extrapolated to large Re. Various predictions, $\mathrm{Rm}_c \propto \mathrm{Re}$ (Novikov *et al.*, 1983), $\mathrm{Rm}_c \to \mathrm{constant} \approx 400$ for steep enough velocity spectra ($p > 3/2$) and no dynamo otherwise (Rogachevskii and Kleeorin, 1997), or dynamo for all possible slope of the velocity spectrum in the range $1 < p < 3$ (Boldyrev & Cattaneo, 2004) have been found. These discrepancies result from non rigorous extrapolation of Kazantsev's model to large Re. The calculation is possible only in the case of a δ–correlated velocity field in time, and $\delta(t - t')$, which has de dimension of the inverse of time, should then be replaced by a finite eddy turn-over frequency in order to describe large Re effects.

A different problem about turbulent dynamos has been considered more recently. It concerns the effect of turbulent fluctuations on a dynamo generated by a mean flow. The problem is to estimate to what extent the dynamo threshold computed as if the mean flow were acting alone, is shifted by turbulent fluctuations. This question has been addressed only recently (Fauve & Pétrélis, 2003) and should not be confused with dynamo generated by random flows with zero mean. It has been shown that weak turbulent fluctuations do not shift the dynamo threshold of the mean flow at first order. In addition, in the case of small scale fluctuations, there is no shift at second order either, if the fluctuations have no helicity. This explains why the observed dynamo threshold in Karlsruhe and Riga experiments has been found in good agreement with the one computed as if the mean flow were acting alone, i.e. neglecting turbulent fluctuations. Recent numerical simulations have shown that in the presence of a mean flow, Rm_c increases with Re at moderate Re, but then seems to saturate at larger Re (Ponty *et al.*, 2005).

2.6.2. Batchelor's predictions for turbulent dynamo threshold and saturation

It may be instructive at this stage to consider the first study on turbulent dynamos made more than half a century ago by Batchelor (1950). Using a questionable analogy between the induction and the vorticity equations, he claimed that the dynamo threshold corresponds to $\mathrm{Pm} = 1$, i.e. $\mathrm{Rm}_c \propto \mathrm{Re}$, using our choice of dimensionless parameters. Pushing the analogy further, he observed that the magnetic field should be generated mostly at the Kolmogorov scale, $\ell_K = L\mathrm{Re}^{-3/4}$, where the velocity gradients are the strongest. He then assumed that saturation of the magnetic field takes place for $\langle B^2 \rangle / \mu_0 \propto \rho v_K^2 = \rho U^2/\sqrt{Re}$, where v_K is the velocity increment at the Kolmogorov scale, $v_K^2 = \sqrt{\nu\varepsilon}$. $\varepsilon = U^3/L$ is the power per unit mass, cascading from L to ℓ_K in the Kolmogorov description of turbulence.

It is now often claimed that Batchelor's criterion $\mathrm{Pm} > 1$ for the growth of magnetic energy in turbulent flows is incorrect. However, it should be noted that for homogeneous isotropic turbulence without mean flow and helicity, the weaker criterion $\mathrm{Pm} >$ constant or $\mathrm{Rm}_c \propto \mathrm{Re}$, is still considered to be a possible scenario (Schekochihin *et al.*, 2004). It is thus of interest to determine the minimal hypothesis for which Batchelor's predictions for dynamo onset and saturation are obtained using dimensional arguments.

First, $\varepsilon = U^3/L$ being the power per unit mass available to feed the dynamo, it may be a wise choice to keep it, instead of U in our minimal set of parameters, thus becoming B, ρ, ε, L, ν, μ_0 and σ. Then, the predictions of Batchelor can be found using the following simple requirement: let us consider only the dynamo eigenmodes that do not depend on L. This is a reasonable requirement, since we may hope that in a large domain, there exist some class of small scale magnetic fields which are insensitive to the details of boundary conditions. Then, forgetting L in our set of parameters, dimensional analysis gives at once $\mathrm{Pm} = \mathrm{Pm}_c = \mathrm{constant}$ for the dynamo threshold, i.e.

$$\mathrm{Rm}_c \propto \mathrm{Re}\,. \tag{2.45}$$

We also obtain for the mean magnetic energy density

$$\frac{\langle B^2 \rangle}{\mu_0} = \rho\sqrt{\nu\varepsilon}\, G(\mathrm{Pm}) = \frac{\rho U^2}{\sqrt{\mathrm{Re}}}\, G(\mathrm{Pm})\,, \tag{2.46}$$

where $G(\mathrm{Pm})$ is an arbitrary function of Pm. Close to dynamo threshold, we expect $G(\mathrm{Pm}) \propto \mathrm{Pm} - \mathrm{Pm}_c$ if the bifurcation is supercritical. Only the prefactor $\rho U^2/\sqrt{\mathrm{Re}}$ of (2.46), which is the kinetic energy at Kolmogorov scale, was considered by Batchelor and assumed to be in equipartition with magnetic energy. This class of dynamos being small scale ones, it is not surprising that the inertial range of turbulence screens the magnetic field from the influence of integral size, thus L can be forgotten. We emphasise that a necessary condition for Batchelor's scenario is that the magnetic field can grow below the Kolmogorov scale, i.e. its dissipative length ℓ_σ should be smaller than ℓ_K, thus $\mathrm{Pm} > 1$.

2.6.3. A KOLMOGOROV TYPE SCALING IN THE LIMIT $\mathrm{Re} \gg \mathrm{Rm} \gg \mathrm{Rm}_c$

The simplest argument in the limit where both Rm and Re are very large is, as usual, to assume that the transport coefficients ν and σ become negligible. We are left with one dimensionless parameter and

$$\frac{\langle B^2 \rangle}{\mu_0} \propto \rho U^2\,. \tag{2.47}$$

We thus obtain equipartition of energy, an assumption often made in the early dynamo literature. The scaling of the mean square magnetic field does not involve the size L without any further assumption. Note however that this result will not subsist if global rotation is important. The right hand side of expression (2.47) will then involve an a priori arbitrary function of the Rossby number, thus leading to a possible dependence of B on Ω and L.

Assuming that the above argument is correct means that the magnetic field is a large scale quantity in the phenomenology of turbulence. There is obviously a strong discrepancy between (2.47) and (2.46). These two laws are the upper and lower limits of a continuous family of scalings that are obtained by balancing the magnetic energy with the kinetic energy at one particular length scale within the Kolmogorov spectrum. It is not known if one of them is selected by turbulent dynamos.

We finally consider the case $\mathrm{Pm} \ll 1$ i.e. $\mathrm{Re} \gg \mathrm{Rm}$. We know from the Karlsruhe and Riga experiments that dynamo action is possible in this range above a moderate value of Rm_c provided that the mean flow is appropriately chosen. As said above, the problem is still open in the absence of mean flow, although some models predict a much larger, but finite, Rm_c in the limit of large Re.

Assuming that a dynamo is generated, we want to give a possible guess for the power spectrum $|\widehat{B}|^2$ of the magnetic field as a function of the wave number k and the parameters ρ, ε, L, ν, μ_0 and σ. Since $\mathrm{Re} \gg \mathrm{Rm} \gg \mathrm{Rm}_c$, the dissipative lengths are such that $\ell_K \ll \ell_\sigma \ll L$. For k in the inertial range, i.e. $k\ell_\sigma \ll 1 \ll kL$, we may use a Kolmogorov type of argument and discard L, σ and ν. Then, only one dimensionless parameter is left and, not too surprisingly, we get

$$|\widehat{B}|^2 \propto \mu_0 \rho\, \varepsilon^{2/3}\, k^{-5/3}\,. \tag{2.48}$$

This is only one possibility among many others proposed for MHD turbulent spectra within the inertial range, but it is the simplest. Integrating over k obviously gives equipartition law (2.47) for the magnetic energy. It is now interesting to evaluate Ohmic dissipation. Its dominant part comes from the current density at scale ℓ_σ. We have

$$\frac{\mathbf{j}^2}{\sigma} = \frac{1}{\sigma}\int |\widehat{j}|^2\, dk \propto \frac{1}{\mu_0^2 \sigma}\int k^2 |\widehat{B}|^2\, dk \propto \frac{\rho}{\mu_0 \sigma}\, \varepsilon^{2/3}\, \ell_\sigma^{2/3} \propto \rho \frac{U^3}{L}\,. \tag{2.49}$$

We thus find that Ohmic dissipation is proportional to the total available power, which corresponds to some kind of optimum scaling law for Ohmic dissipation. However, this does not give any indication that this regime is achieved. The discrepancies between plausible laws given in this section show that the problem of turbulent dynamos still deserves a lot of studies.

2.7. NONLINEAR EFFECTS IN MEAN FIELD DYNAMO THEORY

Let us now consider the limit of large magnetic Reynolds number Rm. The majority of research into astrophysical dynamos (see Chapter 6 and Chapter 7) has been performed within the framework of mean field electrodynamics. It can also be a useful approach in geodynamo models (see Section 4.5.1), but here there has also been much recent work on solving the full three dimensional equations (see

Section 4.5.3). Mean field electrodynamics, conceived in the 1960's by Steenbeck, Krause and Rädler (see Krause & Rädler, 1980 for full references), is an extremely elegant theory – and is, in many ways, extremely successful. By judicious choice of the various parameters in the theory, it is possible to model a vast range of dynamo-generated magnetic fields (see, for example, the review by Rosner, 2000). It should however always be borne in mind that mean field electrodynamics is a theory of MHD turbulence, and, as in all theories of turbulence (magnetic or non-magnetic), it involves approximations and assumptions. The aim of this chapter is to discuss the various approaches that have been taken towards understanding the nonlinear behaviour of mean field dynamos, concentrating mainly on astrophysical modelling (i.e. high values of the Reynolds numbers Rm and Re). Of particular significance is that the power of present-day computers is now allowing realistic simulations of turbulence – though by no means at the extreme parameter values that pertain in astrophysical situations – and that it is therefore becoming possible to compare theoretical predictions with results from numerical simulations.

The aim of mean field electrodynamics is to provide a mathematical theory for the evolution of magnetic fields on scales large compared with that of the driving turbulent velocity field. Its formulation has already been described in depth in Section 1.5, and so will not be reproduced in detail here. There are though two key points to note, namely:

(i) that the formulation is essentially linear – being based on the induction equation for given flow statistics, and

(ii) that, often, progress is possible only for the case of low Rm. Only in this case is it possible to make any rigorous statement about the relationship between the fluctuating and mean magnetic fields.

Typically though, astrophysical plasmas are both nonlinear in their behaviour and possess extremely high values of Rm [$\mathcal{O}(10^{11})$ in the solar convection zone, for example. See Table IV]. We therefore need to examine just how far the formulation and attendant consequences of mean field electrodynamics can be carried over into the regime of astrophysical relevance.

There are essentially three ways of making progress with understanding the nonlinear evolution of a large-scale magnetic field in a turbulent flow:

(i) through the incorporation into the mean field formalism of plausible nonlinearities, based on physical arguments;

(ii) via other MHD turbulence theories, which put the induction equation and momentum equation on an equal footing;

(iii) via direct numerical simulations of the full governing MHD equations.

In this section, we shall consider each of these areas in turn, and try to give a picture of just where the subject stands at present – to discuss which are the areas of agreement, and which are those of contention. It is intended as an introductory text; it is, deliberately, far from exhaustive, and the work we shall describe has been chosen for illustrative purposes. A much fuller list of references can be found, for example, in the review of galactic magnetic fields by Beck *et al.* (1996) and the recent review of the solar dynamo by Ossendrijver (2003).

For turbulent MHD flows there are two important nonlinearities in the momentum equation. One is the inertial $(\mathbf{u} \cdot \nabla)\mathbf{u}$ term – the crucial nonlinearity in hydrodynamic turbulence, responsible for energy transfer between different spatial scales; the other is the Lorentz force $(\mathbf{j} \times \mathbf{B})$, which provides the back-reaction of the magnetic field on the velocity field. In the following sections, we shall explain how these nonlinearities are accounted for in the three approaches outlined above.

2.7.1. NONLINEAR EFFECTS IN THE MEAN FIELD FORMALISM

THE INCORPORATION OF PLAUSIBLE NONLINEARITIES

As described in Section 1.5, the standard formulation of mean field electrodynamics leads to a mean induction equation in which the large-scale field evolves under the influence of the tensors α_{ij} and β_{ijk}, and a large-scale flow (or differential rotation, ω). The simplest means of introducing nonlinear effects into the theory – which are, of course, necessary to prevent unlimited growth of the magnetic field – is to modify one or more of α_{ij}, β_{ijk} or ω in a manner that reflects the underlying physics. It is though important to point out that such modifications typically do not arise from some self-consistent theory, but are merely physically *plausible*.

For simplicity, let us for the moment consider the case when α_{ij} and β_{ijk} are isotropic tensors, namely $\alpha_{ij} = \alpha\delta_{ij}$ and $\beta_{ijk} = \beta\varepsilon_{ijk}$; we need then concern ourselves only with the pseudo-scalar α and the scalar β. The induction equation for the mean magnetic field $\mathbf{B}_0$ then takes the form (see Section 1.5.2):

$$\partial_t \mathbf{B}_0 = \nabla \times (\mathbf{U} \times \mathbf{B}_0) + \nabla \times (\alpha \mathbf{B}_0) + \nabla \times [(\eta + \beta)\nabla \times \mathbf{B}_0] \, . \qquad (2.50)$$

At low values of Rm one may interpret the α–effect in terms of the physical picture first put forward by Parker (1955), of rising and twisting loops of field giving rise to a mean current anti-parallel to the large-scale field. On physical grounds it is entirely reasonable to argue that this process becomes less effective as the field strength increases – the Lorentz force resisting the tendency to twist field lines – and that therefore α should be a monotonically decreasing function of the large-scale field B_0. The most widely used formulation is that proposed by Jepps (1975), with α taking the form

$$\alpha = \frac{\alpha_0}{1 + B_0^2/\mathcal{B}^2}, \tag{2.51}$$

where α_0 represents the kinematic value of α and $\mathcal{B}^2$ represents some reference magnetic energy. At high values of Rm there is no clear physical picture of even the kinematic (linear) α–effect, and thus it is not at all surprising that the precise nature of the Lorentz force is much harder to understand. Formulae of the form (2.51) are commonly advanced, but there is considerable controversy over which value of $\mathcal{B}$ is appropriate.

If α is "quenched" in the manner suggested by expression (2.51) one may argue that the turbulent diffusivity β should be similarly reduced, the general argument being that a stronger field resists shredding and hence that the process of turbulent diffusion is inhibited. Dynamo models therefore sometimes adopt a prescription for β of the form

$$\beta = \frac{\beta_0}{1 + B_0^2/\mathcal{B}^2}, \tag{2.52}$$

where the reference energy $\mathcal{B}^2$ in expression (2.52) may – or may not – take the same value as in (2.51). There is, of course, a tremendous amount of physics hidden away in the formulae (2.51) and (2.52), and we shall return to this issue in later sections; the aim here however is simply to discuss the general nature of the nonlinearities that are typically incorporated into mean field electrodynamics, and to examine their consequences.

An alternative to equation (2.51) is to formulate a separate equation for α – a so-called *dynamic* α–effect. Schmalz & Stix (1991) postulate that α may be expressed as the difference between a kinematic and dynamic component, $\alpha = \alpha_k - \alpha_d$, where α_d obeys the relation

$$\partial_t \alpha_d = \mathcal{D}(\alpha_d) + \mathcal{F}(AB), \tag{2.53}$$

where $\mathcal{D}$ represents a damping term and the function $\mathcal{F}(AB)$ is chosen to represent the quenching of the α–effect by the Lorentz force, whilst maintaining the pseudo-scalar nature of α. Yet another possible approach is that of Yoshimura (1978), who argues that the reaction of the field on the driving flow does not occur instantaneously, but only after a certain time t_d has elapsed. This is built into his formulation of the mean field equations through specifying that α depends *not* on the magnetic field at the present time t, but instead on the magnetic field at an earlier time $t - t_d$. Such formulations, of either a dynamic α–effect or an α–effect that depends on the field at an earlier time, can be justified through rather non-specific physical arguments, as indeed can (2.51); they are though all somewhat arbitrary.

The Lorentz force, via the momentum equation, of course acts not only on the small-scale turbulence – and hence influences the transport coefficients α and β – but also on the large-scale flow (i.e. on the differential rotation). This can be taken into account through a simple ω–quenching model of the form

$$\omega = \frac{\omega_0}{1 + B_0^2/\mathcal{B}^2}, \tag{2.54}$$

based on fairly non-specific arguments that the stress exerted by the small-scale magnetic field inhibits the differential rotation. The large-scale magnetic field also has a *direct* dynamic effect on the large-scale flow; this process, first investigated by Malkus & Proctor (1975), is accounted for by an additional equation for the large-scale velocity.

A third, and rather different, mechanism of dynamo saturation is that due to loss of flux from the region of field generation. This is typically ascribed to an upward escape of flux via magnetic buoyancy, a consequence of the magnetic pressure supporting more gas than would be possible in its absence (see, for example, the review by Hughes & Proctor, 1988). The process is independent of the sign of the magnetic field, and so the simplest prescription is to add a term of the form $-B^3$ to the right hand side of the mean induction equation (2.50). Again it should be stressed that although this is a reasonable parametrisation, the true physics of magnetic buoyancy instabilities is considerably more complex (see, for example, Hughes, 1991). Indeed, the real picture may be quite subtle; magnetic buoyancy instabilities in a rotating frame – which can lead to an upward transport of magnetic flux – yield helical motions which may, through an α–effect, be conducive to field generation (see Moffatt, 1978; Thelen, 2000a,b). So magnetic buoyancy may play a role not only in the loss of field, but also, indirectly, in its generation.

NONLINEAR MODELS

In modelling a stellar dynamo, the obvious interpretation of the averaging process in the mean-field formulation is as an azimuthal average, leading, in spherical geometry, to equations dependent on radius r, meridional angle θ and time t. Although less daunting than the full, three-dimensional MHD equations, solution of the axisymmetric mean field equations is still a non-trivial task. It can often therefore be instructive to consider further simplifications. The most drastic is to reduce the governing partial differential equations in r, θ and t to a low-order set of ordinary differential equations in t. One of the earliest such models is that of Weiss, Cattaneo & Jones (1984) who, via a severe truncation of a modal expansion of the mean induction and momentum equations, derived the following seventh order system, which may be regarded as a complex generalisation of the Lorenz equations:

$$\begin{aligned}\dot{A} &= 2\mathrm{D}(1+\kappa|B|^2)^{-1}B - A\,, && (2.55a)\\ \dot{B} &= \mathrm{i}(1+\omega_0)A - \tfrac{1}{2}\,\mathrm{i}\,A^*\omega - (1+\lambda|B|^2)B\,, && (2.55b)\\ \dot{\omega}_0 &= \tfrac{1}{2}\,\mathrm{i}\,(A^*B - AB^*) - \nu_0\omega_0\,, && (2.55c)\\ \dot{\omega} &= -\mathrm{i}AB - \nu\omega\,, && (2.55d)\end{aligned}$$

where A and B represent the (complex) poloidal flux function and toroidal field, ω_0 (real) and ω (complex) represent the spatially uniform and spatially varying components of the differential rotation, ν and ν_0 are real constants related to an eddy viscosity. There are three forms of nonlinearity in the above set of equations; α–quenching in the A equation, represented through a term of the form (2.51) (κ being a positive real constant), a buoyancy loss term in the B equation (λ a positive real constant), and the feedback of the Lorentz force on the differential rotation in the ω_0 and ω equations.

Weiss *et al.* (1984) concentrated on the case of $\kappa = \lambda = 0$, for which the nonlinear feedback acts only on the differential rotation, and found that solutions of the seventh order system fall into two classes, depending on whether the nonlinear saturation is dominated by ω_0 or ω. The former can be accommodated within the fifth order system obtained by letting $\nu \to \infty$, the latter within the sixth order system formed by letting $\nu_0 \to \infty$. For $\mathrm{D} > 1$ there is an exact nonlinear solution of the seventh order system, corresponding to dynamo waves. For the fifth order system this solution remains stable for all D; for the sixth order system, however, it loses stability and more mathematically interesting behaviour ensues. As D increases, successive Hopf bifurcations, leading to quasi-periodic behaviour, are followed by a period-doubling cascade to chaos. The magnetic field in the chaotic regime has epochs of cyclic activity interspersed with quiescent episodes during which the field amplitude is reduced and varies on a much slower timescale (see Figure 2.1); such behaviour is, of course, suggestive of the time trace of the Sun's magnetic field measured, for example, by the sunspot number.

The natural extension to low-order ODE models – which are local in nature – is to incorporate full spatial variation in one dimension, the most astrophysically relevant way of achieving this being to consider thin-shell dynamos, in which averaging over the radial direction leads to a set of PDEs in θ and t. Such models naturally allow, for example, interactions between the fields in each hemisphere. This approach was adopted by Belvedere, Pidatella & Proctor (1990) who considered a model in which the only manifestation of the Lorentz force is to modify the large-scale velocity. Increases in the dynamo number lead to solutions of increasing spatial and temporal complexity, from simple periodic solutions to quasi-periodic and "pulsed" solutions in which relatively long periods of stasis are interrupted by interludes of cyclic behaviour. The model of Belvedere *et al.* (1990) also allows for the possibility of

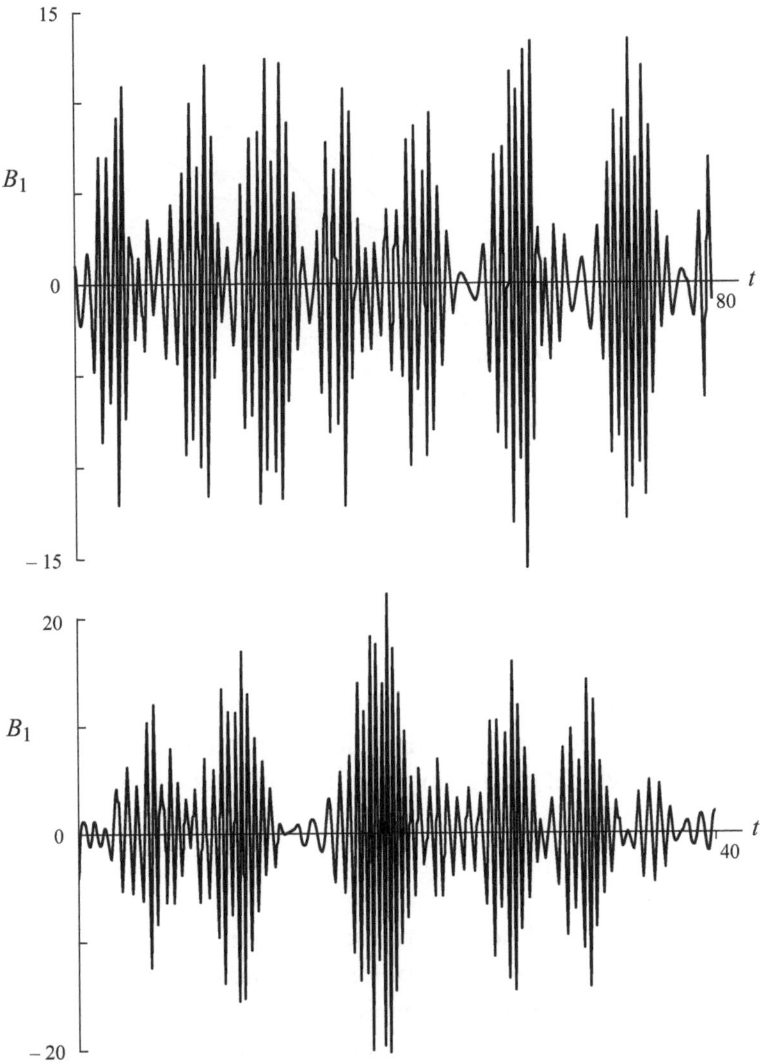

Figure 2.1 - Aperiodic oscillations of the sixth-order system [derived from (2.55a)–(2.55d) with $\nu_0 \to \infty$], modulated to give episodes of reduced activity; $B_1(t)$ (the real part of B) for (a) D $= 8$ and (b) D $= 16$ (from Weiss *et al.*, 1984).

multiple stable solutions for the same parameter values.

Jennings & Weiss (1991) also considered a one-dimensional model, "flattened" into Cartesian geometry ($\theta \to x$), governed by the equations

$$\partial_t A = \frac{\mathrm{D}\cos x}{1+\tau B^2}\, B + \partial_{xx} A\,, \qquad \partial_t B = \frac{\sin x}{1+\kappa B^2}\, \partial_x A + \partial_{xx} B - \lambda B^3\,, \quad \text{(2.56a,b)}$$

which may be regarded as a nonlinear modification to equations (1.112a,b). Their model differs from that of Belvedere *et al.* (1990) not just in the geometry, but also

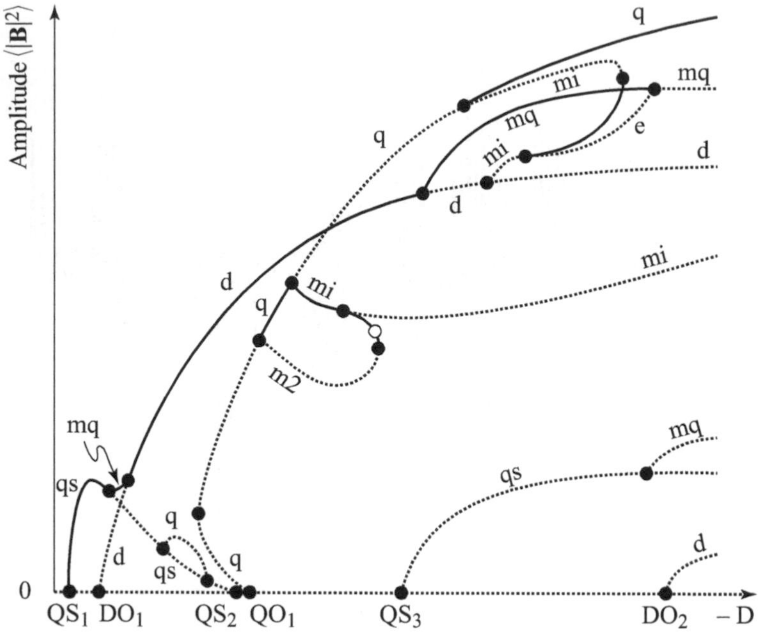

Figure 2.2 - Bifurcation diagram for the system (2.56a,b) with $\kappa = \lambda$, $\tau = 0$ and $\mathrm{D} < 0$; d, q and m refer to dipole, quadrupole and mixed mode solutions, s to steady solutions. Local and global bifurcations are indicated by • and ∘ respectively (from Jennings & Weiss, 1991).

through a different choice of nonlinearities; equations (2.56a,b) contain terms representing α- and ω–quenching and flux loss by magnetic buoyancy, but no direct feedback on the large-scale velocity. Jennings & Weiss (1991) were particularly interested in the phenomenon of symmetry-breaking between the northern and southern hemispheres; via a fairly low-order truncation of equations (2.56a,b), which enabled them to locate both stable and unstable solutions, they were able to construct the bifurcation diagram demonstrating the transitions between dipole, quadrupole and mixed modes. Figure 2.2 shows the bifurcation diagram for the case of $\kappa = \lambda$, $\tau = 0$. Of significance is the existence of different types of stable solution (e.g. dipole and quadrupole) at the same value of the dynamo number D.

Two-dimensional models, in which the variables depend on r and θ (and time) have also received considerable attention, with investigation of all the different types of nonlinearity discussed above (see, for example, Covas *et al.*, 1998, who performed a comparison between algebraic and dynamic α–quenching). Just as for the models with zero or one spatial dimension a wide variety of behaviour can be found through the incorporation of different nonlinearities.

Equation (2.50), with $\mathbf{U}$, α, β and η dependent on (at most) r, θ and t, clearly supports non-axisymmetric solutions – with $\mathbf{B}$ proportional to $\exp(im\phi)$ in the linear regime. Furthermore it may even be the case that the mode of maximum growth rate is non-axisymmetric. However, one has to exercise a certain amount of caution over the interpretation of non-axisymmetric solutions of (2.50). If the mean-field procedure is that of averaging over the azimuthal angle then – for logical consistency – the magnetic field in equation (2.50) must also be axisymmetric. Similarly, an ensemble average – itself somewhat hard to justify for any isolated stellar object – leads to a reduction in the number of spatial dimensions. One other interpretation is that the average represents a filtering between large scales ($m < m_*$, say, for some m_*) and small scales ($m \geq m_*$) – this allows for the possibility of non-axisymmetric modes of equation (2.50), but begs the question as to why α etc. should, in this case, not depend on *all* $m < m_*$. Consequently the most consistent interpretation of three-dimensional (non-axisymmetric) solutions of equation (2.50) is as some sort of temporal average, where ∂_t represents the rate of change over timescales long compared to that involved in the averaging procedure.

THE ROBUSTNESS OF NONLINEAR MODELS

Through the choice of the various nonlinearities discussed above, it is possible to obtain a considerable range of solutions to the mean induction equation (2.50), with differing spatial symmetries and a range of temporal complexity, an excellent agreement being possible with observed cosmical fields. The different types of possible solution and their relation to stellar magnetic fields are discussed further in Chapter 6. However, one has to exercise a certain degree of caution in interpreting the results of mean field models. It is well known that the dynamics of low-order systems may be critically dependent on the severe truncation performed to obtain them, and that, for example, chaotic behaviour may disappear in corresponding higher-order systems. Schmalz & Stix (1991) found such a phenomenon in their dynamic α models. Furthermore, the qualitative behaviour may be sensitively dependent on the precise form of the nonlinearity chosen. Covas *et al.* (1997) re-examined the Schmalz & Stix model by considering different functional forms of the driving term for α_d in (2.53), and found that significant changes in the chaotic nature of the solutions could result. Tobias (1998) has examined the dependence of dynamo cycle periods on the various nonlinearities that may be included in an interface dynamo model (described in more detail in Section 6.4) and, from the time series of the various models, concluded that it is difficult to discriminate between different nonlinearities.

2.7.2. MHD TURBULENCE THEORIES

The essential principle behind the models discussed in Section 2.7.1 is that the mean induction equation (2.50) is pre-eminent, and that nonlinear effects can be incorporated either through parametrisations of the form (2.51) or through auxiliary equations for α or ω. There are clearly advantages to such an approach; the full horrors of the momentum equation are avoided and, importantly, it appears that most astrophysical dynamos can, at some level, be modelled in this way. Clearly though, there is always the worry that some of the essential physics of the problem, contained in the momentum equation, is missing. The aim of this section therefore is to review some of the attempts that have been made to model the turbulent dynamo problem through treating the induction and momentum equations on a more equal footing.

One approach is to apply the ideas of mean field averaging not just to the induction equation but also to the momentum equation, on the grounds that the same turbulence occurs in both. On neglecting the magnetic field for the moment, and splitting the velocity field into its mean and fluctuating components, $\mathbf{U} = \mathbf{U}_0 + \mathbf{u}$, the (dimensionless) mean momentum equation for an incompressible flow may be written as

$$\partial_t \mathbf{U}_0 + (\mathbf{U}_0 \cdot \boldsymbol{\nabla})\mathbf{U}_0 = -\boldsymbol{\nabla} P - \boldsymbol{\nabla} \cdot \langle \mathbf{uu} \rangle + \mathrm{Re}^{-1} \Delta \mathbf{U}_0, \tag{2.57}$$

where the effects of the turbulence are contained in the Reynolds stress tensor

$$Q_{ij} = \langle u_i u_j \rangle. \tag{2.58}$$

Just as in classical mean field electrodynamics, in which the aim is to express the mean EMF $\boldsymbol{\mathcal{E}} = \langle \mathbf{u} \times \mathbf{b} \rangle$ in terms simply of the mean magnetic field, the aim here is to express the tensor Q_{ij} in terms of the mean velocity field. This is, however, an even more daunting task than for the induction equation; whereas, at least for weak fields, the latter represents a problem linear in the magnetic field, the momentum equation for fully turbulent flows is inherently nonlinear in the velocity field. The closure of equation (2.57) is thus fraught with even more uncertainties than that of (2.50).

If, however, the assumption is made that Q_{ij} depends only linearly on the mean part of the velocity field and its first spatial derivatives, then the hydrodynamic analogue of equation (1.102) of Section 1.5.2 may be expressed as

$$Q_{ij} = L_{ijk} \, U_{0k} - N_{ijkl} \, \partial_l U_{0k}. \tag{2.59}$$

Furthermore, if the mean flow takes the form solely of a differential rotation, i.e. $\mathbf{U}_0 = s\omega \mathbf{e}_\phi$, then (2.59) takes the form

$$Q_{ij} = \Lambda_{ijk} \, \omega_k - N_{ijkl} \, \partial_l (\omega \mathbf{e}_\phi \times \mathbf{e}_\mathbf{r})_k. \tag{2.60}$$

The tensor Λ_{ijk} must be symmetric in i and j, and hence anisotropic; the first term on the right hand side of (2.60) – the so-called Λ–effect – therefore represents the contribution towards the differential rotation arising from the interaction between global rotation and anisotropic turbulence (Rüdiger, 1989). The second term denotes the contribution stemming from turbulent diffusivity (analogous to β for the mean induction equation). Of course, just as for α and β, there are no rigorous theories available to calculate Λ_{ijk} and N_{ijkl}; these must come from physically plausible, though to some degree arbitrary, considerations. Mean field dynamos involving the Λ–effect, but with the only nonlinearity that of the large-scale magnetic field on the differential rotation, have been considered by, for example, Brandenburg *et al.* (1991).

The magnetic field though may also influence the differential rotation through modifying, or quenching, the Λ–effect, the turbulent driver of the differential rotation. In the presence of a small-scale magnetic field $\mathbf{b}$ the total stress tensor becomes

$$Q_{ij}^{\mathrm{tot}} = \langle u_i u_j \rangle - \langle b_i b_j \rangle . \tag{2.61}$$

Formal expressions for Q_{ij}^{tot} have been calculated, for a particular turbulence model, by Kitchatinov, Rüdiger & Küker (1994), who also consider the consequences of such a nonlinear Λ–effect for a simple one-dimensional dynamo model. Küker, Arlt & Rüdiger (1999) considered an axisymmetric dynamo model with three different manifestations of the Lorentz force; the Malkus-Proctor mechanism, α–quenching and Λ–quenching. They found that α–quenching leads to temporally periodic solutions, whereas the Malkus-Proctor mechanism and Λ–quenching both yield complicated time series with irregular grand minima.

The approach above, couched solely in terms of mean quantities, may be thought of as a one-point closure model. To study small-scale properties for which correlation functions are of crucial importance it is however necessary to consider higher-order moments of the governing equations. Suppose the system of governing equations is written symbolically as

$$\frac{\mathrm{d}u_i}{\mathrm{d}t} + \nu_i u_i = \sum_{jk} M_{ijk} u_j u_k \, , \tag{2.62}$$

where the $\{u_i\}$ represent the variables of the system (e.g. $\{\mathbf{U}, \mathbf{B}\}$ for incompressible MHD), the ν_i are the dissipation coefficients, and M_{ijk} are the nonlinear coupling coefficients (no implicit summation convention is used here). Then, the equation for the two–point correlation function takes the form

$$\frac{\mathrm{d}}{\mathrm{d}t}\langle u_i u_j \rangle + \nu_i \langle u_i u_j \rangle = \sum_{mn} \left(M_{imn} \langle u_j u_m u_n \rangle + M_{jmn} \langle u_i u_m u_n \rangle \right) \, , \tag{2.63}$$

which clearly involves the triple-correlation function. Continuing in this vein leads to an infinite hierarchy of moment equations; to make progress it is therefore necessary somehow to close the system. One of the infinite number of ways in which this may be done – leading to the only MHD turbulence model that has been used to address the dynamo problem – is to adopt what is known as the eddy–damped quasi–normal Markovian approximation (EDQNM), formulated for hydrodynamics by Orszag (1970) and extended to MHD by Pouquet, Frisch & Léorat (1976). Roughly speaking, closure is achieved by assuming that the joint probability distributions are close to normal, allowing the neglect of all cumulants of order greater than three. An eddy damping, the choice of which allows considerable freedom, is then introduced to determine the decay of the triple correlation, and hence close the system.

From the point of view of mean field dynamo theory, the key result of Pouquet *et al.* (1976) is the derivation of an expression for α of the form

$$\alpha = -\frac{\tau_c}{3}\left(\langle \mathbf{u}\cdot\boldsymbol{\nabla}\times\mathbf{u}\rangle - \frac{\langle \mathbf{j}\cdot\mathbf{b}\rangle}{\rho}\right), \tag{2.64}$$

where τ_c is a typical coherence time of the hydrodynamic turbulence. The result (2.64) provides an extremely appealing description of the saturation of the α–effect, suggesting that the generation of field through kinetic helicity ($\langle \mathbf{u}\cdot\boldsymbol{\nabla}\times\mathbf{u}\rangle$) is nullified through the manifestation of the Lorentz force through small scale current helicity ($\langle \mathbf{j}\cdot\mathbf{b}\rangle$); as such it has been widely used in studies of nonlinear dynamo action. It is however worth bearing in mind that this is a result born of a number of approximations and assumptions, and it is therefore important to discuss the implications of these. The result may be regarded, in some sense, as the nonlinear extension of the quasi-linear result (1.107), a result that follows from approximating integrals of the form

$$\int_0^\infty \langle \mathbf{u}(\mathbf{x},t)\cdot\boldsymbol{\nabla}\times\mathbf{u}(\mathbf{x},t-\tau)\rangle\,\mathrm{d}\tau \qquad \text{by} \qquad \tau_c\langle \mathbf{u}\cdot\boldsymbol{\nabla}\times\mathbf{u}\rangle. \tag{2.65}$$

However, the nature of the correlation time in MHD turbulence, including its dependence on Rm and B_0, remains an important unanswered question (discussed further in the following section). The fact that it is essentially a free parameter of the problem is thus a weakness of the model.

It is also important to consider how the result (2.64) fits in with the classical α–effect picture, as described in Section 1.5.2. The quasi-linear approximation leads, solely from the induction equation for the fluctuating field, to the expression (1.107) for α:

$$\alpha = -\frac{\tau_c}{3}\langle \mathbf{u}\cdot\boldsymbol{\nabla}\times\mathbf{u}\rangle. \tag{2.66}$$

However, as discussed by Proctor (2003), the fact that the induction equation remains linear in the magnetic field – even though in the dynamic regime the flow is of course affected by the field – leads formally – even in the nonlinear regime – again to the result (2.66). Any non-linearity will simply be manifested in a change to the kinetic helicity distribution. So what is the origin of the second term in (2.64)? If, instead of the classical picture of $\mathbf{b}$ being solely dependent on $\mathbf{B}_0$, we consider the introduction of a large-scale field $\mathbf{B}_0$ into a *pre-existing* state of MHD turbulence with a small-scale velocity $\mathbf{u}$ and a small-scale field $\mathbf{b}$ – leading to further perturbations $\mathbf{u}'$ and $\mathbf{b}'$ – then, under the quasi-linear approximation, the EMF may be approximated by

$$\boldsymbol{\mathcal{E}} \approx \langle \mathbf{u} \times \mathbf{b}' \rangle + \langle \mathbf{u}' \times \mathbf{b} \rangle . \tag{2.67}$$

Now, using both the induction *and* momentum equations for the fluctuating quantities, the result (2.64) follows (Pouquet *et al.* 1976; Kleeorin & Ruzmaikin, 1982; Gruzinov & Diamond 1994, 1996; Proctor, 2003). It is though vitally important to be clear about the exact meanings of $\mathbf{u}$ and $\mathbf{b}$ in this formula.

To obtain a further insight into the α–effect it is instructive to write $\mathbf{B} = \boldsymbol{\nabla} \times \mathbf{A}$ and to consider the ideal topological invariant $\chi = \langle \mathbf{A} \cdot \mathbf{B} \rangle$, the magnetic helicity (Gruzinov & Diamond, 1994, 1996). From the induction equation, the equations for $\mathbf{a}$ and $\mathbf{b}$, the small-scale fluctuations of the vector potential and the magnetic field, are

$$\partial_t \mathbf{a} = (\mathbf{u} \times \mathbf{B}_0) + (\mathbf{u} \times \mathbf{b}) - \boldsymbol{\nabla}\chi - \eta \boldsymbol{\nabla} \times \mathbf{b} , \tag{2.68a}$$

$$\partial_t \mathbf{b} = \boldsymbol{\nabla} \times (\mathbf{u} \times \mathbf{B}_0) + \boldsymbol{\nabla} \times (\mathbf{u} \times \mathbf{b}) + \eta \Delta \mathbf{b} . \tag{2.68b}$$

Forming the scalar product of (2.68a) with $\mathbf{b} = \boldsymbol{\nabla} \times \mathbf{a}$, (2.68b) with $\mathbf{a}$, adding, and adopting boundary conditions such that the ensuing surface terms vanish, yields the following equation:

$$\frac{1}{2}\frac{\mathrm{d}}{\mathrm{d}t}\langle \mathbf{a} \cdot \mathbf{b} \rangle = -\mathbf{B}_0 \cdot \boldsymbol{\mathcal{E}} - \eta \, \langle \mathbf{b} \cdot \boldsymbol{\nabla} \times \mathbf{b} \rangle \tag{2.69}$$

[see (1.109)], where the angle brackets denote a spatial average and $\boldsymbol{\mathcal{E}} = \langle \mathbf{u} \times \mathbf{b} \rangle$. For the case of stationary turbulence we may average over time to obtain

$$\mathbf{B}_0 \cdot \boldsymbol{\mathcal{E}} = -\eta \, \mu_0 \, \langle \mathbf{j} \cdot \mathbf{b} \rangle . \tag{2.70}$$

Consequently we have the *exact* result, dependent only on stationarity and suitable boundary conditions, that, for isotropic turbulence

$$\alpha = -\frac{\eta \, \mu_0}{3 \, B_0^2} \, \langle \mathbf{j} \cdot \mathbf{b} \rangle , \tag{2.71}$$

[cf. (1.110)], where $\mathbf{b}$ is the *entire* small-scale magnetic field and where angle brackets here are to be understood as denoting a space *and* time average. The result (2.71) though involves the small-scale field and current, whereas a true mean field theory must involve only large-scale variables. One approach to eliminating the small-scale behaviour is to equate the two expressions for $\langle \mathbf{j} \cdot \mathbf{b} \rangle$ from (2.64) and (2.71) (Gruzinov & Diamond, 1994), thereby leading to what is known as the formula for strong (or even "catastrophic") suppression,

$$\alpha = \frac{\alpha_0}{1 + \mathrm{Rm}\,(B_0^2/\mu_0\,\rho)/\langle \mathbf{u}^2 \rangle}\,. \tag{2.72}$$

It is though worth reiterating the different natures of the expressions (2.71) and (2.64). In expression (2.71) – which is *exact* – $\mathbf{b}$ refers to the total small-scale field, whereas in (2.64) – which is only an approximate result – it refers to a pre-existing small-scale field.

The astrophysical consequences of (2.72), if it is correct, are highly significant in that it implies that the α–effect ceases to be effective at an extremely low value of the large-scale magnetic field (see Vainshtein & Cattaneo, 1992). This issue, which remains very controversial, is now also being addressed through numerical simulations, described in the following section. We shall therefore delay further discussion of (2.72) to the following section.

2.7.3. DIRECT NUMERICAL SIMULATIONS

It is worth stating, from the outset, that direct numerical simulations cannot provide a complete answer to the astrophysical dynamo problem; it is simply not possible to solve the governing equations at the extreme parameter values ($\mathrm{Re} \gg 1$, $\mathrm{Rm} \gg 1$) that pertain astrophysically. With the most powerful computational facilities now available, it is feasible to simulate flows with $\mathrm{Re} \approx \mathrm{Rm} \approx 10^3$ and that possess a reasonable scale separation between that of the driving flow and the largest scale available to the magnetic field. However, given that spatial resolution increases, in each direction, as the inverse square root of the dissipation, and also that the time step decreases in inverse proportion to the resolution, a comparable calculation with $\mathrm{Re} \approx \mathrm{Rm} \approx 10^9$ requires 10^{12} times as many operations. Even with a doubling in computer speed every few years we are clearly nowhere near being able to solve the full problem merely by what Roberts & Soward (1992) term the "brute force" approach. Indeed, even a truly realistic *simulation* of a physical process does not, of itself, constitute a true *understanding* of the process. That said, a computational approach, properly used, can help us to gain an understanding of nonlinear MHD processes, can verify – or refute – existing theories, and can help point the way to new theoretical approaches.

The most ambitious global models of stellar dynamos remain those of Glatzmaier (1985a,b), who investigated self-consistent (i.e. nonlinear) dynamo action driven by thermal convection in a rotating spherical shell. Glatzmaier considered the case of an anelastic gas, thereby filtering out short timescale sound waves whilst retaining the effects of a large density stratification, following on from earlier Boussinesq models of Gilman & Miller (1981) and Gilman (1983). Glatzmaier's models employed subgrid-scale eddy diffusivities, but otherwise contained essentially no parametrisation. In particular, there was no freedom to specify α or ω; these simply emerged, as properties of the convective motions, through a self-consistent solution of the governing equations. Glatzmaier (1985a) considered the case of an everywhere superadiabatic atmosphere; for the parameter values adopted he found that the convection took the form of north-south rolls, as suggested by the Proudman–Taylor theorem for rapidly rotating fluids, with the angular velocity decreasing with increasing latitude at the surface. The magnetic field was antisymmetric about the equator (as for the Sun) but, unlike the Sun, was found to propagate towards the poles. This is sometimes viewed as a failure of the model, in that it differs in this respect from the observed solar field. It is though, as discussed earlier, not practicable to model the Sun in terms of adopting realistic parameter values, and it is (even now) premature to expect self-consistent models that reproduce solar features. The simulations of Glatzmaier represent an extremely important success, demonstrating conclusively the feasibility of a nonlinear dynamo *with minimum parametrisation* (see also the discussion in Section 4.5). Glatzmaier (1985b) did address the question of the direction of propagation of the field, by undertaking a further calculation with a different convective stratification, with the outer two thirds (in radius) superadiabatic and the inner third subadiabatic, the premise being that the helicity and differential rotation in the region of overshooting convection would be such as to drive the dynamo waves towards the equator. The results suggested that this may be the case, but were inconclusive, suffering from the lack of numerical resolution in the inner half of the shell.

Since the studies by Glatzmaier – and in contrast to the path pursued in modelling the geodynamo – attention has shifted away from direct numerical simulations of the entire global dynamo process in a spherical geometry, either towards local, Cartesian models of nonlinear dynamos, or towards "stripped down" simulations aimed at understanding isolated specific aspects of the dynamo mechanism. The former avenue has been pursued by Brandenburg and his co-workers, who have investigated both convectively driven dynamos (Brandenburg *et al.* 1996) and dynamos driven by helical forcing (Brandenburg 2001). The latter approach has been aimed principally at obtaining a more complete understanding of the nonlinear behaviour of the transport coefficients α and β in a turbulent flow at high Rm; for example, does formula (2.51) correctly describe the saturation of α and, if so, what is the appropriate value for $\mathcal{B}^2$ at which the energy of the large-scale field becomes significant? Cattaneo & Vainshtein (1991) considered the (guaranteed) decay of a co-planar, large-scale field in two-dimensional turbulence, in order to calculate the dependence of the turbulent diffusivity on the strength of the large-scale field B_0. With $\mathrm{Rm} = \mathcal{O}(10^2)$, and by varying B_0, they found that the decay of the field could be considered to be kinematic only for extremely weak fields, with $B_0^2 \lesssim \langle \mathbf{u}^2 \rangle / \mathrm{Rm}$, and that the turbulent magnetic diffusion time for a large-scale field of characteristic length L is well-represented by the formula

$$\tau_T \left(= L^2/\beta\right) = \frac{L^2}{\eta} \left(\frac{1}{\mathrm{Rm}} + \frac{1}{\widehat{\mathrm{M}}^2 + 1} \right) , \tag{2.73}$$

where $\widehat{\mathrm{M}}$ is the Alfvénic Mach number, the ratio of the flow speed to the Alfvén speed of the large-scale field. The key physical process behind the suppression of turbulent diffusion is that the field becomes strong (i.e. of equipartition strength) on the scale of the flow whilst remaining weak at large scale, with $\langle |\mathbf{B}|^2 \rangle \approx \mathrm{Rm} B_0^2$. The strong small-scale field resists the rapid deformation necessary for turbulent diffusion, which is thus inhibited. Alternatively, one may consider the problem from a Lagrangian perspective, based on the ideas of Taylor (1921). Turbulent diffusion is achieved by the exponential separation of fluid particles trajectories; the presence of a strong small-scale field provides the fluid particle with a long-term "memory" – their separation is inhibited and the diffusion reduced (Cattaneo, 1994). Clearly any correlation time will be dependent on the magnetic field, and this needs to be brought out in models of MHD turbulence.

The two-dimensional diffusion problem is though rather special, for a number of reasons. Geometrically, there is no possibility of interchange motions, which can bring together oppositely directed field lines without bending them – this suggests that any suppression of diffusion for three-dimensional flows should be weaker. Furthermore, in two dimensions, field decay is guaranteed [Zeldovich's (1957) theorem], following from the fact that one component of the magnetic potential satisfies the heat equation. The question of the suppression (if any) of the turbulent magnetic diffusivity for general, three-dimensional flows remains completely open. It is an extremely difficult question to attack numerically, for two reasons. One is simply a question of computational resources, in that one needs to accommodate a magnetic field that varies on a large scale whilst still resolving the small-scale turbulence. The second, and more difficult, problem is conceptual, arising from the fact that turbulent three-dimensional flows are almost certainly small-scale dynamos at sufficiently high Rm and, for flows lacking reflectional symmetry, may be large-scale dynamos also. It is thus not a straightforward matter to determine how the role of β should be disentangled from that of field amplification.

Calculating the α–effect numerically is more clear-cut since it can be determined unambiguously by measuring the correlation in a turbulent flow between an imposed *uniform* magnetic field and the resulting EMF, $\langle \mathbf{u} \times \mathbf{b} \rangle$. Such calculations are not dynamo simulations – since they have an imposed field with non-zero mean – but are aimed at addressing the one particular issue of the nonlinear nature of the α–effect.

Cattaneo & Hughes (1996) and Cattaneo, Hughes & Thelen (2002) have investigated forced helical, incompressible turbulence, in the presence of an imposed mean field B_0, in order to measure the dependence of α on Rm and B_0. As for the case of β in two dimensions, α is quenched at very weak values of B_0, the results being approximated by a formula of the form

$$\alpha = \frac{\alpha_0}{1 + \mathrm{Rm}^{\gamma}\,(B_0^2/\mu_0\,\rho)/\langle \mathbf{u}^2 \rangle}\,, \tag{2.74}$$

for some $\mathcal{O}(1)$ constant γ (see Figure 2.3). The physics behind the suppression of α can be understood, at least in a rather general manner, in an analogous fashion to the suppression of β; namely that a weak large-scale field gives rise, for large Rm, to a very strong small-scale field which inhibits α. It should be pointed out that this is a more subtle issue than simply a reduction in kinetic helicity; Cattaneo & Hughes (1996) showed that a suppression of α by a factor of order Rm is achieved with only a halving of the kinetic helicity. Clearly, as for diffusion, it must be tied to the ideas of the fluid particles becoming imbued with a "memory". However, the microphysics underlying α at high Rm is not at all well understood, even in the kinematic regime. A formal analysis of the case of perfect electrical conductivity (Rm infinite) leads to the following expression for α in terms of the Lagrangian displacement $\boldsymbol{\xi}$ (Moffatt, 1974):

$$\alpha = -\frac{\mathrm{d}}{\mathrm{d}t}\langle \boldsymbol{\xi} \cdot \boldsymbol{\nabla} \times \boldsymbol{\xi} \rangle, \tag{2.75}$$

and one may speculate that a reduction in the separation of fluid trajectories will lead to a reduction in the average in (2.75). There are though doubts as to the validity of (2.75) even in the kinematic regime for large but finite Rm and certainly, at the moment, there is no proper theory of the suppression of α when Rm is large. The whole issue of the nonlinear behaviour of the transport coefficients of mean field theory is discussed at much greater length in the recent review by Diamond, Hughes & Kim (2004).

As mentioned above, the result (2.74), assuming that it carries through to the astrophysical Rm regime, poses a severe problem for the generation of large-scale fields, in that it implies that the α–effect ceases to be effective once the energy of the large-scale field becomes comparable with the equipartition energy divided by Rm. As such, the result has been criticised, although in a somewhat time-dependent and self-contradictory fashion. Field, Blackman & Chou (1999) claimed that the strong suppression result (2.74) was incorrect, despite its excellent agreement with numerical experiments, but gave no indication as to where they thought the error lay. In a later work, Blackman & Field (2000) underwent an abrupt change of direction, arguing instead that the result was, after all, correct, but was inapplicable to astrophysical situations, their argument being that the dynamics would be dominated by the flux of magnetic helicity through the boundaries, a quantity that is of course

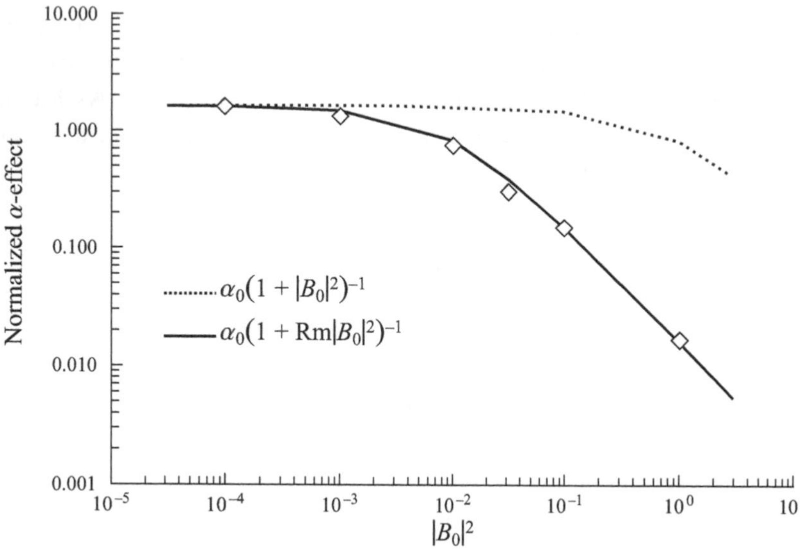

Figure 2.3 - The results of numerical simulations (diamonds) determining α in a forced, helical, chaotic flow, and their relation to two competing theoretical results (from Cattaneo & Hughes, 1996).

zero in periodic domains, such as used by Cattaneo & Hughes (1996). It is indeed true that a formal derivation of the α–effect, via manipulation of equations (2.68a) and (2.68b), leads to the presence of surface terms in the expression for α [i.e. extra surface terms in equations (2.69) and (2.70)] – terms which vanish not only for periodic boundary conditions but also for a number of choices of reasonable boundary conditions. What is totally unclear though is the importance of such terms in an astrophysical context. The issue of α-suppression therefore remains a controversial and important topic.

The aim of this section has been to give an introduction to what, broadly speaking, are the three possible approaches to understanding the behaviour of nonlinear large-scale dynamo action. (The issue of the nonlinear evolution of small-scale dynamos, in which the field exists on scales comparable with or smaller than that of the driving velocity, is also a fascinating and relevant topic, though beyond the scope of this review.) Each has its strengths and weaknesses. Parametrisations of mean field theory, of the sort discussed in Section 2.7.1, are computationally tractable and thus allow an in-depth study of the dependence of a particular model on the governing parameters. Signatures of stellar magnetic fields – such as the solar butterfly diagram – can be faithfully reproduced through parametrised mean field models. The drawback of such models comes though from the somewhat arbitrary choice of parametrisation and the difficulties in assigning particular behaviour to specific physical causes. As such, one must be very careful in asserting that astrophysical magnetic fields can really be *understood* on the basis of such models, and even more careful before making predictions about future magnetic behaviour. Theories of MHD turbulence (such as the EDQNM model discussed above) have their roots more firmly attached to the Navier-Stokes equation, but still rely on a number of assumptions in order to obtain a tractable set of governing equations. It is in formulating these assumptions that all the difficulties arise. Numerical approaches, on the other hand, are able to solve the full nonlinear governing equations, without approximation, but – even with the most powerful computational facilities currently available – only in parameter regimes still far removed from those that describe most astrophysical phenomena. Given all these drawbacks, what is the best hope for progress? Probably the most promising avenue is to improve our understanding of specific, rather narrowly defined questions – such as, for example, the dependence on Rm and B_0 of α and β – via bespoke computational models, and then to incorporate these findings into improved turbulence theories. Today's massively parallel computers are able to model turbulent flows at moderate (from an astrophysical view) values of the Reynolds numbers; from such models we must seek scalings and other information to lead us into the true astrophysical regime. It is a fascinating though formidable challenge.

2.8. PHYSICALLY-REALISTIC FARADAY-DISC SELF-EXCITED DYNAMOS

In this final section, we will highlight how nonlinearities can yield a chaotic dynamical behaviour of dynamo action by returning to the matter of disc dynamos introduced in Section 1.2.1. Owing to the intractablity of the governing nonlinear *partial* differential equations (PDEs) in four independent variables (space and time) in the investigation of generic nonlinear processes in such dynamos, it is not yet possible to exploit numerical models of MHD systems. As a research strategy these processes are better studied in the first instance by analysing the more tractable nonlinear *ordinary* differential equations (ODEs) in just one independent variable (time) that govern the behaviour of simpler systems, such as electro-mechanical devices based, for example, on a steadily forced Faraday disc dynamo.

We summarise the main findings of recent mathematical investigations of the simplest imaginable Faraday disc dynamo systems that are both physically realistic and provide a basis for investigating generic nonlinear effects MHD dynamos. Unlike most systems discussed in the extensive literature on disc dynamos, the governing equations take into account the re-distribution of kinetic energy within the system by Lorentz forces, and the equations are "structurally stable" because they include, in addition to terms representing dissipation due to ohmic heating equally-crucial terms representing mechanical friction. Over wide ranges of conditions these forces give rise to "nonlinear quenching" of dynamo fluctuations, a process which has already been invoked by Raymond Hide as the basis for explaining possibly the most striking feature of the long-term behaviour of the main geomagnetic field, namely "superchron" intervals as long as 30 Ma when the polarity reversals disappear from the palaeomagnetic record (see Chapter 4).

2.8.1. HISTORICAL SURVEY

In the 1860's, three decades after Faraday's invention of a dynamo incorporating a stationary permanent magnet, Varley, Wheatstone & von Siemens independently conceived and applied the self-excitation principle, replacing the permanent magnet of the Faraday dynamo with a stationary coil through which the dynamo current could be diverted. Mathematical models of self-excited homopolar dynamos, which came much later, have been analysed (mainly) by theoretical geophysicists and astrophysicists interested in low-dimensional analogues of MHD self-excited dynamos.

These mathematical investigations began in the 1950's with the pioneering work of Bullard & Rikitake. Bullard treated the simplest-imaginable case of all (as introduced in Section 1.2.1), when (see below for full explanations of the various parameters):

(a) there is no motor in the system [corresponding to $H = 0$, so that the ω together with (2.77d) are therefore redundant];

(b) the disc resistance $\widehat{R}$ is infinite [so that $\widehat{I}$ and equation (2.77b) are redundant];

(c) mechanical friction retarding the motion of the disc is negligible [so that $K = 0$ in (2.77c)].

By coupling two Bullard–type systems together, Rikitake introduced the much-studied two-disc dynamo system governed by an autonomous set of three nonlinear ODEs, the minimum number for chaotic solutions to be possible.

The neglect of mechanical friction seemed at the time to be a reasonable assumption to make, but it is now known that the assumption has the unfortunate consequence of rendering the equations governing the original Bullard & Rikitake systems structurally unstable and their solutions, except as transients, physically unrealistic (see Hide, 1995; Moroz *et al.*, 1998a).

In the original Bullard (1955) dynamo $\breve{\alpha}$ is the only non-zero control parameter, for there is no series motor, the disc conductance is zero and the sliding contacts at the rim and axle of the disc are assumed to be frictionless. Persistent solutions are found with characteristics that depend on $\breve{\alpha}$ and the initial conditions. They represent periodic (but non-harmonic) relaxation oscillations in which the dimensionless electric current generated in the system, x, never changes sign.

The long-held view (see Rikitake, 1966) that the addition of mechanical friction [in our notation (2.81b), $\breve{\kappa} \neq 0$] would make no qualitatitive difference to this behaviour is untenable. Hide (1995) has shown that the mathematical equations governing the Bullard single-disc system, as well as all other friction-free multiple-disc dynamo systems based upon it [including the influential Rikitake (1958) double-disc system], are "structurally unstable". In the presence of mechanical friction the Bullard system eventually becomes steady after initial transients have died away (see also Moroz *et al.*, 1998). When friction is weak these transients certainly resemble "friction-free" fluctuations, notably periodic Bullard-type non-reversing fluctuations in the single-disc case and Rikitake-type chaotic fluctuations with reversals, but they die away. It is noteworthy however that persistent chaotic fluctuations with reversals can occur in a Rikitake system consisting of two coupled identical Bullard dynamos when mechanical friction is added, provided that the two coefficients of mechanical friction are not the same (Ershov *et al.*, 1989; Hide, 1997a; Moroz *et al.*, 1998).

Noting that dynamo action is impossible in the limiting case when the electrical resistance of the disc vanishes (for the magnetic flux linkage of a perfect conductor cannot change) Moffatt (1979) extended the Bullard (friction-free) model by considering the case of non-zero disc conductance, thereby allowing eddy currents to flow. This is the case when, in addition to $\breve{\alpha}$, the control parameters ξ, χ and $\breve{\nu}$ required to specify the electrical properties of the disc are also non-zero.

When mechanical friction in the disc is also taken into account, so that $\breve{\lambda} \neq 0$, we have the case analysed in detail by Knobloch (1981) and later by Plunian *et al.* (1998), who also treated a double-disc system, thereby extending the Ershov study to cases of non-zero disc conductance. In the Knobloch (1981) case, the governing equations are transformable into the celebrated Lorenz set, which can of course have chaotic solutions. We note here that Malkus (1972; see also Robbins, 1976) realised that by adding an electrical shunt to the Bullard system and taking mechanical friction into account he could obtain governing equations of the Lorenz type.

Hide *et al.* (1996) extended the Bullard system by placing a capacitor in series with the coil and including mechanical friction in the disc and then demonstrated the mathematical equivalence of this system to one obtained by replacing the capacitor with a linear motor, with (unavoidable) mechanical friction in the motor equivalent to (unavoidable) leakage resistance in the capacitor.

2.8.2. CHARACTERISTICS OF SELF-EXCITED DYNAMOS

The salient characteristics of all self-excited dynamos can be summarised as follows (Hide, 2000):

(a) the mechanical-to-magnetic energy conversion process is due to motional induction (represented in the equations governing MHD dynamos by the nonlinear motional induction term $\mathbf{u} \times \mathbf{B}$, where $\mathbf{u}$ denotes the Eulerian flow velocity at a general point and $\mathbf{B}$ the magnetic field), and it starts with the amplification of any infinitesimally weak adventitious magnetic field;

(b) for the amplification process to work, motional induction must overcome ohmic losses, implying that the electrical resistance of the system must be sufficiently low (in MHD dynamos this means a sufficiently high magnetic Reynolds number-defined as the product of a characteristic flow speed, a characteristic length, the magnetic permeability of the fluid, and its electrical conductivity);

(c) for the magnetic field to diffuse into the surrounding medium, the electrical resistance must not be *too* low and this sets an *upper* limit on the magnetic Reynolds number in MHD dynamos;

(d) ponderomotive (Lorentz) forces (as represented by the nonlinear term $\mathbf{j} \times \mathbf{B}$ in MHD dynamos, where $\mathbf{j}$ is the electric current density) re-distribute kinetic energy within the system (thereby retarding the buoyancy-driven eddies in typical MHD dynamos such as the geodynamo and accelerating motions in other parts of the eddy spectrum);

(e) no matter how weak, mechanical friction viscosity in MHD dynamos, which *inter alia* dissipates kinetic energy, is never negligible;

(f) internal coupling and feedback (as represented by the terms $\mathbf{u} \times \mathbf{B}$ and $\mathbf{j} \times \mathbf{B}$ in MHD dynamos) give rise to behaviour characteristic of nonlinear systems, i.e. sensitivity to initial conditions leading to non-uniqueness (sometimes called "multiple solutions"), large amplitude fluctuations (including "deterministic chaos"), hysteresis, nonlinear stability, etc.

A strategy advocated in Hide (2000) for discovering generic processes in self-excited dynamos is to start by investigating the temporal behaviour of simple (but not over-simplified) systems (e.g. Faraday disc homopolar generators) governed by tractable ODEs in the single independent variable time, T (say), and then, in the light of the results thus obtained, formulating and executing suitable diagnostic tests of less tractable MHD systems governed by nonlinear PDEs, in four independent space-time variables. Apart from the undoubted mathematical interest of the solutions of the governing ODEs, the findings of those investigations that treat physically-realistic systems –and we must emphasise here this requirement excludes all the friction-free systems that have been treated in the literature (Hide, 1995) including the much-discussed pioneering studies of Bullard (1955) and Rikitake (1958), *cf.* characteristic (e) above– provide general insights into the likely behaviour of the more complex MHD systems, such as the geodynamo operating within the Earth's liquid metallic outer core.

In hydrodynamics the governing mathematical equations express the laws of mechanics and thermodynamics, to which the laws of electrodynamics must be added in the case of MHD. The equations owe their nonlinearity largely to advective terms such as $(\mathbf{u} \cdot \mathbf{\nabla})\mathbf{u}$, $(\mathbf{u} \cdot \mathbf{\nabla})\mathbf{B}$, $(\mathbf{B} \cdot \mathbf{\nabla})\mathbf{u}$, $(\mathbf{B} \cdot \mathbf{\nabla})\mathbf{B}$, etc., which can in some circumstances promote order and stability, as in the case of solitons and in others disorder, instability and sensitivity to initial conditions.

In mathematical analyses, such sensitivity can give rise to multiple solutions associated with "unfoldings" in phase space near co-dimension-two bifurcations, so that steady solutions are able to co-exist at the same point in "control parameter" space with oscillatory and chaotic solutions. In laboratory (and numerical) work sensitivity to initial conditions is manifested as non-uniqueness, chaos, and hysteresis at regime transitions found, for example, in experiments on sloping convection (see e.g. Hide *et al.*, 1994) and Taylor-Couette flow (see e.g. Fenstermacher *et al.*, 1979).

2.8.3. GOVERNING EQUATIONS IN DIMENSIONAL FORM

A Faraday disc homopolar dynamo system that satisfies all the criteria listed above comprises a single disc and coil arrangement with a crucial additional element in the circuit, namely an electric motor with torque characteristics that are not necessarily linear connected in series with the coil (Hide, 1997a,1997b), see Figure 2.4. The motor enables Lorentz forces to re-distribute kinetic energy within the system, where feedback and coupling also contribute to its nonlinear characteristics.

The disc is driven into rotation with angular speed $\Omega(T)$ by a steady applied couple G, where for the rest of this chapter T denotes dimensional time. Retarding the motion of the disc is a frictional couple $-K(T)$ as well as a Lorentz couple $-I(MI+\widehat{L}\widehat{I})$. Here $I = I(T)$ is the main electric current generated by the dynamo and $\widehat{I}(T)$ is the eddy current circulating azimuthally in the plane of the disc (hereafter just "eddy current"), that is induced when $\mathrm{d}I/\mathrm{d}T \neq 0$. The factor $(MI + \widehat{L}\widehat{I})$ thus represents the magnetic flux linkage of the disc if $2\pi M$ is the mutual inductance between the disc and coil and $2\pi\widehat{L}$ is the self inductance of the disc. In the absence of Lorentz forces, friction alone retards the motion of the disc, and when G is steady –the case of interest here– the disc rotates with steady angular speed $\Omega = G/K$.

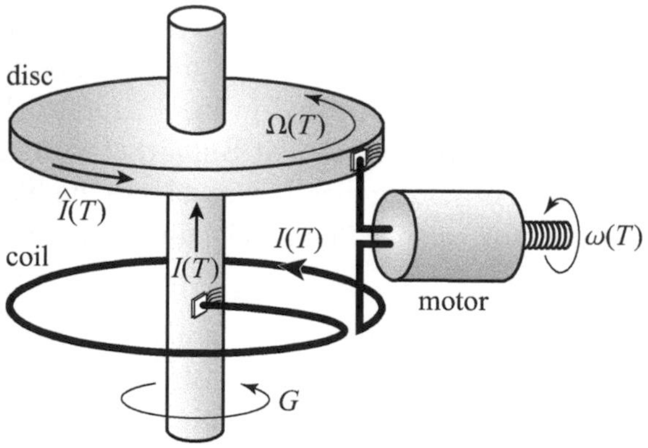

Figure 2.4 - Single-disc dynamo with nonlinear series motor (*cf.* Hide, 1997a,b).

The armature of the motor is driven into rotation with angular speed ω relative to the stationary ambient magnetic field within the motor by a Lorentz couple $HIf(I)$, in general a quadratic function of I, produced by the dynamo current, and it is retarded by a linear frictional couple $-D\omega$. Here H is such that $Hf(I)\omega$ is the back EMF due to the presence of the motor in the dynamo circuit [see equation (2.77a) below], where

$$f(I) = (1 - \varepsilon) + \varepsilon SI, \tag{2.76}$$

and $0 \leq \varepsilon \leq 1$. $f(I)$ specifies the stationary ambient magnetic field within the motor and depends on the design of the motor. The parameter ε measures the nonlinearity of the motor's electro-mechanical characteristics, which vanishes only in the special case when $\varepsilon = 0$. The contribution to the stationary field $\propto \varepsilon SI$ is produced by diverting the dynamo current through stationary field windings (S being a measure of the mutual inductance between the armature and the field windings). This is complemented by the contribution proportional to $(1 - \varepsilon)$ provided by an "outside source". From a geophysical and astrophysical point of view, it is important here to note that this outside source need not necessarily be a permanent magnet, for the magnetic field produced by the current in the coil of a second dynamo would do just as well (see Hide, 2000).

It will be convenient in this section to use the term "linear motor" when $\varepsilon = 0$ and "nonlinear motor" when $0 < \varepsilon \leq 1$ (unless otherwise stated), and also to distinguish two sub-classes of nonlinear motor, namely "quadratic motor" when $0 < \varepsilon < 1$ and "square motor" when $\varepsilon = 1$. The governing 4–mode dimensional set of nonlinear

ODEs in the $(I, \Omega, \omega, \widehat{I})$ is given by (see Hide, 1998)

$$L\,\frac{\mathrm{d}I}{\mathrm{d}T} + M\,\frac{\mathrm{d}\widehat{I}}{\mathrm{d}T} + RI + Hf(I)\omega = \Omega(MI + \widehat{L}\widehat{I})\,, \tag{2.77a}$$

$$L\,\frac{\mathrm{d}\widehat{I}}{\mathrm{d}T} + M\,\frac{\mathrm{d}I}{\mathrm{d}T} + \widehat{R}\widehat{I} = 0\,, \tag{2.77b}$$

$$A\,\frac{\mathrm{d}\Omega}{\mathrm{d}T} = G - I(MI + \widehat{L}\widehat{I}) - K\Omega\,, \tag{2.77c}$$

$$B\,\frac{\mathrm{d}\omega}{\mathrm{d}T} = HIf(I) - D\omega\,, \tag{2.77d}$$

where $2\pi L$ is the self-inductance of the coil, R is the total resistance of the dynamo circuit (including the coil and the armature of the motor), A is the moment of inertia of the disc and B that of the armature of the motor.

Equations (2.77a,b) respectively express Kirchhoff's laws applied to the dynamo current, I, flowing in the main circuit and to the eddy current, $\widehat{I}$, in the disc, $\widehat{R}$ being the azimuthal resistance of the disc (hereafter "disc resistance", the reciprocal of "disc conductance"). Equations (2.77c,d) express angular momentum considerations applied to the motion of the disc and to the motion of the armature of the motor respectively.

The equations can be studied by standard methods involving stability and bifurcation analysis and direct numerical integration. We note here in passing that if $(I, \Omega, \omega, \widehat{I})$ is a solution to (2.77b) then so is $(-I, \Omega, -\omega, -\widehat{I})$ when $\varepsilon = 0$ and $(-I, \Omega, \omega, -\widehat{I})$ when $\varepsilon = 1$. However exact reversal is not a property of any of the solutions when $0 < \varepsilon < 1$. This does not imply that cases when $\varepsilon \neq 1$ can have no geophysical or astrophysical significance. On the contrary, for the "external" contribution to the stationary ambient magnetic field within the motor could be due solely to the current in the coil of a second self-excited dynamo. It is readily shown that the combined system has the requisite symmetry properties.

2.8.4. ENERGETICS AND EQUILIBRIUM SOLUTIONS

Before introducing dimensionless variables and control parameters and thereby abandoning a physically clear but mathematically cumbersome notation, it is instructive to discuss both the energetics of the system and equilibrium solutions on the basis of the dimensional equations (2.76) and (2.77). From these equations it is readily shown that the time rates-of-change of the total magnetic energy and the total mechanical energy of the system satisfy

$$\frac{\mathrm{d}}{\mathrm{d}T}\left[\frac{1}{2}\left(LI^2 + 2MI\widehat{I} + \widehat{L}\widehat{I}^2\right)\right] = -RI^2 - \widehat{R}\widehat{I}^2 + \left\{\Omega I\left(MI + \widehat{L}\widehat{I}\right) - \omega HIf(I)\right\}\,, \tag{2.78a}$$

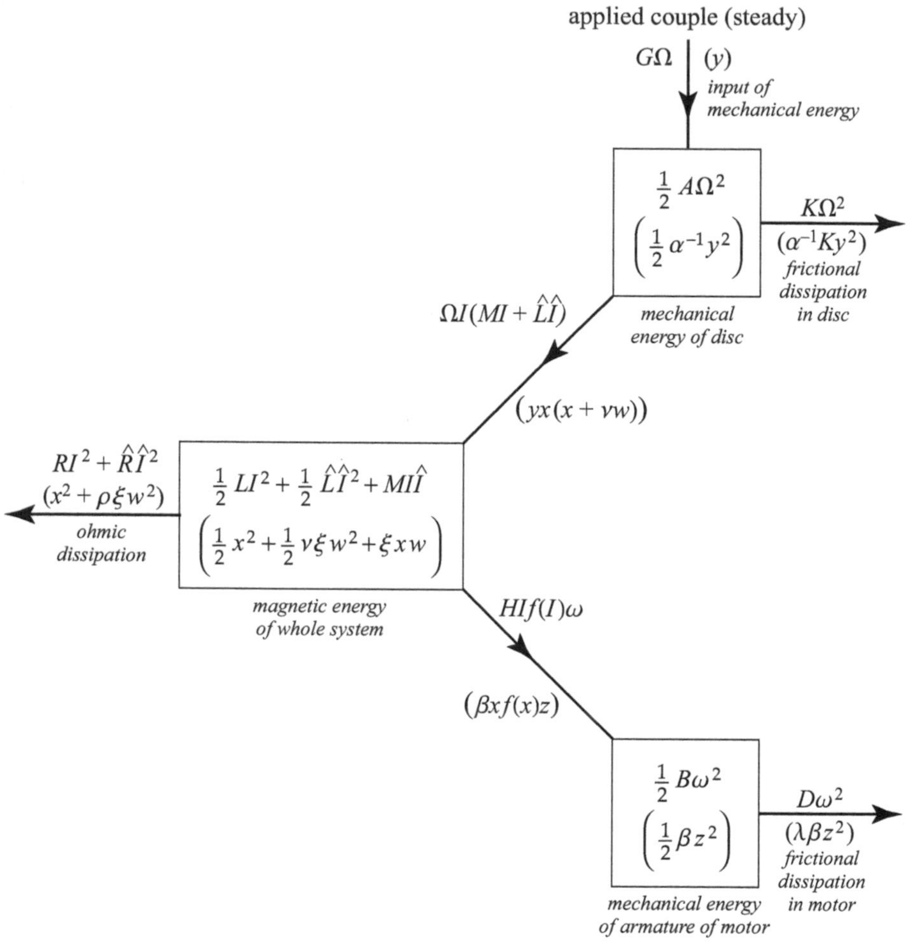

Figure 2.5 - Energetics of the single-disc dynamo with nonlinear series motor.

$$\frac{\mathrm{d}}{\mathrm{d}T}\left[\tfrac{1}{2}\left(A\Omega^2+B\omega^2\right)\right]=G\Omega-K\Omega^2-D\omega^2-\left\{\Omega I\left(MI+\widehat{L}\widehat{I}\right)-\omega HIf(I)\right\}. \tag{2.78b}$$

These equations have an obvious physical interpretation in terms of rates of working of mechanical and Lorentz forces and rates of dissipation by ohmic resistance to the flow of currents and by mechanical friction in the disc and motor. The nonlinear feedback and coupling terms in curly brackets represent the re-distribution of kinetic energy within the system brought about by Lorentz forces, and they cancel out when the equations are added together to give the equation for the rate of change of the total energy of the whole system.

Because we are considering the (important) special case when the applied couple, G, driving the system is steady, there are steady equilibrium solutions –albeit not always stable, as we shall see below in Section 2.8.6– for which the energy equations are given by equations (2.78) with their left hand sides equal to zero. The governing equations (2.76) and (2.77) are then autonomous and have steady equilibrium solutions satisfying

$$I\left[\left(\frac{MG}{K}-R\right)-\left(\frac{M^2I^2}{K}+\frac{H^2f(I)^2}{D}\right)\right]=0\,, \tag{2.79a}$$

$$\Omega=\frac{G-MI}{K}\,,\qquad \omega=HIf(I),\quad \widehat{I}=0\,. \tag{2.79b,c,d}$$

These equations always possess one "trivial" equilibrium solution

$$(I,\Omega,\omega,\widehat{I})=(0,G/K,0,0), \tag{2.80}$$

and this is the only possible equilibrium solution when the dimensionless quantity GM/KR [see (2.82a) below] –which is analogous to the magnetic Reynolds number in MHD dynamos– is so small that the term in square brackets in (2.79a) is negative for all real values of I. Otherwise, when GM/KR is sufficiently large, there are two further equilibrium solutions with $I\neq 0$, obtained by substituting (2.79b) and (2.79c) into (2.79a) [*cf.* equation (2.84e) below].

2.8.5. DIMENSIONLESS EQUATIONS

The electro-mechanical characteristics of the system can be specified in terms of a set of dimensionless control parameters. Various combinations are possible, depending on the choice of scaling of the dependent and independent variables. Following Hide (1997a,1997b) (see also Hide & Moroz, 1999, and Hide, 2000) we take

$$\check{\alpha}=\frac{GLM}{AR^2}\,,\qquad \check{\kappa}=\frac{KL}{AR}\,,\qquad \xi=\frac{M}{L}\,, \tag{2.81a,b,c}$$

$$\chi=\frac{R\widehat{L}}{\widehat{R}L}\,,\qquad \check{\nu}=\frac{\widehat{L}}{M}\,, \tag{2.81d,e}$$

to specify the characteristics of the disc, and

$$\check{\beta}=\frac{H^2L}{R^2B}\,,\qquad \check{\lambda}=\frac{DL}{RB}\,,\qquad \check{\sigma}=S(G/M)^{1/2} \tag{2.81f,g,h}$$

to specify the characteristics of the series motor. Parameters (2.81a-h) as others in this section are noted with a "check" symbol $(\check{\ })$. These variables will be used in the remaining of this chapter to describe the characteristics of the disc-motor setup. They should not be confused with MHD variables $(\alpha,\ \beta,\ \kappa,\ \lambda,\ \mu,\ \nu,\ \rho,\ \sigma)$ used elsewhere in the book.

It is convenient to make use of certain combinations of these basic control parameters, namely

$$\overline{\alpha} = \frac{\breve{\alpha}}{\breve{\kappa}} = \frac{GM}{KR}, \qquad \overline{\beta} = \frac{\breve{\beta}}{\breve{\lambda}} = \frac{H^2}{RD}, \tag{2.82a,b}$$

$$\breve{\mu} = \frac{(\xi/\breve{\nu})}{(1-\xi/\breve{\nu})} = \frac{M^2}{L\widehat{L} - M^2}. \tag{2.82c}$$

These control parameters are all essentially non-negative (including $\breve{\mu}$, since $L\widehat{L} > M^2$) in systems of direct physical interest, but there may, of course, be mathematical interest in solutions of the governing equations in cases when some of the parameters are negative.

We introduce the dimensionless independent variable t and the dimensionless dependent variables $[x(t), y(t), z(t), w(t)]$ where

$$T = (L/R)t, \qquad I = (G/M)^{1/2}x, \qquad \Omega = (R/M)y, \tag{2.83a,b,c}$$

$$\omega = (LH/RB)(G/M)^{1/2}z, \qquad \widehat{I} = (G/M)^{1/2}w. \tag{2.83d,e}$$

Then using equations (2.81)–(2.83) in equations (2.76) and (2.77) gives[4]

$$\dot{x} + \xi\dot{w} = -x - \breve{\beta}\, f(x)\, z + y(x + \breve{\nu}\, w), \tag{2.84a}$$

$$\dot{w} + \dot{x}/\breve{\nu} = -w/\chi, \tag{2.84b}$$

$$\dot{y} = \breve{\alpha}[1 - x\,(x + \breve{\nu}\, w)] - \breve{\kappa}\, y, \tag{2.84c}$$

$$\dot{z} = xf(x) - \breve{\lambda}\, z, \tag{2.84d}$$

where

$$f(x) = 1 - \varepsilon + \varepsilon\,\breve{\sigma}\, x. \tag{2.84e}$$

This formulation is identical to that given in Hide & Moroz (1999) and Moroz & Hide (2000), with a slight redefinition of the control parameters.

The nontrivial equilibrium states are now given by

$$(y, z, w) = (\overline{\alpha}\,(1-x^2), x\, f(x)/\breve{\lambda}, 0), \qquad \overline{\alpha}-1-(\overline{\alpha}\, x^2+\overline{\beta}\, f(x)^2) = 0, \tag{2.85a,b}$$

while the trivial equilibrium state becomes $(x, y, z, w) = (0, \overline{\alpha}, 0, 0)$.

Equations (2.84a-e) can be transformed into other sets of equations, some mathematically more convenient (see Hide & Moroz, 1999; Moroz & Hide, 2000). One such reformulation not considered previously is obtainable by introducing $X = x + \breve{\nu}\, w$, thereby eliminating the parameter $\breve{\nu}$. If, in addition, one introduces the variable $Y = x + \chi\, w$, then one recovers the 4–mode dynamo model investigated by Hide & Moroz (1999); Moroz & Hide (2000). The two new variables X and Y are identifiable as flux variables. The reader is referred to those papers for further details. All of the numerical integrations described in the later subsections of this section are based upon this alternative Moroz & Hide reformulation.

4 We use a dot to denote differentiation with respect to t.

2.8.6. GENERIC SOLUTIONS

Nonlinearity means that the solutions in which we are mainly interested, namely those that persist after transients have died away, can be very sensitive to the initial conditions and/or parameter choices. A comprehensive investigation of the 4–mode dynamo equations is not feasible because of the large numbers of parameters involved. While the control parameters in any given case represent one point in an eight-dimensional parameter space, for many purposes a two-dimensional regime diagram with $\overline{\beta}$ as abscissa and $\overline{\alpha}$ as the ordinate was established at an early stage of the investigations reviewed here.

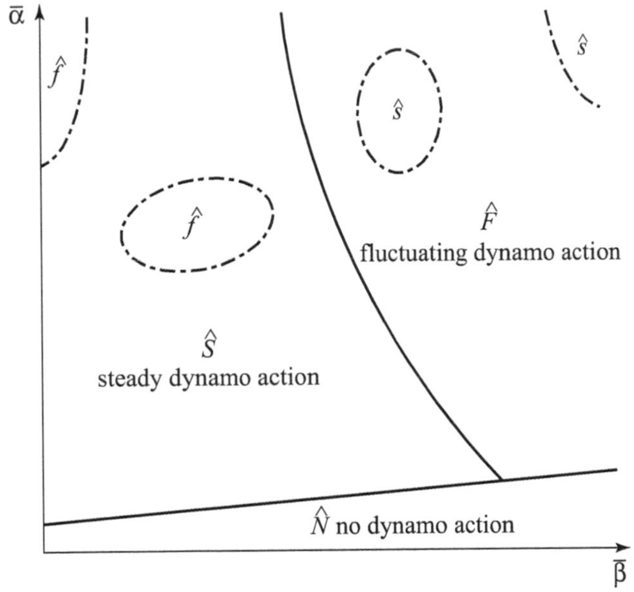

Figure 2.6 - Typical schematic regime diagram in the $(\overline{\beta}, \overline{\alpha}))$ plane (figure 1 of Hide, 2000, reproduced by kind permission of the Royal Society). Within the region labelled $\widehat{N}$, the parameter $\overline{\alpha}$ (the effective "magnetic Reynolds number") is too small for dynamo action to occur. At higher values of $\overline{\alpha}$, steady dynamo action occurs within the main region $\widehat{S}$ and sub-regions $\widehat{s}$; fluctuating dynamo action occurs within the main region $\widehat{F}$ and the sub-regions $\widehat{f}$. The sub-regions disappear when the electrical properties of the disc are such that the azimuthal component of the current in the disc is negligible.

Figure 2.6 shows but one possibility amongst a wide variety of different regime diagrams that have been obtained in studies of the transformed version of equations (2.84a-e) (see Hide & Moroz, 1999; Moroz & Hide, 2000). The trivial equilibrium state $(x, y, z, w) = (0, \overline{\alpha}, 0, 0)$ is the only stable equilibrium state within those regions $\widehat{N}$ (say) of parameter space for which $\overline{\alpha} < \overline{\alpha}^*$, where $\overline{\alpha}^*$ is determined from whichever bifurcation curve forms that segment of the stability boundary (see, for example, Figure 1 of Hide & Moroz, 1999, or Figure 7 of Moroz & Hide, 2000). Persistent dynamo action cannot occur within N. It is throughout the rest of parameter space, in regions Y, say, where $\overline{\alpha} > \overline{\alpha}^*$ that the trivial solution is unstable, that persistent dynamo action takes place. Within these Y regions there are two general possibilities, namely *steady* dynamo action and *fluctuating* dynamo action. The first occurs within regions labelled as $\widehat{S}$ and $\widehat{s}$ in Figure 2.6.

Fluctuating dynamo action occurs throughout the rest of Y, within regions labelled $\widehat{F}$ and $\widehat{f}$ in Figure 2.6 (or in the more explicitly labelled regime diagrams of Hide & Moroz, 1999, and Moroz & Hide, 2000), where the non-trivial equilibrium solutions lose their stability to large amplitude fluctuations of varying degrees of complexity, including multiple solutions and chaos (see below).

Self-excited dynamos, be they disc or MHD dynamos, satisfy essentially nonlinear equations, with generic solutions that are multiple and much more varied and interesting than just reflectionally- symmetric pairs [in MHD cases $(\mathbf{u}, \mathbf{B})$ and $(\mathbf{u}, -\mathbf{B})$], corresponding to an unaltered velocity field and a completely reversed magnetic field. We note here in passing (see below) that when $0 < \varepsilon < 1$, bias is automatically introduced into the fluctuating time series, regardless of its length. When $\varepsilon = 0$ or $\varepsilon = 1$, the symmetry properties of the governing equations suggest that one can define the length of time $\widetilde{T}$ taken for any bias in the time series to vanish. $\widetilde{T}$ is clearly infinite in the case of stable steady persistent solutions. On the other hand, for fluctuating persistent solutions $\widetilde{T}$ can of course be finite. Time series exhibiting these asymmetry/symmetry characteristics are presented in a later subsection.

2.8.7. SURVEY OF BEHAVIOUR

VARIATIONS ON A THEME

Hide *et al.* (1996) extended the Bullard system by placing a capacitor in series with the coil and including mechanical friction in the disc and then demonstrated the mathematical equivalence of this system to one obtained by replacing the capacitor with a linear motor, with (unavoidable) mechanical friction in the motor equivalent to (unavoidable) leakage resistance in the capacitor. In this case the only non-zero parameters are $\breve{\alpha}$, $\breve{\beta}$, $\breve{\kappa}$ and $\breve{\lambda}$.

We find it useful to employ the notation used in Section 2.8.5 above to summarise the results of extensions to the Hide *et al.* (1996) dynamo, including the one described by (2.84a-e).

(a) a linear motor ($\varepsilon = 0$) and no disc eddy currents ($\chi = 0$);
Case (a) is the original Hide *et al.* (1996) study in which the nonlinear dynam-

ics was found to be controlled by the presence of a codimension-two Takens-Bogdanov double-zero bifurcation. The linear stability curves for steady and oscillatory dynamo action for both the trivial and the non-trivial states all emerge from one bifurcation point in $(\overline{\beta}, \overline{\alpha})$–space. Steady and fluctuating (periodic and chaotic) solutions are possible, with chaotic dynamics being confined to a small region of parameter space, near the (subcritical) Hopf stability boundary for the onset of oscillatory solutions associated with the nontrivial equilibria, provided $\check{\lambda} > \check{\kappa}$. When $\check{\kappa} > \check{\lambda}$, no chaotic solutions were observed.

(b) a square motor ($\varepsilon = 1$) and no disc eddy currents ($\chi = 0$);
Hide (1997b) considered the case of a square series motor so that $\varepsilon = 1$ and found parameter space to be dominated by steady dynamo action. He termed this phenomenon "nonlinear quenching".
According to Hide (2000), nonlinear quenching is associated with the redistribution of kinetic energy within the system by Lorentz forces (see item (d) of Section 2.8.2 above), and if, as seems likely, the process is generic and therefore occurs in MHD dynamos, it could provide the basis of testable theory of geomagnetic polarity reversals, the most striking property of which is the *absence* of reversals during very long intervals of time, the so-called "polarity superchrons".
Mathematically, nonlinear quenching arises because, as noted by Moroz (2002), the Takens-Bogdanov double-zero bifurcation, responsible for the oscillatory solutions in the Hide *et al.* (1996) dynamo, now occurs at infinity.

(c) a linear motor ($\varepsilon = 0$) with azimuthal eddy currents ($\chi \neq 0$);
The extent to which this picture is changed when eddy currents are allowed to flow in the disc has been considered by treating systems for which the control parameters χ, ξ and $\check{\nu}$ are no longer zero (Hide & Moroz, 1999). For a linear motor, the dynamics of the system is much richer than in the absence of eddy currents

$$\check{\rho} = \frac{\check{\nu}}{\chi(\check{\nu} - \xi)}, \quad \check{\mu} = \frac{\xi}{\check{\nu} - \xi}, \tag{2.86a,b}$$

then four scenarios are possible:

- when $\check{\rho} < \check{\lambda}(1 + \check{\mu})$, only steady solutions are possible and nonlinear quenching occurs;
- when $\check{\lambda}(1 + \check{\mu}) < \check{\rho} < \check{\rho}_L$ (where $\check{\rho}_L$ denotes the critical value of $\check{\rho}$ for the existence of the Lorenz subcritical Hopf bifurcation for $\check{\beta} = 0$), the scenario resembles that of Hide *et al.* (1996);
- when $\check{\rho} > \check{\rho}_L$, parameter space is dominated by fluctuating solutions,

either periodic or chaotic, with steady states occupying only a small region;

- when $\breve{\lambda}(1+\breve{\mu}) > \breve{\rho} > \breve{\rho}_L$, no double-zero bifurcation is possible and partial nonlinear quenching occurs. Oscillatory solutions are confined to small values of $\breve{\beta}$ and large values of $\breve{\alpha}$ and emanate from the subcritical Lorenz bifurcation point on the $\breve{\beta} = 0$ axis.

(d) a square motor ($\varepsilon = 1$) with azimuthal eddy currents ($\chi \neq 0$);
In the cases when the motor is square (i.e. e=1) and eddy currents are allowed to flow in the disc, nonlinear quenching is still a key process, but it is again partial rather than complete in the sense defined in (c) above (Hide & Moroz, 1999).

(e) a quadratic motor ($0 < \varepsilon < 1$) with no azimuthal eddy currents ($\chi = 0$);
Moroz (2002) extended the analyses of Hide *et al.* (1996) and Hide (1997a,b) to the case of a nonlinear series motor with $0 < \varepsilon < 1$ in the absence of eddy currents. The double-zero bifurcations for the trivial and the nontrivial equilibria no longer coincide. There are multiple steady state bifurcation curves, as well as an additional Hopf bifurcation curve, which result in additional (non-degenerate) codimension-two Hopf-steady bifurcations. This yields a much richer range of behaviour. The continuous range of chaotic solutions, a feature of case (a), now fragments and gives rise to a structure of interleaving chaotic and periodic behaviour of differing oscillatory patterns.

(f) a quadratic motor ($0 < \varepsilon < 1$) with azimuthal eddy currents ($\chi \neq 0$);
When $0 < \varepsilon < 1$ and in the presence of eddy currents, the two double-zero bifurcations again become non-coincident and multiple steady and Hopf bifurcation curves generate a greater diversity of nonlinear behaviour than that found in case (e). Depending upon the parameter values, Moroz & Hide (2000) also found chaos occurring not far above the transitional curve for the onset of nontrivial dynamo action. Multiple solutions are possible and the nonlinear and linear stability thresholds are subject to hysteresis effects (see also the following subsection).

OTHER EXTENSIONS

Since the seminal work of Hide *et al.* (1996), other extensions to the basic dynamo systems have been investigated. Moroz *et al.* (1998a,b) investigated the behaviour of two coupled dynamo units with linear motors and in the absence of eddy currents. The first study confirmed the work of Hide (1995) on the structural instability of the Rikitake dynamo in the presence of even a small amount of friction, while the second study focused upon establishing general criteria for the existence of phase locked states. Moroz (2001a) extended this study of synchronisation to a three dynamo configuration. The general problem of two coupled dynamos with nonlinear series motors was addressed in Moroz (2002), who also reviewed the research to date on the Hide family of dynamos to which the interested reader is also referred.

A start was made by Goldbrum *et al.* (2000) to analyse dynamo models, biased by immersion in a background magnetic field and/or by connecting a battery in series with the motor and coil (*cf.* the so-called "Biermann" battery of astrophysics), as given in Hide (1997a). The initial study was for the battery only, while Moroz (2001b) investigated both the battery and magnetic field.

Finally, Moroz (2003, 2004a, 2004b) returned to the original Malkus–Robbins dynamo, extended to incorporate both a linear and a quadratic series motor, but in the absence of azimuthal eddy currents, to find different types of regime diagrams and different transition sequences between nonlinear states.

2.8.8. SOME NUMERICAL INTEGRATIONS

In addition to the regime diagrams and behaviours described in Hide & Moroz (1999) and Moroz & Hide (2000), we present a selection of phase portraits, time series and bifurcation diagrams, which represent slices of parameter space for specific choices of the various parameters of the four-mode dynamo of Section 2.8.5 when re-written in the flux-formulation of Hide & Moroz (1999). In all of our integrations we chose $\breve{\alpha} = 100$, $\breve{\kappa} = 1$, $\breve{\lambda} = 1.2$, $\breve{\mu} = 0.5$ and $\breve{\rho} = 16$, where $\breve{\mu}$ and $\breve{\rho}$ are defined in equation (2.86a,b). In so doing we shall demonstrate the existence of multiple solutions, as well as bias in the time series when $0 < \varepsilon < 1$.

$\varepsilon = 0$ AND $\varepsilon = 1$

The two cases reported here should be viewed in conjunction with Figures 1 and 2 of Hide & Moroz (1999). When $\varepsilon = 0$, chaotic solutions persist for the range of $\breve{\beta}$ that we investigated, namely $0 \leq \breve{\beta} \leq 25$. Figure 2.7 shows the time series of X and Figure 2.8 shows that corresponding phase portrait in the (X, w)–plane for $\breve{\beta} = 5$.

When $\varepsilon = 1$ and as described above, chaotic solutions are confined to much smaller regions of parameter space. Figure 2.9 shows a section of the $X(t)$ time series and Figure 2.10 the phase portrait in (X, w)–space, for $\breve{\beta} = 0.4$ which is close to the transition from chaotic to steady dynamo action, when $\breve{\beta}$ is increased.

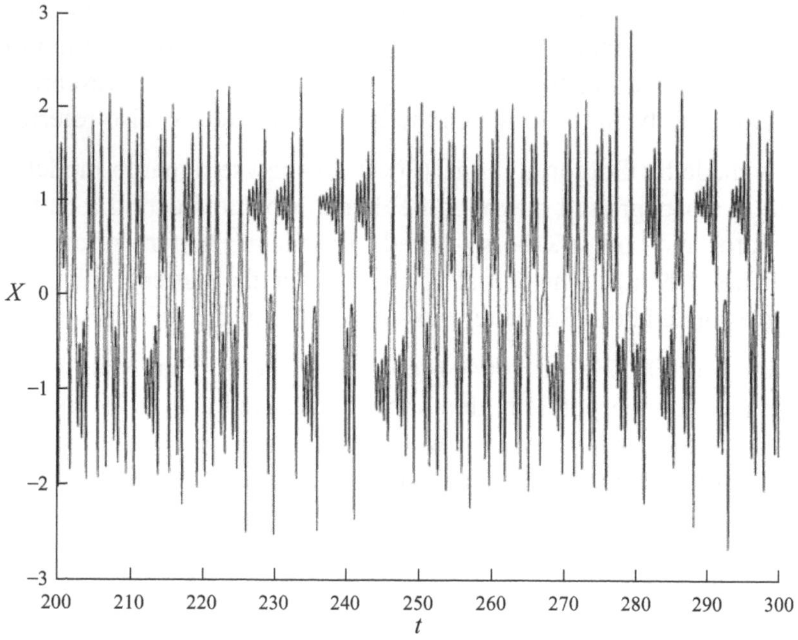

Figure 2.7 - Time series of X for $\varepsilon = 0$ and $\check{\beta} = 5$.

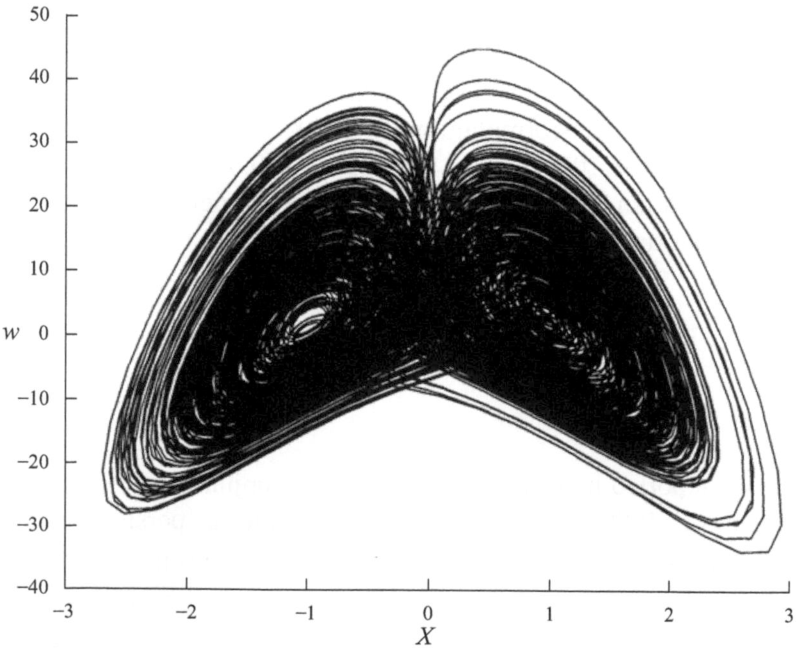

Figure 2.8 - The phase portrait in the (X, w)–plane for the same parameter values as in Figure 2.7.

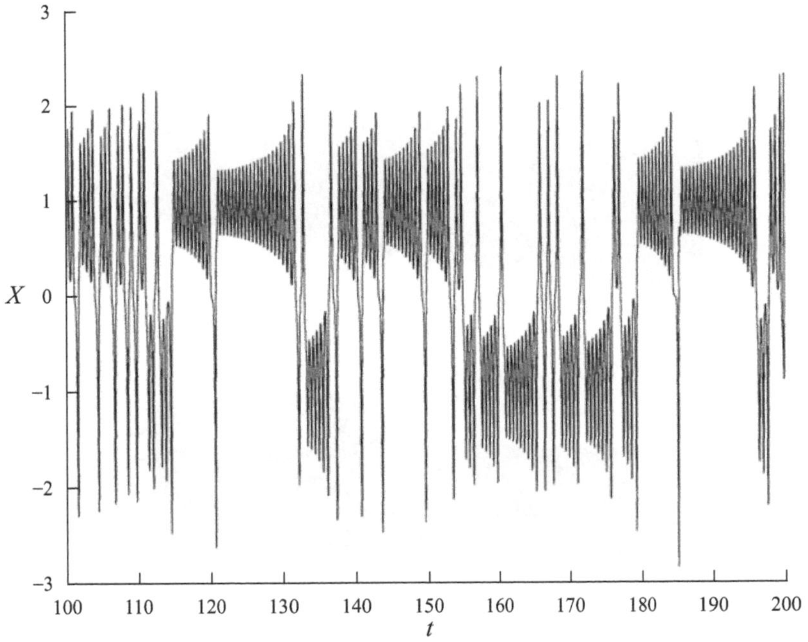

Figure 2.9 - Times series of X for $\varepsilon = 1$ and $\breve{\beta} = 0.4$.

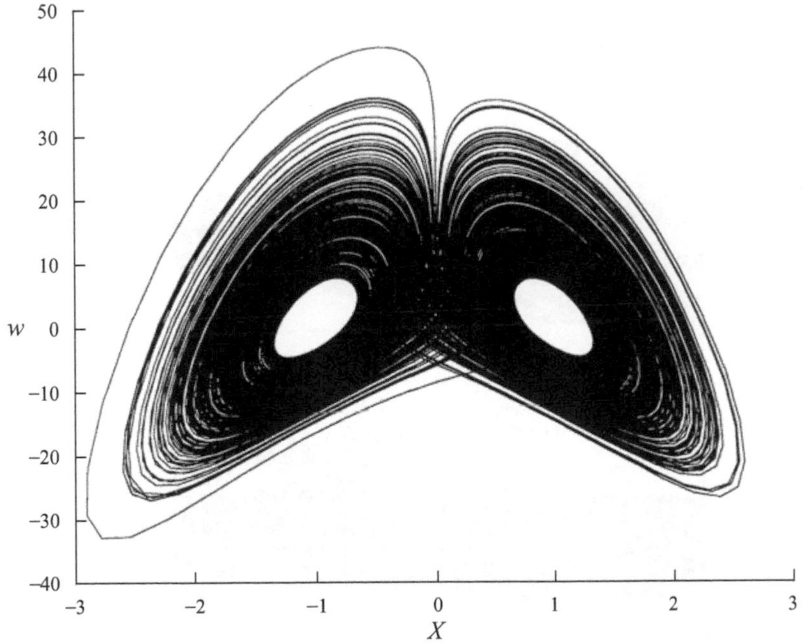

Figure 2.10 - The phase portrait in the (X, w)–plane for the same parameter values as in Figure 2.9.

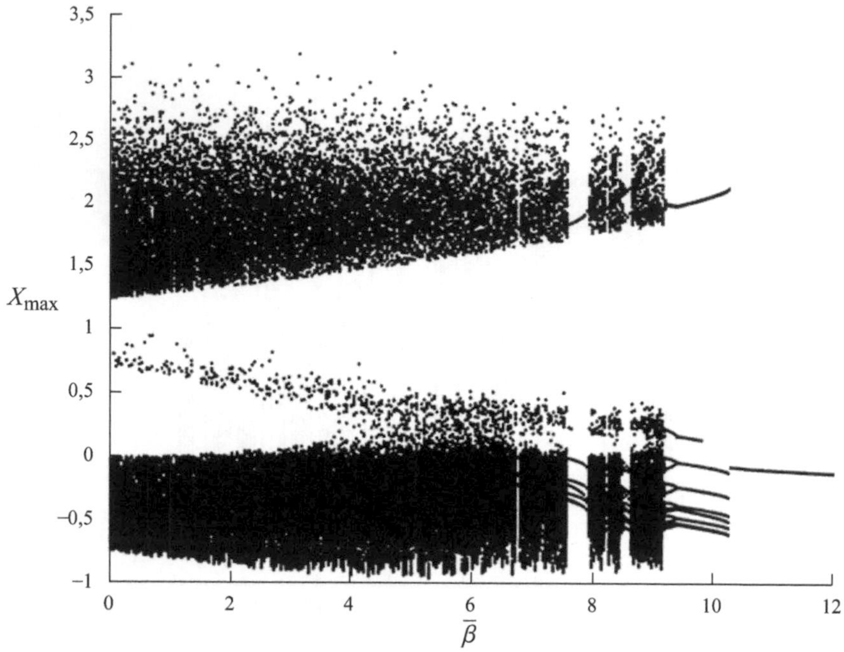

Figure 2.11 - Plot of the local maximum value of X when $\varepsilon = 0.4$, as a function of $\overline{\beta}$, for $\check{\beta}$ increasing.

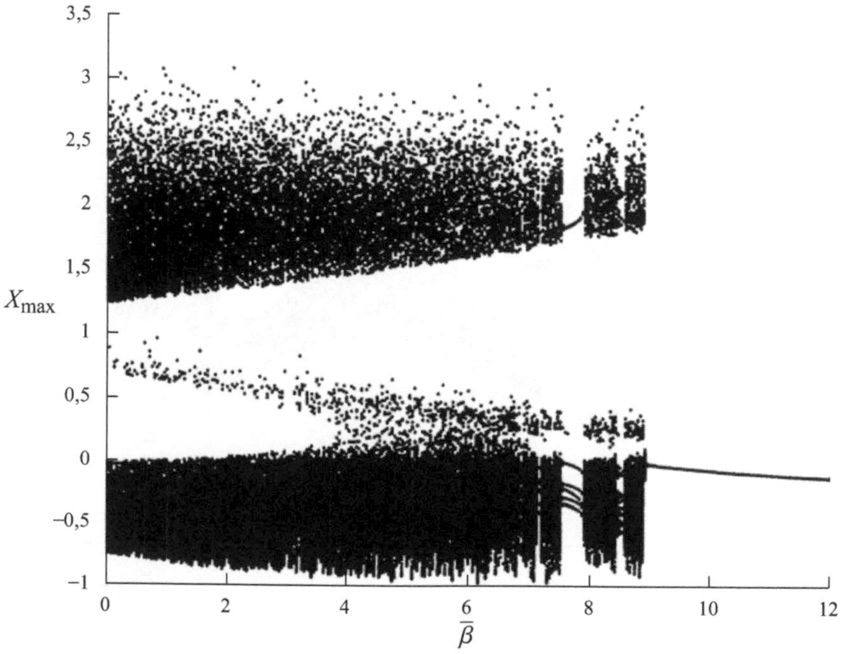

Figure 2.12 - As in Figure 2.11, but for $\check{\beta}$ decreasing.

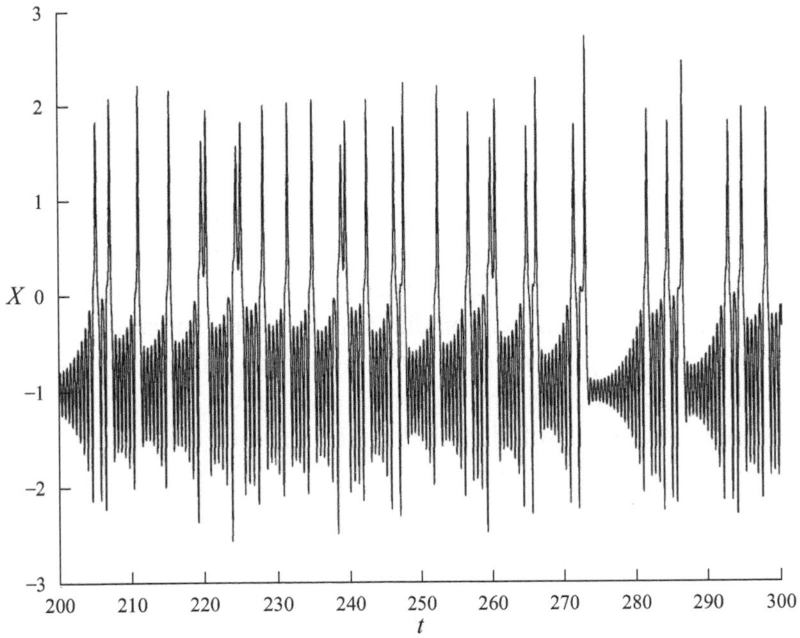

Figure 2.13 - A section of the time series for X when $\varepsilon = 0.4$ and $\check{\beta} = 6$.

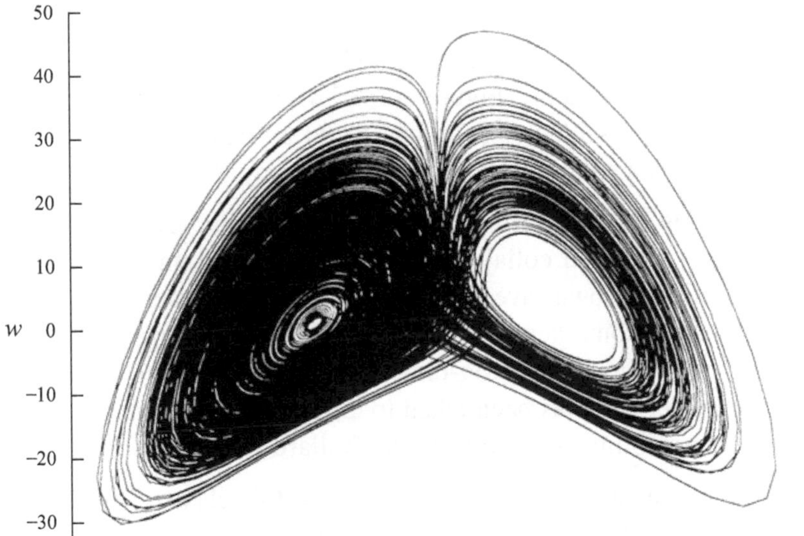

Figure 2.14 - The phase portrait in the (X, w)–plane for the same parameter values as in Figure 2.13.

$0 < \varepsilon < 1$

We now amplify the results depicted in Figure 8 of Moroz & Hide (2000). Figure 2.11 shows the plot of local maximum values of X as a function of $\overline{\breve{\beta}}$ for $\varepsilon = 0.4$ as $\breve{\beta}$ is increased, while Figure 2.12 shows the corresponding plot when $\breve{\beta}$ is decreased. The procedure is as follows. The initial value of $\breve{\beta}$ is chosen and the maximum values of X are recorded after transients have decayed. Then $\breve{\beta}$ is increased/decreased and the final state is used as the initial condition for the next integration. This results in a bifurcation diagram, as a slice in parameter space, which affords a direct and simple way of identifying where different types of oscillatory behaviour may be found. Note the presence of windows of periodic solutions, separated by bands of chaotic solutions before the solution loses stability to a simple periodic solution when $\overline{\breve{\beta}} \approx 10.3$.

Figure 2.13 shows part of a time series for X when $\breve{\beta} = 6$ and $\varepsilon = 0.4$, while Figure 2.14 shows the corresponding phase portrait in the (X, w)–plane. Immediately apparent is the bias, introduced when ε differs from 0 or 1. The system spends more time oscillating (irregularly) around one of the (unstable) equilibrium states than it does around the other. A reversal in the time series occurs after a gradual build up in the maximum and minimum amplitudes.

When $\varepsilon = 0.535$, chaotic solutions persist until $\overline{\breve{\beta}} \approx 4$, when the system loses stability to a simple periodic limit cycle. This also persists until $\overline{\breve{\beta}} \approx 5.2$, when steady dynamo action obtains. Figure 8(c) of Moroz & Hide (2000) suggests that the disappearance of oscillatory solutions could be caused by the presence of the branch H_3 of periodic solutions.

In conclusion, this last section has presented a brief survey of some recent work which the authors and their collaborators have conducted on self-excited dynamos. As well as placing our own investigations into a historical context, we have made an effort to identify some key features of naturally-occurring MHD systems with their counterparts in the much lower-dimensional (and more tractable) Faraday-disc dynamos. Moreover care has been taken to ensure that the dynamo models studied exhibit structural stability, in contrast to the Bullard and Rikitake models.

In Section 2.8.6, we saw that features, generic to this class of dynamo are regions of parameter space in which no dynamo action, steady dynamo action and fluctuating dynamo action occur. The precise details as to where and which type of persistent behaviour dominates is, however, model and parameter dependent (see Section 2.8.7). Bifurcation transition sequences between different finite amplitude states are possible, in which chaotic and simple periodic behaviour interleave (see, for example, Figure 2.11). In addition, the nonlinear regime exhibits hysteresis with multiple solutions possible.

Other studies, referenced in Section 2.8.7, have introduced terms into the basic model, such as the effects of an external battery, which break the symmetry of the finite amplitude steady state solutions, as well as creating additional codimension-two bifurcations (Moroz, 2001b).

It is clear that this class of low order dynamo is capable of producing a rich range behaviours, depending upon both the parameters and the specific dynamo model chosen. What is required is some way of distinguishing between the whole gamut of possibilities. One such approach involves the identification of the underlying basis of unstable periodic orbits (UPOs), specific to a given model. Further investigations along these lines should prove rewarding.

CHAPTER 3

DYNAMICS OF ROTATING FLUIDS

Friedrich Busse, Emmanuel Dormy,
Radostin Simitev & Andrew Soward

In this chapter we introduce basic concepts concerning the dynamics of rotating fluids. The effect of rapid rotation on a flow can significantly alter its nature. These effects are particularly important for planetary applications. In Section 3.1 we introduce steady boundary layers and shear flows, which occur in rotating fluids, and show how these layers can have a strong influence on the mainstream flow outside them. In Section 3.2, we consider the combined influence of rotation and magnetic field on these layers. Time dependent effects, particularly waves propagating in rotating and electrically conducting fluids, are considered in Section 3.3. Finally, in Section 3.4, we address the particular case of thermal Rossby wave, which is the preferred mode of convection in a rapidly rotating sphere such as planetary interiors. We shall devote Section 3.4.2 to a description of the simpler rotating cylindrical annulus model, since it offers the simplest access to the spherical problem. The basic equations for the spherical problem are introduced in Section 3.4.3 and the onset of columnar convection in spherical shells is discussed in Section 3.4.4. In Section 3.4.5 the onset of inertial mode convection is described which prevails at very low Prandtl numbers. In Section 3.4.6 the properties of finite amplitude convection are outlined for moderately low Prandtl numbers Pr, while convection at higher values of Pr is considered in Section 3.4.7. In Section 3.4.8 equatorially attached convection is considered which evolves from inertial mode convection. The problem of penetrative convection is addressed in Section 3.4.9 where also some aspects of convection in the presence of thermal as well as chemical buoyancy are discussed. The chapter concludes with some remarks on applications among which the dynamo action of convection is of special importance.

3.1. BOUNDARY AND SHEAR LAYERS IN ROTATING FLOWS

When a fluid has a low viscosity, the ensuing laminar flow may be largely inviscid in character in the bulk of the flow domain, which we refer to as the mainstream. At a boundary $\partial\mathcal{D}$, with unit normal $\mathbf{n}$ directed into the fluid, the velocity $\mathbf{U}$ of inviscid flow satisfies the impermeable boundary condition[5]

$$\mathbf{n}\cdot\mathbf{U} = 0 \qquad \text{on} \quad \partial\mathcal{D}\,. \tag{3.1a}$$

However, viscous flow, velocity $\mathbf{u}$, also satisfies further conditions dependent on the nature of the boundary. For stationary rigid boundaries considered throughout this section, the viscous fluid satisfies, in addition to (3.1a), the no-slip boundary condition

$$\mathbf{n}\times\mathbf{u} = \mathbf{0} \qquad \text{on} \quad \partial\mathcal{D}\,, \tag{3.1b}$$

namely that the tangential component of the velocity vanishes. To meet that condition there is a thin region adjacent to the boundary, where viscosity is important, which we call the boundary layer. Inside it, approximations to the equation of motion may be made based on the assumption that the typical length scale parallel to the boundary L is long compared with the boundary layer thickness, δ. The most important consequence of this is that the fluid pressure, p, inside the boundary layer is independent, correct to leading order in δ/L, of the coordinate z (say) normal to the wall and determined by the mainstream pressure just outside. The components parallel to the boundary of the equation of motion are then solved to determine the boundary layer velocity $\mathbf{u}$. Having solved the boundary layer equations the results are used to obtain reduced boundary conditions on the mainstream flow velocity $\mathbf{u}$. In the case of a fluid of constant density ρ, to which we restrict attention throughout this section, the impermeable boundary condition $\mathbf{n}\cdot\mathbf{u} = 0$ on the mainstream is replaced by

$$\mathbf{n}\cdot\mathbf{u} = \mathbf{n}\cdot\boldsymbol{\nabla}\times(\mathbf{Q}\times\mathbf{n})\,, \qquad \text{where} \qquad \mathbf{Q} = \int_0^{z/\delta\to\infty}(\mathbf{u}_{\|}-\mathbf{U})\,\mathrm{d}z \tag{3.2a,b}$$

is the volume flux deficit carried by the boundary layer; here we assume that the components of $\mathbf{u}$ parallel to the boundary, namely $\mathbf{u}_{\|}$, and $\mathbf{U}$ merge on leaving the boundary layer:

$$\mathbf{u}_{\|} - \mathbf{U} \to \mathbf{0} \qquad \text{as} \qquad z/\delta \to \infty\,. \tag{3.3}$$

The result (3.2) follows from mass continuity integrated over a a thin "penny-shaped disc" sitting on the boundary, yet thick enough to encompass the boundary layer. It

5 In the following two sections (dealing with asymptotic developments), we will use capital letters for mainstream solutions.

says that $-\mathbf{n}\cdot\mathbf{u} = \boldsymbol{\nabla}_{\|}\cdot\mathbf{Q}$, namely the two-dimensional divergence of $\mathbf{Q}$ on the boundary surface. We offer no proof here but establish the result in a simple Ekman layer case with planar boundary in the following subsection.

In some cases boundary layers become detached from the boundary but retain a boundary layer structure. We call them shear layers. Though a good account of Ekman layers is given by Greenspan (1968), the then topical shear layer theory was surveyed with a clearer perspective later by Moore (1978).

3.1.1. EKMAN LAYERS

Relative to a Cartesian reference frame (x, y, z) rotating with angular velocity $\boldsymbol{\Omega} = (0, 0, \Omega)$, the slow, steady flow, velocity $\mathbf{U} = \mathbf{u}_G$, of inviscid fluid is in geostrophic balance and satisfies

$$2\boldsymbol{\Omega}\times\mathbf{u}_G = -\boldsymbol{\nabla}(p/\rho)\,, \tag{3.4}$$

where p is the pressure. In consequence, the flow satisfies the Proudman–Taylor theorem and is independent of the co-ordinate z parallel to the rotation axis.[6]

The essential properties of an Ekman layer can be understood in terms of the following simple model. Consider a viscous fluid, kinematic viscosity ν, moving with uniform geostrophic $\mathbf{u}_G = (u_G, v_G, 0)$ in the region $z > 0$ above the plane boundary $\partial\mathcal{D} : \{z = 0\}$. The fluid velocity $\mathbf{u} = (u(z), v(z), 0)$ in the Ekman layer is a function of z alone and satisfies

$$-2\Omega v = -\frac{\partial}{\partial x}\left(\frac{p}{\rho}\right) + \nu\frac{\mathrm{d}^2 u}{\mathrm{d}z^2}\,, \qquad -2\,\Omega v_G = -\frac{\partial}{\partial x}\left(\frac{p}{\rho}\right), \tag{3.5a,b}$$

$$2\Omega u = -\frac{\partial}{\partial y}\left(\frac{p}{\rho}\right) + \nu\frac{\mathrm{d}^2 v}{\mathrm{d}z^2}\,, \qquad 2\,\Omega u_G = -\frac{\partial}{\partial y}\left(\frac{p}{\rho}\right). \tag{3.5c,d}$$

By multiplying the second equation by i and adding to the first, this set of equations of equations reduces to the single complex equation

$$\delta_E^2\frac{\mathrm{d}^2 Z}{\mathrm{d}z^2} - 2\,(\operatorname{sgn}\Omega)\,\mathrm{i}\,(Z - Z_G) = 0\,, \quad \text{in which} \quad \delta_E = \sqrt{\nu/|\Omega|} \tag{3.6a,b}$$

and $Z \equiv u + \mathrm{i}v$, $Z_G \equiv u_G + \mathrm{i}v_G$. The solution satisfying the boundary conditions (3.1b) and (3.3) is

$$Z = Z_G\{1 - \exp[-(1 + (\operatorname{sgn}\Omega)\,\mathrm{i})\,z/\delta_E]\}\,, \tag{3.7}$$

where we identify δ_E with the Ekman layer thickness. The path $Z(z) : 0 \le z < \infty$ traced by Z on the Argand Diagram (equivalently $\mathbf{u}(z)$ in the (x, y)–plane) is referred to as the "Ekman Spiral" (see Figure 3.1).

6 Indeed, the curl of (3.4) yields immediately $(\boldsymbol{\Omega}\cdot\boldsymbol{\nabla})\,\mathbf{u}_G = \mathbf{0}$.

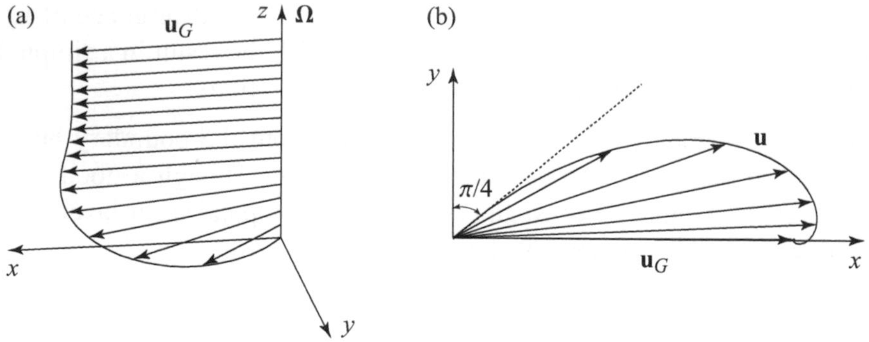

Figure 3.1 - Ekman layers ($\Omega > 0$). (a) The velocity profile $\mathbf{u}(z)$. (b) The Ekman spiral projection on the (x, y)–plane.

The tangential stress $\boldsymbol{\tau} = \rho\nu(\mathrm{d}\mathbf{u}/\mathrm{d}z)|_{z=0}$ on the the boundary is determined by differentiating (3.7). Its components (τ_x, τ_y) satisfy

$$\tau_x + \mathrm{i}\,\tau_y = \rho\nu \left.\frac{\mathrm{d}u}{\mathrm{d}z}\right|_{z=0} = \rho\sqrt{2\nu|\Omega|}\, Z_G \exp[(\operatorname{sgn}\Omega)\,\mathrm{i}\,\pi/4]\,. \tag{3.8}$$

Likewise the components (Q_x, Q_y) of the volume flux deficit (3.2b) determined by integrating (3.7) satisfy

$$Q_x + \mathrm{i}\,Q_y = (\delta_E/\sqrt{2})\, Z_G \exp[(\operatorname{sgn}\Omega)\,\mathrm{i}\,3\pi/4] \tag{3.9a}$$

or explicitly

$$\mathbf{Q} = \tfrac{1}{2}\,\delta_E\,[(-u_G\,,\,-v_G) + (\operatorname{sgn}\Omega)\,(-v_G\,,\,u_G)]\,. \tag{3.9b}$$

To help interpret the above results, take axes such that $\mathbf{u}_G = (u_G, 0, 0)$. Then as z increases from zero the flow velocity $\mathbf{u}$ increases from zero where the Ekman Spiral has tangent $\mathrm{d}\mathbf{u}/\mathrm{d}z|_{z=0}$ at $(\operatorname{sgn}\Omega)\,45°$ to the x–axis [see (3.8)]. The velocity has a substantial positive y–component v until the velocity vector $\mathbf{u}$ spirals in tightly about $\mathbf{u}_G$ (see also Figure 3.1b). This process is quantified by the mass transport $\rho\, Q_y = (\operatorname{sgn}\Omega)\,(\delta_E/2)\,\rho\, u_G$ in the y–direction [see (3.9)] perpendicular to the mainstream geostrophic flow. The point is that with Ωu_G positive a negative y–component of pressure gradient is set up [see (3.5d)] which is unbalanced by the weaker Coriolis acceleration in the Ekman layer leading to the positive (negative) y–mass transport ρQ_y, when Ω is positive (negative).

When the geostrophic flow is no longer uniform but varies on a length scale L large compared with δ_E, the velocity $\mathbf{u}_G$ continues to have no z–component and its mass continuity implies that $\partial_x u_G + \partial_y v_G = 0$. It follows that the boundary layer form of the mass continuity equation for $\mathbf{u} = (u, v, w)$ can be expressed in the form

$$\partial_z w = -\partial_x(u - u_G) - \partial_y(v - v_G)\,. \tag{3.10}$$

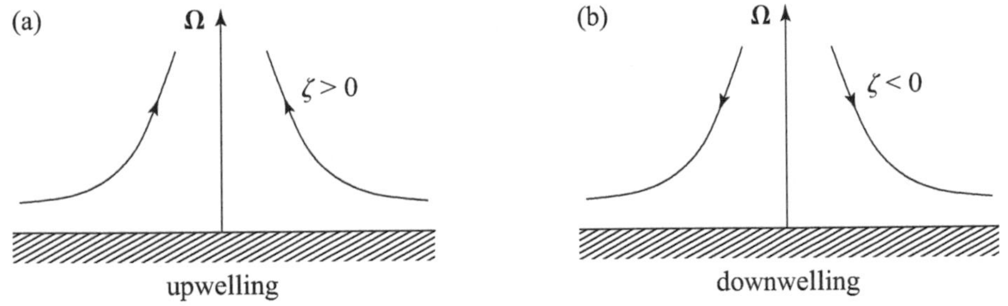

Figure 3.2 - Ekman pumping ($\Omega > 0$). (a) Upwellings $\boldsymbol{\Omega} \cdot \boldsymbol{\nabla} \times \mathbf{u}_G > 0$. (b) Downwellings $\boldsymbol{\Omega} \cdot \boldsymbol{\nabla} \times \mathbf{u}_G < 0$.

Integration with respect to z and application of the impermeable boundary condition (3.1a), namely $w|_{z=0} = 0$, determines

$$w_E = -\partial_x Q_x - \partial_y Q_y, \tag{3.11}$$

where w_E is the value of w as $z/\delta_E \uparrow \infty$. Upon substitution of (3.9b) and use again of the continuity condition $\partial_x u_G + \partial_y v_G = 0$, we obtain

$$w_E = \tfrac{1}{2}\delta_E (\operatorname{sgn}\Omega)\,\zeta, \qquad \text{where} \qquad \zeta = \partial_x v_G - \partial_y u_G \tag{3.12a,b}$$

is the z–component of the vorticity of the Geostrophic flow. The process by which fluid is transferred out of (into) the Ekman layer is known as Ekman pumping (suction). When gravity identifies the positive (negative) z–direction as upwards (downwards), it leads to upwelling (downwellings) when the vorticity ζ has the same (opposite) sign to Ω [see Figure 3.2a(b)].

When the rotation vector $\boldsymbol{\Omega}$ is no longer parallel to the unit normal $\mathbf{n}$, the Ekman layer structure is essentially the same but with $\boldsymbol{\Omega}$ replaced by $\Omega_\perp \mathbf{n}$, where $\Omega_\perp = \mathbf{n} \cdot \boldsymbol{\Omega}$; the components of $\boldsymbol{\Omega}$ parallel to the boundary do not influence the laminar Ekman layer dynamics. Thus the volume flux deficit (3.9b) becomes

$$\mathbf{Q} = \tfrac{1}{2}\delta_{E\perp}\left[-\mathbf{u}_G + (\operatorname{sgn}\Omega_\perp)\,\mathbf{n} \times \mathbf{u}_G\right], \tag{3.13a}$$

where

$$\delta_{E\perp} = \sqrt{\nu/|\Omega_\perp|} = \delta_E \Big/ \sqrt{|\cos\theta|} \tag{3.13b}$$

is the Ekman boundary layer thickness and $\Omega_\perp = \mathbf{n} \cdot \boldsymbol{\Omega} = \Omega\cos\theta$. The Ekman pumping velocity $\mathbf{n} \cdot \mathbf{u}$ is now obtained from (3.2) in the form given on p. 46 of Greenspan (1968). At a planar boundary the result reduces to

$$\mathbf{n} \cdot \mathbf{u} = \tfrac{1}{2}\delta_{E\perp} (\operatorname{sgn}\Omega_\perp)\,\mathbf{n} \cdot \boldsymbol{\nabla} \times \mathbf{u}_G, \tag{3.14}$$

which relates the normal flux to the vorticity $\boldsymbol{\nabla} \times \mathbf{u}_G$ of the mainstream geostrophic flow (see also Pedlosky, 1979) and generalises (3.12) above.

3.1.2. SIDEWALL $\mathrm{E}^{1/3}$–LAYERS

When the boundary is parallel to the rotation vector ($\Omega_\perp = 0$), the boundary layer is no longer governed by the system (3.5) and we have to consider the steady, linear equation of motion

$$2\boldsymbol{\Omega} \times \mathbf{u} = -\boldsymbol{\nabla}(p/\rho) + \nu\Delta\mathbf{u} \qquad \text{with} \qquad \boldsymbol{\nabla}\cdot\mathbf{u} = 0 \qquad (3.15\text{a,b})$$

together with the vorticity equation

$$-\,2\boldsymbol{\Omega}\cdot\boldsymbol{\nabla}\mathbf{u} = \nu\delta\boldsymbol{\zeta} \qquad \text{with} \qquad \boldsymbol{\zeta} = \boldsymbol{\nabla}\times\mathbf{u} \quad (3.16\text{a,b})$$

obtained by taking its curl.

Throughout this section we will take $\boldsymbol{\Omega} = (0, 0, \Omega)$ with $\Omega > 0$ and consider 2–dimensional flows

$$\mathbf{u} = (u,\, v,\, w) = (-\partial_z\chi,\, v,\, \partial_x\chi)\,, \qquad \boldsymbol{\zeta} = (-\partial_z v,\, -\Delta\chi,\, \partial_x v)\,, \quad (3.17\text{a,b})$$

in which the streamfunction χ and velocity v are independent of the coordinate y.

Our objective now is to consider boundary layers tangent to the rotation axis for which the x–length scale is small compared to the z–length scale. So on making the approximations $\partial_x \gg \partial_z$, the y–components of (3.15a) and (3.16a) reduce to

$$-2\,\partial_z\chi = \delta_E^2\,\partial_{xx}v \qquad \text{and} \qquad -\,2\,\partial_z v = -\,\delta_E^2\,\partial_{xxxx}\chi\,, \quad (3.18\text{a,b})$$

respectively, where δ_E is the boundary layer thickness (3.6b). Then boundary layer solutions, which decay to zero as $|x| \to \infty$ in one of the half-spaces $x > 0$ and $x < 0$ and are spatially periodic on a length $4H$ in the z–direction, have the complex form

$$\{w,\, v\} = \widetilde{w}\left\{\cos\left(\frac{\pi z}{2H}\right),\, -(\mathrm{sgn}\,x)\,\alpha^3\sin\left(\frac{\pi z}{2H}\right)\right\}\exp\left[-\,\alpha\left(\frac{\pi}{\delta_E^2 H}\right)^{1/3}|x|\right], \quad (3.19\text{a})$$

where $\widetilde{w}$ is a complex constant and α can take each of the three values

$$\alpha = [1\,, (1 \pm \mathrm{i}\sqrt{3})/2] \quad \text{and} \quad \alpha^3 = [1\,, -1]\,. \qquad (3.19\text{b,c})$$

Evidently the boundary layer width described by (3.19a) is

$$\delta_S = (\delta_E^2 H)^{1/3} = \mathrm{E}^{1/3}H\,, \quad \text{where} \quad \mathrm{E} = (\delta_E/H)^2 = \nu/H^2|\Omega| \qquad (3.20\text{a,b})$$

is the Ekman number; such structures are often called $\mathrm{E}^{1/3}$–Stewartson layers. Simple scale analysis of the original equation (3.18a,b) would yield this $\mathrm{E}^{1/3}H$ length scale.

To illustrate the idea, we consider fluid confined between two plane boundaries $z = \pm H$. Each half-plane $\operatorname{sgn} x > 0$ moves rigidly parallel to itself with velocity $\pm(\operatorname{sgn} x)V_0(0,1,0)$. This is a slight variant of one of Stewartson's (1957) original split-disc problems. The anti-symmetry in z ensures that the geostrophic velocity vanishes: $\mathbf{u}_G = \mathbf{0}$. As a consequence the Ekman volume flux determined from (3.13) in the associated boundary layers is uniform and takes the value $\mathbf{Q} = \pm(\operatorname{sgn} x)(\delta_E/2)V_0(1,1,0)$. Since the Ekman fluxes are uniform on each half-plane the Ekman pumping velocity vanishes everywhere except at $x = 0$. There on the bottom $z = -H$ (top $z = H$) boundary the Ekman boundary layer fluxes converge (diverge), where they emerge as source (sink) flows with volume fluxes $\delta_E V_0$. This leads to the boundary conditions

$$w = \delta_E V_0\, \delta(x) \qquad (\Omega > 0) \qquad \text{on} \qquad z = \pm H\,, \tag{3.21a}$$

where $\delta(x)$ is the Dirac δ–function. Equivalent conditions are

$$\begin{aligned} w &= 0 && \text{on} \quad z = \pm H\,, && x \neq 0\,, && (3.21\text{b})\\ \chi &\to \pm\tfrac{1}{2}\,\delta_E V_0 && \text{on} \quad -H \le z \le H\,, && \text{as} \quad x \to \pm\infty\,, && (3.21\text{c}) \end{aligned}$$

which accounts for the volume flux between the source-sink combination. The Fourier transform solution of (3.18) subject to this boundary condition (3.21a) is

$$\{w,\, v\} = \frac{\delta_E V_0}{\pi}\int_0^\infty \frac{\{[\cos(kx)\,\cosh(\frac{1}{2}\delta_E^2 k^3 z)]\,,\,[\sin(kx)\,\sinh(\frac{1}{2}\delta_E^2 k^3 z)]\}}{\cosh\frac{1}{2}\delta_E^2 k^3 H}\,dk\,. \tag{3.22}$$

These expressions may be turned into infinite sums by application of contour integration in the complex plane and use of the residue theorem. The result for $w = \partial_x\chi$ yields

$$\chi = \delta_E V_0\,(\operatorname{sgn} x)\left[\frac{1}{2} - \sum_{n=0}^{\infty}\frac{2(-1)^n}{(2n+1)\pi}F(X_n)\,\cos\left((2n+1)\frac{\pi z}{2H}\right)\right], \tag{3.23a}$$

where

$$X_n = [(2n+1)\pi]^{1/3}\,x/\delta_S\,, \tag{3.23b}$$

$$F(X_n) = \tfrac{1}{3}\left[\exp\left(-|X_n|\right) + 2\exp\left(-\tfrac{1}{2}|X_n|\right)\cos\left(\tfrac{1}{2}\sqrt{3}|X_n|\right)\right] \tag{3.23c}$$

and we have taken advantage of the finite Fourier transform identity

$$\frac{1}{2} = \sum_{n=0}^{\infty}\frac{2(-1)^n}{(2n+1)\pi}\cos\left[(2n+1)\frac{\pi z}{2H}\right] \qquad \text{on} \quad -H < z < H\,. \tag{3.23d}$$

Rather than do the contour integrations to obtain the results (3.23), it is perhaps more straightforward and illuminating to check directly that it meets the conditions of the problem. Since $w = \partial_x \chi$ obtained from (3.23a) is composed of the eigenmodes (3.19), it solves (3.18a,b) whenever $x \neq 0$, and evidently meets the boundary conditions (3.21b,c). Finally the continuity of (3.23a) and all its x–derivatives across $x = 0$ may be checked by expanding the exponentials in (3.23c) as power series. Then we may recast (3.23a) in the form

$$\chi = \delta_E V_0 \sum_{n=0}^{\infty} \frac{(-1)^n}{(2n+1)\pi} \left[\sum_{m=0}^{\infty} \left((-1)^m + \tfrac{1}{3}\right) \frac{(X_n)^{2m+1}}{(2m+1)!} \right] \cos\left((2n+1)\frac{\pi z}{2H}\right), \tag{3.23e}$$

from which continuity is self-evident.

The solution (3.23a) shows that w is small $\mathcal{O}((\delta_E/\delta_S)V_0) = \mathcal{O}(\mathrm{E}^{1/6}V_0)$. Similar expressions exist for v which is likewise small $\mathcal{O}(\mathrm{E}^{1/6}V_0)$. Nevertheless, as the source at $(x, z) = (0, -H)$ is approached the strengths of these flows increase and their asymptotic values determined by (3.22), on retaining only the dominant exponential in each of the hyperbolic functions, are

$$\{w,\, v\} = \frac{\delta_E V_0}{\pi} \int_0^\infty \{\left[\cos(kx)\,\exp(-\tfrac{1}{2}\delta_E^2 k^3 Z)\right], \left[-\sin(kx)\,\exp(-\tfrac{1}{2}\delta_E^2 k^3 Z)\right]\}\,\mathrm{d}k\,, \tag{3.24a}$$

where $Z = z + H$ and $0 \leq Z \ll H\,.$ (3.24b,c)

Moreover these solutions are of similarity form

$$\{w,\, v\} = V_0 \left(\frac{\delta_E}{Z}\right)^{1/3} \{W(\xi)\,, V(\xi)\}\,, \quad \text{where} \quad \xi = \frac{x}{(\delta_E^2 Z)^{1/3}} = \mathcal{O}(1) \tag{3.25a,b}$$

is the similarity variable. Furthermore by (3.24a) W and V are given by the real and imaginary parts of

$$W - \mathrm{i}V = \frac{1}{\pi} \int_0^\infty \exp(\mathrm{i}K\xi - \tfrac{1}{2}K^3)\,\mathrm{d}K \tag{3.26a}$$

[Moore & Saffman, 1969, equations (3.23–26)]. Direct substitution shows that it satisfies the inhomogeneous complex Airy equation

$$(W - \mathrm{i}V)'' + \tfrac{2}{3}\mathrm{i}\xi\,(W - \mathrm{i}V) = -\tfrac{2}{3}\pi^{-1}\,, \tag{3.26b}$$

where the prime denotes the ξ–derivative. It is readily verified that the ξ–derivative of (3.26b) is consistent with (3.18). This integrated form, which incorporates the boundary conditions in a non-trivial way, is useful as it is shows that $V \approx -1/(\pi\xi)$ (equivalently $v \approx -\delta_E V_0/(\pi x)$) as $|\xi| \to \infty$. Note also that, since $\chi = \delta_E V_0 \Upsilon(\xi)$

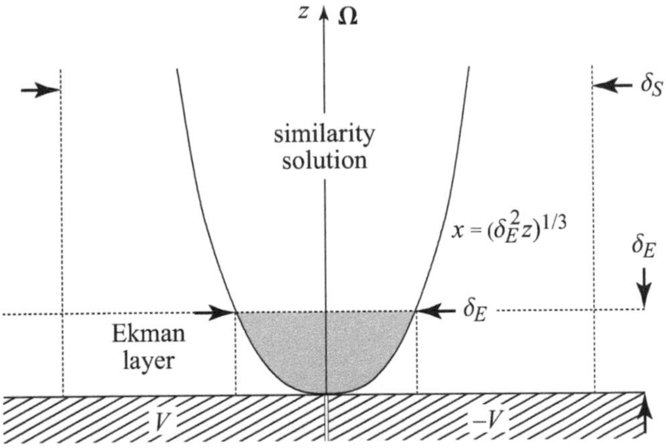

Figure 3.3 - Boundary singularity in the neighbourhood of $x = 0$, $Z = z + H = 0$. The location of the $\mathrm{E}^{1/3}$–layer, similarity region and Ekman layer are indicated.

where $\Upsilon = \int_0^\xi W \mathrm{d}\xi$, the curves $\xi =$ constant are streamlines. The similarity solution is valid in the "similarity" region where $\xi = \mathcal{O}(1)$ (see Figure 3.3) but is unreliable outside it when $|x| = \mathcal{O}(\delta_S) = \mathcal{O}(\mathrm{E}^{1/3}H)$. In that part of the $\mathrm{E}^{1/3}$–Stewartson layer, where w and v are $\mathcal{O}(\mathrm{E}^{1/6}V_0)$, the solution is properly determined by (3.23a).

The essential point is that within the similarity region w and v increase as $(\delta_E/Z)^{1/3}$ with decreasing Z along curves $\delta_E^2 Z = (x/\xi)^3$, where $\xi =$ constant. Nevertheless they remain small outside the Ekman layer, where $Z \gg \delta_E$. Once $Z = \mathcal{O}(\delta_E)$, the width of the similarity region has also shrunk to $\mathcal{O}(\delta_E)$. Here, in a small region centred at $(x, Z) = 0$ of radius $\mathcal{O}(\delta_E)$, both the x and z length scales are comparable. Accordingly the scale separations assumed in both the Ekman and similarity layers are no longer valid and the solutions fail.

Similar results pertain to the neighbourhood of the sink at $(x, z) = (0, H)$. Together they confirm that, since $\int_{-\infty}^{\infty} W \mathrm{d}\xi = 1$, all the Ekman flux passes through the $\mathrm{E}^{1/2} \times \mathrm{E}^{1/2}$– regions at $(x, z) = (0, \pm H)$ and they provide the source-sink combination for the $\mathrm{E}^{1/3}$–Stewartson layer. The ensuing streamline pattern defined by curves $\chi =$ constant in the (x, z)–plane is sketched in Figure 3.4.

3.1.3. SIDEWALL $\mathrm{E}^{1/4}$–LAYERS

The situation investigated in the previous subsection was unusual in the sense that the geostrophic velocity was uniform everywhere and did not suffer a jump across the boundary layer region. In more general situations that is not the case and the jump is smoothed out in a wider boundary layer region, where the flow is quasi-geostrophic.

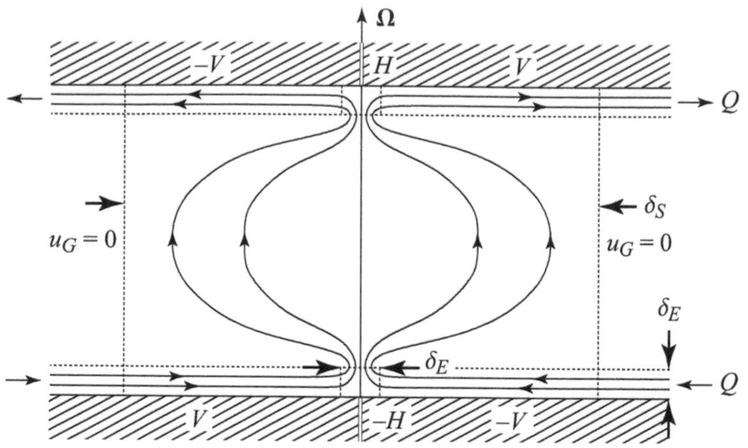

Figure 3.4 - The $\mathrm{E}^{1/3}$–shear layer for the solution (3.22) illustrating the streamlines $\chi =$ constant of the ageostrophic flow driven by the Ekman layer fluid flux.

In order to appreciate the nature of these boundary layers, we consider the geostrophic motion $\mathbf{u}_G = (u_G(x,y), v_G(x,y), 0)$ between stationary parallel plane boundaries $z = \pm H$ with $\boldsymbol{\Omega} = (0,0,\Omega)$ and $\Omega > 0$. According to (3.5b,d) that flow may be expressed in terms of a streamfunction in the form

$$u_G = -\partial_y \psi\,, \qquad v_G = \partial_x \psi\,, \qquad \text{where} \qquad \psi = p/2\rho\Omega\,. \tag{3.27a,b}$$

The associated vorticity $\boldsymbol{\zeta} = (0,0,\zeta)$ given by (3.12b) is given by

$$\zeta = \nabla_\perp^2 \psi\,, \qquad \text{where} \qquad \nabla_\perp^2 \equiv \partial_{xx} + \partial_{yy}\,. \tag{3.28a,b}$$

The streamfunction is at this stage undefined (p is as yet an unknown function of x and y), leading to a situation often referred to as geostrophic degeneracy. It may be resolved here by first noting that this is only a first approximation to the true solution $\mathbf{u} = (u,v,w)$ of the Navier-Stokes equation (3.15a), in the sense that $|\mathbf{u} - \mathbf{u}_G| \ll |\mathbf{u}_G|$. Since w is small, the dominant balance in the z–component of the vorticity equation (3.16a) in the mainstream is between two small terms:

$$-2\,\partial_z w = \delta_E^2 \nabla_\perp^2 \zeta\,. \tag{3.29}$$

Integrating this equation with respect to z across the mainstream determines

$$-(w_E|_{z=H} - w_E|_{z=-H}) = H\delta_E^2 \nabla_\perp^2 \zeta\,, \tag{3.30a}$$

where $w_E|_{z=-H}(= -w_E|_{z=H}) = \delta_E \zeta/2$ is the Ekman pumping velocity on the lower (upper) boundary. It shows that the vorticity ζ satisfies

$$\zeta = \delta_E H \nabla_\perp^2 \zeta \qquad (\Omega > 0)\,. \tag{3.30b}$$

For unidirectional geostrophic flows $\mathbf{u}_G = (0, v_G(x), 0)$, the vorticity equation (3.30b) can be integrated once giving

$$v_G - V_G = \delta_E H \frac{\mathrm{d}^2 v_G}{\mathrm{d}x^2}, \tag{3.31}$$

where V_G is an arbitrary constant. This equation allows boundary layer structures of the form

$$v_G - V_G \propto \exp[\pm(x/\Delta)], \quad \text{where} \quad \Delta = \sqrt{\delta_E H} = \mathrm{E}^{1/4} H, \tag{3.32a,b}$$

which are often called $\mathrm{E}^{1/4}$–Stewartson layers.

To illustrate the idea, we adopt the split plane boundary geometry of the previous section but assume instead that both the top ($z = H$) and bottom ($z = -H$) boundaries move together with velocity $-(\operatorname{sgn} x) V_0(0, 1, 0)$. This is the planar version of the other Stewartson (1957) split-disc problem. We assume that, far from the discontinuity, the fluid moves with the boundary, specifically $v_G \to -(\operatorname{sgn} x) V_0$ as $|x| \to \infty$. The appropriate combination of solutions (3.32a) that meet that condition and have v_G and $\mathrm{d}v_G/\mathrm{d}x$ continuous at $x = 0$ is

$$v_G = -(\operatorname{sgn} x) V_0 \left[1 - \exp(-|x|/\Delta)\right]. \tag{3.33}$$

The corresponding Ekman suction velocities, which take account of the velocity jumps at the boundaries, are

$$w_E\big|_{z=\mp H} = \pm \delta_E V_0 \left[-\frac{1}{2\Delta} \exp\left(-\frac{|x|}{\Delta}\right) + \delta(x)\right]. \tag{3.34a}$$

Since (3.29) implies that the axial velocity w is linear in z, the velocity jumps (3.34a) show that

$$w = \tfrac{1}{2} \mathrm{E}^{1/4} V_0 \, (z/H) \exp(-|x|/\Delta) \tag{3.34b}$$

outside the $\mathrm{E}^{1/3}$–layer, where $|x| \gg \delta_S$. The corresponding streamfunction χ for $u = -\partial_z \chi$ and $w = \partial_x \chi$ is

$$\chi = -\tfrac{1}{2} \delta_E V_0 \, (z/H) \, (\operatorname{sgn} x) \exp(-|x|/\Delta). \tag{3.34c}$$

It is important to appreciate that (3.34c) determines the jump

$$[\chi]_-^+ = \chi_{(x/\Delta)\downarrow 0} - \chi_{(x/\Delta)\uparrow 0} = -\delta_E V_0 \, (z/H) \tag{3.34d}$$

across $x/\Delta = 0$. It means that jets emanate from sources at $(0, -H)$ and $(0, H)$ of equal magnitude $\delta_E V_0$ but both directed towards the mid-plane $z = 0$. This flux along $x/\Delta = 0$ escapes from the $\mathrm{E}^{1/3}$–layer at a constant rate to provide the uniform outflow $u = \frac{1}{2}(\operatorname{sgn} x) \mathrm{E}^{1/2} V_0$ as $x/\Delta \to 0$. This efflux is returned symmetrically

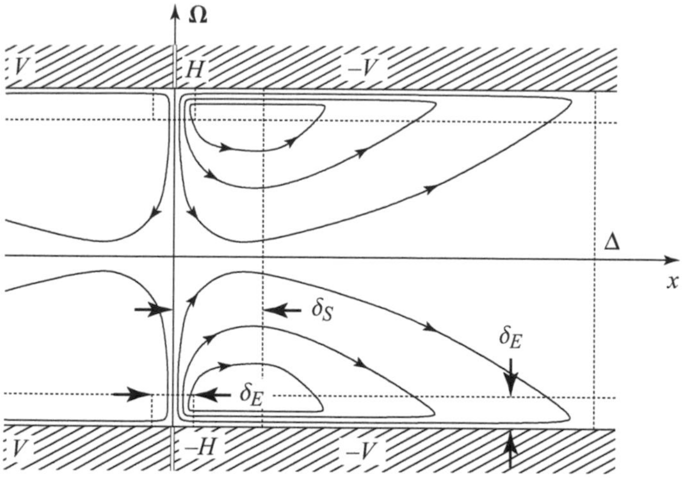

Figure 3.5 - The $E^{1/4}$–shear layer streamlines $\chi =$constant for the solution (3.34c) valid in the regions $\delta_S \ll |x| \ll \Delta$. The streamlines for the Ekman layer flow that drain it and the $E^{1/3}$–shear layer that feeds it are also indicated. Note that all streamlines pass through the $E^{1/2} \times E^{1/2}$–regions at $(x, z) = (0, \pm H)$.

through the $E^{1/4}$–layer and reabsorbed by the upper and lower boundary Ekman layers from whence it is recirculated through the sources at $(0, \mp H)$, as indicated by the streamline pattern in Figure 3.5.

In the thinner $E^{1/3}$–layer, flow is made up in part by the quasi-geostrophic contribution (3.33), which has the expansion

$$v_G = -V_0 \left[E^{1/12}(x/\delta_S) - \tfrac{1}{2}(\operatorname{sgn} x)E^{1/6}(x/\delta_S)^2 + \mathcal{O}(E^{1/4})\right], \qquad (3.35)$$

as well as an additional z–dependent ageostrophic part. The velocity in this layer is dominated by the leading term $\mathcal{O}(E^{1/12}V_0)$ of (3.35). The second term $\mathcal{O}(E^{1/6}V_0)$ turns out to be the same size as the $E^{1/3}$–layer ageostrophic velocities predicted in the previous subsection. This is significant because, though we have ensured that v_G and $\mathrm{d}v_G/\mathrm{d}x$ are continuous across $x = 0$, that is evidently not the case for $\mathrm{d}^2 v_G/\mathrm{d}x^2$, which suffers a jump in value and must be eliminated by the lowest order solution for that layer. The safest way to compute that solution is to solve the governing equations (3.15) using the Fourier transform method in conjunction with the Ekman-layer jump condition across both the $E^{1/3}$ and $E^{1/4}$–layers simultaneously as executed by Greenspan (1968). Nevertheless a detailed boundary layer analysis more in the spirit of the development here is given by Moore & Saffman (1969).

3.1.4. DIFFERENTIALLY ROTATING SPHERES: THE PROUDMAN–STEWARTSON PROBLEM

An important example which may have relevance to motion in the Earth's fluid core is the steady axisymmetric slow motion in a shell induced by rotating the outer, radius r_o, and inner, radius r_i, spherical boundaries at slightly different rates $\Omega\ (>0)$ and $\Omega + \omega_i$ about a common rotation axis. Relative to cylindrical polar coordinates (s, ϕ, z) rotating in the frame fixed in the outer sphere, the fluid occupies the region $r_i < r < r_o$, where $r = (s^2 + z^2)^{1/2}$, and meets the boundary conditions $\mathbf{u} = 0$ on $r = r_i$ and $\mathbf{u} = (0, s\omega_i, 0)$ on $r = r_i$.

The only admissible geostrophic flow in a shell is azimuthal and independent of z which we write in the form $\mathbf{u}_G = (0, s\omega_G(s), 0)$. According to (3.13), the θ–directed (θ is the co-latitude) Ekman layer volume fluxes are

$$Q_{\theta i} = \frac{\delta_E\, s}{2\sqrt{|\cos\theta_i|}}(\omega_i - \omega_G) \quad \text{and} \quad Q_{\theta o} = -\frac{\delta_E\, s}{2\sqrt{|\cos\theta_o|}}\omega_G \qquad (3.36\text{a,b})$$

on the inner and outer spheres respectively, where

$$\cos\theta_i = \sqrt{1 - (s/r_i)^2}\,, \quad \text{and} \quad \cos\theta_o = \sqrt{1 - (s/r_o)^2}\,. \qquad (3.36\text{c,d})$$

We define the cylinder $s = r_i$ tangent to the equator of the inner sphere to be the tangent cylinder (see Figure 3.6).

The geostrophic velocity $s\omega_G$ behaves very differently inside and outside this tangent cylinder. So if we consider a cylinder of radius s inside the tangent cylinder, which intersects the inner and outer spheres, then the total radial fluid flux across it vanishes. Since there is no radial component of the geostrophic velocity the values of $Q_{\theta i}(s)$ and $Q_{\theta i}(s)$ must be equal and opposite and that determines the Proudman (1956) solution

$$\omega_i - \omega_G = \omega_i \sqrt{|\cos\theta_i|}\Big/\Big[\sqrt{|\cos\theta_i|} + \sqrt{|\cos\theta_o|}\Big] \quad \text{for} \quad s < r_i\,. \qquad (3.37\text{a})$$

Outside the tangent cylinder, a cylinder $s =$ constant only intersects the upper and lower boundaries of the outer sphere and that determines

$$\omega_G = 0 \qquad \text{for} \qquad s > r_i\,. \qquad (3.37\text{b})$$

We sketch ω_G *versus* s in Figure 3.7a. This profile can be recovered in full numerical simulations at finite E (Dormy *et al.*, 1998).

We consider the complete axisymmetric flow velocity in the form

$$\mathbf{u} = (u,\, v,\, w) = (-s^{-1}\partial_z\chi,\, s\omega,\, s^{-1}\partial_s\chi)\,. \qquad (3.38\text{a,b})$$

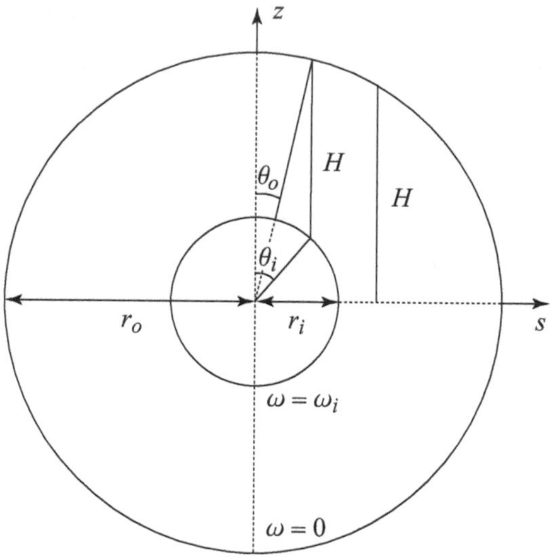

Figure 3.6 - The shell geometry of the Proudman–Stewartson problem. The height $H(s)$ [see (3.41b)] both inside $(s < r_i)$ and outside $(s > r_i)$ the tangent cylinder $s = r_i$ are indicated.

For $\omega_i > 0$ the Ekman flux $Q_{\theta o}(< 0)$ on the outer sphere is inwards towards the axis. Since the total influx $-2\pi s Q_{\theta o}$ is a decreasing function of s, fluid continually escapes from the outer sphere Ekman layer and returns to the inner sphere as a purely axial flow characterised by $\chi(s)$. Its value is determined by the total flux balance $2\pi s Q_{\theta o} = 2\pi \int_0^s sw\, \mathrm{d}s = 2\pi\chi$, which yields

$$\chi(s) = -\tfrac{1}{2}\,\delta_E\omega_i\, s^2 \Big/ \left[\sqrt{|\cos\theta_i|} + \sqrt{|\cos\theta_o|}\right] \qquad \text{for} \qquad s < r_i\,. \quad (3.39a)$$

On entering the inner sphere Ekman layer the fluid is transported outwards $Q_{\theta i}(> 0)$. On reaching $s = r_i$ we find that

$$\chi(r_i) = -\tfrac{1}{2}\,\delta_E\omega_i\, r_i^2\, r_o^{1/2}(r_o^2 - r_i^2)^{-1/4} \qquad (3.39b)$$

but outside the tangent cylinder we have

$$\chi(s) = 0 \qquad \text{for} \qquad s < r_i\,. \qquad (3.39c)$$

That means that there is a return flow jet along the tangent cylinder. The fluid outside the tangent cylinder is stagnant. The flow that we have described is the Proudman (1956) solution of the problem. We sketch the meridional streamline pattern $\chi =$ constant in Figure 3.7b.

Evidently the boundary layer structure on the tangent cylinder is very complicated and was the problem addressed by Stewartson (1966). We first note that the ϕ–

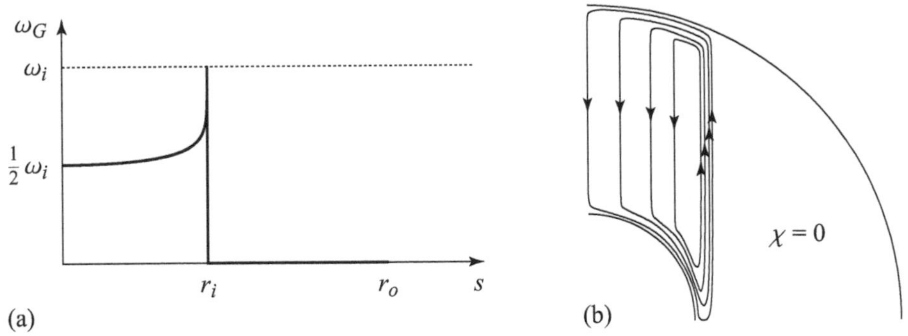

Figure 3.7 - The Proudman solution. (a) The geostrophic angular velocity ω_G (3.37) plotted versus s. (b) The meridionanal streamlines χ = constant (3.39a). Note the location of the jet along the tangent cylinder $s = r_i$, which returns the volume flux $-2\pi\chi(r_i)$ (3.39b) from the Ekman layer on the inner sphere at its equator to the outer sphere Ekman layer.

components of the equation of motion (3.15a) and vorticity (3.16a) reduce to

$$-2\,\partial_z(\chi/s) = \delta_E^2\,(\Delta - s^{-2})(s\omega)\,, \tag{3.40a}$$

$$-2\,\partial_z(s\omega) = -\,\delta_E^2\,(\Delta - s^{-2})^2(\chi/s) \tag{3.40b}$$

respectively [*cf.* their planar counterparts (3.18)]. The quasi-geostrophic layers can be analysed on the basis that $\omega \approx \omega_G$ is independent of z. Then (3.40a) can be integrated with respect to z between the Ekman layer boundaries. It determines

$$H\,\delta_E^2\left[\frac{1}{s}\frac{\mathrm{d}}{\mathrm{d}s}\left(s\frac{\mathrm{d}}{\mathrm{d}s}\right) - \frac{1}{s^2}\right](s\omega_G) = \begin{cases} -2(Q_{\theta o} + Q_{\theta i}) & \text{for } s < r_i\,; \\ -2Q_{\theta o} & \text{for } s > r_i\,, \end{cases} \tag{3.41a}$$

where

$$H(s) = \begin{cases} (r_o\cos\theta_o - r_i\cos\theta_i) & \text{for } s < r_i; \\ r_o\cos\theta_o & \text{for } s > r_i. \end{cases} \tag{3.41b}$$

Here and henceforth we restrict $\theta_i(s)$ and $\theta_o(s)$ to the range 0 up to $\pi/2$.

In the vicinity of the tangent cylinder, it is sufficient to take local Cartesian coordinates (x, y, z), where $x = s - r_i$ and the y–axis points in the ϕ–direction. Locally the inner sphere boundary is $z \approx \sqrt{-2r_i x}$ giving $\cos\theta_i \approx \sqrt{-2x/r_i}$ for $x \le 0$. Thus correct to lowest order (3.41a) reduces to

$$\Delta^2\,\frac{\mathrm{d}^2\omega_G}{\mathrm{d}x^2} - \frac{\omega_G}{\sqrt{\cos\theta_o}} = \begin{cases} (-2x/r_i)^{-1/4}\,(\omega_G - \omega_i) & \text{for } s < r_i; \\ 0 & \text{for } s > r_i, \end{cases} \tag{3.42}$$

in which $H(s)$, $\Delta(s) = \sqrt{\delta_E H}$ and $\theta_o(s)$ are all evaluated on the tangent cylinder $s = r_i$. The structure of (3.42) is most informative. Outside the tangent cylinder the

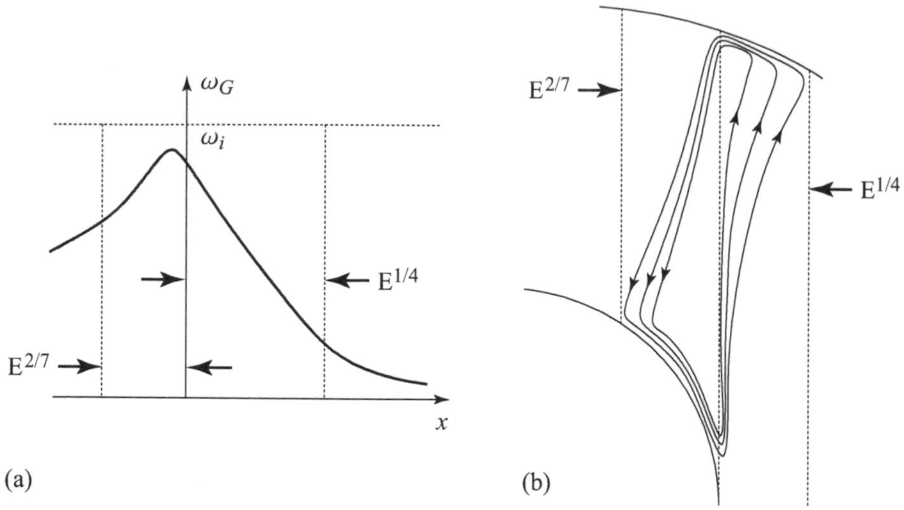

Figure 3.8 - The quasi-geostrophic Stewartson shear layers. (a) The geostrophic angular velocity ω_G, which solves (3.42), plotted versus $x = s - r_i$ in both the $E^{1/4}$ and $E^{2/7}$–layers exterior and interior to the tangent cylinder. (b) The corresponding meridional streamlines $\chi =$constant.

quasi-geostrophic layer is a standard $E^{1/4}$–layer of width $\mathcal{O}(\Delta(\cos\theta_o)^{1/4})$. Inside, however, the mass flux $Q_{\theta i}$ is linked to the weak singularity of $(r_i/x)^{1/4}$ which is manifest by the large term on the right of (3.42). That dominates over the influence of the mass flux $Q_{\theta o}$. The remaining two terms in (3.42) show that the boundary layer inside the tangent cylinder has width $\mathcal{O}(\Delta^{8/7}/r_i^{1/7})$, which is the width of the $E^{2/7}$–Stewartson layer. Stewartson (1966) showed that the jump in ω_G is largely eliminated in the exterior and wider $E^{1/4}$–layer. This is affected by the mass flux of the Proudman tangent cylinder jet escaping into it to be reabsorbed into the outer sphere Ekman layer. The remaining jump in $\mathrm{d}\omega_G/\mathrm{d}x$ is met across the thinner interior $E^{2/7}$–layer. The resulting profile of ω_G in the $E^{2/7}$ and $E^{1/4}$–layers is sketched in Figure 3.8a, while the resulting meridional streamline pattern $\chi =$ constant is illustrated in Figure 3.8b.

Stewartson (1966) also analysed the ageostrophic $E^{1/3}$–layers width $\delta_S = (\delta_E^2\, H)^{1/3}$. Now the equator $(x, z) = 0$ of the inner sphere is even more singular than the Proudman solution suggested. The difficulty may be traced to the Ekman-flux term $-2Q_{\theta i}/(\delta_E r_i) = (-2x/r_i)^{-1/4}(\omega_G - \omega_i)$ in (3.42). In the case of the Proudman solution, this tends to a finite limit as $x \uparrow 0$ by demanding that $\omega_G \to \omega_i$ simultaneously. In the case of the Stewartson extension the difference $\omega_i - \omega_G$ though small is finite tending to zero with E. Thus the volume flux in the Ekman layer exceeds the $\mathcal{O}(\delta_E r_i \omega_i)$ estimate of Proudman and leads to the streamline deflection in the $E^{2/7}$–layer indicated in Figure 3.8b. The Ekman flux must be eventually ejected

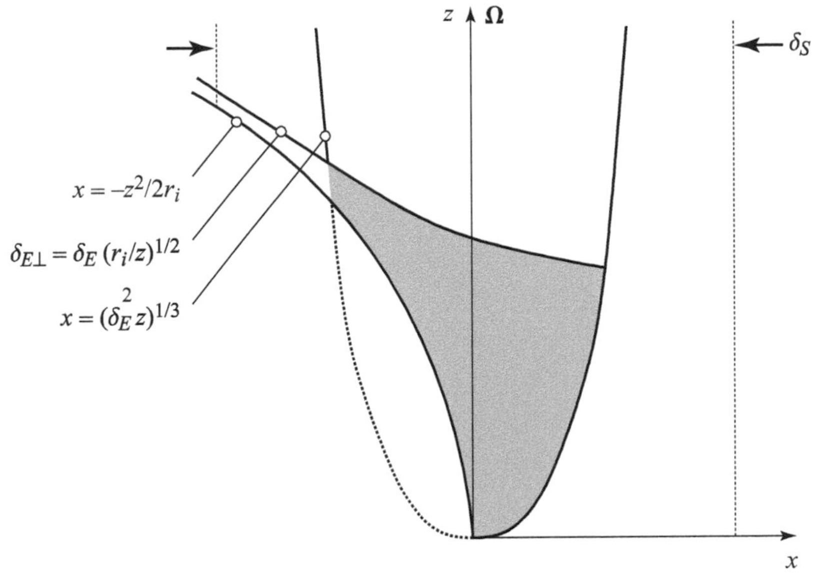

Figure 3.9 - The boundary layer structure on the tangent cylinder $x = s - r_i = 0$ in the vicinity of the equator $z = 0$ of the inner sphere. The inner sphere boundary is $x = -\frac{1}{2}z^2/r_i$. The intersection of the Ekman layer thickness $\delta_{E\perp} = \delta_E(r_i/z)^{1/2}$ and the similarity region $|x| < \mathcal{O}((\delta_E^2 z)^{1/3})$ determines the equatorial Ekman layer identified by the shaded region. Eventually, as z is increased, the singularity fills the Stewartson $\mathrm{E}^{1/3}$–layer of width $\delta_S = (\delta_E^2 H)^{1/3}$.

at the equator into the similarity sublayer of x–width $\mathcal{O}((\delta_E^2 z)^{1/3})$ [see (3.25b)] at distance z from the equator along the tangent cylinder. The similarity sublayer intersects the inner sphere $z \approx \sqrt{-2r_i x}$, when $z = \mathcal{O}(r_i^{3/5}\delta_E^{2/5})$, where it has x–width $\mathcal{O}(r_i^{1/5}\delta_E^{4/5})$. From another point of view the Ekman layer thickens as the equator is approached exhibiting according to (3.13b) an x–width $\mathcal{O}(\delta_{E\perp}) = \mathcal{O}(\delta_E\sqrt{r_i/z})$. So on decreasing z the Ekman layer thickens and merges with the similarity layer at the same order of magnitude of z as that layer attaches itself to the inner sphere as illustrated on Figure 3.9. Whereas the Proudman flux escapes into the exterior $\mathrm{E}^{1/4}$–layer (see Figure 3.8b), the additional flux fixed by the interior $\mathrm{E}^{2/7}$–layer is returned to the inner sphere Ekman layer in an eddy close to the equatorial Ekman layer of latitudinal extent $\mathcal{O}(r_i^{3/5}\delta_E^{2/5}) = \mathcal{O}(\mathrm{E}^{1/5})$ and width $\mathcal{O}(r_i^{1/5}\delta_E^{4/5}) = \mathcal{O}(\mathrm{E}^{2/5})$. Though Stewartson (1966) was unable to provide the solution in this layer, which remains a challenging unsolved problem, he does indicate the nature of the similarity solution in its vicinity, which determines the eddy structure illustrated in Figure 3.10. The O–type eddy stagnation point is located inside the equatorial Ekman layer. Stewartson's asymptotic analysis of the various boundary layers reveals a most exotic collection of powers of E.

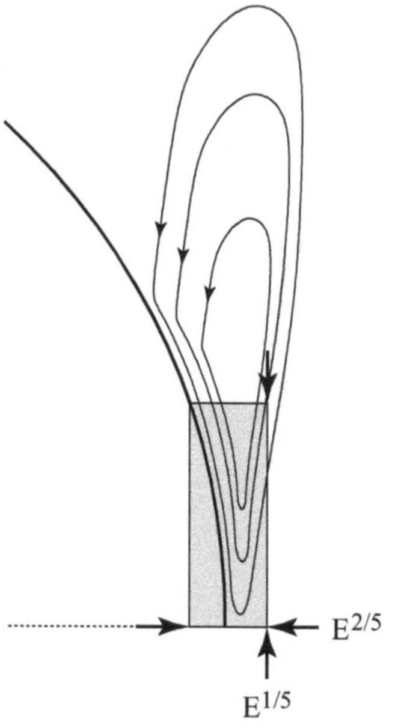

Figure 3.10 - The meridional streamlines χ = constant of the similarity solution stemming from the smaller $E^{2/5} \times E^{1/5}$–equatorial Ekman layer.

3.2. BOUNDARY AND SHEAR LAYERS IN ROTATING MHD FLOWS

In this section we investigate the role of magnetic fields $\mathbf{B}$ in boundary layer theory. The important new ingredient is the presence of the Lorentz force $\mathbf{j} \times \mathbf{B}$, where $\mu\mathbf{j} = \boldsymbol{\nabla} \times \mathbf{B}$ and μ is the magnetic permeability (as derived in Section 1.1.1). As in the case of non-magnetic boundary layer theory a key objective is to obtain jump conditions across the layer. So for example, the analogous result for electric current flow, which corresponds to the Ekman jump condition (3.2), is

$$\left[\mathbf{n}\cdot\mathbf{j}\right]_-^+ \equiv \mathbf{n}\cdot\mathbf{j}\big|_{z/\delta\to\infty} - \mathbf{n}\cdot\mathbf{j}\big|_{z=0} = \mathbf{n}\cdot\boldsymbol{\nabla}\times(\mathbf{J}\times\mathbf{n}) = -\boldsymbol{\nabla}_{\|}\cdot\mathbf{J}\,, \quad (3.43a)$$

where

$$\mathbf{J} = \int_0^{z/\delta\to\infty} (\mathbf{j}_{\|} - \mathbf{j}_{0\|})\,\mathrm{d}z \quad (3.43b)$$

in which $\mathbf{j}_0$ is the mainstream electric current and the subscript $\|$ denotes the component parallel to the boundary. The magnitude of the electrical conductivity of the boundary plays a crucial role in determining the strength of the boundary layer. Essentially insulating boundaries lead to the most intense layers as electric currents remain in the layers being unable to leak into the boundaries: $\mathbf{n}\cdot\mathbf{j}|_{z=0} = 0$. As the boundary conductivity increases, the boundary layer electric current $\mathbf{J}$ decreases and is almost completely absent in the case of perfectly conducting boundaries. All strong electric currents reside as surface currents in the solid conductor. We explore these ideas in the non-rotating context (see, for example, the books by Roberts, 1967a; Müller & Bühler, 2001) before considering rotating MHD flows (see, for example, Acheson & Hide, 1973).

3.2.1. THE HARTMANN LAYER

The nature of the Hartmann boundary layer is most readily appreciated in plane layer geometry. Relative to Cartesian coordinates (x, y, z) we consider fluid of constant density in the half-space $z > 0$. The fluid, which is electrically conducting with conductivity σ, permeability μ and resistivity $\eta = 1/\sigma\mu$, is permeated by a uniform transverse applied magnetic field $(0, 0, B_0)$. Away from the boundary the fluid is in uniform steady motion $\mathbf{u}_0 = (u_0, 0, 0)$, which generally perturbs the magnetic field leading to the form $\mathbf{B}_0 = (b_0(z), 0, B_0)$. In the presence of the pressure gradient $\boldsymbol{\nabla} p_0 = (-G, 0, \partial_z p_0)$, the uniform component $-G$ applied in the x–direction must be balanced by the Lorentz force $\mathbf{j}_0 \times \mathbf{B}_0$, in which $\mathbf{j}_0 = (0, j_0, 0)$ is a uniform electric current satisfying $j_0 B_0 = -G$. The z–component of the force balance shows that the total pressure $p_{T0} \equiv p_0 + p_{m0}$ is independent of z, where $p_{m0} = |\mathbf{B}_0(z)|^2/2\mu$ is the magnetic pressure. According to Ampère's law (1.3) $\mu\mathbf{j}_0 = \boldsymbol{\nabla} \times \mathbf{B}_0$, we have $\mathrm{d}b_0/\mathrm{d}z = \mu j_0$ and so $b_0(z) = b_0(0) + \mu j_0 z$. The realised magnitude of the flow is related to the uniform electric field $\mathbf{E}_0 = (0, E_0, 0)$. According to Ohm's Law (1.7) $\mathbf{E}_0 = -\mathbf{u}_0 \times \mathbf{B}_0 + \mathbf{j}_0/\sigma$ and this determines $E_0 = u_0 B_0 + j_0/\sigma$.

To consider the Hartmann layer we set

$$\mathbf{u} = (u(z),\, 0,\, 0)\,, \qquad \mathbf{b} = (b(z),\, 0,\, B_0)\,, \tag{3.44a,b}$$

$$\mathbf{j} = (0,\, j(z),\, 0)\,, \qquad \mathbf{E} = (0,\, E_0,\, 0)\,. \tag{3.44c,d}$$

Then the x–component of the equation of motion, the y–components of Ampère's and Ohm's laws yield

$$0 = G + B_0 j + \rho\nu\,\frac{\mathrm{d}^2 u}{\mathrm{d}z^2}\,, \qquad j = \frac{1}{\mu}\frac{\mathrm{d}b}{\mathrm{d}z} = \sigma\,(E_0 - B_0 u) \tag{3.45a,b}$$

respectively. These combine to give the single equation

$$\delta_H^2\,\frac{\mathrm{d}^2 u}{\mathrm{d}z^2} - (u - u_0) = 0\,, \qquad \text{where} \qquad \delta_H = \frac{\sqrt{\rho\nu\mu\eta}}{|B_0|} \tag{3.46a,b}$$

$$\text{and} \quad u_0 = \frac{E_0}{B_0} + \delta_H^2\,\frac{G}{\rho\nu}\,, \qquad j_0 = \sigma(E_0 - B_0 u_0)\,\left(= -\frac{G}{B_0}\right). \tag{3.46c,d}$$

Here $u = u_0$ is the mainstream velocity outside the Hartmann layer, which significantly is independent of z by analogy with the Proudman–Taylor theorem for rotating fluids (see Section 3.1.1).

If the boundary is at rest with $u(0) = 0$, the Hartmann boundary layer solution satisfying $u \to u_0$ as $z/\delta_H \uparrow \infty$ is

$$u - u_0 = -u_0 \exp(-z/\delta_H)\,, \tag{3.47a}$$

which identifies δ_H as the Hartmann layer width. It determines the magnetic field

$$b - b_0(z) = -(u_0 B_0/\eta)\delta_H \exp(-z/\delta_H) = (\delta_H B_0/\eta)\,(u - u_0)\,. \tag{3.47b}$$

It shows that the magnetic field and velocity jumps across the boundary layer are related by

$$b_0(0) - b(0) = \frac{\delta_H B_0}{\eta}\, u_0 = (\mathrm{sgn}\, B_0)\, \mathrm{Pm}^{1/2} \sqrt{\rho\mu}\, u_0\,, \tag{3.48a}$$

where $\mathrm{Pm} \equiv \nu/\eta$ is the magnetic Prandtl number.

Since the magnetic field jump (3.48a) depends on Pm, its value is sensitive to both ν and η even in the perfect fluid limit $\nu \downarrow 0$ and $\eta \downarrow 0$. Furthermore we also can determine the electric current deficit $J = \int_0^\infty (j - j_0)\mathrm{d}z$ in terms of the volume flux deficit $Q = \int_0^\infty (u - u_0)\mathrm{d}z$ from (3.48a). It is

$$J = -\sigma B_0 Q\,, \quad \text{where} \quad J = [b_0(0) - b(0)]/\mu\,, \quad Q = -\delta_H u_0\,. \tag{3.48b,c,d}$$

Other than the nature of the boundary layer structure, the most striking feature of the solution (3.47) is the dependence of the mainstream velocity u_0 on the magnitude of the electric field E_0 [see (3.46c)]. Its value is sensitive to the electromagnetic boundary conditions. If, for example, the boundary is a perfect conductor then $E_0 = 0$ and whence $u_0 = \delta_H^2 G/\rho\nu$. Note also that (3.45b), used in conjunction with the no-slip boundary condition $u(0) = 0$, yields the perfect conductor boundary condition $\mathrm{d}b/\mathrm{d}z(0) = 0$. If on the other hand the boundary is an insulator, no electric current can escape into the region $z < 0$ and we must consider carefully how the currents return elsewhere. Indeed in the well known case of channel flow (Müller & Bühler, 2001), where there is a second stationary insulating boundary at $z = 2H\ (\gg \delta_H)$, we may form the symmetric ($u(2H - z) = u(z)$, $b(2H - z) = -b(z)$) composite solution

$$u \approx u_0 \{1 - [\exp(-z/\delta_H) + \exp((z - 2H)/\delta_H)]\}\,, \tag{3.49a}$$

$$b \approx \mu j_0 (z - H) - (u_0 B_0/\eta)\delta_H [\exp(-z/\delta_H) - \exp((z - 2H)/\delta_H)]\,. \tag{3.49b}$$

The requirement that no net electric current leaves the system ($\int_0^{2H} j\mathrm{d}z = 0$) is effected by the insulating boundary conditions $b(0) = b(2H) = 0$, which is met when

$$j_0 \approx -\sigma u_0 B_0/\mathrm{M}\,, \quad \text{where} \quad \mathrm{M} = H/\delta_H = H\,|B_0|\,/\sqrt{\rho\nu\mu\eta} \tag{3.50a,b}$$

is the Hartmann number. Accordingly the electric current j_0 in Ohm's law (3.46d) is small implying $E_0 \approx u_0 B_0$. Furthermore the dynamical property $j_0 = -G/B_0$, used in conjunction with (3.50a), determines the mainstream velocity $u_0 \approx \mathrm{M}\delta_H^2 G/\rho\nu$. Significantly this is a factor M larger than the value $u_0 \approx \delta_H^2 G/\rho\nu$ for the case of perfectly conducting boundaries (see (3.46c) with $E_0 = 0$ and also Roberts, 1967a).

This sensitivity of the magnitude of the flow to the boundary conditions is remarkable. Essentially strong boundary layers are possible at insulating boundaries because the electric currents required to support the Hartmann layer cannot leak into the boundary, as they can at conducting boundaries. For given G the flow rate decreases as the conductivity of the boundary increases.

Another pair of examples that illustrates the dependence on boundary conductivity is the following. Firstly, consider the motion caused by moving insulating boundaries at $z = 0$ and $2H$ with velocities $u = 0$ and $2U_0$ respectively, but without a pressure gradient, $G = 0$. For this case we have the anti-symmetric ($u(2H - z) - U_0 = -[u(z) - U_0]$, $b(2H - z) = b(z)$) composite solution

$$u - U_0 \approx -U_0 \left[\exp(-z/\delta_H) - \exp((z - 2H)/\delta_H)\right], \tag{3.51a}$$

$$b \approx (U_0 B_0/\eta)\, \delta_H \left\{1 - \left[\exp(-z/\delta_H) + \exp((z - 2H)/\delta_H)\right]\right\}, \tag{3.51b}$$

in which we note that $j_0 = 0$, $E_0 = u_0 B_0$ and $u_0 = U_0$. Secondly, we consider instead the asymmetric configuration, in which the stationary boundary at $z = 0$ is perfectly conducting, while the moving boundary at $z = 2H$ is an insulator. This problem has the composite solution

$$u \approx 2U_0 \exp((z - 2H)/\delta_H), \tag{3.52a}$$

$$b \approx 2(U_0 B_0/\eta)\delta_H \left[1 - \exp((z - 2H)/\delta_H)\right], \tag{3.52b}$$

which satisfies, up to the neglect of exponentially small terms, the magnetic boundary conditions $\mathrm{d}b/\mathrm{d}z(0) = b(2H) = 0$; we note that $E_0 = j_0 = u_0 = 0$. This example shows the dominating influence of the conducting boundary, which essentially controls the electric field. In turn its effect is to lock the mainstream flow onto the conducting boundary.

3.2.2. DIFFERENTIALLY ROTATING SPHERES

The example cited at the end of the previous section relates to the geophysical and planetary configurations of a shell of electrically conducting fluid confined between an electrically conducting inner solid core and an outer insulating mantle. The x–directed flow in our example is analogous to an azimuthal flow forced by differentially rotating the inner and outer boundaries. Planar extensions in the rapidly

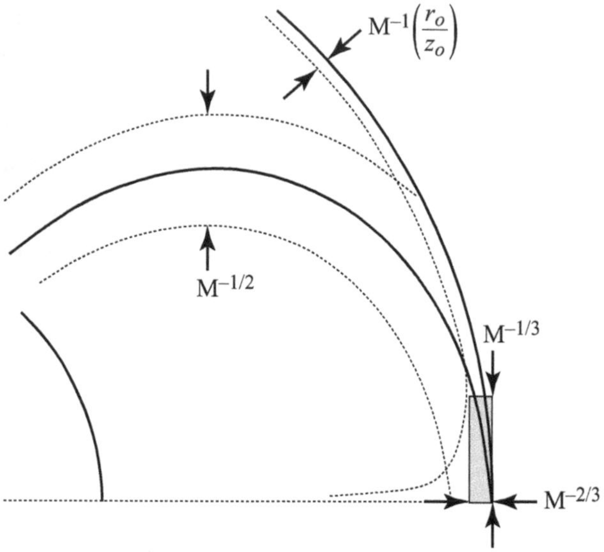

Figure 3.11 - The dipole magnetic field permeating the spherical shell. The magnetic field line touching the outer sphere $r = r_o$ is the location of the super-rotating MHD-shear layer width $\mathcal{O}(L\mathrm{M}^{-1/2})$. The dimensions of the equatorial Hartmann layer are also indicated.

rotating limit that relate to the planetary problem are discussed in Sections 3.2.3 and 3.2.4 below.

Here we consider a shell model studied by Dormy *et al.* (1998) and Dormy *et al.* (2002) in the opposite slow rotation limit; though to simplify our outline development, we modify some of the details of their model. We suppose that the inner solid core radius $r = r_i$ is a stationary $\omega = 0$ perfect conductor, while the outer solid mantle of inner radius $r = r_o$ is an insulator rotating at angular velocity $\omega = \omega_o$. Fluid of finite electrical conductivity occupies the shell $r_i < r < r_o$. We suppose that the entire system is permeated by an axisymmetric magnetic dipole $\mathbf{B}_p$ whose source is at the centre of the solid conductor (see Figure 3.11). Relative to the cylindrical polar coordinates of Section 3.1.4, it is

$$\mathbf{B}_P = s^{-1}\boldsymbol{\nabla}A \times \mathbf{e}_\phi = L^3 B_0 \left(\frac{3sz}{2r^5}\,,\ 0\,,\ \frac{2z^2 - s^2}{2r^5}\right), \qquad (3.53a)$$

where $L = r_o - r_i$ (say) and B_0 are typical length and magnetic field strengths,

$$A = \tfrac{1}{2} L^3 B_0 r^{-3} s^2 \qquad \text{and} \qquad \mathbf{e}_\phi = (0\,,1\,,0) \qquad (3.53b,c)$$

is the unit vector in the azimuthal direction.

For slow motion we may linearise the governing equations and, for the flow velocity, magnetic field and electric current, write

$$\mathbf{u} = s\omega\,\mathbf{e}_\phi\,, \qquad \mathbf{b} = \mathbf{B}_P + b\,\mathbf{e}_\phi\,, \qquad \mathbf{j} = (\mu s)^{-1}\,\boldsymbol{\nabla}(sb) \times \mathbf{e}_\phi\,, \tag{3.54}$$

where $\omega = \omega(s,z)$, $b = b(s,z)$ and $\mathbf{j} = \mathbf{j}(s,z)$. Ohm's law determines the electric field

$$\mathbf{E} = -\mathbf{u} \times \mathbf{B}_P + \sigma^{-1}\mathbf{j} = -\omega\boldsymbol{\nabla}A + \eta\, s^{-1}\,\boldsymbol{\nabla}(sb) \times \mathbf{e}_\phi\,. \tag{3.55a}$$

Using $\boldsymbol{\nabla} \times \mathbf{E} = \mathbf{0}$, we obtain the steady magnetic induction equation

$$\eta\left(\Delta - s^{-2}\right)b + s\mathbf{B}_P \cdot \boldsymbol{\nabla}\omega = 0\,. \tag{3.55b}$$

Since the Lorentz force is $\mathbf{j} \times \mathbf{B}_p = [(\mu s)^{-1}\mathbf{B}_P \cdot \boldsymbol{\nabla}(sb)]\mathbf{e}_\phi$ and there is no azimuthal pressure gradient, the ϕ–component of the equation of motion becomes

$$\nu\left(\Delta - s^{-2}\right)(s\omega) + (\rho\mu s)^{-1}\,\mathbf{B}_P \cdot \boldsymbol{\nabla}(sb) = 0\,. \tag{3.56}$$

For the coupled system of equations (3.55b), (3.56), the natural definition of the Hartmann number is

$$\mathrm{M} = L/\delta_H\,, \qquad \text{where as usual} \quad \delta_H = \sqrt{\rho\nu\mu\eta}/|B_0| \tag{3.57a,b}$$

[*cf.* (3.50b)].

Our system is very close to the situation described by the solution (3.52). There is no electric field in the central solid conductor and so in the large Hartmann number limit, the flow is virtually locked to the conductor. According to the magnetic induction equation (3.55b) we have that $\mathbf{B}_P \cdot \boldsymbol{\nabla}\omega \approx 0$ in the mainstream. Hence, since $\omega \approx 0$ at the edge of the inner boundary Hartmann layer, it remains so throughout the mainstream. In consequence Hartmann layer on the outer boundary must support the magnetic field jump $-(\mathrm{sgn}\,z_o)s_o(\omega_o B_0/\eta)\delta_H \equiv -[\mathrm{sgn}\,(z_o B_0)](\nu/\eta)^{1/2}\sqrt{\rho\mu}\,s_o\omega_o$ (see (3.68) below) between the insulator and the mainstream. Now in the mainstream the Lorentz force in (3.56) vanishes ($\mathbf{B}_P \cdot \boldsymbol{\nabla}(sb) \approx 0$) and so sb is almost constant on field lines. Since $b = 0$ in the insulator, the magnetic field jump condition shows that sb is proportional to s_o^2 as $r \uparrow r_o$ from the mainstream. Since A is also proportional to s_o^2 on $r = r_o$, it follows that sb is proportional to A everywhere in the mainstream. Consequently, b takes the value

$$b \approx (\mathrm{sgn}\,z)\,\frac{r_o^3\omega_o B_0\delta_H}{\eta}\,\frac{s}{r^3} \quad \text{with} \quad \mathbf{j} \approx (\mathrm{sgn}\,z)\;2\sigma\,\frac{r_o^3\delta_H}{L^3}\,\omega_o\,\mathbf{B}_p \tag{3.58a,b}$$

at all points on field lines that pass through the outer sphere. (Since $\delta_H|B_0| = \sqrt{\rho\nu\mu\eta}$, these results do not depend on the magnitude of B_0 but only its sign.) This region is bounded by the field line $r_o s^2 = r^3$, which is tangent to the outer sphere at the equator. On all field lines on the other side of it which close within the fluid, the symmetry conditions on the equator show that there $b \approx 0$ and $\mathbf{j} \approx \mathbf{0}$.

We consider carefully the electric current flow in the system. The current flow (3.58b) enters the outer Hartmann layer, where it then flows consistent with (3.43) to the equator. It is returned along the tangent line $r_o s^2 = r^3$ as a current sheet to the perfectly conducting inner sphere. There is becomes a surface current, which flows towards the poles but simultaneously feeds the mainstream current flow (3.58b).

The current sheet on the tangent line has finite width $\mathcal{O}(L\mathrm{M}^{-1/2})$. Across it the magnetic field must jump from zero to the value determined by (3.58a). Dormy *et al.* (2002) studied this layer in detail and showed that the ensuing Lorentz force made the fluid move $\omega \neq 0$ in this shear layer and in parts at angular velocities in excess of the outer sphere. This interesting result provides a powerful illustration of the fact that Lenz's law, which states that the Lorentz force opposes motion, is only true in some mean sense but does not necessarily apply everwhere.

Finally we comment on the nature of the Hartmann layer on the outer boundary, whose unit normal directed into the fluid is $\mathbf{n} = -\mathbf{r}/r$. The planar analysis of Section 3.2.1 continues to apply provided that only the normal component of the applied magnetic field is employed in the boundary layer calculation. Thus the Hartmann layer thickness (3.46b) is replaced by

$$\delta_{H\perp} = \sqrt{\rho\nu\mu\eta}\,/|\mathbf{n}\cdot\mathbf{B}_P|_{r=r_o}\,, \tag{3.59}$$

which, in view of the definition (3.53a) of $\mathbf{B}_P$, takes the value $\delta_{H\perp} = (r_o/z_o)\delta_{He}$, where $\delta_{He} = (r_o/L)^3\delta_H$ is the Hartmann layer width based on $(L/r_o)^3 B_0$, which is twice the magnitude of the equatorial magnetic field. Thus the Hartmann boundary layer thickness $\delta_{H\perp}$ is proportional to r_o/z_o. It thickens as the equator is approached becoming singular on the equator $z_o = 0$. It is this singularity which triggers the tangent field line shear layer, just as the equatorial Ekman layer does in the non-magnetic Proudman–Stewartson problem of Section 3.1.4. Indeed Dormy *et al.* (2002), following Roberts (1967b), determine the solution in the ensuing equatorial Hartmann layer of width $\mathcal{O}((r_o\delta_{He}^2)^{1/3})$ and latitudinal extent $\mathcal{O}((r_o^2\delta_{He})^{1/3})$.

3.2.3. THE EKMAN–HARTMANN LAYER

We now extend the analysis of Section 3.2.1 to include the effect of vertical rotation $\mathbf{\Omega} = \Omega\mathbf{n} = (0, 0, \Omega)$ and to investigate the Ekman–Hartmann boundary layer that forms above a stationary plane $z = 0$ (Gilman & Benton, 1968). In the mainstream, the pressure gradient $\nabla p_0 = (-G_x, -G_y, \partial_z p_0)$ has constant x and y–components and a steady electric current $\mathbf{j}_0 = (j_{x0}, j_{y0}, 0)$ drives the uniform steady magnetogeostrophic flow $\mathbf{u}_0 = (u_0, v_0, 0)$. The realised steady magnetic field has the form

$\mathbf{B}_0 = (b_{x0}(z), b_{y0}(z), B_0)$, while the electric field is $\mathbf{E}_0 = (E_{x0}, E_{y0}, -u_0 b_{y0} + v_0 b_{x0})$ has constant x and y–components too ($\boldsymbol{\nabla} \times \mathbf{E}_0 = \mathbf{0}$). The equation of motion, Ampère's and Ohm's law give the mainstream balance

$$2\rho\Omega\mathbf{n} \times \mathbf{u}_0 = \mathbf{G} + B_0\mathbf{j}_0 \times \mathbf{n}\,, \qquad \mathbf{j}_0 \times \mathbf{n} = \frac{1}{\mu}\frac{\mathrm{d}\mathbf{b}_0}{\mathrm{d}z} = \sigma(\mathbf{E}_0 \times \mathbf{n} - B_0\mathbf{u}_0)\,, \tag{3.60a}$$

where $\mathbf{G} = (G_x, G_y, 0)$ and $\mathbf{b}_0 = (b_{x0}(z), b_{y0}(z), 0)$. Accordingly, $\mathbf{u}_0$ is related to $\mathbf{G}$ and $\mathbf{E}_0$ by

$$2\,(\mathrm{sgn}\,\Omega)\,\mathbf{n} \times \mathbf{u}_0 + \Lambda\mathbf{u}_0 = \frac{\mathbf{G}}{\rho|\Omega|} + \Lambda\frac{\mathbf{E}_0 \times \mathbf{n}}{B_0}\,, \tag{3.60b}$$

where
$$\Lambda = \frac{\sigma B_0^2}{\rho|\Omega|} = \frac{\delta_E^2}{\delta_H^2} \tag{3.60c}$$

is the Elsasser number. The limits $\Lambda = 0$ and $1/\Lambda = 0$ recover the earlier relations (3.5b,d) and (3.46c) respectively.

To resolve the Ekman–Hartmann layer we set

$$\mathbf{u} = (u(z),\, v(z),\, 0)\,, \quad \mathbf{B} = (b_x(z),\, b_y(z),\, B_0)\,, \quad \mathbf{j} = (j_x(z),\, j_y(z),\, 0)\,. \tag{3.61}$$

Then the (x, y)–components of the equation of motion, Ampère's and Ohm's laws yield

$$2\,(\mathrm{sgn}\,\Omega)\,\mathbf{n} \times (\mathbf{u} - \mathbf{u}_0) = -\,\Lambda\,(\mathbf{u} - \mathbf{u}_0) + \delta_E^2\,\frac{\mathrm{d}^2\mathbf{u}}{\mathrm{d}z^2}\,, \tag{3.62a}$$

$$\sigma^{-1}(\mathbf{j} - \mathbf{j}_0) \times \mathbf{n} = \eta\,\frac{\mathrm{d}}{\mathrm{d}z}(\mathbf{b} - \mathbf{b}_0) = -\,B_0\,(\mathbf{u} - \mathbf{u}_0)\,. \tag{3.62b}$$

From (3.62b) we can determine the electric current deficit $\mathbf{J} = \int_0^\infty(\mathbf{j} - \mathbf{j}_0)\mathrm{d}z$ in terms of the volume flux deficit $\mathbf{Q} = \int_0^\infty(\mathbf{u} - \mathbf{u}_0)\mathrm{d}z$. It is

$$\mu\mathbf{J} \times \mathbf{n} = \mathbf{b}_0(0) - \mathbf{b}(0) = -\,B_0\mathbf{Q}/\eta\,. \tag{3.63}$$

To determine the solution of (3.62a), we proceed as in the Ekman layer case and write $Z \equiv u + \mathrm{i}v$, $Z_0 \equiv u_0 + \mathrm{i}v_0$. Then the corresponding extension of (3.6) is

$$\delta_E^2\,\frac{\mathrm{d}^2 Z}{\mathrm{d}z^2} - (\Lambda + 2\,(\mathrm{sgn}\,\Omega)\,\mathrm{i}\,)\,(Z - Z_0) = 0\,. \tag{3.64}$$

The solution which satisfies $\mathbf{u} = 0$ on $z = 0$, and decays to zero as z tends to infinity is

$$Z = Z_0\,\{1 - \exp[-(1 + \mathrm{i}\,(\mathrm{sgn}\,\Omega)\,\tan\Upsilon)z/\delta_{EH}]\} \quad (0 \le \Upsilon \le \pi/4)\,, \tag{3.65a}$$

where $\delta_E^2/\delta_{EH}^2 = \cot\Upsilon = \tfrac{1}{2}\left(\Lambda + \sqrt{\Lambda^2 + 4}\right)\,, \quad 2\cot 2\Upsilon = \Lambda\,.$ (3.65b,c)

The solution (3.65a) may be used to calculate the volume flux deficit

$$\begin{aligned} Q_x + \mathrm{i}Q_y &= -\tfrac{1}{2}\,\delta_{EH} Z_0\,\{1 + \exp[-2\mathrm{i}\,(\operatorname{sgn}\Omega)\,\Upsilon]\} \\ &= [-\,\delta_Q^{\parallel} + \mathrm{i}\,(\operatorname{sgn}\Omega)\,\delta_Q^{\perp}]\,Z_0\,. \end{aligned} \tag{3.66a}$$

where $$\delta_Q^{\parallel} = \frac{\delta_E^2}{\delta_{EH}\sqrt{\Lambda^2+4}}\,, \qquad \delta_Q^{\perp} = \frac{\delta_{EH}}{\sqrt{\Lambda^2+4}}\,. \tag{3.66b,c}$$

These two length scales have the limiting forms

$$\delta_Q^{\parallel} \approx \begin{cases} \delta_H \Lambda^{1/2}/2 & \text{for } \Lambda \ll 1; \\ \delta_H & \text{for } \Lambda \gg 1\,; \end{cases} \qquad \delta_Q^{\perp} \approx \begin{cases} \delta_E/2; \\ \delta_E/\Lambda^{3/2}. \end{cases} \tag{3.66d,e}$$

Expressed in terms of real variables, (3.66a) gives

$$\mathbf{Q} = -\,\delta_Q^{\parallel}\,\mathbf{u}_0 + (\operatorname{sgn}\Omega)\,\delta_Q^{\perp}\,\mathbf{n}\times\mathbf{u}_0 \tag{3.67a}$$

and for the current flux deficit (3.63) gives

$$\mathbf{J} = \mu^{-1}\mathbf{n}\times[\mathbf{b}_0(0)-\mathbf{b}(0)] = \sigma B_0\left[\delta_Q^{\parallel}\,\mathbf{n}\times\mathbf{u}_0 + (\operatorname{sgn}\Omega)\,\delta_Q^{\perp}\,\mathbf{u}_0\right]. \tag{3.67b}$$

Note that the possibilities that Ω and B_0 are negative is accommodated by our formulation. In addition, (3.65a) shows that we continue to have an Ekman type spiral, while (3.66a) demonstrates that there is mass flux in the boundary layer transverse to the mainstream geostrophic flow. Importantly in the non-rotating limit $\Lambda \uparrow \infty$, (3.67b) generalises (3.48a) to give the tangential magnetic field jump

$$\mathbf{b}_0(0) - \mathbf{b}(0) = (\delta_H B_0/\eta)\,\mathbf{u}_0 = (\operatorname{sgn} B_0)\ \mathrm{Pm}^{1/2}\sqrt{\rho\mu}\,\mathbf{u}_0 \tag{3.68}$$

across the Hartmann layer. Note the dependence of the magnetic field jump on the magnetic Prandtl number Pm (3.48b), which should be compared with the Alfvén velocity scaling in (3.117b) below.

Evidently the complete solution of any particular problem is quite complicated and involves the considerations appropriate to both Ekman and Hartmann flows. Thus, whereas we were free to specify the pressure gradient $-\mathbf{G}$ in the Hartmann case, its value is part of the solution in the Ekman case. In general, for Ekman–Hartmann case, we are not free to specify $-\mathbf{G}$ in its entirety. Of course, we have to take due care with the electrical boundary conditions as well. Interestingly when the mainstream flow $\mathbf{u}_0$ varies on length scales large compared to δ_{EH}, while the applied magnetic field B_0 is held constant, the obvious generalisations of (3.14), which determine the Ekman-Hartmann pumping velocity and the normal electric current density jumps from (3.2) and (3.43), are

$$\mathbf{n}\cdot\mathbf{u} = \delta_Q^{\perp}(\operatorname{sgn}\Omega)\,\mathbf{n}\cdot\boldsymbol{\nabla}\times\mathbf{u}_0\,, \qquad [\mathbf{n}\cdot\mathbf{j}]_-^+ = \sigma B_0\,\delta_Q^{\parallel}\,\mathbf{n}\cdot\boldsymbol{\nabla}\times\mathbf{u}_0 \tag{3.69a,b}$$

(see Acheson & Hide, 1973). As Λ increases, the suction decreases and vanishes in the non-rotating limit, while as Λ decreases, the normal electric current jump decreases and vanishes in the non-magnetic limit.

When $\mathbf{\Omega}$ and $\mathbf{B}_0$ are not parallel to $\mathbf{n}$, the boundary layer calculation of this section still holds but with the B_0 and Ω replaced by the normal components $B_{0\perp} = \mathbf{n} \cdot \mathbf{B}_0$ and $\Omega_\perp = \mathbf{n} \cdot \mathbf{\Omega}$. Accordingly δ_E becomes $\delta_{E\perp}$ [see (3.13b)], δ_H becomes $\delta_{H\perp}$ [see (3.59)] and in turn, Λ becomes $\Lambda_\perp = \delta_{E\perp}^2/\delta_{H\perp}^2$ [see (3.60c)]. With these values used in (3.65b), its solution δ_{EH} is replaced by $\delta_{EH\perp}$ and, in turn, the solutions $\delta_Q^{\|}$ and $\delta_Q^{\perp}$ of (3.66b,c) are replaced by $\delta_{Q\perp}^{\|}$ and $\delta_{Q\perp}^{\perp}$. Then all our formulae (3.67) to (3.69a,b) remain valid provided all the new $\perp$–values are employed. When $B_{0\perp}$ and $\Omega_\perp$ are functions of position, we must return to the primitive formulae (3.2), (3.43) and take proper account of their spatial derivatives; Loper (1970) investigated the case of a spherical boundary.

3.2.4. ROTATING MHD FREE SHEAR LAYERS; $\Lambda \ll 1$

Our main interest in this section is with the MHD extension of the Proudman-Stewartson problem of the flow driven by differentially rotating two concentric spheres, which we discussed in Section 3.1.4, with particular consideration of the free shear layers on the tangent cylinder. The MHD extension of Stewartson's (1957) axisymmetric split disc problem was investigated by Vempaty & Loper (1975, 1978). They considered the case of $\mathbf{B}_0$ parallel to the rotation axis $\mathbf{\Omega}$. Since the magnetic field has no component transverse to the tangent cylinder, the model fails to capture important features of the geophysical system. This aspect, however, has been addressed by Hollerbach (1994a) and Kleeorin *et al.* (1997), who considered an applied axisymmetric dipole field. Furthermore non-axisymmetric flows driven by an asymmetric forcing (as opposed to differentially rotating the inner and outer boundaries) have also been investigated by Hollerbach (1994b) and Soward & Hollerbach (2000).

The key ingredient in the spherical model is the radial magnetic field crossing the tangent cylinder. For that reason Hollerbach (1996) considered a plane layer geometry with a transverse magnetic field because that removes many complications associated with the spherical geometry.

In order to identify the main features of the shear layer structure found by Kleeorin *et al.* (1997), we take Hollerbach's (1996) plane layer geometry with $\mathbf{\Omega} = (0, 0, \Omega)$ and $\Omega > 0$. We outline how aspects of the ageostrophic and quasi-geostrophic analyses of Sections 3.1.2 and 3.1.3 generalise to the case of an applied transverse magnetic field $(B_0, 0, 0)$ in the small Elsasser number limit. We consider a duct geometry $-H < z < H$ and $-L < x < L$ but emphasise that we continue to base our Ekman number E (3.20b) and Hartmann number M (3.50b) on the length H. We will assume that the region exterior to the duct is split across the plane $x = 0$. The exterior region $x < 0$ is a stationary solid perfect conductor, while that in $x > 0$ is a solid insulator moving with slow velocity $(0, V_0, 0)$, in the sense that we may linearise the governing equations. There is no applied pressure gradient $\mathbf{G} = \mathbf{0}$ and so motion is driven entirely by the moving boundary.

From a general point of view, we may use the 2–dimensional representation (3.17) for the flow velocity introduced in Section 3.1.2 with the corresponding forms

$$\mathbf{B} - \mathbf{B}_0 = (-\partial_z a,\, b,\, \partial_x a)\,, \qquad \mathbf{E} = (-\partial_x \varphi,\, 0,\, -\partial_z \varphi) \qquad (3.70a,b)$$

for the perturbation magnetic and electric fields. Then the y–component of the linearised equations of motion and vorticity become

$$-2\,\partial_z \chi = \Lambda\,(\eta/B_0)\,\partial_x b + \delta_E^2\,\Delta v\,, \qquad (3.71a)$$

$$-2\,\partial_z v = -\Lambda\,(\eta/B_0)\,\partial_x \Delta a - \delta_E^2\,\Delta\,\Delta\chi\,. \qquad (3.71b)$$

The x and z–components of the linearised Ohm's law give

$$-\partial_x \varphi = -\eta\,\partial_z b\,, \qquad -\partial_z \varphi - B_0 v = -\eta\,\partial_x b\,. \qquad (3.72a,b)$$

Ellimination of the electric potential φ in (3.72a,b) and the y–component of Ohm's law give

$$0 = B_0\,\partial_x v + \eta\,\Delta b \qquad \text{and} \qquad 0 = B_0\,\partial_x \chi + \eta\,\Delta a \qquad (3.73a,b)$$

respectively.

When $\Lambda \ll 1$ the steady mainstream velocity away from any boundary or shear layers is quasi-geostrophic dependent only on one coordinate, namely x. Under this 1–dimensional assumption, the quasi-geostrophic flow is $\mathbf{u}_G = (0, v_G(x), 0)$. The induced magnetic field b however is 2–dimensional and according to (3.73a) satisfies the Poisson equation

$$\Delta b = -\frac{B_0}{\eta}\frac{\mathrm{d}v_G}{\mathrm{d}x} \qquad (3.74)$$

subject to the appropriate boundary condititions on the duct boundary. Use of the Green's function solution for b, enables an intergro-dfferential equation to be obtained for v_G as Kleeorin *et al.* (1997) explain..

Since the boundary in the region $x < 0$ is a perfect conductor, the electric field in this region is virtually zero ($\varphi \approx 0$) and there is no fluid motion ($v_G \approx 0$). So in the vertical shear layer at $x = 0$, we may make the usual boundary layer approximation that $\partial/\partial x \gg \partial/\partial z$. Accordingly we neglect the vertical component of electric field $-\partial_z \varphi$ in (3.72b) to obtain

$$\eta\,\partial_x b = -B_0 v_G\,, \qquad (3.75a)$$

in which $\partial_x b$, like v_G, is a function of x alone. Substitution of this result into the y–component of the equation of motion (3.71a) yields

$$-\frac{2}{\delta_E^2}\,\partial_z\chi = -\frac{1}{\delta_H^2}\,v_G + \frac{\mathrm{d}^2 v_G}{\mathrm{d}x^2} \qquad (\Omega > 0)\,, \tag{3.75b}$$

which shows that the x–component of velocity $u = -\partial_z\chi$ is also only a function of x. In fact the symmetry of our system about the mid–plane indicates that $\chi(x,z) = -zu(x)$. Evidently (3.75b) is simply the Hartmann balance (3.46a) with the Coriolis acceleration added.

We now assume that there are Ekman layers on the top and bottom boundaries $z = \pm H$ of the shear layer. According to (3.9b), the x–directed volume flux in the bottom boundary $z = -H$ is

$$Q = \begin{cases} -\tfrac{1}{2}\,\delta_E\, v_G & \text{for } x < 0\,; \\ -\tfrac{1}{2}\,\delta_E\,(v_G - V_0) & \text{for } x > 0\,. \end{cases} \tag{3.76}$$

A similar flux is found on the top boundary. The condition that there is no total x–directed mass flux is that $Q + Hu = 0$ or equivalently $\chi(x, \pm H) = \pm Q(x)$. Used in conjunction with (3.75), we obtain the MHD extension of (3.31), namely

$$\frac{\mathrm{d}^2 v_G}{\mathrm{d}x^2} - \frac{v_G}{\Delta_M^2} = \begin{cases} 0 & \text{for } x < 0\,; \\ -\dfrac{V_0}{\Delta^2} & \text{for } x > 0\,, \end{cases} \tag{3.77a}$$

where

$$\frac{1}{\Delta_M^2} = \frac{1}{\Delta^2} + \frac{1}{\delta_H^2} = \frac{1}{\mathrm{E}^{1/2}H^2}\left(1 + \frac{\Lambda}{\mathrm{E}^{1/2}}\right) \tag{3.77b}$$

and $\Delta = \sqrt{\delta_E H}$ is the $\mathrm{E}^{1/4}$–Stewartson layer thickness (3.32b). The solution of (3.77a) analogous to (3.33) is

$$v_G \approx \begin{cases} \tfrac{1}{2}\,V_M\,\exp(x/\Delta_M) & \text{for } x < 0\,; \\ V_M\left[1 - \tfrac{1}{2}\,\exp(-x/\Delta_M)\right] & \text{for } x > 0\,, \end{cases} \tag{3.78a}$$

where

$$V_M = V_0 \Big/ \left(1 + \Lambda \mathrm{E}^{-1/2}\right)\,. \tag{3.78b}$$

It describes a free shear layer of width Δ_M. The corresponding ageostrophic velocity component $u = -Q/H$, namely

$$u = \begin{cases} \tfrac{1}{2}\,\mathrm{E}^{1/2}\,v_G & \text{for } x < 0\,; \\ \tfrac{1}{2}\,\mathrm{E}^{1/2}\,(v_G - V_0) & \text{for } x > 0\,. \end{cases} \tag{3.78c,d}$$

is readily obtained from (3.76). As $x/\Delta_M \downarrow -\infty$, we have $v_G \to 0$ and $u \to 0$ compatable with the absense of motion in the mainstream $x < 0$. As $x/\Delta_M \uparrow \infty$, we have $v_G \to V_M$ and $u \to -\Lambda V_M/2$ and they provide boundary conditions for the mainstream problem in $x > 0$. A further shear layer of width Δ_M exists at $x = L$, but its nature will depend on the magnitude of the vertical electric field $-\partial_z\varphi$ there.

For $\Lambda \ll \mathrm{E}^{1/2}$ the shear layer at $x = 0$ reduces to an $\mathrm{E}^{1/4}$–Stewartson layer. As the Elsasser number Λ increases, the free shear layer width Δ_M decreases in concert with the magnetogeostrophic velocity V_M. For $\Lambda \gg \mathrm{E}^{1/2}$ the layer becomes a weak Hartmann layer with $V_M \approx (\mathrm{E}^{1/2}/\Lambda)V_0\,(\ll V_0)$. As in the case of the $\mathrm{E}^{1/4}$–Stewartson layer, some important physical quantities derived from the solution (3.78a) suffer unacceptable discontinuities at $x = 0$ (e.g. $\mathrm{d}^2 v_G/\mathrm{d}x^2$). Provided that $\Lambda \ll \mathrm{E}^{1/3}$ these are resolved in ageostrophic $\mathrm{E}^{1/3}$–Stewartson sublayers (see Section 3.1.2). This restriction on Λ stems from the requirement that the quasi-geostrophic shear layer width Δ_M be large compared to $\mathrm{E}^{1/3}H$ (see Figure 3.12a). When $\Lambda = \mathcal{O}(\mathrm{E}^{1/3})$ the two sets of shear layers merge and for $\Lambda \gg \mathrm{E}^{1/3}$ the quasi-geostrophic layers are lost completely. Similar remarks apply to the boundary layer structures on the sidewall $x = L$, $-H \le z \le H$.

When $\Lambda \gg \mathrm{E}^{1/3}$ the ageostrophic apparatus of Section 3.1.2 may be adapted. Indeed the governing equations (3.18a,b) are simply supplemented by the Lorentz force and according to (3.71) and (3.73) become

$$-2\,\partial_z\chi = -\Lambda v + \delta_E^2\,\partial_{xx}v \qquad \text{and} \qquad -2\,\partial_z v = \Lambda\,\partial_{xx}\chi - \delta_E^2\,\partial_{xxxx}\chi\,. \tag{3.79a,b}$$

Just as before, the streamfunction χ is coupled to the velocity v by the Ekman suction boundary condition. The main point to appreciate is that the mainstream velocity V_M defined by (3.78b) is small: $V_M/V_0 = \mathrm{E}^{1/2}/\Lambda$ and that v just outside the top and bottom Ekman layers is small too for $|x| \gg \mathrm{E}^{1/2}$, as confirmed by (3.85) below. Accordingly the boundary condition (3.21) is replaced by

$$w = \pm\tfrac{1}{2}\,\delta_E V_0\,\delta(x) \qquad \text{on} \qquad z = \pm H\,. \tag{3.80}$$

The solution corresponding to (3.22) is

$$w = \frac{\delta_E V_0}{2\pi}\int_0^\infty \frac{\cos(kx)\,\sinh\left[\frac{1}{2}k(\Lambda + \delta_E^2k^2)z\right]}{\sinh\frac{1}{2}k(\Lambda+\delta_E^2k^2)H}\,\mathrm{d}k\,. \tag{3.81a}$$

By performing a binomial expansion of the denominator this may expressed in the alternative form

$$w = -\,\delta_E V_0\sum_{n=-\infty}^{\infty}\mathcal{W}(x, Z_n)\,, \qquad Z_n = z - (2n+1)H\,, \tag{3.81b}$$

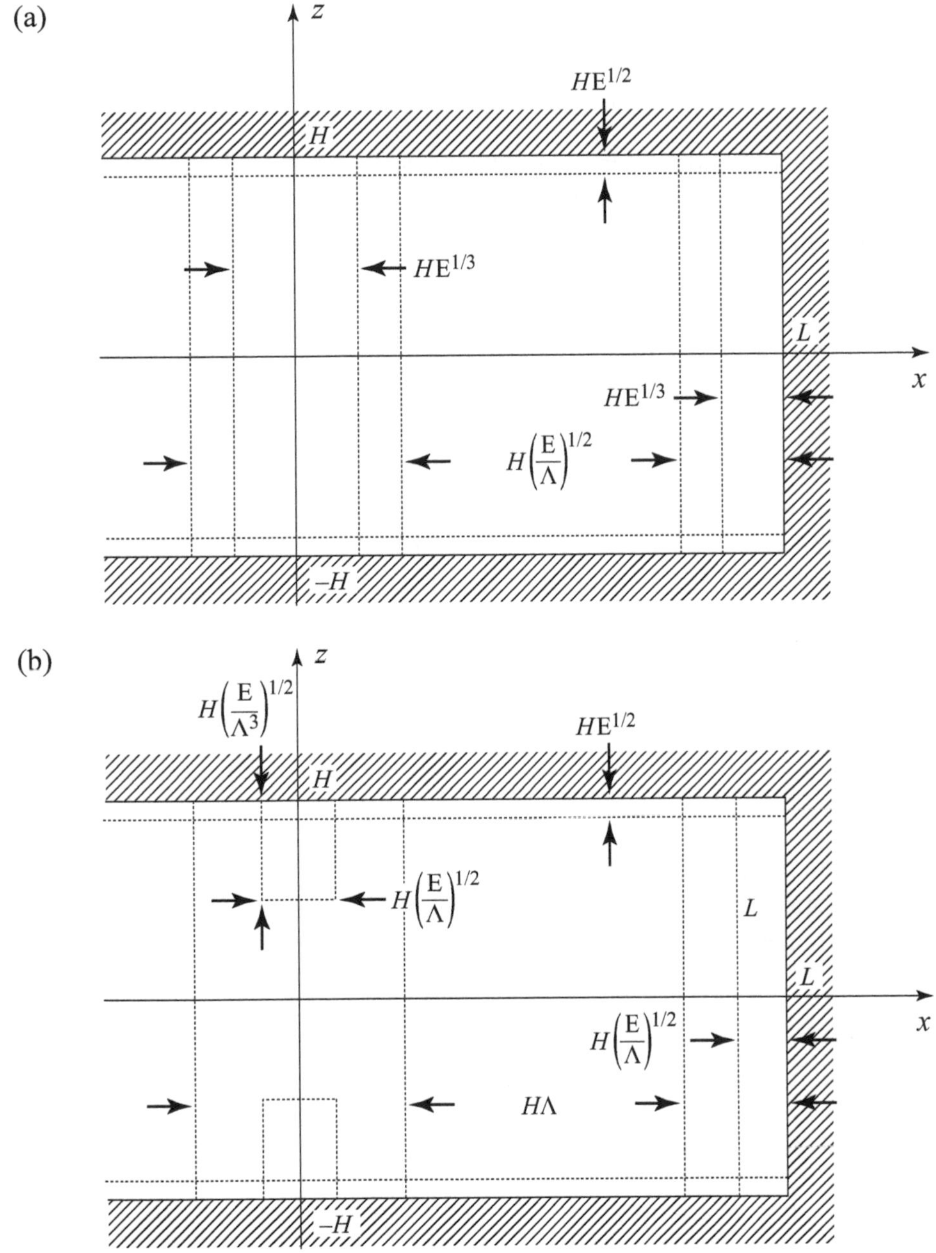

Figure 3.12 - The duct shear layer structure including the horizontal $H\mathrm{E}^{1/2}$–Ekman layers. The vertical layers are for: (a) the case $\mathrm{E}^{1/2} \ll \Lambda \ll \mathrm{E}^{1/3}$, the quasi-geostrophic $H(E/\Lambda)^{1/2}$–Hartmann layers and the ageostrophic $H\mathrm{E}^{1/3}$–Stewartson layers; (b) the case $\mathrm{E}^{1/3} \ll \Lambda \ll 1$, the $H\Lambda$–magnetogeostrophic layers, the hybrid $\mathrm{E}^{1/3}$–Hartmann layer stumps and the $H(E/\Lambda)^{1/2}$–Hartmann layer at the sidewall $x = L$.

where

$$\mathcal{W}(x,Z) \;=\; \frac{\operatorname{sgn} Z}{2\pi}\int_0^\infty \cos(kx)\,\exp\left[-\tfrac{1}{2}k(\Lambda+\delta_E^2k^2)|Z|\right]\mathrm{d}k\,. \tag{3.81c}$$

This constitutes the flow due to an image system composed of sinks strength $\delta_E V_0$ at $z=(2n+1)H$, all integer n.

Inside the flow domain $-H<z<H$, a scale analysis of (3.79) suggests the existence of a Hartmann layer width $x=\mathcal{O}(H\mathrm{E}^{1/2}/\Lambda^{1/2})=\mathcal{O}(\delta_H)$. However, the solution (3.81a) exposes this suggestion as over simplistic. Instead a hybrid $\mathrm{E}^{1/3}$ and Hartmann boundary layer of Hartmann layer width $\mathcal{O}(\delta_H)$ is restricted to stump like regions, aspect ratio Λ^{-1}, stemming from the sinks $x=0$, $z=\pm H$ of height $H\mp z=\mathcal{O}(H\mathrm{E}^{1/2}/\Lambda^{3/2})=\mathcal{O}(\delta_H/\Lambda)$, in which motion is dominated by $w\approx-\delta_E V_0\mathcal{W}(x,z\mp H)$ (see Figure 3.12b). We may distinguish two limiting cases. First, when $H\mp z\ll H\mathrm{E}^{1/2}/\Lambda^{3/2}$, the magnetic forces are unimportant [set $\Lambda=0$ in (3.81c)] and so, close to the sink at $z=-H$, we are left with the viscous similarity solution (3.25) modified by a factor $-1/2$. Second, when $H\mathrm{E}^{1/2}/\Lambda^{3/2}\ll H\mp z\ll H$, the viscous forces are unimportant. Then, on setting $\delta_E=0$, integration of (3.81c) yields

$$\mathcal{W}(x,z\mp H) \;=\; \frac{1}{2\pi}\,\frac{(\Lambda/2)(z\mp H)}{x^2\,+\,(\Lambda/2)^2(z\mp H)^2}\,. \tag{3.82}$$

Furthermore, in the magnetogeostrophic shear layer of width $\mathcal{O}(\Lambda H)$ exterior to the stump regions, we may neglect viscosity. Thus all $\mathcal{W}(x,Z_n)$ in the sum (3.81a) may be approximated by (3.82), so yielding

$$w \;=\; \partial_x\chi \;=\; -\,\frac{\delta_E V_0}{2\pi}\sum_{n=-\infty}^{n=\infty}\frac{(\Lambda/2)[z-(2n+1)H]}{x^2\,+\,(\Lambda/2)^2[z-(2n+1)H]^2}\,. \tag{3.83}$$

Indeed with $\delta_E=0$, the shear layer equations (3.79a,b) reduce to

$$2\,u \;=\; -\,2\,\partial_z\chi \;=\; -\,\Lambda\,v \qquad\text{and}\qquad -\,2\,\partial_z v \;=\; \Lambda\,\partial_{xx}\chi\,. \tag{3.84a,b}$$

This defines an elliptic problem $(\Lambda/2)^2\partial_{xx}v+\partial_{zz}v=0$. Use of (3.83) shows that the solution, which meets the boundary conditions $v\to\mathrm{E}^{1/2}V_0/\Lambda$, $u\to-\mathrm{E}^{1/2}V_0/2$ as $x/\Delta_M\uparrow\infty$ and $v\to0$, $u\to0$ as $x/\Delta_M\downarrow-\infty$, is

$$\frac{\Lambda}{2}\,v \;=\; u \;=\; -\,\frac{\mathrm{E}^{1/2}V_0}{4}\left[1+\frac{1}{\pi}\sum_{n=-\infty}^{n=\infty}\frac{\Lambda H x}{x^2\,+\,(\Lambda/2)^2[z-(2n+1)H]^2}\right]. \tag{3.85}$$

Similar layers are to be expected on the boundary $x = L$. The only difference is that the Hartmann sub–layer covers the entire boundary $-H < z < H$ rather than being simply stumps as they are for the free shear layers at $x = 0$.

As Λ increases to unity, the Λ–magnetogeostrophic shear layers at $x = 0$ and L expand to fill the entire mainstream domain. Since the velocity is small $\mathcal{O}(\mathrm{E}^{1/2}V_0)$ [see (3.85)], the moving boundaries $z = \pm H$, $0 < x < L$ support strong Ekman layers which carry fluid flux between the sink–source combinations at $x = 0$ and L. These in turn drive the magnetogeostrophic flow, which according to (3.71) and (3.73) is governed by

$$-2\,\partial_z\chi = \Lambda\,(\eta/B_0)\,\partial_x b\,, \qquad -2\,\partial_z v = \Lambda\,\partial_{xx}\chi\,, \qquad (3.86\text{a,b})$$

while b itself satisfies (3.73a). The mainstream equations (3.86a,b) and (3.73a) must be solved in the region $x > 0$ subject to the vanishing on the boundaries of b and the normal velocity (χ =constant), except in the corners, where the sink–source combinations are located. The magnetic boundary conditions in $x < 0$ are of course those that correspond to perfect electrical conductors.

3.3. WAVES

Asymptotic methods can also be very useful in capturing time dependent processes such as waves. A large variety of waves can propagate in rotating and magnetic fluids. We shall restrict our attention here to waves propagating in incompressible fluids ($\boldsymbol{\nabla}\cdot\mathbf{u} = 0$).

3.3.1. INERTIAL WAVES

The fundamental wave motion in an inviscid, rotating, non-magnetic system is the inertial wave. To understand its basic properties, we consider coplanar motion $\mathbf{u}(z,t) = (u,\,v,\,0)$, relative to Cartesian co-ordinates $(x,\,y,\,z)$. This flow automatically satisfies the solenoidal condition $\boldsymbol{\nabla}\cdot\mathbf{u} = 0$. While the equation of motion reduces to

$$\partial_t\mathbf{u} + 2\boldsymbol{\Omega}\times\mathbf{u} = -\boldsymbol{\nabla}(p/\rho) \qquad [p = p(z,t)]\,. \qquad (3.87)$$

The x and y components of motion are simply

$$\partial_t u - 2\Omega_z v = 0\,, \qquad \partial_t v + 2\Omega_z u = 0\,, \qquad (3.88\text{a,b})$$

where $\boldsymbol{\Omega} = (\Omega_x,\,\Omega_y,\,\Omega_z)$. The equations can be combined to obtain

$$\partial_t(u + \mathrm{i}v) + 2\mathrm{i}\Omega_z(u + \mathrm{i}v) = 0 \qquad (3.88\text{c})$$

with solution

$$u + \mathrm{i}v = (u_0 + \mathrm{i}v_0)\exp(-\mathrm{i}\omega t) \qquad \omega = 2\Omega_z\,, \tag{3.89}$$

where $u_0 = u(z,0)$ and $v_0 = v(z,0)$. The pressure distribution $p(z,t)$ is determined from the z component of the equation of motion. Fluid particles move in circles on the planes z =constant with constant velocity and constant angular velocity $-\omega = -2\Omega_z$.

From a more specific point of view, we can consider the travelling wave

$$[\mathbf{u},\, \boldsymbol{\zeta},\, p] \;=\; \mathrm{Re}\Big\{[\widetilde{\mathbf{u}},\, \widetilde{\boldsymbol{\zeta}},\, \widetilde{p}]\, \mathrm{e}^{\mathrm{i}(\mathbf{k}\cdot\mathbf{x}-\omega t)}\Big\} \qquad \text{with} \qquad \boldsymbol{\zeta} \;=\; \boldsymbol{\nabla}\times\mathbf{u}\,, \tag{3.90}$$

where $\widetilde{\mathbf{u}}$, $\widetilde{\boldsymbol{\zeta}}$ are constant complex vectors and $\widetilde{p}$ is a complex constant. According to (3.87), they are related by

$$-\,\mathrm{i}\omega\widetilde{\mathbf{u}} + 2\boldsymbol{\Omega}\times\widetilde{\mathbf{u}} \;=\; -\,\mathrm{i}\mathbf{k}\widetilde{p}/\rho\,, \qquad \text{with} \qquad \mathbf{k}\cdot\widetilde{\mathbf{u}} \;=\; 0\,, \tag{3.91a,b}$$

which in turn yields

$$\widetilde{p} \;=\; -\,2\rho\boldsymbol{\Omega}\cdot\widetilde{\boldsymbol{\zeta}}/k^2 \qquad \text{with} \qquad \widetilde{\boldsymbol{\zeta}} \;=\; \mathrm{i}\mathbf{k}\times\widetilde{\mathbf{u}} \qquad (k = |\mathbf{k}|)\,. \tag{3.91c,d}$$

Significantly (3.91a,b) imply that $\widetilde{\mathbf{u}}\cdot\widetilde{\mathbf{u}} = 0$. Compatible with the result (3.89), the fluid velocity has constant amplitude $|\mathbf{u}|$, where $|\mathbf{u}|^2 = \widetilde{\mathbf{u}}\cdot\widetilde{\mathbf{u}}^*/2$ (the star denotes complex conjugate), and the wave is circularly polarised. We may also deduce from (3.91), that the fluid velocity $\mathbf{u}$ and vorticity $\boldsymbol{\zeta}$ satisfy

$$\omega\boldsymbol{\zeta} \;=\; -\,2(\boldsymbol{\Omega}\cdot\mathbf{k})\mathbf{u}\,, \qquad \mathbf{k}\cdot\mathbf{u} \;=\; 0\,, \tag{3.92a,b}$$

in which the frequency is given by

$$\omega = \pm\omega_C \;=\; \pm 2\boldsymbol{\Omega}\cdot\mathbf{e}_k\,, \qquad \mathbf{e}_k = \mathbf{k}/k\,. \tag{3.93a}$$

Accordingly the helicity of the wave

$$\mathbf{u}\cdot\boldsymbol{\zeta} \;=\; \mp k|\mathbf{u}|^2 \tag{3.93b}$$

has the opposite sign to the frequency. It is easy to show that the phase velocity $\mathbf{c}_p = (\omega/k)\mathbf{e}_k$ and the group velocity $\mathbf{c}_g = \boldsymbol{\nabla}_{\mathbf{k}}\,\omega$ are related by

$$\mathbf{c}_g + \mathbf{c}_p \;=\; \pm 2\boldsymbol{\Omega}/k \qquad \text{with} \qquad \mathbf{c}_p\cdot\mathbf{c}_g = 0\,. \tag{3.93c,d}$$

Essentially $\mathbf{c}_p$, $\mathbf{c}_g$ and $\pm 2\boldsymbol{\Omega}$ form a right-angled triangle.

From a more general point of view, we may consider modes of given frequency ω of the form

$$[\mathbf{u}, p] = \mathrm{Re}\left\{[\widetilde{\mathbf{u}},\, \widetilde{p}](\mathbf{x})\, \mathrm{e}^{-\mathrm{i}\omega t}\right\}. \tag{3.94a}$$

These have the poloidal–toroidal decomposition (see Section 1.3.4)

$$\widetilde{\mathbf{u}} = \boldsymbol{\nabla} \times [\boldsymbol{\nabla} \times (\widetilde{\phi}\, \mathbf{e}_\Omega)] - \boldsymbol{\nabla} \times (\widetilde{\psi}\, \mathbf{e}_\Omega) , \qquad \mathbf{e}_\Omega = \boldsymbol{\Omega}/|\boldsymbol{\Omega}| . \tag{3.94b}$$

This satisfies (3.87) when

$$2\boldsymbol{\Omega} \cdot \boldsymbol{\nabla}\widetilde{\phi} = \mathrm{i}\omega\, \widetilde{\psi} , \qquad 2\boldsymbol{\Omega} \cdot \boldsymbol{\nabla}\widetilde{\psi} = -\mathrm{i}\omega\, \Delta\widetilde{\phi} , \tag{3.95a,b}$$

from which we deduce that both $\widetilde{\psi}$ and $\widetilde{\phi}$ satisfy

$$\omega^2 \Delta\widetilde{\psi} - 4(\boldsymbol{\Omega} \cdot \boldsymbol{\nabla})^2 \widetilde{\psi} = 0 . \tag{3.95c}$$

Furthermore, $\widetilde{\psi}$ is related to $\widetilde{p}$ by

$$(4\Omega^2 - \omega^2)\widetilde{\psi} = 2\Omega\widetilde{p}/\rho . \tag{3.96}$$

Relative to Cartesian coordinates $(x,\, y,\, z)$ for which the z axis is parallel to the rotation vector $\boldsymbol{\Omega} = (0,\, 0,\, \Omega)$, (3.95c) becomes

$$\omega^2\, \nabla_\perp^2 \widetilde{\psi} - (4\Omega^2 - \omega^2)\partial_{zz}\widetilde{\psi} = 0 , \tag{3.97a}$$

where $\nabla_\perp^2 \equiv \partial_{xx} + \partial_{yy}$ as in (3.28b). Moreover since $2\Omega \geq |\omega|$ by (3.93a) this equation is hyperbolic. So for example y independent solutions have have characteristic lines

$$(4\Omega^2 - \omega^2)^{1/2}\; x \pm \omega\, z = \text{constant} , \tag{3.97b}$$

which generate the planes with unit normal

$$\mathbf{e}_k = \left[\left(1 - \left(\frac{\omega}{2\Omega} \right)^2 \right)^{1/2} ,\, 0,\, \pm \frac{\omega}{2\Omega} \right] . \tag{3.97c}$$

Since any plane inertial wave (3.90) with wave vector $\mathbf{k} = (k_x,\, 0,\, k_z)$ and frequency ω has this property [see (3.93a)], the characteristic planes coincide with the wavefronts.

We note that geostrophy, namely $\phi = \phi(x, y)$ and $\psi = \psi(x, y)$, is recovered from (3.95a.b) in the limit $\omega = 0$.

REFLECTION AT A PLANE BOUNDARY

In this section we consider the reflection of a plane wave in a plane boundary with unit normal $\mathbf{n}$. A more complete description is given by Stewartson (1978) which in turn is a summary of Phillips (1963). To that end, we adopt Cartesian coordinates $(x,\, y,\, z)$, chosen such that $\mathbf{n} = (0, 0, 1)$ and $\mathbf{k} = (k \sin\alpha,\, 0,\, k\cos\alpha)$ for some α.

Accordingly, the most general form of $\widetilde{\mathbf{u}}$ satisfying $\mathbf{k}\cdot\widetilde{\mathbf{u}} = \widetilde{\mathbf{u}}\cdot\widetilde{\mathbf{u}} = 0$, (3.92a) and (3.93a) is

$$\widetilde{\mathbf{u}} = \widetilde{\mathbf{u}}_{\pm}(\alpha,\varphi_0) \;=\; u_0\,(\cos\alpha,\,\mp\mathrm{i},\,-\sin\alpha)\,\exp(\mathrm{i}\varphi_0)\,, \qquad \omega \;=\; \pm 2\Omega\cos\alpha\,, \tag{3.98a,b}$$

where the amplitude $u_0 = |\mathbf{u}|$ and phase φ_0 are arbitrary real constants. On the boundary $z = 0$, the velocity (3.90) defined by (3.98a) is

$$\mathbf{u} \;=\; u_0\,\mathrm{Re}\{(\cos\alpha,\,\mp\mathrm{i},\,-\sin\alpha)\,\exp[\mathrm{i}(k(\sin\alpha)\,x - \omega t + \varphi_0)]\} \tag{3.98c}$$

[for the case $\alpha = 0$ *cf.* (3.89) and (3.93a)]. When $\alpha \neq 0$, the z–component of velocity is non-zero and then an inertial wave is reflected that must eliminate that component for all x and t.

If we denote the incident and reflected waves by the superscripts i and r, it then follows from (3.93a) and (3.98c) that $\mathbf{e}_k^r$ and $\mathbf{e}_k^i$ must satisfy the relations

$$|\boldsymbol{\Omega}\cdot\mathbf{e}_k^i| \;=\; |\boldsymbol{\Omega}\cdot\mathbf{e}_k^r|\,, \qquad \mathbf{n}\times\mathbf{k}^i \;=\; \mathbf{n}\times\mathbf{k}^r\,, \qquad |\mathbf{n}\times\mathbf{e}_k^i|\,u_0^i \;=\; |\mathbf{n}\times\mathbf{e}_k^r|\,u_0^r\;; \tag{3.99a,b,c}$$

the former two ensure that the incident and reflected wave have the same phase on the boundary, while the latter is required to ensure that $\mathbf{u}\cdot\mathbf{n}$ vanishes there. Any ambiguity in the sign of the $\mathbf{k}$'s is resolved by the requirement that energy propagates towards (away) from the boundary for the incident (reflected) waves. Thus the group velocities satisfy

$$\mathbf{n}\cdot\mathbf{c}_g^i < 0 \qquad \text{and} \qquad \mathbf{n}\cdot\mathbf{c}_g^r > 0\,. \tag{3.99d,e}$$

A simple illustrative example is provided by the special case for which the rotation vector is normal to the boundary: $\boldsymbol{\Omega} = (0,0,\Omega)$ with $\Omega > 0$. Then all the conditions (3.99a-e) are met by

$$\widetilde{\mathbf{u}}^i \;=\; \widetilde{\mathbf{u}}_-(\alpha,\varphi_0)\,, \qquad \widetilde{\mathbf{u}}^r \;=\; \widetilde{\mathbf{u}}_+(\pi-\alpha,\pi+\varphi_0)\,, \qquad \omega \;=\; -2\,\Omega\,\cos\alpha\,(<0)\,, \tag{3.100a,b,c}$$

where $|\alpha| < \pi/2$. The ensuing motion is

$$\mathbf{u} \;=\; 2u_0\,\mathrm{Re}\{\,(\cos\alpha\,\cos Kz,\;\mathrm{i}\cos Kz,\;-\mathrm{i}\sin\alpha\,\sin Kz)\;\mathcal{E}(x,t)\}\,, \tag{3.100d}$$

in which

$$\mathcal{E}(x,t) \;=\; \exp[\mathrm{i}(k(\sin\alpha)\,x - \omega t + \varphi_0)] \qquad K \;=\; k\cos\alpha\,(>0) \qquad (k>0)\,. \tag{3.100e,f}$$

When $\boldsymbol{\Omega}$ is not parallel to $\mathbf{n}$, the reflection is more complicated. Of particular interest is the case when the reflected wave is almost normal to the boundary

$$|\mathbf{n}\times\mathbf{e}_k^r| \ll 1\,, \qquad |\mathbf{n}\times\mathbf{e}_k^i| \;=\; \mathcal{O}(1)\,. \tag{3.101a,b}$$

Then, according to (3.99b,c), the magnitude of the reflected wave is large, while its wavelength is short:

$$u_0^r \gg u_0^i \qquad\qquad |\mathbf{k}^r| \gg |\mathbf{k}^i|\,. \qquad (3.101\text{c,d})$$

In the case of bounded systems, the hyperbolic equation (3.97a) must be solved subject to the impermeable boundary condition $\mathbf{n}\cdot\widetilde{\mathbf{u}} = 0$ on the boundary $\partial\mathcal{D}$ of the fluid. This leads to a badly posed problem. In the case of a spherical cavity, however, the solution in terms of normal modes with distinct frequencies (eigenvalues) is well known (see, for example, Greenspan, 1968; Zhang *et al.*, 2001). Nevertheless, the occurrence of a wave vector normal to the boundary is linked to the tangency of a characteristic line with the boundary. The issue is particularly pertinent to inertial waves in a shell. Evidently singular behaviour is triggered on the inner sphere at latitudes $\pm\sin^{-1}(|\omega|/2\Omega)$, where they characteristics touch the inner sphere. This singular behaviour [relatively large amplitude and short length scale (3.101d)], follows the characteristics to the outer sphere where it is reflected. This reflection proceeds indefinitely. Indeed it is far from clear that normal modes even exist in the usual sense as the numerical results of Rieutord & Valdettaro (1997) indicate.

BOUNDARY LAYERS

In view of the difficulties mentioned above it is of some interest to investigate the role of small viscosity. When the Ekman number is small we may anticipate that most of the dissipation occurs in boundary layers. So if we consider wave motion of the form (3.94a) adjacent to a plane boundary, we can resolve the boundary layer structure by simply adapting the Ekman analysis of Section 3.1.1.

We adopt a local Cartesian coordinate system with $\mathbf{n} = (0,0,1)$. Within the framework of the boundary layer approximation we assume that the flow just outside the boundary layer is uniform $\widetilde{\mathbf{u}}_C = (\widetilde{u}_C, \widetilde{v}_C, 0)$, where $\widetilde{u}_C$ and $\widetilde{v}_C$ are complex constants. The motion in the boundary layer is governed by

$$-\mathrm{i}\,\omega\widetilde{u} - 2\Omega_z\widetilde{v} = -\partial_x(\widetilde{p}/\rho) + \nu\partial_{zz}\widetilde{u}\,, \qquad -\mathrm{i}\,\omega\widetilde{u}_C - 2\Omega\widetilde{v}_C = -\partial_x(\widetilde{p}/\rho)\,, \qquad (3.102\text{a,b})$$

$$-\mathrm{i}\,\omega\widetilde{v} + 2\Omega_z\widetilde{u} = -\partial_y(\widetilde{p}/\rho) + \nu\partial_{zz}\widetilde{v}\,, \qquad -\mathrm{i}\,\omega\widetilde{v}_C + 2\Omega\widetilde{u}_C = -\partial_y(\widetilde{p}/\rho)\,, \qquad (3.102\text{c,d})$$

where $\boldsymbol{\Omega} = (\Omega_x,\, \Omega_y,\, \Omega_z)$. They can be combined to form the single equation

$$\nu\partial_{zz}\widetilde{Z}^{\pm} \mp \mathrm{i}(2\Omega_z \mp \omega)\left(\widetilde{Z}^{\pm} - \widetilde{Z}_C^{\pm}\right) = 0\,, \qquad (3.103)$$

where $\widetilde{Z}^{\pm} \equiv \widetilde{u} \pm \mathrm{i}\widetilde{v}$ and $\widetilde{Z}_C^{\pm} \equiv \widetilde{u}_C \pm \mathrm{i}\widetilde{v}_C$. This gives rise to a boundary layer with a double structure composed of two Ekman spiral solutions

$$\widetilde{Z}^{\pm} = \widetilde{Z}_C^{\pm}\left\{1 - \exp\left[-\left(1 \pm \mathrm{i}\,\mathrm{sgn}\,(2\mathbf{n}\cdot\boldsymbol{\Omega} \mp \omega)\right) z/\delta_C^{\pm}\right]\right\}, \qquad (3.104\text{a})$$

each characterised by the boundary layer lengths

$$\delta_C^{\pm} = \sqrt{2\nu / |2\mathbf{n}\cdot\boldsymbol{\Omega} \mp \omega|}\,. \tag{3.104b}$$

Evidently, (3.103) and its solution (3.104a,b) reduce to (3.6a,b) and (3.7) with $\delta_C^{\pm} = \delta_E$ in the geostrophic limit $\omega = 0$.

As a simple example, we may consider the boundary layer induced from the reflection of the inertial wave at the boundary with normal $\mathbf{n}$ parallel to $\boldsymbol{\Omega}$ described by (3.100) with $\Omega = \mathbf{n}\cdot\boldsymbol{\Omega} > 0$. For that model, we have $\widetilde{u}_C = 2u_0 \cos\alpha\, \mathcal{E}(x,t)$, $\widetilde{v}_C = 2\mathrm{i}u_0\, \mathcal{E}(x,t)$ and $\omega = -2\Omega\cos\alpha$, which gives

$$\widetilde{Z}_C^{\pm} = 2u_0(\cos\alpha \mp 1)\, \mathcal{E}(x,t)\,, \qquad \delta_C^{\pm} = \delta_E/\sqrt{1 \pm \cos\alpha}\,. \tag{3.105a,b}$$

Furthermore, when $2KH = n\pi$ for integer n, the solution (3.100d) provides a wave guided solution confined between boundaries $z = 0$ and $2H$. This solution is lightly damped with ω now complex and possessing a small negative imaginary part $\mathrm{Im}\{\omega\}$. The small damping rate $-\mathrm{Im}\{\omega\}$ may be estimated by equating the rate of decay of the kinetic energy $\mathcal{O}(-2\,\mathrm{Im}\{\omega\}H\rho u_0^2)$ of the guided inertial wave to the viscous dissipation $\mathcal{O}(2\delta_E\rho\nu(u_0/\delta_E)^2)$ in the Ekman layers on the boundaries $z = 0$ and $2H$. They determine the decay rate

$$-\mathrm{Im}\{\omega\} = \mathcal{O}(\sqrt{\nu\Omega}/H) = \mathcal{O}(\mathrm{E}^{1/2}\Omega)\,. \tag{3.106}$$

Note that $[-\mathrm{Im}\{\omega\}]^{-1} = \mathcal{O}(\mathrm{E}^{-1/2}\Omega^{-1})$ is the so-called "spin-up" (or "spin-down") time for the readjustment of geostrophic flow, which results from the impulsive change of velocity of horizontal boundaries moving in their own planes. The important point is that the spin-up time $[-\mathrm{Im}\{\omega\}]^{-1}$ is a much shorter time than the viscous decay time $\mathcal{O}((k^2\nu)^{-1}) = \mathcal{O}((kH)^{-2}\mathrm{E}^{-1}\Omega^{-1})$ based on the viscous dissipation in the main stream. Essentially the main stream viscous dissipation $\mathcal{O}(2H\rho\nu(ku_0)^2)$ is smaller than Ekman layer dissipation by the factor $\mathcal{O}((kH)^2\mathrm{E}^{1/2})$. This factor places a lower limit $L \gg H\mathrm{E}^{1/4}$ on the inertial wave length scale $L = \mathcal{O}(k^{-1})$ for which our estimate (3.106) of the decay rate $-\mathrm{Im}\{\omega\}$ is applicable [*cf.* (3.135) below, where a comparable estimate is made concerning the decay rate of a quasi-geostrophic flow].

Another interesting feature of the boundary layer structure (3.104) for our particular case of $\boldsymbol{\Omega} = \Omega\mathbf{n}$ ($\Omega > 0$) is that the boundary layer thickness δ_C^- (3.105b) tends to ∞ as $\alpha \to 0$. Simultaneously, the amplitude $\widetilde{Z}_C^+$ of the remaining boundary layer structure tends to zero. Consequently, the viscous boundary layer dissipation associated with the reflection is small. Nevertheless, the limit $\alpha \to 0$ does illustrate a degeneracy that occurs when $\omega = \pm\mathbf{n}\cdot\boldsymbol{\Omega}$. For the more general configuration, in which $\boldsymbol{\Omega}$ is no longer parallel to $\mathbf{n}$, we see that the degeneracy identified occurs whenever the boundary is tangent to a characteristic, for then one of the boundary layer widths

$\delta_C^{\pm}$ is infinite. So for example in the case of a shell, the boundary layer on the inner sphere, radius r_i, becomes singular at the critical latitudes $\pm\sin^{-1}(|\omega|/2\Omega)$, where the tangent cones exhibit similar singularities to the tangent cylinder for geostrophic flow. At these critical latitudes the boundary layer thickens to $\mathcal{O}(r_i\delta_{\mathrm{E}}^{4/5})$ proportional to $\mathrm{E}^{2/5}$, as explained at the end of Section 3.1.4. Note, however, that some double limits are involved here. For not only does the boundary layer thicken but the length scale of the reflected wave on the characteristic tangent, specifically in the direction normal to it [see (3.101)], shortens leading to increased viscous dissipation all along the characteristic and its subsequent reflections. Note, however, that this feature is absent in our illustrative example (3.100), for which $k^i = k^r = k$.

All the issues discussed here pertain to the matter of instabilities in processing spheroidal cavities and shells (see, for example, Aldridge, 2003; Lorenzani & Tilgner, 2003).

LOW FREQUENCY MODES

Whereas inertial modes of frequency $\omega = \mathcal{O}(\Omega)$ are pertinent to issues like precession, low frequency modes with

$$|\omega| \ll |\Omega| \tag{3.107}$$

may be excited as part of the convective process in the Earth's core (of outer radius r_o). Accordingly, relative to the cylindrical polar coordinates (s, ϕ, z) of Section 3.1.4, solutions of (3.97a), which vary on the long length scale r_o in the z–direction, must vary on the much shorter length scale $r_o\omega/|\Omega|$ in the orthogonal s and ϕ–directions.

To investigate the properties of inertial modes in the limit (3.107), we will consider, for simplicity, a spherical cavity radius r_o and ignore the fact that the true Earth's core has an inner boundary. For many inertial modes, the inner boundary is an irrelevance because they lie in the region outside the inner core tangent cylinder $s = r_i$. Thus, we restrict attention to modes with a short radial s–length scale and seek WKBJ type solutions, for which the poloidal–toroidal decomposition (3.94b) has the local form

$$[\widetilde{\phi},\, \widetilde{\psi}] = [\widetilde{\Phi},\, \widetilde{\Psi}](z)\exp[\mathrm{i}(\textstyle\int k\,\mathrm{d}s + m\phi)] \qquad (kr_o \gg 1)\,. \tag{3.108}$$

Accordingly we approximate the axial z–component of velocity $\widetilde{W}$ by

$$\widetilde{W} = -\nabla_\perp^2\widetilde{\Phi} \approx a^2\widetilde{\Phi}\,, \qquad \text{where} \qquad a^2 = k^2 + m^2/s^2\,, \tag{3.109a,b}$$

while the orthogonal components s and ϕ–components $(\widetilde{U},\widetilde{V})$ are dominated by their toroidal parts and in particular

$$\widetilde{U} \approx -\frac{im}{s}\widetilde{\Psi} \approx -\frac{2m\Omega}{\omega s a^2}\frac{\mathrm{d}\widetilde{W}}{\mathrm{d}z}, \tag{3.109c}$$

where use has been made of (3.95a). Hence, the impermeable boundary condition $\mathbf{u}\cdot\mathbf{n} = 0$ on the sphere $(s, z_o(s))$ becomes

$$m\frac{\mathrm{d}\widetilde{W}}{\mathrm{d}z} - \frac{\omega a^2}{2\Omega} z_o \widetilde{W} = 0. \tag{3.110}$$

This provide the boundary condition which must be satisfied by the solutions of (3.97a), which, correct to lowest order, becomes

$$\frac{\mathrm{d}^2\widetilde{W}}{\mathrm{d}z^2} + \frac{\omega^2 a^2}{4\Omega^2}\widetilde{W} = 0. \tag{3.111}$$

The odd and even solutions of (3.111) are

$$\widetilde{W} = \widetilde{W}_0^o \sin\left(\frac{\omega a z}{2\Omega}\right) \qquad \text{and} \qquad \widetilde{W} = \widetilde{W}_0^e \cos\left(\frac{\omega a z}{2\Omega}\right). \tag{3.112a,b}$$

They satisfy the boundary condition (3.110) when

$$\tan\left(\frac{\omega a z_o}{2\Omega}\right) = \frac{m}{a z_o} \qquad \text{and} \qquad \cot\left(\frac{\omega a z_o}{2\Omega}\right) = -\frac{m}{a z_o} \tag{3.112c,d}$$

respectively. For any given azimuthal wavenumber m and frequency ω, there is a spectrum of odd and even modes determined by the solutions $a z_o$ of (3.112c,d). In fact there is a critical radius s_c at which the corresponding value of k vanishes and there $a z_o = m z_o(s_c)/s_c$. Beyond that radius, this formula with (3.109b) shows that k satisfies

$$k^2 = \frac{m^2}{s_c^2}\left[\left(\frac{z_o(s_c)}{z_o(s)}\right)^2 - \left(\frac{s_c}{s}\right)^2\right] \qquad \text{for} \qquad s \geq s_c, \tag{3.113}$$

while inside $(s < s_c)$ k is pure imaginary and the mode is evanescent. Outside $(s > s_c)$ there are two families of solutions determined by (3.113) and distinguished by wave vectors $(k, m/s)$ and $(-k, m/s)$. The wave fronts for both are initially radial at $s = s_c$ but then twist either prograde or retrograde for $s > s_c$. In fact, the requirement of evanescence for $s < s_c$ leads to a reflection condition at $s = s_c$ that the prograde and retrograde waves are of equal amplitude. Furthermore their amplitude deceases algebraically with increasing $s - s_c$.

The main interest in these inertial waves is their occurrence at the onset of instability in a self-gravitating sphere containing a uniform distribution of heat sources.

There the lowest order odd modes with $m = \mathcal{O}(\mathrm{E}^{-1/3})$ are readily identified with a prograde twist. The role of thermal and viscous dissipation is crucial in that problem and leads to the localisation of the convection pattern (see Jones *et al.*, 2000). Nevertheless, in the low Prandtl number limit $\mathrm{Pr} \ll 1$, the modes do extend almost out to the outer sphere equatorial boundary $s = r_o$, as in our pure wave theory.

In this context, the lowest order odd modes are of particular interest when $|m/az_o| \ll 1$. Then, together with (3.109c), the solution (3.112a,c) determines

$$\widetilde{U} \approx -\widetilde{W}_0^o \frac{m}{as}, \qquad \widetilde{W} \approx \widetilde{W}_0^o \frac{\omega a z}{2\Omega}, \tag{3.114a,b}$$

$$\text{where} \qquad \omega \approx \frac{2\Omega m}{a^2 z_o^2}. \tag{3.114c}$$

These modes are characterised by a z–independent toroidal part; they propagate in the prograde direction and are referred to as Rossby waves. They form the underlying idea behind Busse's annulus model of thermal convection (see Section 3.4.2 and Busse, 1970).

The extension of the above ideas to a spherical cavity, of inner radius r_i, is beset with difficulties, except when the critical radius lies outside the inner core tangent cylinder (i.e. $s_c > r_i$). There the low frequency modes again pertain to the onset of convective instability (Dormy *et al.*, 2004). Nevertheless, even inside the tangent cylinder (i.e. $s < r_i$) local modes can be investigated in much the same way as before (see, e.g., Gubbins & Roberts, 1987) and they may have some relevance to the convective process.

We should also note that low frequency equatorially trapped inertial modes have been identified at low Prandtl number by Zhang (1995). Nevertheless the double limit $\mathrm{E} \to 0$ and $\mathrm{Pr} \to 0$ is difficult to unravel and is yet to be properly resolved (see Section 3.4.5).

3.3.2. ALFVÉN WAVES

Let us now investigate the role of magnetic fields and begin by outlining the nature of Alfvén waves. We consider an inviscid, incompressible, non-rotating, perfectly electrically conducting fluid permeated by a uniform magnetic field $\mathbf{B} = (B_x, B_y, B_z)$ at rest with constant total pressure (sum of fluid and magnetic pressure). This rest state is perturbed by a slow velocity $\mathbf{u}$, a small magnetic field $\mathbf{b}$ and a small perturbation total pressure p_T. They satisfy the linearised equations of motion

$$\rho \partial_t \mathbf{u} = -\boldsymbol{\nabla} p_T + \mu^{-1}(\mathbf{B}\cdot\boldsymbol{\nabla})\mathbf{b}, \qquad \boldsymbol{\nabla}\cdot\mathbf{u} = 0 \tag{3.115a}$$

and magnetic induction

$$\partial_t \mathbf{b} = (\mathbf{B}\cdot\boldsymbol{\nabla})\mathbf{u}, \qquad \boldsymbol{\nabla}\cdot\mathbf{b} = 0. \tag{3.115b}$$

Note that together they imply that

$$\Delta p_T = 0. \tag{3.115c}$$

For waves in an unbounded region, the solution of (3.115c) is necessarily $p_T = 0$.

If we take axes (x, y, z) such that $\mathbf{u} = (0, u(x,t), 0)$, $\mathbf{b} = (0, b(x,t), 0)$ and assume that $p_T = 0$, then (3.115a,b) are satisfied when

$$\partial_t u = (\rho\mu)^{-1} B_x \partial_x b \qquad \text{and} \qquad \partial_t b = B_x \partial_x u \tag{3.116a,b}$$

respectively. They possess travelling wave solutions

$$\mp (\rho\mu)^{-1/2} b = u = f_\pm(x \mp V_x t), \tag{3.117a}$$

where $f_\pm$ are arbitrary functions of their arguments and

$$\mathbf{V} = (\rho\mu)^{-1/2}(B_x, B_y, B_z) = (\rho\mu)^{-1/2}\mathbf{B} \tag{3.117b}$$

is the Alfvén velocity. Here $f_+(x - V_x t)$ and $f_-(x + V_x t)$ define transverse Alfvén waves travelling to the right and left respectively.

3.3.3. MHD WAVES IN ROTATING FLUIDS

When the fluid rotates, the equation of motion (3.115a) is replaced by

$$\rho\partial_t \mathbf{u} + 2\rho\boldsymbol{\Omega} \times \mathbf{u} = -\boldsymbol{\nabla} p_T + \mu^{-1}(\mathbf{B}\cdot\boldsymbol{\nabla})\mathbf{b}, \qquad \boldsymbol{\nabla}\cdot\mathbf{u} = 0, \tag{3.118a}$$

while $\mathbf{b}$ continues to be determined by the magnetic induction equation (3.115b). Furthermore, if we introduce the fluid particle displacement $\boldsymbol{\eta}$ (hopefully not to be confused with the magnetic diffusivity η), we may express solution $\mathbf{b}$ of (3.115b) in terms of its "frozen" field representation

$$\mathbf{b} = (\mathbf{B}\cdot\boldsymbol{\nabla})\boldsymbol{\eta} \qquad \text{with} \qquad \mathbf{u} = \partial_t \boldsymbol{\eta}. \tag{3.118b}$$

This enables us to construct the single equation

$$\partial_{tt}\boldsymbol{\eta} + 2\boldsymbol{\Omega} \times \partial_t\boldsymbol{\eta} = -\boldsymbol{\nabla}(p_T/\rho) + (\mathbf{V}\cdot\boldsymbol{\nabla})^2\boldsymbol{\eta}, \qquad \boldsymbol{\nabla}\cdot\boldsymbol{\eta} = 0. \tag{3.119}$$

Suppose, as in our preliminary discussion in Section 3.3.1, we choose our axes so that we are considering waves propagating in the z–direction. Accordingly, the fluid particle displacement is $\boldsymbol{\eta} = (\eta_x(z,t), \eta_y(z,t), 0)$ and likewise the total pressure perturbation p_T is a function of z and t alone. Now we form the complex displacement

$$\eta_x + \mathrm{i}\,\eta_y = \Upsilon(z,t)\exp(-\mathrm{i}\,\Omega_z t) \tag{3.120}$$

as in (3.89), where Υ is a complex function of z and t. The inclusion of the extra factor $\exp(-\mathrm{i}\,\Omega_z t)$ essentially rotates the frame such that the z–component Ω_z of rotation is removed. By this device the natural extension of (3.88c) is

$$\partial_{tt}\Upsilon + \Omega_z^2\Upsilon = V_z^2\partial_{zz}\Upsilon\,. \qquad (3.121)$$

Here the term $\Omega_z^2\Upsilon$ is essentially the pressure gradient in the original rotating frame, which there balanced the centrifugal acceleration. Evidently in our new rotating frame plane polarised waves exist of the form

$$\Upsilon = \Upsilon_0\cos(kz \pm \omega_p t)\,, \qquad \text{where} \qquad \omega_p = \sqrt{\Omega_z^2 + (V_z k)^2}\,. \qquad (3.122)$$

Note however that this is a linear combination of two circularly polarised waves

$$\eta_x + \mathrm{i}\,\eta_y = \Upsilon_0\exp\{\mathrm{i}\,[kz - (\Omega_z \mp \omega_p)t]\}\,, \qquad (3.123)$$

where here k may be both positive and negative. At a particular level $z =$ constant the particle paths described by (3.123) are circles provided that $V_z \neq 0$ for then $\omega_p > |\Omega_z|$. Curiously, the above discussion continues to hold for the non-magnetic case $V_z = 0$, for which $\omega_p = |\Omega_z|$. Then the choice $\mp\omega_p = \Omega_z$ in (3.123) recovers the inertial wave solution (3.89), while the choice $\mp\omega_p = -\Omega_z$ in (3.123) determines a stationary displacement $\boldsymbol{\eta}$, which corresponds to the trivial solution $\mathbf{u} = \mathbf{0}$.

The alternative plane wave solution formulation, which is not dependent on the choice of coordinate axes, is

$$[\boldsymbol{\eta},\, \mathbf{u},\, \mathbf{b},\, p_T] = \mathrm{Re}\Big\{\Big[\widetilde{\boldsymbol{\eta}},\, \widetilde{\mathbf{u}},\, \widetilde{\mathbf{b}},\, \widetilde{p}_T\Big]\,\mathrm{e}^{\mathrm{i}(\mathbf{k}\cdot\mathbf{x}-\omega t)}\Big\}\,. \qquad (3.124)$$

They satisfy (3.115b) and (3.118) when

$$\mp\mathrm{i}\,\omega_C\widetilde{\mathbf{u}} + 2\boldsymbol{\Omega}\times\widetilde{\mathbf{u}} = -\,\mathrm{i}\,\mathbf{k}\widetilde{p}_T/\rho\,, \qquad \mathbf{k}\cdot\widetilde{\mathbf{u}} = 0 \qquad (3.125\text{a,b})$$

with

$$\widetilde{\mathbf{b}} = -\,(\mathbf{k}\cdot\mathbf{B})\widetilde{\boldsymbol{\eta}}\,, \qquad \widetilde{\boldsymbol{\eta}} = \mathrm{i}\omega^{-1}\widetilde{\mathbf{u}}\,, \qquad \pm\omega_C = \omega - \omega_M^2/\omega\,, \qquad (3.125\text{c,d,e})$$

where

$$\omega_C = 2\boldsymbol{\Omega}\cdot\mathbf{e}_k\,, \qquad \omega_M = \mathbf{k}\cdot\mathbf{V} \qquad (3.125\text{f,g})$$

are the inertial (3.93a) and Alfvén wave frequencies respectively. In this way the structure of the solution has been reduced to that attributed to inertial waves of frequency $\pm\omega_C$ [see (3.91) to (3.93a)], albeit their actual frequency ω is determined by the roots of (3.125e) (see, for example, Moffatt, 1978), namely

$$\omega = \pm\tfrac{1}{2}\omega_C + \omega_p \qquad \text{and} \qquad \pm\tfrac{1}{2}\omega_C - \omega_p\,, \qquad (3.126\text{a,b})$$

where

$$\omega_p = \sqrt{\left(\tfrac{1}{2}\omega_C\right)^2 + \omega_M^2}\,, \qquad (3.126\text{c})$$

consistent with (3.123).

ALFVÉN WAVES: $\omega_C = 0$

When the inertial frequency vanishes ($\omega_C = 0$), it follows from (3.125f) that $\boldsymbol{\Omega}$ is perpendicular to $\mathbf{k}$ and so lies in the plane of the wavefront. That means motion is independent of the coordinate parallel to the rotation axis and thus satisfies the Proudman–Taylor condition. In summary therefore motion is geostrophic and according to (3.125) satisfies

$$2\boldsymbol{\Omega} \times \widetilde{\mathbf{u}} = -\mathrm{i}\mathbf{k}\widetilde{p}_T/\rho \qquad \mathbf{k}\cdot\widetilde{\mathbf{u}} = 0\,, \qquad \mathbf{k}\cdot\boldsymbol{\Omega} = 0\,. \quad \text{(3.127a,b,c)}$$

Though geostrophic, the motion that ensues is a pure Alfvén wave of the type (3.117a) with frequency $\omega = \pm\omega_M$, as the equation (3.121) and result (3.122) with $\Omega_z = 0$ and $\omega_p = |V_z k|$ illustrate.

An interesting degenerate mode of zero frequency occurs when both $\omega_C = 0$ and $\omega_M = 0$. Then according to (3.125f,g) motion is not influenced by either Coriolis or Lorentz forces and lies in the plane containing both $\mathbf{B}$ and $\boldsymbol{\Omega}$, i.e.

$$\mathbf{u}\cdot(\mathbf{B} \times \boldsymbol{\Omega}) = 0 \qquad \text{with} \qquad \mathbf{k} \parallel \mathbf{B} \times \boldsymbol{\Omega}\,. \qquad \text{(3.128a,b)}$$

Motion of this type with arbitrary length scale L in the $\mathbf{B} \times \boldsymbol{\Omega}$–direction is possible. This type of motion is believed to play an important role in rotating MHD turbulence. In that case with dissipation (particularly ohmic) included, there is evidence of plate like motion of this type with a small value of L and determined by the diffusion length scale (Braginsky & Meytlis, 1964; St. Pierre, 1996).

LEHNERT WAVES: $\omega_C \neq 0$

When $\omega_C \neq 0$, the both families of waves, with frequencies (3.126a,b), are circularly polarised exactly as illustrated by (3.123).

For geophysical parameter values, we generally have Alfvén velocities $|\mathbf{V}|$ which are small compared to $L\Omega$, where $L = \mathcal{O}(k^{-1})$ is the length scale of the waves. So, unless $\mathbf{k}$ is almost orthogonal to $\boldsymbol{\Omega}$, we have

$$|\omega_M| \ll |\omega_C| \qquad \text{(3.129a)}$$

and then the two families of waves identified by the roots (3.126a,b) of the dispersion relation (3.125e) have very different characters. One corresponds to the fast inertial wave with frequency $\omega \approx \pm\omega_C$ for which the Lorentz force only provides a small perturbation. The other is the slow hybrid MC–wave (Lehnert, 1954) with frequency

$$\omega = \mp\omega_{MC}\,, \qquad \text{where} \qquad \omega_{MC} \approx \omega_M^2/|\omega_C| = \mathcal{O}\left(B^2/L^2\mu\rho|\Omega|\right). \qquad \text{(3.129b,c)}$$

In this approximation, the inertial term $\partial_{tt}\boldsymbol{\eta}$ is neglected in the equation of motion (3.119) leaving the magnetogeostrophic balance

$$2\boldsymbol{\Omega} \times \partial_t\boldsymbol{\eta} = -\boldsymbol{\nabla}(p_T/\rho) + (\mathbf{V}\cdot\boldsymbol{\nabla})^2\boldsymbol{\eta}\,, \qquad \boldsymbol{\nabla}\cdot\boldsymbol{\eta} = 0\,. \qquad (3.130\text{a,b})$$

For our earlier axis choice with $\mathbf{k} = (0, 0, k)$, the slow MC–wave solution expressed in the form (3.123) is

$$\eta_x + \mathrm{i}\,\eta_y = \Upsilon_0 \exp\left\{\mathrm{i}\left[kz + \tfrac{1}{2}\left(B_z^2k^2/\rho\mu\Omega_z\right)t\right]\right\}. \qquad (3.131)$$

Of course, they are circularly polarised, but curiously the sense of rotation of fluid particles at given z is opposite to that for the fast inertial oscillations. This is simply a reflection of the fact that $|\omega_p|$ is always greater than $|\omega_C|/2$, when $\omega_M \neq 0$.

Note that the slow MC–wave timescale defined by the inverse of (3.129b) is comparable to the magnetic diffusion time L^2/η, when the Elsasser number

$$\Lambda = B^2/\mu\eta\rho|\Omega| \qquad (3.132)$$

[see (3.60c)] is of order unity. Significantly the dimensionless number Λ is independent of the length L. These waves are believed to play an important role in the geodynamo process as the Elsasser number is estimated to be of roughly order unity in the Earth's core.

TORSIONAL OSCILLATIONS WITH DISSIPATION

The low frequency Alfvén waves relate to the short period length of day variation for the Earth and are manifest as a geostrophic torsional oscillation (see Section 4.3.4). Thus geostrophic circular cylinders with their generators aligned to the rotation axis rotate rigidly but independently. They are coupled by the meridional magnetic field which threads them and permits a cylindrical Alfvén wave to propagate.

In order to appreciate the nature of the mechanism, we adopt the parallel plane geometry of Section 3.1.3; a more complete description in cylindrical geometry is given by Gubbins & Roberts (1987). Relative to Cartesian coordinates (x, y, z), we consider fluid confined between to plane boundaries $z = \pm H$. The system rotates about the z–axis with angular velocity $\boldsymbol{\Omega} = (0, 0, \Omega)$ and $\Omega > 0$, while the electrically conducting fluid is permeated by the uniform magnetic field $\mathbf{B} = (B, 0, 0)$. The idea is that the y–direction is to be identified with the azimuthal direction in the sphere and so we consider velocity and magnetic field perturbations $\mathbf{u} = (0, v(x,t), 0)$ and $\mathbf{b} = (0, b(x,t), 0)$ respectively.

We take into account small viscosity and finite ohmic diffusion. We therefore assume that there are thin Ekman layers of width $\delta_E = \sqrt{\nu/\Omega}$ adjacent to each boundary, in which the Lorentz force is unimportant. Since the applied magnetic field is aligned with the boundary, this approximation is reasonable provided that the magnetic field is not too large. The solution in the mainstream is the time dependent extension of (3.31), namely

$$\partial_t v + \mathrm{E}^{1/2}\Omega v = (B/\mu\rho)\,\partial_x b + \nu\,\partial_{xx} v\,, \quad \partial_t b = B\,\partial_x v + \eta\,\partial_{xx} b\,, \quad (3.133\text{a,b})$$

where $\mathrm{E} = \nu/H^2\Omega$ is the Ekman number. The dispersion relation for damped torsional waves proportional to $\exp[\mathrm{i}(kx - \omega t)]$ is

$$\left(-\mathrm{i}\omega + \mathrm{E}^{1/2}\,\Omega + \nu k^2\right)\left(-\mathrm{i}\omega + \eta k^2\right) = -V^2 k^2 \qquad \left(V^2 = B^2/\mu\rho\right). \quad (3.134\text{a,b})$$

Assuming that the damping is not too strong, the decay rate characterising the damping is

$$-\operatorname{Im}\{\omega\} = \tfrac{1}{2}\left[\mathrm{E}^{1/2}\,\Omega + (\nu + \eta)k^2\right]. \quad (3.135)$$

Since $\nu \ll \eta$ in the Earth's core, whether the damping of the waves is by Ekman suction or Ohmic diffusion in the mainstream depends on whether the horizontal length scale $L\ (= \mathcal{O}(k^{-1}))$ is much greater or less than $(\eta/\nu)^{1/2}\mathrm{E}^{1/4}H = (\eta H)^{1/2}/(\nu\Omega)^{1/4}$ respectively.

The model may be made more geophysically relevant by placing side walls at $x = \pm L$ as we did in Section 3.2.4. On solving (3.134a,b) at given ω, we obtain $k \approx \pm\omega/V$ and $k \approx \pm \mathrm{i}V/\sqrt{\nu\eta}$. The former with k real, determines the complex frequency ω (3.135) of our lightly damped wave. The latter indicates that the boundary layers to be found on the walls $x = \pm L$ are Hartmann layers as in Section 3.2.4.

Though we capture important features of the torsional waves by our model, it does possess degenerate features. They are that the top and bottom boundaries are parallel to the applied magnetic field $\mathbf{B}$ and that the side walls $x = \pm L$ are parallel to the rotation vector $\boldsymbol{\Omega}$. As we explained in Section 3.2.4, neither of these features are typical in the geophysical situation. Indeed the Earth's meridional magnetic field will generally have a component normal to both the inner core boundary $r = r_i$ and the core mantle boundary $r = r_o$. There we expect an Ekman-Hartmann boundary layer to play a role rather than the Ekman layer advocated in (3.133a,b). The meridional magnetic field, which permeates the boundaries, may lead to an important dynamical coupling. For even though the mantle may be a relatively poor electrical conductor, the weak electric currents that leak into the mantle may produce forces on it far larger than the viscous stress at the boundary. Nevertheless that is a complicated and involved matter which is outside the scope of our present analysis (see, e.g., Jault, 2003).

3.3.4. STRATIFIED ROTATING MHD WAVES

We now briefly discuss the effects of gravity $\mathbf{g} = -g\mathbf{n}$, where the unit vector $\mathbf{n}$ points upwards, in a horizontally stratified Boussinesq fluid density $\rho_0(\mathbf{x})$, for which $\nabla\rho_0$ is parallel to $\mathbf{g}$. The governing equations for the full dissipative system are, as derived in Chapter 1,

$$\partial_t\mathbf{u} + 2\boldsymbol{\Omega} \times \mathbf{u} = -\nabla(p_T/\rho) + \vartheta\mathbf{g} + (\mu\rho)^{-1}\mathbf{B}\cdot\nabla\mathbf{b} + \nu\Delta\mathbf{u}\,, \qquad (3.136a)$$

$$\nabla\cdot\mathbf{u} = 0\,, \qquad \nabla\cdot\mathbf{b} = 0\,, \qquad (3.136b,c)$$

$$\partial_t\mathbf{b} = \mathbf{B}\cdot\nabla\mathbf{u} + \eta\Delta\mathbf{b}\,, \qquad (3.136d)$$

$$\partial_t\vartheta = -{\rho_0}^{-1}\mathbf{u}\cdot\nabla\rho_0 + \kappa\Delta\vartheta \qquad (\mathbf{g}\times\nabla\rho_0 = \mathbf{0})\,, \qquad (3.136e,f)$$

where, p_T denotes the total pressure and ϑ corresponds to the density perturbation (i.e. $\vartheta = \alpha\,\Theta$, where Θ is the temperature perturbation used in Chapter 1, page 16).

The dispersion relation for waves of the form (3.124) is

$$\Sigma - \frac{\omega_C^2}{\Sigma} - \frac{\omega_A^2}{\omega + \mathrm{i}\kappa k^2} = 0\,, \qquad (3.137a)$$

in which

$$\Sigma = \omega + \mathrm{i}\nu k^2 - \frac{\omega_M^2}{\omega + \mathrm{i}\eta k^2}\,, \qquad (3.137b)$$

where $\qquad \omega_A^2 = N^2\,|\mathbf{e}_k \times \mathbf{n}|^2 \qquad$ and $\qquad N^2 = {\rho_0}^{-1}\,\mathbf{g}\cdot\nabla\rho_0 \qquad$ (3.137c,d)

is the Brunt-Väisälä frequency. Note that the fluid is stably (unstably) stratified when $N^2 \geq 0$ ($N^2 < 0$). Here ω_A (possibly pure imaginary) is the gravity wave frequency, while ω_C and ω_M are the inertial and Alfvén frequencies (3.125f,g).

When all the diffusions are negligible, we may set $\nu = \eta = \kappa = 0$ and take advantage of the particle displacement approach (3.118b) and write

$$\vartheta = -{\rho_0}^{-1}\boldsymbol{\eta}\cdot\nabla\rho_0 \qquad (3.138)$$

for the pertubation buoyancy. Accordingly the equation of motion (3.136a) becomes

$$\partial_{tt}\boldsymbol{\eta} + 2\boldsymbol{\Omega}\times\partial_t\boldsymbol{\eta} = -\nabla(p_T/\rho) - N^2(\boldsymbol{\eta}\cdot\mathbf{n})\mathbf{n} + (\mathbf{V}\cdot\nabla)^2\boldsymbol{\eta}\,, \qquad \nabla\cdot\boldsymbol{\eta} = 0\,, \qquad (3.139a,b)$$

which is the buoyancy driven extension of (3.119). To obtain a simple and informative picture of the nature of our waves, we choose Cartesian coordinate axes as follows. The z–axis is parallel to the wave vector $\mathbf{k}$ but its direction $\mathbf{e}_z = (0, 0, 1)$ is chosen so that $\mathbf{e}_z\cdot\mathbf{n} \geq 0$. The y–axis is horizontal and so we write $\mathbf{n} = (-\sin\alpha,\, 0,\, \cos\alpha)$, where α is the tilt angle of the wave front to the horizontal. The x–axis points in the direction of steepest descent (ascent) on the wave front for

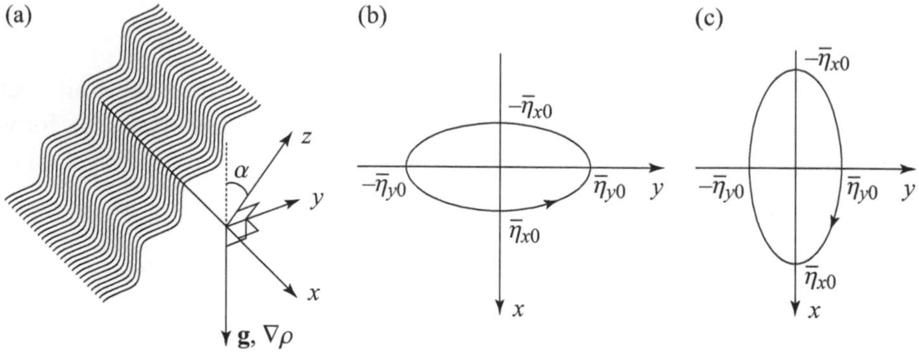

Figure 3.13 - Elliptically polarised plane waves for $N^2 > 0$ and $\Omega_z > 0$. (a) The geometry of the wave front $z = 0$. The particle paths and their direction are indicated for two cases: (b) Slow waves with $\overline{\eta}_{x0} > \overline{\eta}_{y0} > 0$, $\mathcal{R} > 0$; (c) Fast waves with $0 < \overline{\eta}_{x0} < \overline{\eta}_{y0}$, $\mathcal{R} < 0$.

$0 \leq \alpha \leq \pi/2$ $(-\pi/2 \leq \alpha \leq 0)$ (see Figure 3.13a). The fluid particle displacement $\boldsymbol{\eta} = (\eta_x(z,t), \eta_y(z,t), 0)$ lies in the plane of the wave front and, when ω is real, possesses elliptically polarised travelling wave solutions of the form

$$\eta_x = \overline{\eta}_{x0} \cos(kz - \omega t), \qquad \eta_y = \overline{\eta}_{y0} \sin(kz - \omega t), \tag{3.140a}$$

where the real constants $\overline{\eta}_{x0}$ and $\overline{\eta}_{y0}$ measure the lengths of the principal axes of the elliptical particle paths. Note that according to (3.140a) the sense of rotation of the particle paths about the ellipse is determined by

$$\mathcal{R} = \operatorname{sgn}\big((\boldsymbol{\eta} \times \mathbf{u}) \cdot \mathbf{e}_z\big), \qquad \text{where} \qquad \boldsymbol{\eta} \times \mathbf{u} = (0,\, 0,\, -\omega\overline{\eta}_{x0}\overline{\eta}_{y0}) \tag{3.140b}$$

is a constant vector independent of the sign of k. Note, however, that the sense of rotation is also linked to the helicity $\mathbf{u} \cdot \boldsymbol{\nabla} \times \mathbf{u}$ because the wave (3.140a) satisfies the relation

$$\omega^2 \boldsymbol{\eta} \times \mathbf{u} = (\mathbf{u} \cdot \boldsymbol{\nabla} \times \mathbf{u})\, \mathbf{c}_p\,, \tag{3.140c}$$

where as in (3.93b) $\mathbf{c}_p = (\omega/k)\mathbf{e}_k$ is the phase velocity.

The x and y–components of (3.139a,a) are satisfied when

$$(\omega^2 - \omega_M^2 - \omega_A^2)\, \overline{\eta}_{x0} = 2\Omega_z \omega\, \overline{\eta}_{y0}\,, \tag{3.142a}$$

$$(\omega^2 - \omega_M^2)\, \overline{\eta}_{y0} = 2\Omega_z \omega\, \overline{\eta}_{x0}\,, \tag{3.142b}$$

where $$\omega_M^2 = (kB_z)^2/(\rho\mu)\,, \qquad \omega_A^2 = N^2 \sin^2 \alpha\,. \tag{3.142c}$$

From (3.142a,b), we may deduce that the square of the ellipse aspect ratio is

$$\left(\frac{\overline{\eta}_{y0}}{\overline{\eta}_{x0}}\right)^2 = \frac{\omega^2 - \omega_M^2 - \omega_A^2}{\omega^2 - \omega_M^2}\,, \tag{3.143a}$$

while the dispersion relation for ω is

$$(\omega^2 - \omega_M^2 - \omega_A^2)(\omega^2 - \omega_M^2) - \omega_C^2\omega^2 = 0 \qquad (\omega_C^2 = 4\Omega_z^2) \quad (3.143b)$$

consistent with (3.137a-f). The rotational sense $\mathcal{R} = \mathrm{sgn}(-\omega\overline{\eta}_{x0}\overline{\eta}_{y0})$ of the particle paths determined by (3.142a,b) is

$$\mathcal{R} = \mathrm{sgn}\left(\Omega_z(\omega_M^2 - \omega^2)\right) = \mathrm{sgn}\left(\Omega_z(\omega_M^2 + \omega_A^2 - \omega^2)\right). \quad (3.143c)$$

The four roots $\omega = \pm\omega_S$ and $\pm\omega_F$ of (3.143b), where $0 \le \omega_S \le \omega_F$, identify slow ω_S and fast ω_F waves with opposite rotational senses

$$\mathcal{R}_S = -\mathcal{R}_F = \mathrm{sgn}\,\Omega_z\,. \quad (3.144)$$

Here it is important to appreciate that our wave front plane with normal $\mathbf{e}_z$ has been orientated relative to the direction of gravity $-g\mathbf{n}$ such that $\mathbf{n}\cdot\mathbf{e}_z \ge 0$ and not the direction $(\mathrm{sgn}\,k)\mathbf{e}_z$ of the wave vector. So it is in this sense that $\Omega_z = \mathbf{n}\cdot\mathbf{e}_\Omega$ enters the formula (3.144). As an example, the inertial waves of Section 3.3.1 have the fast wave rotational sense $-\mathrm{sgn}\,\Omega_z$, as illustrated by (3.89).

In the absence of stratification $\omega_A^2 = 0$ ($N^2 = 0$), both the slow and fast waves are circularly polarised $|\overline{\eta}_{y0}| = |\overline{\eta}_{x0}|$ and we recover the results of Section 3.3.3. Furtermore the result (3.123) illustrates the rotational sense rule (3.144), since $\omega_p \ge |\Omega_z|$.

In the absence of the Coriolis acceleration $\omega_C = 0$, plane waves are recovered. One class of waves is horizontal $\overline{\eta}_{x0} = 0$ Alfvén waves $\omega^2 = \omega_M^2$, for which the buoyancy does no work. The other class is gravity waves modified by the magnetic field $\omega^2 = \omega_M^2 + \omega_A^2$, which are confined in the direction of steepest descent/ascent $\overline{\eta}_{y0} = 0$. This latter class consists of fast (slow) waves in the case of stable (unstable) stratification $\omega_A^2 > 0$ ($0 \le -\omega_A^2 \le \omega_M^2$). For sufficiently large unstable stratification, namely $-\omega_A^2 > \omega_M^2$, stability is lost ($\omega_S^2 < 0$).

With the inclusion of both the stratification $\omega_A^2 \ne 0$ and the Coriolis acceleration $\omega_C \ne 0$, the particle paths are elliptical. In the case of stable stratification $\omega_A^2 > 0$, the slow (fast) waves $\omega_S^2 < \omega_M^2$ ($\omega_F^2 > \omega_M^2 + \omega_A^2$) are elongated in the horizontal (steepest descent/ascent) direction with $|\overline{\eta}_{y0}| > |\overline{\eta}_{x0}|$ ($|\overline{\eta}_{y0}| < |\overline{\eta}_{x0}|$) (see Figure 3.13b,c). By contrast, in the case of unstable stratification $\omega_A^2 < 0$, the slow (fast) waves $\omega_S^2 < \omega_M^2 + \omega_A^2$ ($\omega_F^2 > \omega_M^2$) are elongated in the steepest descent/ascent (horizontal) direction with $|\overline{\eta}_{y0}| < |\overline{\eta}_{x0}|$ ($|\overline{\eta}_{y0}| > |\overline{\eta}_{x0}|$).

As remarked the limit $\omega_M \ll |\omega_C|$ (3.129a) is geophysically interesting. For that the fast waves are pertubations of inertial waves with the inertial wave frequency $\omega_F = \pm|\omega_C|$. In the case of the slow waves inertia is negligable and the particle displacements satisfy

$$-(\omega_M^2 + \omega_A^2)\,\overline{\eta}_{x0} = 2\Omega_z\omega\,\overline{\eta}_{y0}\,, \qquad -\omega_M^2\,\overline{\eta}_{y0} = 2\Omega_z\omega\,\overline{\eta}_{x0}\,, \quad (3.145a,b)$$

in which

$$\omega = \pm\omega_S \approx \pm\omega_{MAC}\,, \qquad \text{where} \quad \omega_{MAC} = |\omega_M| \sqrt{\omega_M^2 + \omega_A^2} \Big/ |\omega_C| \tag{3.145c}$$

is the so-called MAC–wave frequency (Braginsky, 1967). Like the slow MC–wave (3.131) of Section 3.3.3 it exhibits the particle path rotational sense $\mathcal{R}_S = \operatorname{sgn}\Omega_z$ (3.144). Furthermore, the principal axis ratio of the MAC–wave ellipse is

$$|\overline{\eta}_{y0}| / |\overline{\eta}_{x0}| = \omega_{MAC}/\omega_{MC}\,, \tag{3.145d,e}$$

where $\omega_{MC} = \omega_M^2/|\omega_C|$ is the MC–wave frequency (3.129).

We describe briefly the role of dissipation as it appears in the general formulation (3.136). In geophysical situations, inertia and viscocity are unimportant. With those terms neglected, we can make the approximation

$$\Sigma = -\omega_M^2/(\omega + \mathrm{i}\eta k^2) \tag{3.146}$$

in (3.137b) and simplifications follow. Since the magnetic diffusion timescale is arguably comparable to the MC–wave timescale, the modification involving (3.146) is an attractive direction of investigation. Furthermore, the excitation of MAC–waves with both the magnetic and thermal diffussions included involves a comprehensive study of convective and resistive instabilities which lies outside the scope of this chapter but is reviewed extensively elsewhere (see, for example, Acheson & Hide, 1973; Fearn, Roberts, & Soward, 1988; Gubbins & Roberts, 1987).

3.4. CONVECTION IN ROTATING SPHERICAL FLUID SHELLS

In the previous section we addressed free waves in a non dissipative system, we will now address dissipative systems for which an energy source is necessary to maintain motions. In most natural dynamos, the relevant source is buoyancy and the resulting motions are then referred to as "convection".

3.4.1. PHYSICAL MOTIVATIONS

Convection driven by thermal buoyancy in rotating spherical bodies of fluid has long been recognised as a fundamental process in the understanding of the properties of planets and stars. Since these objects are rotating in general and since their evolution is associated with the transport of heat from their interiors convection influenced by the Coriolis force does indeed play a dominant role in the dynamics of their fluid parts. In the case of the Earth it is the generation of the geomagnetic field by motions in the molten outer iron core which has stimulated much interest in the subject of convection in rotating spheres. But the zones and belts seen on Jupiter are a just as interesting phenomenon driven by convection in the deep atmosphere of the planet. Similarly, the differential rotation of the Sun and its magnetic cycle are intimately connected with the solar convection zone encompassing the outer 30 percent of the Sun in terms of its radial extent (see Figure 6.1).

Theoretical studies of convection in rotating fluid spheres started about 50 years ago. The attention was restricted to the linear problem of the onset of convection and for simplicity axisymmetric motions were assumed. An account of these early efforts can be found in Chandrasekhar's famous treatise (1961) and in the papers by Bisshopp & Niiler (1965) and by Roberts (1965). A little later it became evident that the preferred forms of convection in the interesting limit of rapid rotation are not axisymmetric, but highly non–axisymmetric (Roberts, 1968). In this later paper, however, the incorrect assumption was made that the preferred mode of convection exhibits a z–component of the velocity field parallel to the axis that is symmetric with respect to the equatorial plane. The correct mode for the onset of convection was found by Busse (1970a) who approached the problem on the basis of the rotating cylindrical annulus model. This model takes advantage of the approximate validity of the Proudman–Taylor theorem (see Section 3.1.1) and the analysis can thus be reduced from three to two spatial dimensions.

3.4.2. CONVECTION IN THE ROTATING CYLINDRICAL ANNULUS

Convection in the fluid filled gap between two rigidly rotating coaxial cylinders is receiving increasing attention because it shares many linear and nonlinear dynamical properties with convection in rotating spherical fluid shells, at least as far as columnar convection is concerned. From the point of view of planetary applications it seems natural to have gravity pointing inward and to keep the inner cylinder at the higher temperature T_2. However, since the centrifugal force is used as a source of buoyancy in laboratory experiments with the higher temperature at the outer cylinder we shall use this latter configuration as shown in Figure 3.14. Since only the product of effective gravity and applied temperature gradient is physically relevant the two cases are equivalent. In experimental realisations of the system ordinary gravity plays a minimal role when a vertical axis of rotation is used and when the rate of rotation is sufficiently high such that the centrifugal force exceeds gravity by at least a factor of two or three. An important ingredient of the geometrical configuration shown in Figure 3.14 are the conical boundaries at top and bottom which

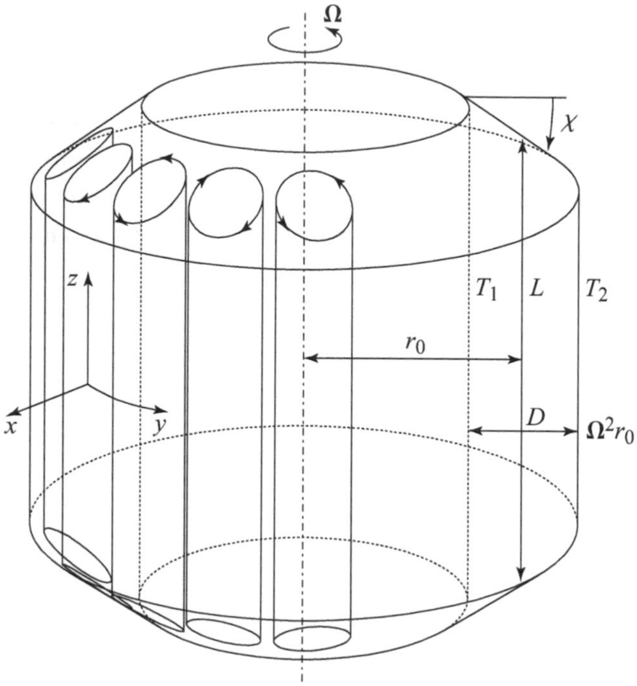

Figure 3.14 - Sketch of the geometrical configuration of the rotating cylindrical annulus.

cause a variation in height with distance from the axis of rotation. Without this variation in height steady two–dimensional convection rolls aligned with the axis will be realised since they obey the Proudman–Taylor condition. The Coriolis force is entirely balanced by the pressure gradient in this case and Rayleigh number for onset of convection in the small gap limit is given by the Rayleigh–Bénard value for a non–rotating layer. Thin Ekman layers at the no–slip top and bottom boundaries exert only a minor influence on the dynamics of convection if the height L of the annulus is sufficiently large in comparison to the gap size. As soon as the height changes in the radial direction any flow involving a radial velocity component can no longer satisfy the geostrophic balance. Instead a weak time dependence is required and the flow assumes the character of Rossby waves as introduced in Section 3.2.4 on page 159. These waves are well known in the meteorological context where the variation of the vertical component of rotation with latitude has the same effect as the variation of height in the annulus of Figure 3.14. The dynamics of Rossby waves can be visualised most readily if the action of the vorticity acquired by the fluid columns displaced radially from the middle of the gap is considered. As indicated in Figure 3.15 columns shifted inward acquire cyclonic vorticity because they are stretched owing to the increasing height. The opposite sign of vorticity is exhibited by columns moving outward. Since their moments of inertia are increased they must

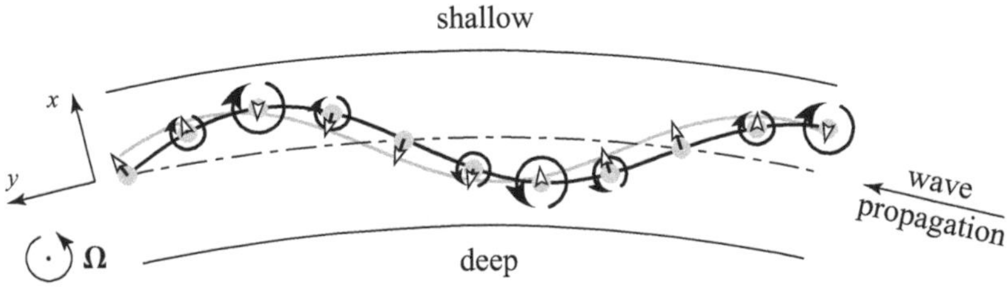

Figure 3.15 - The mechanism of propagation of a Rossby wave visualised in the equatorial plane of the rotating annulus: Fluid columns originally resting at the mid–surface acquire anti–cyclonic vorticity relative to the rotating system when they are displaced outwards towards the shallow region. Cyclonic vorticity is acquired by the columns displaced inwards. The action of the columnar motion on the neighboring fluid columns is such that an initial sinusoidal displacement propagates in the prograde direction.

rotate anti–cyclonically relative to the rotating system in order to conserve angular momentum. The action of the acquired motion of sinusoidally displaced columns on their neighbors results in the propagation of a wave as shown in Figure 3.15. The phase velocity is in the prograde (retrograde) direction when the height decreases (increases) with distance from the axis.

In the case of convection the dynamics is modified by the presence of thermal buoyancy and the phase velocity is less than that of Rossby waves except in the limit of vanishing Prandtl number. In the laboratory thermal Rossby waves, as the propagating convection waves are called, can be realised relatively easily (Busse & Carrigan, 1974). In Figure 3.16 photographs of a combination of two experiments on convection driven by centrifugal buoyancy in a rotating cylindrical annulus are shown. In the upper part an annular region of constant height is realised and convection occurs as a nearly geostrophic flow (except for the thin Ekman layers at the upper and lower boundaries). Accordingly the Coriolis force is balanced by the pressure gradient and the Rayleigh number for onset of convection is nearly the same as in the case of a fluid layer heated from below. In the lower part of the apparatus an annular region of varying height is realised through conical end boundaries as sketched in Figure 3.14. In this situation the Rayleigh number for onset of convection is increased such that in the left photograph the basic state of pure conduction is still stable. Only after a further increase of the temperature difference across the annular gap does convection set in the lower annular region as well. A movie would be needed to demonstrate the wave like propagation of the columns, but is worth noting that the wavelength is smaller than that of the steady convection in the upper part as is predicted by theory.

The mathematical analysis of thermal Rossby waves is quite simple if the inclination of the cones with respect to the equatorial plane is introduced as a small perturba-

Figure 3.16 - Photographs of convection rolls made visible by small flaky particles which align themselves with the shear of the fluid motion. The cylinder is rotating in a warm water bath and the inner cylindrical wall is cooled by water flowing through the axis. The upper annular region has parallel top and bottom boundaries, while the lower region is bounded by conical boundaries. Since the latter tend to inhibit convection the basic state of pure conduction is still stable in the bottom of the left picture, while convection has already developed in the top part of the apparatus. An increased Rayleigh number is necessary to cause the onset of the thermal Rossby waves as shown in the right picture.

tion. Using the gap width D as length scale and D^2/ν as timescale where ν is the kinematic viscosity of the fluid we may assume the velocity field in the form

$$\mathbf{u} = \nabla\psi(x, y, t) \times \mathbf{k} + \dots\,, \tag{3.147}$$

where $\mathbf{k}$ is the unit vector in the z–direction parallel to the axis of rotation and where the small gap approximation with x as radial coordinate and y as azimuthal coordinate has been assumed. Only the geostrophic component of $\mathbf{u}$ has been denoted explicitly in expression (3.147). Deviations from geostrophy are induced by the condition of vanishing normal velocity at the conical boundaries:

$$u_x\eta_0 \pm u_z = 0 \quad \text{at} \quad z = \pm\frac{L}{2D} \tag{3.148}$$

where the tangent η_0 of the angle χ between the cones and the equatorial plane has been introduced as small parameter. By taking the z–component of the curl of the equation of motion and averaging it over the height of the annulus we can incorporate the boundary condition (3.148) into an equation for the z–component of vorticity, $-\Delta_2\psi$:

$$(\partial_t + \partial_y\psi\partial_x - \partial_x\psi\partial_y)\,\Delta_2\psi - \Delta_2^2\psi - \widehat{\eta}\,\partial_y\psi = -\partial_y\Theta \tag{3.149a}$$

where Δ_2 is the two–dimensional Laplacian, $\Delta_2 = \partial^2/\partial x^2 + \partial^2/\partial y^2$. Equation (3.149a) must be considered together with the heat equation for the deviation Θ from the static temperature distribution of pure conduction:

$$\mathrm{Pr}\,(\partial_t + \partial_y\psi\partial_x - \partial_x\psi\partial_y)\,\Theta + \widehat{\mathrm{Ra}}\,\partial_y\psi = \Delta_2\Theta\,, \tag{3.149b}$$

where Θ is measured in multiples of $(T_2 - T_1)\mathrm{Pr}/\widehat{\mathrm{Ra}}$. T_1 and T_2 are the temperatures prescribed at the inner and outer cylindrical boundaries, respectively. We have used here a non–dimensional form which differs slightly from that of Chapter 1, Section 1.1.3. The Prandtl number ($\mathrm{Pr} = \nu/\kappa$) is unaltered. The strength of the Coriolis term is not measured here by the inverse of the Ekman number, but by the dimensionless Coriolis parameter $\widehat{\eta}$ (not to be confused with the forthcoming Coriolis number τ) defined as

$$\widehat{\eta} = \frac{4\Omega\eta_0 D^3}{\nu L}\,, \tag{3.150}$$

where Ω denotes, as before, the angular velocity of rotation. The Rayleigh number here takes a modified form $\widehat{\mathrm{Ra}}$ (where $r_0\,\Omega^2$ replaces gravity)

$$\widehat{\mathrm{Ra}} = \frac{\alpha(T_2 - T_1)\Omega^2 r_0 D^3}{\nu\kappa}\,, \tag{3.151}$$

where α is as before the coefficient of thermal expansion and r_0 is the mean radius of the annulus.

Assuming stress–free boundaries at the cylindrical walls,

$$\psi = \partial_{xx}\psi = \Theta = 0 \quad \text{at} \quad x = \pm\tfrac{1}{2}, \tag{3.152}$$

we obtain a completely specified mathematical formulation of the problem of centrifugally driven convection in the cylindrical annulus.

The onset of convection is described by the linearised version of (3.149a) which can be solved by

$$\psi = A \sin n\pi \left(x + \tfrac{1}{2}\right) \exp\{\mathrm{i}\alpha y - \mathrm{i}\omega t\}, \qquad \Theta = \frac{-\mathrm{i}\,\alpha\,\widehat{\mathrm{Ra}}\,\psi}{\alpha^2 + (n\pi)^2 - \mathrm{i}\omega}, \tag{3.153a,b}$$

with the following relationships for ω and $\widehat{\mathrm{Ra}}$:

$$\omega = \frac{\widehat{\eta}\,\alpha}{(1+\mathrm{Pr})(n^2\pi^2 + \alpha^2)}, \tag{3.154a}$$

$$\widehat{\mathrm{Ra}} = (n^2\pi^2 + \alpha^2)^3\alpha^{-2} + \left(\frac{\widehat{\eta}\,\mathrm{Pr}}{1+\mathrm{Pr}}\right)^2 (n^2\pi^2 + \alpha^2)^{-1}. \tag{3.154b}$$

As expected the dependence of the Rayleigh number on the wavenumber in the case of Rayleigh–Bénard convection in a non–rotating layer is recovered in the limit $\widehat{\eta} = 0$. The mode corresponding to $n = 1$ is preferred in this case, of course. This property continues to hold for finite $\widehat{\eta}$. But as the limit $\widehat{\eta} \to \infty$ is approached, the values of $\widehat{\mathrm{Ra}}$ and ω do not depend on n in first approximation as can be seen from the following expressions for the critical values in the limit of large $\widehat{\eta}$:

$$\alpha_c = \widehat{\eta}_P^{1/3}\left(1 - \frac{7}{12}\pi^2\widehat{\eta}_P^{-2/3} + \ldots\right), \tag{3.155a}$$

$$\widehat{\mathrm{Ra}}_c = \widehat{\eta}_P^{4/3}\left(3 + \pi^2\widehat{\eta}_P^{-2/3} + \ldots\right), \tag{3.155b}$$

$$\omega_c = \sqrt{2}\mathrm{Pr}^{-1}\widehat{\eta}_P^{2/3}\left(1 - \frac{5}{12}\pi^2\widehat{\eta}_P^{-2/3} + \ldots\right), \tag{3.155c}$$

where the definition $\widehat{\eta}_P = \widehat{\eta}\mathrm{Pr}\sqrt{1/2}(1+\mathrm{Pr})^{-1}$ has been used.

Expressions (3.155a,b,c) have been derived for $n = 1$, but they hold for arbitrary n when π^2 is replaced by $(n\pi)^2$. The weak dependence on the radial coordinate of the problem has two important consequences:

(i) The onset of convection is rather insensitive to the cylindrical boundaries. The analysis can thus be applied to the case of a sphere where these boundaries are missing.

(ii) Modes of different radial dependence correspond to the same critical parameters asymptotically. Secondary bifurcations right above threshold become possible through couplings of these modes.

The latter possibility is indeed realised in the form of the mean flow instability. A transition to a solution of the form

$$\psi = A\sin(\alpha y - \omega t)\sin\pi(x+\tfrac{1}{2}) + B\sin(\alpha y - \omega t + \varphi)\sin 2\pi(x+\tfrac{1}{2}) \quad (3.156)$$

occurs as the Rayleigh number is increased beyond the critical value unless the Prandtl number is rather small (Or & Busse, 1987). A characteristic property of solution (3.156) is the strong mean zonal shear which it generates through its Reynolds stress, $\overline{u_x u_y} \propto AB\sin\pi(x+1/2)\sin\varphi$, where the bar indicates the average over the y–coordinate. Both signs of the shear are equally possible since the sign of B is arbitrary. The mean flow instability corresponds to a tilt of the convection columns as indicated in Figure 3.17. When the columns are slightly tilted in the prograde sense towards the outside prograde momentum is carried outward and retrograde momentum is transported inwards leading to a differential rotation in which the outer fluid rotates faster than the inner one. An equilibrium is reached through viscous stresses which tend to oppose the differential rotation. The reverse situation occurs when the columns are tilted the other way as shown in Figure 3.17(b). The instability occurs because the differential rotation tends to increase the initial tilt and a feedback process is thus initiated. The mean flow instability of convection rolls is also possible in a non–rotating Rayleigh–Bénard layer. But there it is usually preceded by three–dimensional instabilities.

There is another way in which a differential rotation in the annulus can be generated. When curved cones instead of straight cones are used as indicated in Figure 3.18 solutions of the form (3.152) with separating x– and y–dependences are no longer possible. The term $\widehat{\eta}\,\partial_y\psi$ in (3.149a) must now be replaced by $\widehat{\eta}\,(1+\varepsilon x)\,\partial_y\psi$ where positive ε refers to the convex case of Figure 3.18(a) while a negative ε corresponds to the concave cones of Figure 3.18(b). The columns are tilted because the thermal Rossby wave has the tendency to propagate faster on the outside than on the inside when ε is positive and vice versa. A differential rotation prograde on the outside and retrograde on the inside must thus be expected for $\varepsilon > 0$ as shown in Figure 3.18(a) while the opposite results is obtained for $\varepsilon < 0$. The experiment of Busse & Hood (1982) has demonstrated this effect.

There are numerous other interesting features of convection in the cylindrical annulus such as vacillations and relaxation oscillations which appear at higher Rayleigh numbers and which can be related to analogous phenomena of convection in rotating spheres. We refer to the papers by Or & Busse (1987), Schnaubelt & Busse (1992), Brummell & Hart (1993) for details. The influence of non–axisymmetric

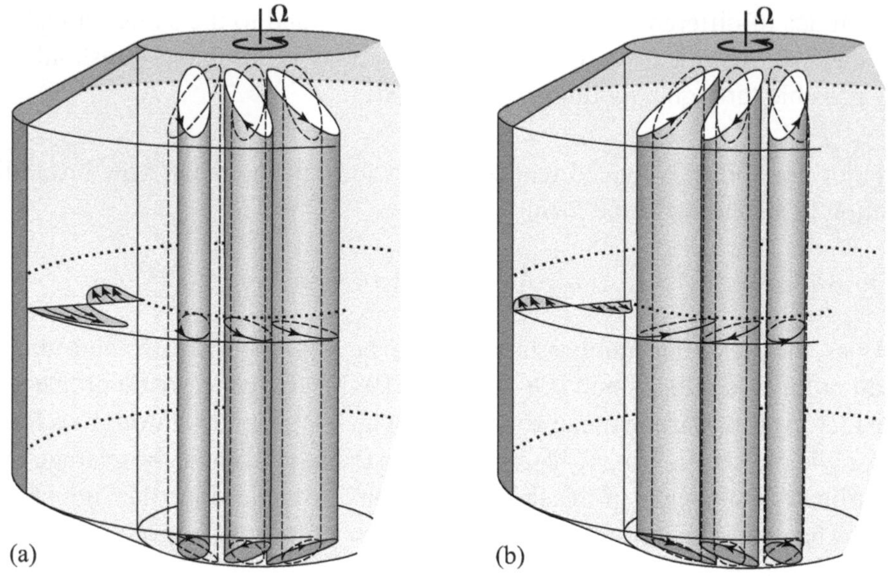

Figure 3.17 - The mean flow instability leading to either outward (a) or inward (b) transport of prograde angular momentum.

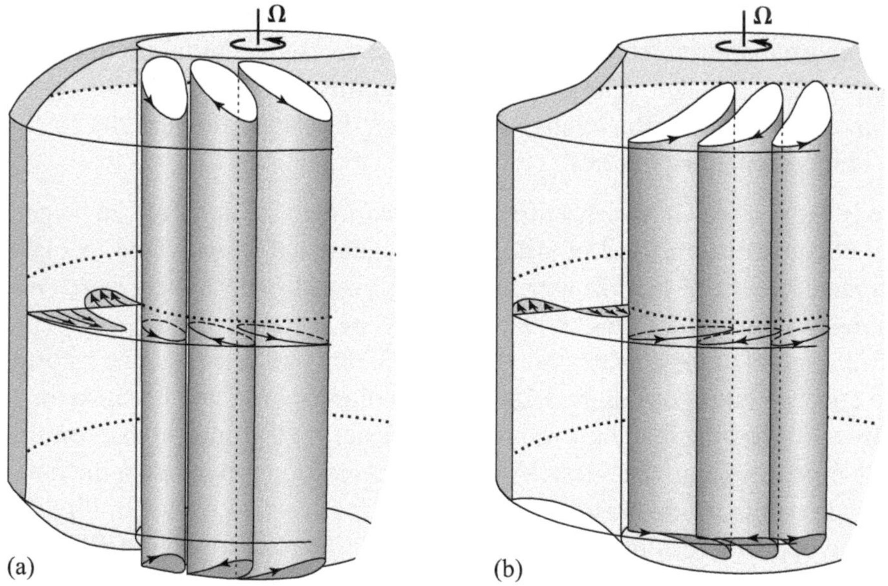

Figure 3.18 - Influence of curved conical boundaries. In the convex case (a) the columns tend to spiral outward in the prograde direction. They thus create a different rotation with higher angular velocity on the outside than on the inside. The opposite situation is found in the concave case (b).

modulations of the boundaries can also easily be investigated in the annulus model as shown by Bell & Soward (1996), Herrmann & Busse (1997) and Westerburg & Busse (2003). Besides the narrow gap limit the finite gap case of the rotating cylindrical annulus system is also of interest. The discrete manifold of realizable wavenumbers gives rise to resonances and changes in the character of secondary instabilities. We refer to the recent papers by Pino *et al.* (2000, 2001) and Chen & Zhang (2002).

3.4.3. MATHEMATICAL FORMULATION OF THE PROBLEM OF CONVECTION IN ROTATING SPHERICAL SHELLS

The presence of the centrifugal force in rotating planets and stars usually causes some deviations from spherical symmetry. The surfaces of constant potential become spheroidal and a basic state of vanishing motion relative to a rotating frame of reference may not exist since surfaces of constant density do not coincide with surfaces of constant potential in general. As a result of this baroclinicity a differential rotation must be expected (see, for example, Busse, 1982). But these effects are usually much smaller than those introduced by convection and it is thus a good approximation to neglect the effects of the centrifugal force and to assume that there exists a basic static solution with spherically symmetric distributions of gravity and temperature.

For the theoretical description of thermal convection in rotating spheres usually the case of a gravity varying linearly with radius, $\mathbf{g} = -\widetilde{g}_0\, d\,\mathbf{r}$, is assumed where $\mathbf{r}$ is the position vector with respect to the center of the sphere $\mathbf{r} = r\,\mathbf{e}_r$, which is made dimensionless with the thickness d of the shell.[7] We assume that a static state exists with the temperature distribution

$$T_S = T_0 - \widetilde{\beta}\, d^2 r^2/2 + \Delta T\, \widetilde{\eta}\, r^{-1}(1-\widetilde{\eta})^{-2}\,, \qquad (3.157)$$

where $\widetilde{\eta}$ denotes the ratio of inner to outer radius of the shell and $\widetilde{\beta}$ is proportional to a uniform density of heat sources in the sphere. In addition an applied temperature difference is admitted such that ΔT is the temperature difference between the boundaries in the special case $\widetilde{\beta} = 0$. Of course, in geophysical and astrophysical applications only the super–adiabatic part of the temperature field must be identified with the temperature distribution given above.

In addition to d (the thickness of the shell), the time d^2/ν and the temperature $\nu^2/\widetilde{g}_0 \alpha d^4$ are used here as scales for the dimensionless description of the problem. We use here the Boussinesq approximation introduced in Chapter 1, Section 1.1.3.

---

7 Note that $\widetilde{g}_0$ does not have the dimension of an acceleration, $\widetilde{g}_0\, d$ does.

We recall the basic equations of motion and the heat equation for the deviation Θ from the static temperature distribution:

$$\partial_t \mathbf{u} + (\mathbf{u}\cdot\boldsymbol{\nabla})\mathbf{u} + \tau\mathbf{k}\times\mathbf{u} = -\boldsymbol{\nabla}\pi + \Theta\mathbf{r} + \Delta\mathbf{u}\,, \qquad \boldsymbol{\nabla}\cdot\mathbf{u} = 0\,, \qquad (3.158\text{a,b})$$

$$\mathrm{Pr}\,(\partial_t\Theta + \mathbf{u}\cdot\boldsymbol{\nabla}\Theta) = \left[\mathrm{Ra}_i + \mathrm{Ra}_e\,\widetilde{\eta}\,r^{-3}\,(1-\widetilde{\eta})^{-2}\right]\mathbf{r}\cdot\mathbf{u} + \Delta\Theta\,. \qquad (3.158\text{c})$$

For convenience, the strength of the Coriolis term is here measured by the Coriolis number

$$\tau \equiv E^{-1} = \frac{2\Omega d^2}{\nu}\,. \qquad (3.159)$$

The Rayleigh numbers Ra_i and Ra_e, follow from Section 1.1.3 (but keeping in mind that $\widetilde{g}_0\,d$ has here the dimension of an acceleration)

$$\mathrm{Ra}_i = \frac{\alpha\widetilde{g}_0\widetilde{\beta}d^6}{\nu\kappa}\,, \qquad \mathrm{Ra}_e = \frac{\alpha\widetilde{g}_0\Delta T d^4}{\nu\kappa}\,. \qquad (3.160\text{a,b})$$

Since the velocity field $\mathbf{u}$ is solenoidal the general representation in terms of poloidal and toroidal components, as introduced Section 1.3.4, can be used:

$$\mathbf{u} = \boldsymbol{\nabla}\times(\boldsymbol{\nabla}u_p\times\mathbf{r}) + \boldsymbol{\nabla}u_t\times\mathbf{r}\,,$$

(see also Appendix B). By multiplying the (curl)^2 and the curl of equation (3.158a) by $\mathbf{r}$ we obtain two equations for u_p and u_t:

$$[(\Delta-\partial_t)L_2 + \tau\partial_\phi]\,\Delta u_p + \tau Q_3 u_t - L_2\Theta = -\mathbf{r}\cdot\boldsymbol{\nabla}\times[\boldsymbol{\nabla}\times(\mathbf{u}\cdot\boldsymbol{\nabla}\mathbf{u})] \quad (3.161\text{a})$$

$$[(\Delta-\partial_t)L_2 + \tau\partial_\phi]\,u_t - \tau Q_3 u_p = \mathbf{r}\cdot\boldsymbol{\nabla}\times(\mathbf{u}\cdot\boldsymbol{\nabla}\mathbf{u}), \qquad (3.161\text{b})$$

where ∂_ϕ denotes the partial derivatives with respect to the angle ϕ of a spherical system of coordinates r, θ, ϕ and where the operator L_2 was defined in (1.80) page 27, we recall

$$L_2 \equiv -r^2\Delta + \partial_r(r^2\partial_r)\,, \qquad (3.162\text{a})$$

and the operator Q_3 is defined by

$$Q_3 \equiv r\cos\theta\,\Delta - (L_2 + r\partial_r)(\cos\theta\partial_r - r^{-1}\sin\theta\partial_\theta)\,. \qquad (3.162\text{b})$$

Stress-free boundaries with fixed temperatures are most often assumed:

$$u_p = \partial_{rr}u_p = \partial_r(u_t/r) = \Theta = 0\,, \qquad (3.163)$$
$$\text{at}\quad r = r_i \equiv \widetilde{\eta}/(1-\widetilde{\eta}) \quad\text{and}\quad r = r_o = (1-\widetilde{\eta})^{-1}.$$

The numerical integration of equations (3.161a,b) together with boundary conditions (3.163) in the general nonlinear case proceeds with the pseudo–spectral method as described by Tilgner & Busse (1997) which is based on an expansion of all dependent variables in spherical harmonics for the θ, ϕ–dependences, i.e.

$$u_p = \sum_{\ell,m} V_\ell^m(r,t) P_\ell^m(\cos\theta) \exp\{im\phi\} \tag{3.164}$$

and analogous expressions for the other variables, u_t and Θ. P_ℓ^m denotes the associated Legendre functions. For the r–dependence expansions in Chebychev polynomials are used. For further details see also Busse *et al.* (1998). For the computations to be reported in Sections 3.4.5 and 3.4.6 a minimum of 33 collocation points in the radial direction and spherical harmonics up to the order 64 have been used. But in many cases the resolution was increased to 49 collocation points and spherical harmonics up to the order 96 or 128.

3.4.4. THE ONSET OF CONVECTION IN ROTATING SPHERICAL SHELLS

The main difficulty in analyzing convection in rotating spherical shells arises from the fact that the role of the Coriolis force varies with the angle between gravity and the vector $\mathbf{\Omega}$ of angular velocity. The geometrical configuration of the polar regions of the shell thus resembles that of a Rayleigh–Bénard layer rotating about a vertical axis while in the equatorial region the model of the rotating cylindrical annulus can be applied. Only at low rotation rates does convection set in in a global fashion and an axisymmetric mode can become preferred in this case (Geiger & Busse, 1981). At higher rotation rates the onset of convection does indeed occur in the form of the columnar modes as predicted on the basis of the annulus model of Section 3.4.2. A visualisation of the convection motion in the form of thermal Rossby waves travelling in the prograde direction is shown in Figure 3.19.

A rough idea of the dependence of the critical Rayleigh number Ra_{ic} for the onset of convection on the parameters of the problem in the case $\mathrm{Ra}_e = 0$ can be gained from the application of expressions (3.155a,b,c) of the annulus model:

$$\mathrm{Ra}_{ic} = 3\left(\frac{\tau\mathrm{Pr}}{1+\mathrm{Pr}}\right)^{4/3} (\tan\theta_m)^{8/3}\, r_m^{-1/3}\, 2^{-2/3}\,, \tag{3.165a}$$

$$m_c = \left(\frac{\tau\mathrm{Pr}}{1+\mathrm{Pr}}\right)^{1/3} (r_m \tan\theta_m)^{2/3}\, 2^{-1/6}\,, \tag{3.165b}$$

$$\omega_c = \left(\frac{\tau^2}{(1+\mathrm{Pr})^2\mathrm{Pr}}\right)^{1/3} 2^{-5/6} \left(\frac{\tan^2\theta_m}{r_m}\right)^{2/3}, \tag{3.165c}$$

where r_m refers to the mean radius of the fluid shell, $r_m = (r_i + r_o)/2$, and θ_m to the corresponding co–latitude, $\theta_m = \arcsin[r_m(1-\widetilde{\eta})]$. The azimuthal wavenumber

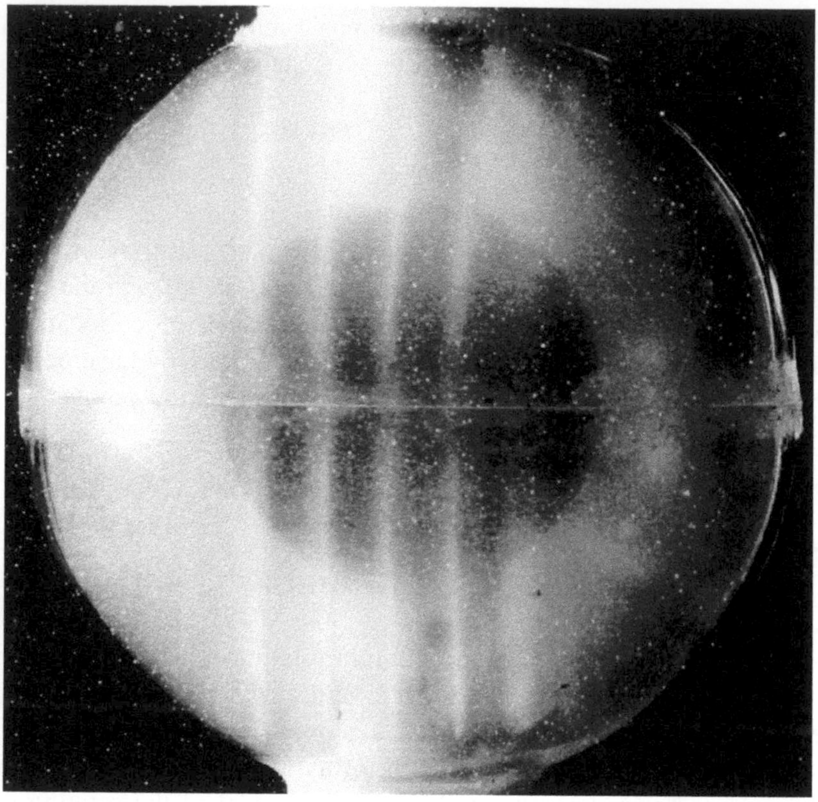

Figure 3.19 - Centrifugally driven convection in a rotating fluid between an inner cooled and an outer heated spherical boundary. The motions are visualised through nearly neutrally buoyant thin platelets which align themselves with the shear. The photograph shows the onset of thermal Rossby waves in the form of convection columns in a thick fluid shell. For details see Carrigan & Busse (1983).

of the preferred mode is denoted by m_c and the corresponding angular velocity of the drift of the convection columns in the prograde direction is given by ω_c/m_c. In Figure 3.20 the expressions (3.165a,c) are compared with accurate numerical values in the case $\mathrm{Ra}_e = 0$ which indicate that the general trend is well represented by expressions (3.165a,c). The same property holds for m_c. In the case $\mathrm{Ra}_i = 0$ the agreement with expressions (3.165a,c) is not quite as good since the onset of convection is more concentrated towards the tangent cylinder touching the inner boundary at its equator because of the higher temperature gradient in that region. Since we shall continue to restrict the attention to the case $\mathrm{Ra}_e = 0$, unless indicated otherwise, we shall drop the subscript i of Ra_i.

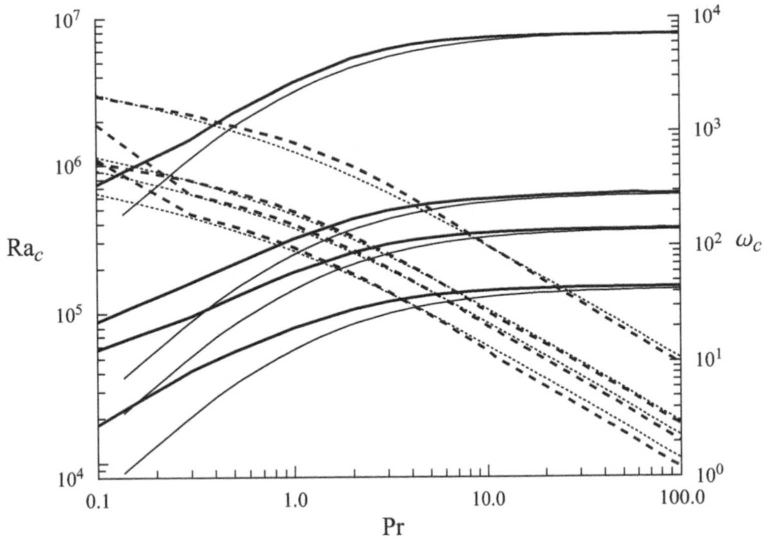

Figure 3.20 - Critical Rayleigh number Ra_{ic} (thick lines) and frequency ω_c (right ordinate, thick dashed lines) as a function of the Prandtl number Pr in the case $\widetilde{\eta} = 0.4$ for the Coriolis numbers $\tau = 5 \times 10^3$, 10^4, 1.5×10^4 and 10^5 (from bottom to top). The thin solid and dotted lines correspond to expression (3.165a) and (3.165c).

For the rigorous analysis of the onset problem in the limit of rapid rotation the dependence of the solution on the distance s from the axis must be considered which has been neglected in the application of the annulus model. It turns out that a finite difference exists between the results of the local analysis and the exact global analysis as was pointed out already by Soward (1977). Using a WKBJ approach with the double turning point method of Soward & Jones (1983b), Yano (1992) has analyzed the asymptotic problem in a refined version of the rotating cylindrical annulus model. He assumes a finite gap and the same dependence on s of the small inclination of the convex conical end surfaces as in the case of the sphere. The component of gravity parallel to the axis of rotation is still neglected. These assumptions have been dropped by Jones *et al.* (2000), who attacked the full spherical problem. They find that their results agree surprisingly well with those of Yano (1992). The asymptotic analysis of Jones *et al.* (2000) has recently been extended to the case of spherical shells with varying radius ratio by Dormy *et al.* (2004). These authors have also considered the case $\mathrm{Ra}_e \neq 0$, $\mathrm{Ra}_i = 0$ and have taken into account the effect of no–slip boundaries.

The analytical findings have been confirmed through numerous numerical studies (Zhang & Busse, 1987; Zhang, 1991, 1992a; Ardes *et al.*, 1997; Sun *et al.*, 1993) of which we show Figure 3.21 as an example. This figure emphasises the difference

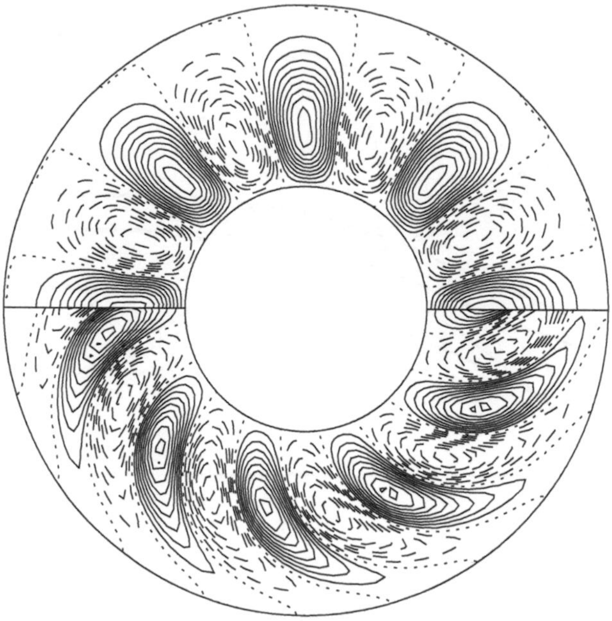

Figure 3.21 - Equatorial streamlines, $r\partial u_p/\partial\varphi = $const., of convection columns in the cases $\mathrm{Pr} = 10^5$, $\tau = 10^5$ and $\mathrm{Ra} = 3 \times 10^6$ (upper half) and $\mathrm{Pr} = 1$, $\tau = 10^4$ and $\mathrm{Ra} = 2 \times 10^5$ (lower half).

between the strongly spiralling nature of the convection columns at Prandtl numbers of the order unity or less and the more radially oriented columns at higher Prandtl numbers. This property is a result of the strong decrease of the frequency ω with increasing Pr.

The interest in convection in rotating spherical shells has motivated a number of laboratory investigations of the problem. The experiment of Hart *et al.* (1986) in which a spherically symmetric electric field acting on a dielectric insulating liquid is used to simulate gravity was carried out in space in order to avoid the interference from laboratory gravity. Less sophisticated experiments (Busse & Carrigan, 1976; Carrigan & Busse, 1983; Cardin & Olson, 1994; Cordero & Busse, 1992; Sumita & Olson, 2000) have used the centrifugal force with a cooled inner and a heated outer sphere to simulate the onset as well as the finite amplitude properties of convection. The main handicap of these experiments is the zonal flow generated as a thermal wind and the associated meridional circulation in the basic axisymmetric state (Cordero & Busse, 1992). But this handicap can be minimised through the use of high rotation rates and the observations correspond quite well to the theoretical expectations as shown by the example of Figure 3.19.

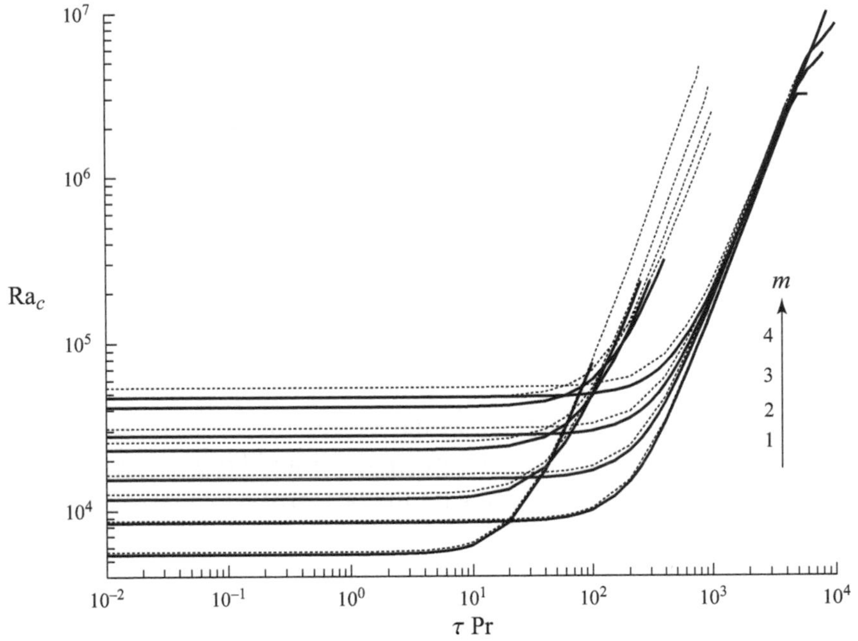

Figure 3.22 - The critical Rayleigh number Ra_c (solid lines) for the onset of the inertial convection as a function of $\tau\,\mathrm{Pr}$ for various wavenumbers m. The dotted lines correspond to the analytical expressions derived in Busse & Simitev (2004). For each m there are two modes corresponding to the two signs of the square root in expression (3.167b). The retrograde mode corresponding to the positive sign is preferred at lower values of $\tau\mathrm{Pr}$. But it turns up before the prograde mode which thus is preferred for larger values of $\tau\mathrm{Pr}$.

3.4.5. ONSET OF INERTIAL CONVECTION AT LOW PRANDTL NUMBERS

Onset of instability in the form of inertial convection is well known from the case of a plane horizontal layer heated from below and rotating about a vertical axis. As has been discussed by Chandrasekhar (1961) convection in the form of modified inertial waves represents the preferred mode at the onset of instability for Prandtl numbers of the order 0.6 or less for sufficiently high values of the Coriolis number τ. A similar situation is found for convection in rotating spherical shells where Zhang & Busse (1987) identified equatorially attached modes of convection as modified inertial waves. This connection has motivated Zhang (1994, 1995) to develop a perturbation approach for analytical description of the equatorially attached convection. Recently this approach has been extended and simplified by Busse & Simitev (2004). Here we shall just present a short introduction to the subject.

It is obvious from (3.161a,b) even in their linearised versions that solutions for which the $r-$ and $\theta-$dependences separate are not admissible in general. Nevertheless for some parameter regimes simple, physically realistic solutions of the linearised version of (3.161a,b) together with conditions (3.163) can be obtained through the separation ansatz

$$u_p = P_m^m(\cos\theta)\,\exp\{im\varphi - i\omega t\}\,f(r)\,, \tag{3.166a}$$
$$u_t = P_{m+1}^m(\cos\theta)\,\exp\{im\varphi - i\omega t\}\,g(r)\,, \tag{3.166b}$$
$$\Theta = P_m^m(\cos\theta)\,\exp\{im\varphi - i\omega t\}\,h(r)\,. \tag{3.166c}$$

Solutions of this form satisfy (3.161a,b) after the right hand sides have been dropped when the term proportional to P_{m+2}^m in the expression for $Q_3 u_t$ can be neglected. This term vanishes exactly in the limit of small Pr and high τ when the convection assumes the form of an inertial wave with the property

$$f(r) = \left(\frac{r}{r_o}\right)^m - \left(\frac{r}{r_o}\right)^{m+2}, \tag{3.167a}$$
$$g(r) = \frac{2im(m+2)(r/r_o)^{m+1}}{(2m+1)\left[\omega(m^2+3m+2) - m\right] r_o}, \tag{3.167b}$$
$$\omega = \frac{-\tau}{m+2}\left(1 \pm \left[1 + m(m+2)(2m+3)^{-1}\right]^{1/2}\right). \tag{3.167c}$$

Since this solution does not satisfy all the boundary conditions (3.163), weak Ekman layers must be added and finite critical Rayleigh numbers for onset have thus been obtained (Zhang, 1994). Results for the Rayleigh number for different values of the azimuthal wavenumber m are shown in figure 3.22. According to these results the mode with $m = 1$ is always preferred in the case $\widetilde{\eta} = 0$ if τPr is sufficiently low (Busse & Simitev, 2004). As τPr increases a transition occurs to the mode propagating in the prograde direction. Besides the equatorially wall attached mode, convection can be described approximately in the form (3.166) for arbitrary Prandtl numbers if τ is less than 10^3 (see Ardes *et al.*, 1997) or if the thin shell limit $\widetilde{\eta} \to 1$ is approached (Busse, 1970b,1973).

3.4.6. EVOLUTION OF CONVECTION COLUMNS AT MODERATE PRANDTL NUMBERS

In general the onset of convection in rotating fluid spheres occurs supercritically. As long as the convection assumes the form of shape preserving travelling thermal Rossby waves as described by linear theory, its azimuthally averaged properties are time independent. In fact, as seen from a frame of reference drifting together with

Figure 3.23 - Time periodic vacillations of convection at $\mathrm{Ra} = 2.8 \times 10^5$ (left) and $\mathrm{Ra} = 3 \times 10^5$ (right) for $\tau = 10^4$, $\mathrm{Pr} = 1$ The streamlines, $r\, \partial u_p / \partial \varphi = \mathrm{const.}$ are shown in one quarter of the equatorial plane. The four quarters are equidistant in time with $\Delta t = 0.015$ ($\Delta t = 0.024$) in the left (right) case in the clockwise sense such that approximately a full period is covered by the circles.

the convection columns the entire pattern is steady. A differential rotation is generated through the action of the Reynolds stress as explained in Section 3.4.2. The latter is caused by the spiralling cross section of the columns which persists as a dominant feature at moderate Prandtl numbers far into the turbulent regime. The plots of the streamlines $r\, \partial u_p / \partial \phi = \mathrm{const.}$ in the equatorial plane shown in any of the quarter circles of Figure 3.23 give a good impression of the spiralling nature of the columns.

A true time dependence of convection develops in the form of vacillations after a subsequent bifurcation. First the transition to amplitude vacillations occurs in which case just the amplitude of convection varies periodically in time as exhibited in the left plot of Figure 3.23. At a somewhat higher Rayleigh number shape vacillations become noticeable which are characterised by periodic changes in the structure of the columns as shown in the right plot of Figure 3.23. The outer part of the columns is stretched out, breaks off and decays. The tendency towards breakup is caused by the fact that the local frequency of propagation varies with distance from the axis according to expression (3.165c) after θ_m has been replaced by the local co–latitude θ.

The two types of vacillations also differ significantly in their frequencies of oscillation. This is evident from the time records of the energy densities of convection which have been plotted in Figure 3.24. This figure gives an overview of the evolution of time dependence in the interval $2.8 \times 10^5 \leq \mathrm{Ra} \leq 10^6$. The various

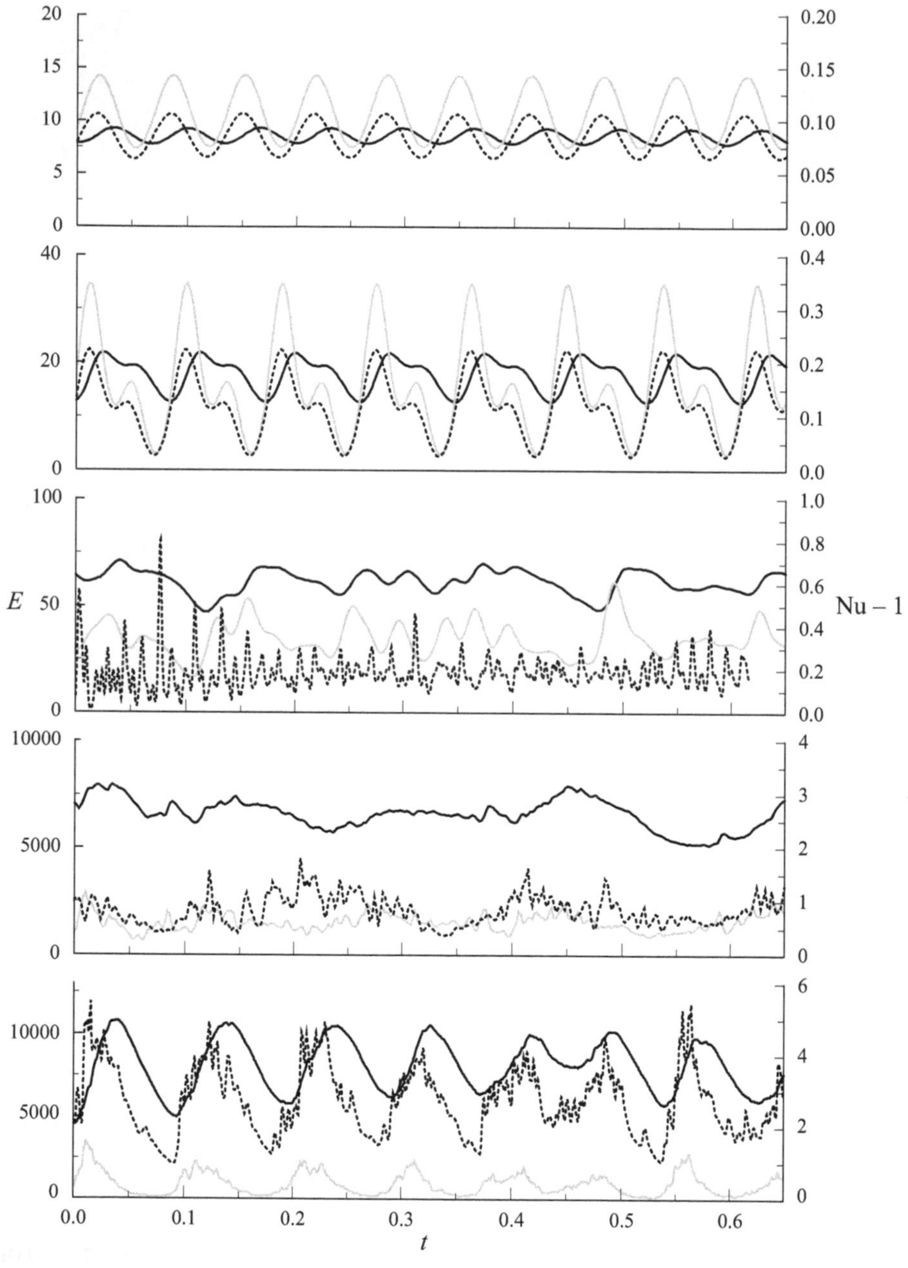

Figure 3.24 - Time series of energy densities E_t^m (thick solid lines), E_t^f (thin solid lines) and Nusselt number (dotted lines, right ordinate) are plotted for $\mathrm{Pr} = 1$, $\tau = 10^4$, and $\mathrm{Ra} = 2.8 \times 10^5$, 3×10^5, 3.5×10^5, 7×10^5 and 12×10^5 (from top to bottom). E_p^m and E_p^f have not been plotted. E_p^m is several orders of magnitude smaller than the other energies and E_p^f always approaches closely $0.4 \times E_t^f$.

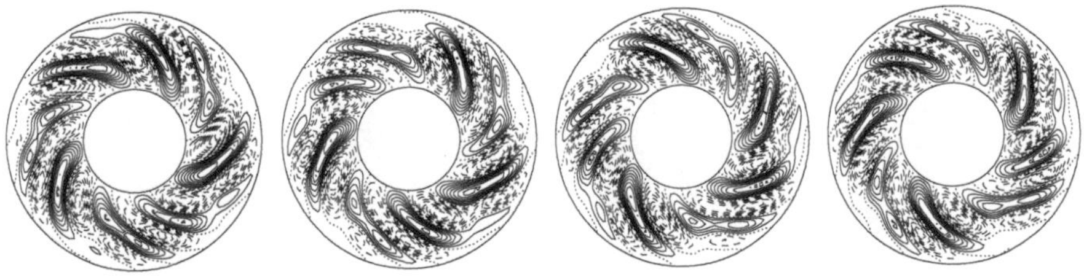

Figure 3.25 - Modulated shape vacillations of convection for $\mathrm{Ra} = 2.9 \times 10^5$, $\tau = 10^4$, $\mathrm{Pr} = 1$. The plots show streamlines, $r\,\partial u_p/\partial\phi = \text{const.}$, in the equatorial plane and are equidistant in time with $\Delta t = 0.04$ so that approximately a full period is covered.

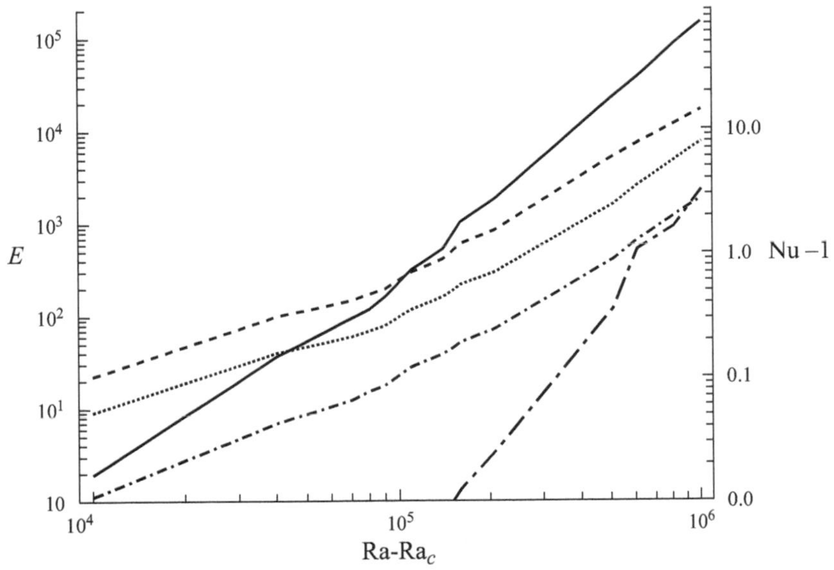

Figure 3.26 - Energy densities E_t^m (solid line), E_p^m (long dash – short dash line, multiplied by the factor 100), E_t^f (dashed line), E_p^f (thick dashed line) and the Nusselt number Nu (dash–dotted line, right ordinate) are plotted as function of $\mathrm{Ra} - \mathrm{Ra}_c$ in the case $\tau = 10^4$, $\mathrm{Pr} = 1$. $\mathrm{Ra}_c = 1.9 \times 10^5$ has been used corresponding to $m_c = 10$.

components of the energy densities are defined by

$$E_p^m = \tfrac{1}{2}\langle|\,\boldsymbol{\nabla}\times(\boldsymbol{\nabla}\overline{u}_p\times\mathbf{r})\,|^2\rangle,\quad E_t^m = \tfrac{1}{2}\langle|\,\boldsymbol{\nabla}\overline{u}_t\times\mathbf{r}\,|^2\rangle, \tag{3.168a,b}$$

$$E_p^f = \tfrac{1}{2}\langle|\,\boldsymbol{\nabla}\times(\boldsymbol{\nabla}\check{u}_p\times\mathbf{r})\,|^2\rangle,\quad E_t^f = \tfrac{1}{2}\langle|\,\boldsymbol{\nabla}\check{u}_t\times\mathbf{r}\,|^2\rangle, \tag{3.168c,d}$$

where $\overline{u}_p$ refers to the azimuthally averaged component of u_p and $\check{u}_p$ is given by $\check{u}_p = u_p - \overline{u}_p$.

With a further increase of the Rayleigh number spatial modulations of the shape

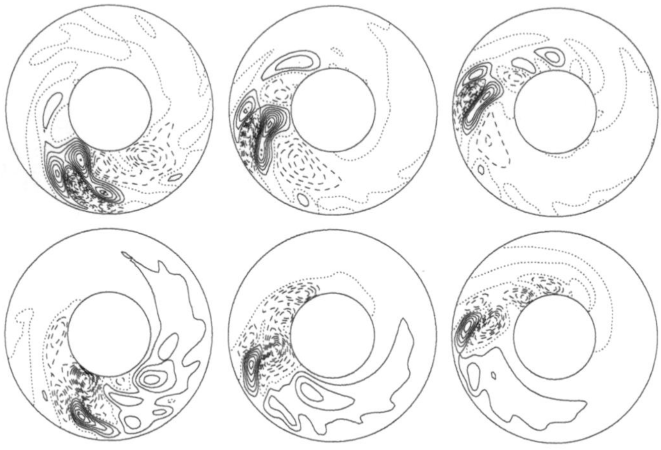

Figure 3.27 - Localised convection for $\mathrm{Ra} = 7 \times 10^5, \tau = 1.5 \times 10^4, \mathrm{Pr} = 0.5$. The streamlines, $r\, \partial u_p/\partial\varphi$ = const. (first row) and the isotherms, Θ = const. (second row), are shown in the equatorial plane for equidistant times (from left to right) with $\Delta t = 0.03$.

vacillations occur as shown in Figure 3.25. These modulations often correspond to a doubling of the azimuthal period but soon contributions with the azimuthal wavenumber $m = 1$ arise as shown in Figure 3.25. The pattern in this particular case is still periodic if appropriately shifted in azimuth. But as further modulations enter convection becomes quasiperiodic and with increasing Ra a chaotic state is reached. We refer to Simitev & Busse (2003a) which includes some movies to demonstrate the time dependence of convection. Figure 3.24 also demonstrates the diminishing fraction of the total kinetic energy that is associated with the poloidal component of motion which carries the convective heat transport. The differential rotation in particular increases much faster with Ra than the amplitude of convection since the Reynolds stress is proportional to the square of the latter. While this is already evident from the sequence of plots in Figure 3.24 it is even more obvious from Figure 3.26. Here it can also be seen that the onset of vacillations and aperiodic time dependence tends to increase the heat transport as indicated by the Nusselt number in contrast to the situation in a planar convection layer with the same Prandtl number (Clever & Busse, 1987). In the latter case the mismatch between the structure of the convection flow and the configuration of the boundary is absent which inhibits the heat transport in rotating spherical fluid shells. The time varying shift in the radial position of the convection columns thus promotes the heat transport. In Figure 3.26 and elsewhere in the paper the Nusselt number, Nu, is defined as the ratio of the average heat transport at the outer boundary divided by the conductive heat transport in the absence of convection.

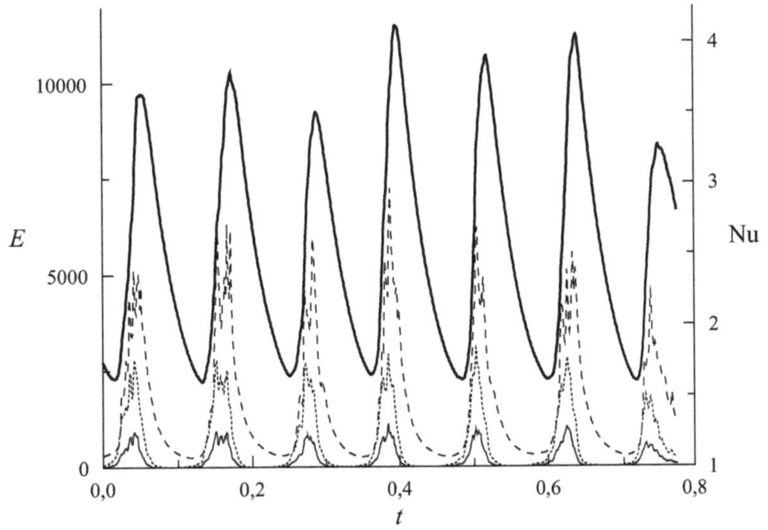

Figure 3.28 - Relaxation oscillation of turbulent convection in the case $\mathrm{Ra} = 10^6$, $\tau = 1.5 \times 10^4$, $\mathrm{Pr} = 0.5$. Energy densities E_t^m (solid line), E_t^f (dotted line), E_p^f (thin solid line) and the Nusselt number (long dashed line, right ordinate) are shown as functions of time t.

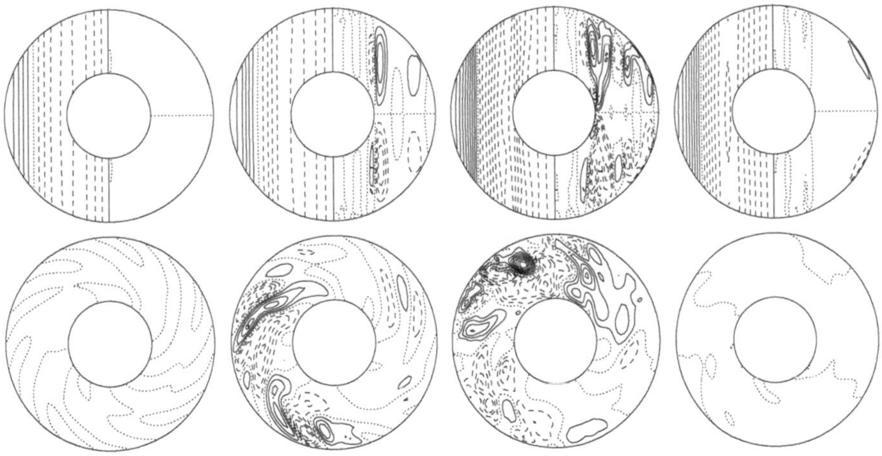

Figure 3.29 - Sequence of plots starting at $t = 0.12015$ and equidistant in time ($\Delta t = 0.016$) for the same case as in Figure 3.28. Lines of constant $\overline{u_\varphi}$ and mean temperature perturbation, $\overline{\Theta} = \text{const.}$ in the meridional plane, are shown in the left and right halves, respectively, of the first row. The second row shows streamlines, $r\,\partial u_p/\partial\varphi = \text{const.}$, in the equatorial plane.

Surprisingly the spatio–temporal randomness of convection columns does not just increase at larger values of Ra, but instead new coherent structures evolve (Grote & Busse, 2001). First there is localised convection as shown in Figure 3.27. The differential rotation has become so strong that its shearing action inhibits convection in most parts of the spherical fluid shell as is evident from the figure. Only in a certain region of longitude is convection strong enough to overcome the shearing action of differential rotation. In the "quiet" zone the basic temperature profile recovers towards the purely conducting state and thus provides the buoyancy in the interior of the shell which sustains the localised convection as it is recirculated into the "active" zone by the differential rotation.

After a further amplification of the differential rotation with increasing Ra the local intensification of convection no longer suffices to overcome the shearing action of the zonal flow. Instead of a spatial separation between "active" and "quiet" zones the system chooses a separation in time which manifests itself in the relaxation oscillations seen in the lowermost plot of Figure 3.24. The fluctuating component of motion is still rather turbulent in the case of the relaxation oscillation as demonstrated in Figures 3.28 and 3.29. When the differential rotation has decayed sufficiently in the near absence of Reynolds stresses generated by convection, a sudden burst of convection activity occurs leading to a sharp peak in the heat transport. But since the Reynolds stress grows just as suddenly as the kinetic energy of convection, the growth of the differential rotation occurs with only a slight delay. The shearing off of the convection columns then leads to their decay almost as quickly as they had set in. The relaxation oscillations occur over a wide region in the parameter space for high Rayleigh and Coriolis numbers and for Prandtl numbers of the order unity or less. Since it is mainly determined by the viscous decay of the differential rotation, the period of the relaxation oscillation does not vary much with these parameters. For the case of $\widetilde{\eta} = 0.4$ which has been used for almost all numerical simulations a period of about 0.1 is usually found. The sequence of bifurcations presented here can also be reproduced using a simple quasi–geostrophic model (inspired by the annulus model of Section 3.4.2). It is then found that these bifurcations occure increasingly close to the onset as the Coriolis number is increased (Morin & Dormy, 2004).

3.4.7. FINITE AMPLITUDE CONVECTION AT HIGHER PRANDTL NUMBERS

The transitions from drifting convection columns to vacillating convection and modulated vacillating convection do not change much as as the Prandtl number tends to high values (Zhang, 1992b). But as Pr increases the influence of the differential rotation which dominates the evolution of the convection columns for Prandtl num-

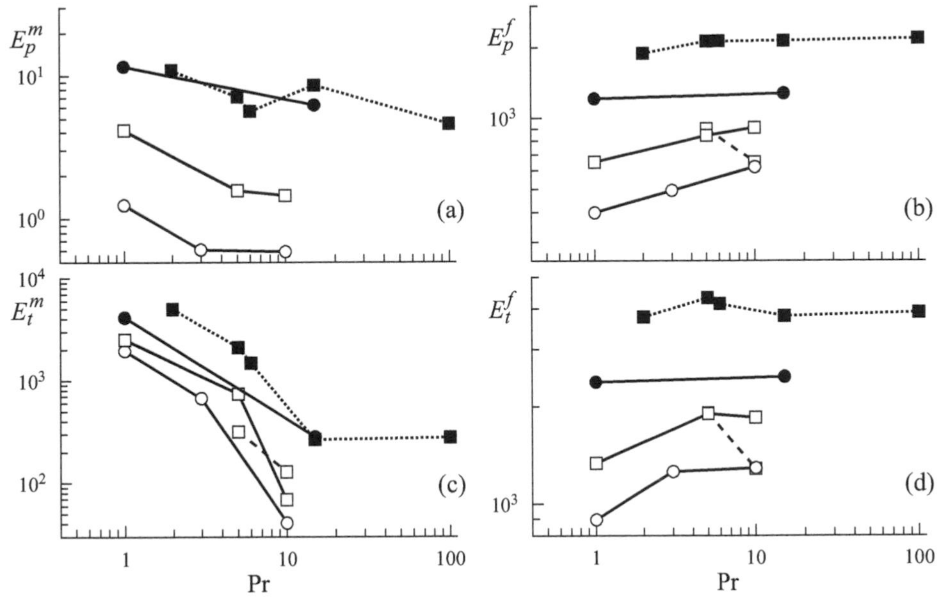

Figure 3.30 - Kinetic energy densities (a) E_p^m, (b) E_p^f, (c) E_t^m and (d) E_t^f all multiplied by Pr^2 as a function of the Prandtl number Pr in the case $\tau = 5 \times 10^3$. The values of the Rayleigh number $R = 5 \times 10^5$, 6×10^5, 8×10^5, 10^6 are denoted by empty circles and squares and full circles and squares, respectively.

bers of the order unity and below diminishes rapidly. The feedback process exhibited most clearly by the mean flow instability discussed in Section 3.4.2 ceases to operate at Prandtl numbers of the order 10. Above this value of Pr the properties of convection become nearly independent of Pr when the thermal timescale instead of the viscous one is used. This property which is familiar from Rayleigh–Bénard convection in plane layers heated from below holds rather generally in convection systems. For this reason we have plotted in Figure 3.30 energy densities as defined by expressions (3.168a,b), but multiplied by the factor Pr^2 in order to demonstrate the tendency towards independence of Pr. It should be noted that most of values used in this figure have been obtained from dynamo computations. But the action of the Lorentz force is rather weak and hardly affects the Prandtl number dependence of convection. Only the lower left plot for the energy densities of the differential rotation shows the expected strong decay with increasing Pr. These energy densities do not decay to zero for $\mathrm{Pr} \to 0$, however. The thermal wind relationship obtained from the azimuthal average of the ϕ–component of the curl of (3.158a)

$$\tau \mathbf{k} \cdot \nabla \overline{u}_\phi = \partial_\theta \overline{\Theta} \tag{3.169}$$

continues to require a finite field $\overline{u}_\phi$ in the limit of high Pr. The right hand side is finite because the azimuthally averaged temperature field $\overline{\Theta}$ deviates strongly from

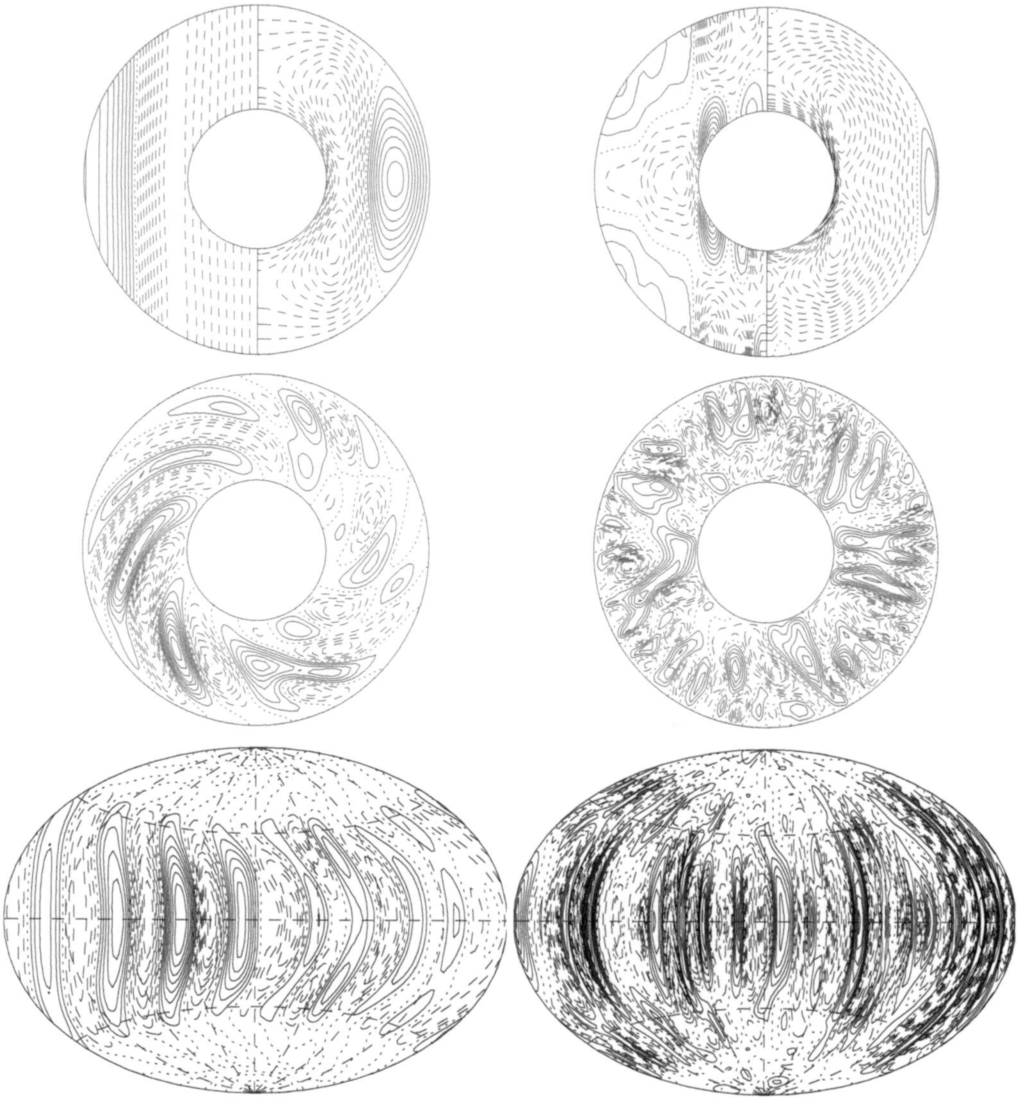

Figure 3.31 - Convection in rotating spherical fluid shells in the cases $\tau = 10^4$, $\mathrm{Ra} = 4 \times 10^5$, $\mathrm{Pr} = 1$ (left column) and $\tau = 5 \times 10^3$, $\mathrm{Ra} = 8 \times 10^5$, $\mathrm{Pr} = 20$ (right column). Lines of constant mean azimuthal velocity $\overline{u}_\varphi$ are shown in the left halves of the upper circles and isotherms of $\overline{\Theta}$ are shown in the right halves. The plots of the middle row show streamlines, $r\,\partial u_p/\partial\varphi$ = const., in the equatorial plane. The lowermost plots indicate lines of constant u_r in the middle spherical surface, $r = r_i + 0.5$.

Figure 3.32 - Streamlines $r\,\partial u_p/\partial\phi = \text{const.}$ in the equatorial plane for the case $\mathrm{Pr} = 0.025$, $\tau = 10^5$ with $\mathrm{Ra} = 3.2 \times 10^5$, 3.4×10^5, 4×10^5 (from left to right).

spherical symmetry as long as the influence of rotation is significant.

The typical differences between convection at Prandtl numbers of the order unity and higher values are exhibited in Figure 3.31 where typical results obtained for $\mathrm{Pr} = 1$ and $\mathrm{Pr} = 20$ are compared at similar values of Ra and τ. The approximate validity of the thermal wind relationship (3.169) can be noticed from a comparison of the plots of $\overline{u}_\phi$ and of $\overline{\Theta}$ in the case $\mathrm{Pr} = 20$. The convection columns retain their alignment with the axis of rotation with increasing Pr, but the spiralling nature of their radial orientation disappears.

3.4.8. FINITE AMPLITUDE INERTIAL CONVECTION

For an analysis of nonlinear properties of equatorially attached convection we focus on the case $\mathrm{Pr} = 0.025$ with $\tau = 10^5$. The critical Rayleigh number for this case is $\mathrm{Ra}_c = 28300$ corresponding to $m = 10$. As Ra is increased beyond the critical value other values of m from 7 to 12 can be realised, but $m = 10$ and lower values are usually preferred. An asymptotic perfectly periodic solution with $m = 10$ or $m = 9$ can be found only for Rayleigh numbers close to the critical value when computations are started from arbitrary initial conditions. On the other hand, perfect periodic patterns appear to be stable with respect to small disturbances over a more extended regime of supercritical Rayleigh numbers. Distinct transitions like the transition to amplitude vacillations and to structure vacillations do not seem to exist for equatorially attached convection. Instead modulated patterns are typically already observed when Ra exceeds the critical value by 10% as can be seen in the plots of Figure 3.32. These modulations are basically caused by the superposition of several modes with neighboring values of the azimuthal wavenumber m which appear to propagate nearly independently. For example, the period of 1.68×10^{-2} visible in the energy densities shown in Figure 3.33 in the case $\mathrm{Ra} = 3.2 \times 10^5$

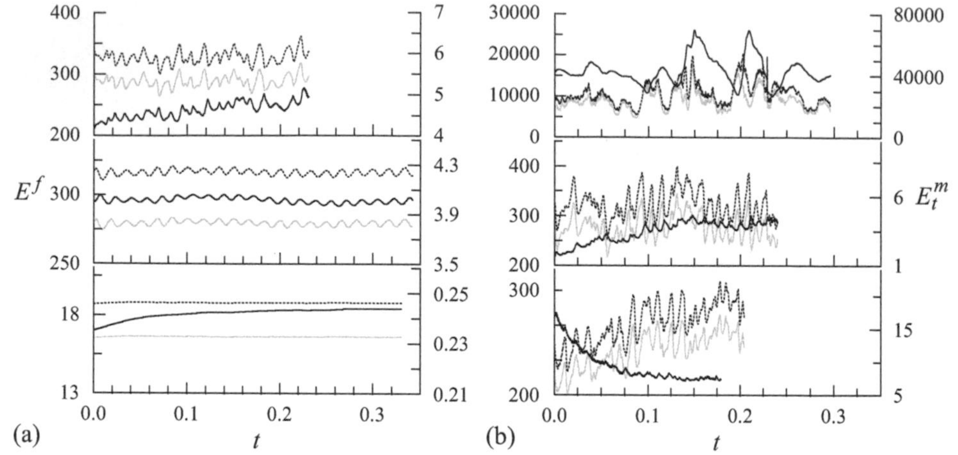

Figure 3.33 - Time series of energy densities of convection in the case $\mathrm{Pr} = 0.025$, $\tau = 10^5$, for $\mathrm{Ra} = 3.1 \times 10^5, 3.2 \times 10^5, 3.4 \times 10^5$ (from bottom to top, left side) and 3.8×10^5, 4×10^5, 8×10^5 (from bottom to top, right side). The solid, dashed and shaded lines correspond to E_t^m, E_t^f, E_p^f, respectively. E_t^m is measured on the right ordinate.

Figure 3.34 - Streamlines $r\,\partial u_p/\partial\phi = \text{const.}$ in the equatorial plane for the case $\mathrm{Pr} = 0.025$, $\tau = 10^5$ with $\mathrm{Ra} = 6 \times 10^5$, 8×10^5, 10^6 (from left to right).

is just about half the difference of 3.34×10^{-2} between the periods of the modes $m = 8$ and $m = 9$ according the inertial wave dispersion relation (3.167b) for prograde modes. This property must be expected if a small component with $m = 8$ participates in the pattern shown in Figure 3.32 which is dominated by the $(m = 9)$–component. The time series of energy densities shown in Figure 3.33 indicate that usually more than two modes contribute to the dynamics of the pattern with the exception of the case just discussed since the time dependence is not periodic as in the case when only two modes interact. The computations of the time series require a high spatial resolution together with a small time step. The time spans indicated in

Figure 3.33 are sufficient for reaching a statistically steady state of the fluctuating components of motion since these equilibrate on the fast thermal timescale of the order Pr^{-1}. Only close to Ra_c the adjustment process takes longer as can be seen in the case $\mathrm{Ra} = 3.1 \times 10^5$ where a ($m = 10$)–pattern approaches its equilibrium state. The pattern corresponding to the other cases of Figure 3.33 are shown in Figures 3.32, and 3.34. The differential rotation represented by E_t^m relaxes on the viscous timescale and therefore takes a long time to reach its asymptotic regime in the examples shown in Figure 3.33. But the differential rotation is quite weak such that it has a negligible effect on the other components of motion except in the case of the highest Rayleigh number of Figure 3.33. Even smaller is the axisymmetric part of the poloidal component of motion which is not shown in the plots of Figure 3.33. At higher Rayleigh numbers the convection eddies spread farther into the interior and in some cases become detached from the equator as can be seen in the plots of Figure 3.34. In this way the convection eddies contribute to the heat transport from the inner boundary. But at the same time they acquire the properties of the convection columns which are characteristic for convection at higher Prandtl numbers. Accordingly the differential rotation is steeply increased at $\mathrm{Ra} = 10^6$ and a tendency towards relaxation oscillation can be noticed in the upper right time series of Figure 3.33. For additional details on low Prandtl number spherical convection we refer to Simitev & Busse (2003a).

3.4.9. PENETRATIVE AND COMPOSITIONAL CONVECTION

Penetrative convection occurs when only part of the fluid layer is unstably stratified and convection flows penetrate to some extent into the stably stratified part. The subject of penetrative convection has been studied in the cases of the planetary boundary layer of the Earth's atmosphere and of the solar convection zone which is bounded from below by a radiative core in which entropy increases with distance from the center. But penetrative convection may also occur in planetary cores. In the outer part of the Earth's core, for example, the adiabatic temperature gradient increases so strongly in absolute value that a subadiabatic temperature gradient may be realised because it is sufficient to carry by thermal conduction the heat flux from the core to the mantle. Penetrative convection could thus occur in an upper stably stratified layer of the outer core.

The effects of a stably stratified sublayer of a convection layer are minimal when the corresponding temperature contrast (subadiabatic temperature difference) is small compared to the temperature contrast (superadiabatic temperature difference) in the adjacent convectively unstable layer. This property reflects the ability of convection to average the local variations of available buoyancy over the height of the layer. On the other hand, strongly stably stratified sublayers may have an effect approaching

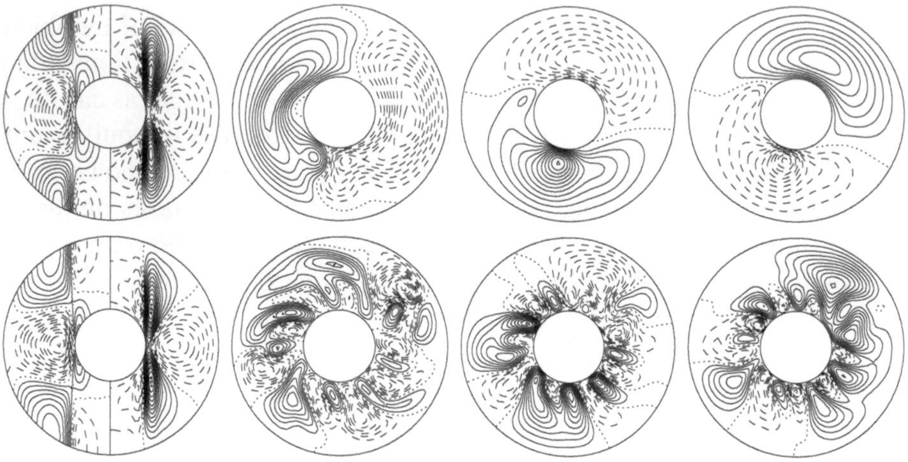

Figure 3.35 - Compositional convection in the presence of thermal stratification. The latter is described by $\mathrm{Ra}_e = -1.45 \times 10^6$ (-1.1×10^6) in the upper (lower) row while the compositional Rayleigh number $\mathrm{Ra}_i^{comp} = 2 \times 10^4$ and Lewis number $L = 80$ are the same in both cases. Lines of $u_\varphi =$const. (left half) and meridional streamlines, $r \sin\theta \partial_\theta \overline{u}_p =$const. (right half) are shown in the left circles. Streamlines, $r\, \partial u_p / \partial \varphi =$const., temperature perturbations, $\Theta =$const., and light element concentration perturbation, all in the equatorial plane, are shown in the subsequent plots (left to right). The values $\mathrm{Pr} = 0.1$ and $\tau = 10^4$ have been used.

that of a rigid boundary (Stix, 1970). Here we like to mention only the effect of a stably stratified layer in a rapidly rotating sphere which has been studied by Zhang & Schubert (1997, 2002). These authors find that convection in an inner part of a fluid shell which is surrounded by a stably stratified outer layer can excite toroidal motions in the outermost part of the shell. They call this phenomenon "teleconvection". This phenomenon disappears, however, as more stable stratifications are considered (Takehiro & Lister, 2001).

In the absence of an unstably stratified layer convection in a thermally stably stratified layer can still occur if there exists a gradient in the composition of the fluid. In the Earth's core, for example, light elements accumulate in the lower part of the outer core because of the growth of the solid inner core. Iron and nickel crystallise at its surface leaving lighter elements in solution. The latter provide a source of buoyancy for convection flows. It differs from the thermal buoyancy in that the diffusivity of the light elements is much lower – by the factor $1/\mathrm{L}$ where L is the Lewis number[8] – than the thermal diffusivity. The competition between the two types of buoyancy gives rise to new types of convection, especially when the ther-

8 The Lewis number is the ratio of the thermal diffusivity κ to the chemical diffusivity.

mal stratification is stabilizing. As has been shown by Busse (2002) in the case of the rotating cylindrical annulus model the Coriolis force may be balanced by one of the two buoyancy forces (which can be stabilizing or destabilizing) while the other buoyancy force (which must be destabilizing) gives rise to convection flows as in the absence of rotation. Figure 3.35 demonstrates that this effect also occurs in rotating spherical shells. In particular close to the onset of convection – as in the upper part of Figure 3.35 – the large scale $m = 1$–mode dominates as must be expected for convection in the absence of rotation. We note that the pattern of shown in Figure 3.35 drifts very slowly in the prograde direction with a frequency proportional to L^{-2}. The corresponding third root of the dispersion relationship for the frequency has not been determined explicitly in the analysis of Busse (2002), but it can easily be shown that it gives rise to $\mathrm{L}Rc + \mathrm{Ra} = a^6/\alpha^2$ in the notation of that paper.

Convection flows visible on the Sun (see Chapter 6) do not seem to exhibit any influence of rotation even though the Coriolis parameter τ based on the molecular value of viscosity is huge. The level of turbulence associated with the convection flows is so high, however, that an eddy viscosity of the order of $3 \times 10^8\,\mathrm{m}^2\,\mathrm{s}^{-1}$ is appropriate. This gives rise to a value of τ of the order unity if the supergranulation length scale (10^7 m) is used. Indeed effects of rotation are exhibited by convection flows on this scale. One of this effects is the drift of hexagon like cells. As shown by Busse (2004) this drift is a nonlinear phenomenon in contrast to the drift of convection columns considered earlier in this chapter. The drift thus assumes opposite directions for hexagonal cells with rising and with descending motion in the center. As an example of this drift we show the retrograde drift of a dodecahedral pattern of convection cells in Figure 3.36.

3.4.10. CONCLUDING REMARKS ON CONVECTION

The review of convection in rotating spherical fluid shells presented in this chapter is necessarily incomplete in view of the vast literature that has been accumulated in the past decades. We refer to the cited papers and the references given therein. A topic that has not been mentioned so far is the possibility of extrapolation of the results to conditions realised in planetary cores and in rapidly rotating stars. Based on molecular diffusivities huge Rayleigh and Coriolis numbers prevail in these situations and it is advantageous to use parameters that are independent of diffusivities. $\mathrm{Ro}^2 \equiv \mathrm{Ra}/(\mathrm{Pr}\,\tau^2)$ is such a parameter. Julien *et al.* (1996) refer to Ro as the convective Rossby number, while Christensen (2002) calls Ro^2 the modified Rayleigh number. In the latter paper it is demonstrated that a modified Nusselt number defined by $\widetilde{\mathrm{Nu}} = \mathrm{Nu}/(\tau\,\mathrm{Pr})$ appears to obey the asymptotic relationship $\widetilde{\mathrm{Nu}} = 0.0031\,\mathrm{Ro}^{5/2}$ in the case of convection in rotating spherical fluid shell with $\mathrm{Pr} = 1, \widetilde{\eta} = 0.35$ and $\mathrm{Ra}_i = 0$.

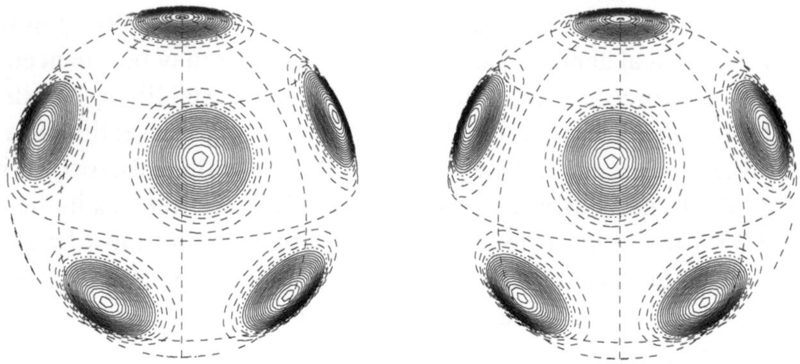

Figure 3.36 - Drifting hexagonal convection cells in a slowly rotating spherical shell. Two plots ($\Delta t = 0.39$ apart) of lines of constant u_r in the mid surface, $r = r_i + 0.5$, are shown for $\mathrm{Pr} = 1$, $\mathrm{Ra}_e = -6000$, $\mathrm{Ra}_i = 3000$, $\tau = 60$, $\widetilde{\eta} = 0.6$. In order to provide a preference for hexagonal cells with rising motion in the center the term $0.05 \times \Theta^2$ – representing a typical deviation from the Boussinesq approximation – has been added on the right hand side of (3.158a).

The most important application of the theory of convection in rotating spherical fluid shells is the dynamo process of the generation of magnetic fields in planets and in stars. This topic will be addressed in Chapters 4, 5 & 6 of this book. Here we just like to emphasise the intimate connection between the style of convection and the structure of the magnetic field generated by it. This property has been the focus of the papers by Grote & Busse (2001) and by Simitev & Busse (2005) and it has been reviewed in the article by Busse *et al.* (2003).

PART II

NATURAL DYNAMOS AND MODELS

CHAPTER 4

THE GEODYNAMO

David Fearn & Paul Roberts

This chapter is devoted to the planet for which we have the best knowledge, namely the Earth. We will review observational facts about the Earth's magnetic field in Section 4.1. We review the equations and parameters relevant to the Geodynamo in Section 4.2. We elaborate on the theoretical results outlined in Chapter 1 to derive the fundamental results appropriate to the Earth's core in Section 4.3. The existing constraints on the control parameters are addressed in Section 4.4. We then discuss the present status of numerical geodynamo models in Section 4.5. We focus then on the difficult problem of the small scale turbulent flow in the Earth's core in Sections 4.6 to 4.10.2. Next, we touch on some relevant geophysical issues (Section 4.7). Then (Section 4.8), we shall expound the traditional method, the insights it gives and the difficulties it faces. This is followed (Section 4.9) by a description of the engineering approach. Finally we discuss future developments on the geodynamo and a critique of turbulence in Section 4.10.

4.1. THE EARTH AND ITS MAGNETIC FIELD

4.1.1. A BRIEF HISTORY

Interest in the Earth's magnetic field goes back some 2000 years, in particular to the ancient Chinese, whose major achievements include the invention of the magnetic compass and the subsequent discovery of *declination*, the angle between magnetic and geographic north. Initially, the compass was believed to be attracted to the pole star. Later, the favoured source of attraction moved to the polar regions of the Earth and subsequently to the interior of the Earth. This followed the discovery in Europe in the XVIth Century of *inclination*, the angle of dip of the field direction. At this time, considerable effort was expended in mapping the declination and inclination as a potential aid to navigation.[9] The year 2000 was the 400th anniversary of William Gilbert's treatise on geomagnetism *De Magnete*. This gave the first rational explanation for the mysterious ability of the compass needle to point north-south; that the Earth itself was magnetic.

Up until this point, the Earth's magnetic field had been assumed to be steady, but it was not long before a series of observations at Greenwich led Henry Gellibrand in 1634 to deduce that the declination changes with time. This was the first observed feature of the so-called *Geomagnetic Secular Variation* (GSV), the slow (on a human timescale) change of the field emanating from the Earth's core. From detailed observations we now know this behaviour in considerable detail over the past few hundred years (see for example Courtillot & Le Mouël, 1988; Bloxham & Jackson, 1992; Jackson *et al.*, 2000).[10]

4.1.2. STRUCTURE OF THE EARTH

A spherical harmonic analysis of the geomagnetic field averaged over a few years shows clearly that the long-time field is essentially entirely of internal origin (see for example the discussion in Backus *et al.*, 1996). A key prerequisite to understanding the generation mechanism of the field is therefore a knowledge of the interior of the Earth. Our principal sources of information are: (A) the composition of meteorites thought to be characteristic of the material from which the Earth was formed, (B) the analysis of seismic waves, and (C) the properties of materials at high pressure determined from high-pressure experiments and, more recently, theoretical calculations. Useful references, explaining the ideas, are Bolt (1982), Melchior (1986), Stacey (1992) and Poirier (2000). A review of the theoretical approach can be found in Alfè *et al.* (2002b).

These three sources give the following picture, see Figure 4.1. The Earth is composed of a core of radius 3485 km surrounded by a rocky mantle of radius 6370 km. On top of that is the thin crust on which we live. The mantle is a good electrical insulator (except perhaps close to its boundary with the core) so the only possible

9 Modern maps can be found at "`http://geomag.usgs.gov/`".

10 Much further information about the Earth's field and its history can be found at the de Magnete website at "`http://www-spof.gsfc.nasa.gov/earthmag/demagint.htm`", see also Chapter 1 of Merrill *et al.* (1996) and Stern (2002).

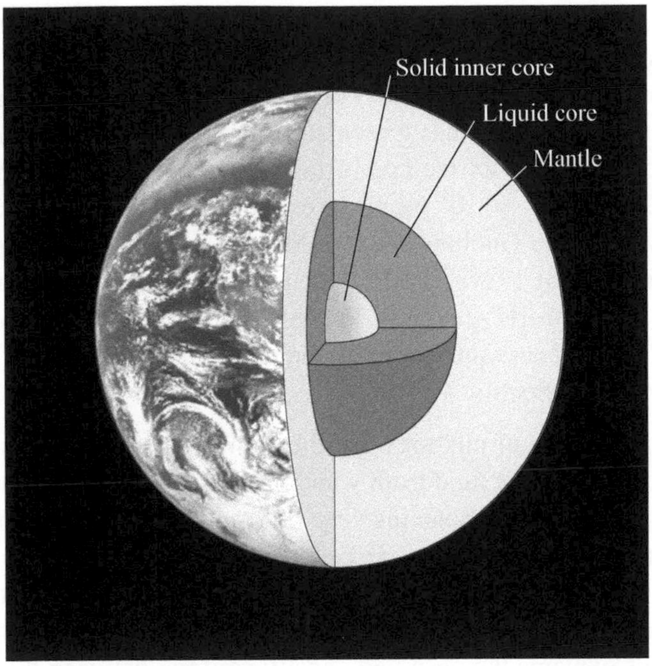

Figure 4.1 - The interior of the Earth. The outer region, the mantle, is composed mainly of silicates, it is a good electrical insulator. Some 2885 km below the surface lies the Earth core, composed mainly of liquid iron. It occupies a little more than half the radius of the Earth. At the center the pressure increase is such that the iron solidifies, this is the solid inner core, occupying 1215 km in radius. (**See colour insert**.)

source of electromagnetic induction that can generate magnetic field is in the core. The core has two distinct parts: an *inner core* of radius 1215 km that is solid and an *outer core* that is fluid. The principal constituent of both is iron. While the density of the inner core is consistent with it being pure iron, the density of the outer core is up to 10% lighter than iron at core pressures. While there remains considerable controversy as to the identity of the lighter element or elements that are mixed with iron in the outer core, it is clear that the outer core is composed of a mixture of iron and some lighter constituents. Possible candidates include sulphur, oxygen and silicon. Recent studies have estimated the core density deficit at closer to 5% (Anderson & Isaac, 2002) and have highlighted the importance of oxygen in the core (Alfè *et al.*, 2002a).

The presence of the solid inner core can be explained by the freezing of the outer core. While the temperature in the interior of the Earth increases with depth, the freezing temperature also increases because of the effect of pressure. Indeed, the latter increases more rapidly with depth. This explains why, as the Earth cools, freezing takes place first at the centre. It is believed that the core was initially completely molten and that a proto inner core nucleated first at the centre of the Earth several billion years ago and has grown steadily since then through the freezing of the outer core. Estimates of the age of the inner core vary. Recent studies put it as at least 3 Gyr (1 Gyr $= 10^9$ years) to explain the paleomagnetic measurements of the Earth's field, see Gubbins *et al.* (2003, 2004) but, for example, Labrosse *et al.* (2001) estimate it at 1 Gyr based on the assumption of no radioactive heating in the core. See also Roberts *et al.* (2001) who come to a similar conclusion but then go on to consider the consequences of all the potassium 40 "missing" from the Earth being in the core (see Section 4.1.4).

It is a common property of mixtures that the composition of the solid which freezes is different from that of the fluid from which it has frozen. For example, if sea water is cooled, the ice that forms contains very little salt, with the remaining fluid being enriched in salt. The same process is believed to take place as the outer core fluid freezes; as the mixture of iron and lighter constituent cools, what freezes is predominantly iron, with most of the light constituent being rejected into the remaining outer core. This explains the observed density contrast between the inner and outer cores which is larger than can be explained by density change upon freezing. Our picture then is of a dense inner core growing steadily as the Earth cools, with the density of the remaining outer core gradually decreasing as the fraction of light constituent in it increases. This picture has major implications for the energy source of the geodynamo which we discuss in Section 4.1.4.

4.1.3. THE GEOMAGNETIC FIELD

While direct measurements of the Earth's field go back only a few hundred years, we have information on its behaviour going back several billion years through *paleomagnetic* measurements, see for example Merrill *et al.* (1996). Though most rock forming minerals are non-magnetic, all rocks exhibit some magnetic properties due to the presence of traces of iron oxides. The magnetisation of these may be used to determine both the local direction and intensity of the Earth's field at the time the rock was formed. Relatively short time-scale behaviour can be determined from sequences of lava flows, and longer time-scale behaviour from sedimentary rocks.

Intensity measurements show that the field has roughly maintained its strength over the past 3.5 Gyr, see Figure 4.2. When compared with the ohmic decay time of the Earth's core

$$\tau_\eta = \frac{r_o^2}{\eta} \approx 3 \times 10^5 \text{ years,} \tag{4.1}$$

there is a clear requirement to explain how the field is maintained and what is its

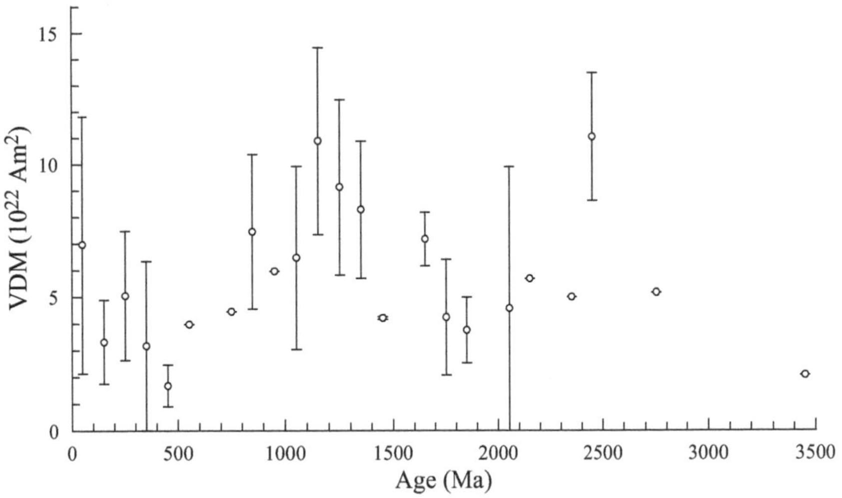

Figure 4.2 - Intensity of the geomagnetic field over the past 3.5 Gyr averaged over 100 Myr intervals (total number of data: 899). The current strength is 8×10^{22} A m^2 (from Kono & Tanaka, 1995).

energy source. A sharper and even more striking estimate can be made by noting that the slowest decaying mode in the sphere is the dipole, with decay time

$$\tau_d = \frac{r_o^2}{\pi^2 \eta} \approx 3 \times 10^4 \text{ years.} \tag{4.2}$$

Permanent magnetism is not a possible explanation because, below a depth of the order of a hundred kilometers, the temperature inside the Earth exceeds the Curie temperature (see for example Stacey, 1992). In (4.1), r_o is the radius of the core and $\eta = 1/\mu\sigma$ is the magnetic diffusivity of the core (see Chapter 1, Section 1.1.1). For the Earth, substituting $r_o = 3.485 \times 10^6$ m and $\sigma \approx 6 \times 10^5\,\mathrm{Sm}^{-1}$ (Merrill *et al.*, 1996) gives $\eta \approx 2\,\mathrm{m}^2\,\mathrm{s}^{-1}$ (see Table II for a summary of the core characteristics).

The principal component of the Earth's field at present is a dipole whose axis is roughly aligned with the geographic axis (the declination measuring the deviation). Indeed, the average field over the past 5 Myr closely approximates a geocentric axial dipole (see Merrill & McFadden, 2003). Directional paleomagnetic measurements show that the field has reversed its direction many times. Reversals are irregular and take place over a time that is short (of the order of 5000 years) compared with the quiescent period between reversals. The last (the Brunhes-Mutuyama reversal) was some 7×10^5 years ago. The reversal frequency has varied over time, see Figure 4.3 and McFadden & Merrill (2000). Typically there have been a few every million years over the past 45 Myr but there was a period (the Cretaceous Superchron) of some 20 Myr ending 86 Myr ago in which hardly any reversals have been found

Table II - Some orders of magnitudes for the geodynamo.

Kinematic viscosity	ν	10^{-6}	$\mathrm{m^2\,s^{-1}}$
Thermal diffusivity	κ	5×10^{-6}	$\mathrm{m^2\,s^{-1}}$
Magnetic diffusivity	η	2	$\mathrm{m^2\,s^{-1}}$
Angular velocity	Ω	7.29×10^{-5}	$\mathrm{s^{-1}}$
Mean core density	ρ_0	1.1×10^{4}	$\mathrm{kg\,m^{-3}}$
Outer core radius	r_o	3.48×10^{6}	m
Inner core radius	r_i	1.22×10^{6}	m
Core surface velocity	$\lvert\mathbf{u}\rvert$	10^{-4}	$\mathrm{m\,s^{-1}}$
Core surface field	$\lvert\mathbf{B}\rvert$	5×10^{-4}	T $(10^{-4}\,\mathrm{T} \equiv 1\,\mathrm{G})$
Ekman number	E	$\nu/\Omega r_o^2$	$\mathcal{O}(10^{-15})$
Magnetic Ekman number	E_η	$\eta/\Omega r_o^2$	$\mathcal{O}(10^{-9})$
Rossby number	Ro	$\lvert\mathbf{u}\rvert/\Omega r_o$	$\mathcal{O}(10^{-6})$
Prandtl number	Pr	ν/κ	$1/5$
Magnetic Prandtl number	Pm	ν/η	$\mathcal{O}(10^{-7})$
Roberts number	q	κ/η	$\mathcal{O}(10^{-6})$
Reynolds number	Re	$\lvert\mathbf{u}\rvert r_o/\nu$	$\mathcal{O}(10^{9})$
Magnetic Reynolds number	Rm	$\lvert\mathbf{u}\rvert r_o/\eta$	$\mathcal{O}(10^{2} - 10^{3})$
Elsasser number	Λ	$\lvert\mathbf{B}\rvert^2/\Omega\mu_0\rho_0\eta$	$\mathcal{O}(1)$

(see Merrill *et al.*, 1996). Heller *et al.* (2002) have investigated the relationship between the frequency of reversals and what is known about the field intensity. They conclude "that there is not a simple correlation between reversal rate and intensity". In addition to reversals, features known as *excursions* have been found. When observed in detail, these start off in a similar way to reversals, with an increase in the declination and typically a decrease in intensity. However, in these events, the field returns to its original polarity rather than the reversed one. Excursions may be "aborted reversals" and may occur ten times more frequently than reversals, see for example Gubbins (1999). They may be due to an intrinsic instability of the dynamo process, see McFadden & Merrill (1993) and Zhang & Gubbins (2000). McFadden & Merrill demonstrate that, following a reversal, there is a reduced probability of a further reversal during a period of some 45000 yrs, a period of the order of τ_η.

The governing equations (4.5a-d) clearly admit $-\mathbf{B}$ as a solution if $\mathbf{B}$ is a solution, so the existence of reversed fields is not a puzzle. However, we do not have a good understanding of what triggers a reversal, what influences their frequency or why some should fail. Simulations are beginning to give some insight into these issues, see Section 4.5. For example, the pattern of heat flux at the *core-mantle boundary* (CMB) has been shown to strongly influence reversal behaviour (Glatzmaier *et al.*, 1999), see Figure 4.5.

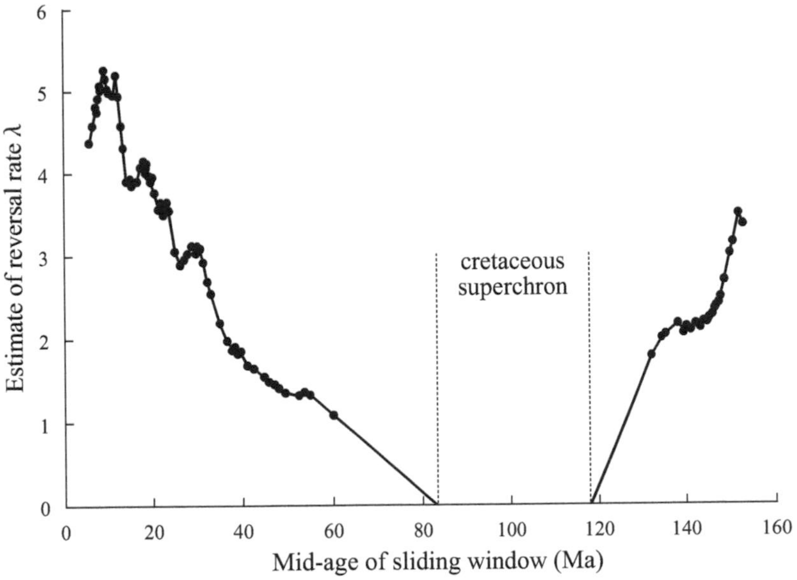

Figure 4.3 - Frequency of reversal of the Earth's magnetic field (from Merrill *et al.*, 1996). The vertical scale indicates the number of reversals per Myr (50 intervals per analysis).

4.1.4. ENERGY SOURCES

The observation in the previous section that the Earth's field has existed at around its present strength for a time four orders of magnitude longer than the ohmic diffusion time (4.1) clearly indicates the need for a mechanism for maintaining the field against ohmic decay, together with an adequate power source.

A rough estimate of the power required can be obtained by setting $\mathbf{u} = \mathbf{0}$ in (4.5a), taking the scalar product with $\mathbf{B}/\mu_0$ and integrating over all space V to give

$$P = \frac{\mathrm{d}}{\mathrm{d}t}\int_V \frac{\mathbf{B}^2}{2\mu_0}\mathrm{d}V = -\int_V \frac{\mathbf{j}^2}{\sigma}\mathrm{d}V\,. \tag{4.3}$$

The standard vector identity (A.26) and the divergence theorem have been used in deriving (4.3). For an insulating mantle, the current $\mathbf{j}$ vanishes outside the core. The left-hand side of (4.3) is the rate of change of magnetic energy. Now, since $\mu_0\mathbf{j} = \boldsymbol{\nabla}\times\mathbf{B}$, the ohmic power dissipation can then be roughly estimated from the right-hand side to give

$$P \approx \frac{4}{3}\pi\, r_o^3\, \frac{|\mathbf{B}|^2}{\sigma\,\mu_0^2\,\mathcal{L}^2}\,, \tag{4.4}$$

where $|\mathbf{B}|$ is a typical field strength and $\mathcal{L}$ an appropriate length scale (which cannot be greater than r_o). If we choose $\mathcal{L} = 10^6$ m, then (4.4) gives an ohmic power dissipation of $1.4\times10^{14}|\mathbf{B}|^2$ W. For a field strength of 6 mT this equates to 5×10^9 W. Of course, this estimate depends crucially on our (rather arbitrary) choice of $\mathcal{L}$. Shorter length scales lead to higher dissipation. Loper & Roberts (1983) reviewed the various estimates; these give $P/|\mathbf{B}|^2$ in the range from 0.7×10^{14} to 200×10^{14} WT^{-2}. Loper & Roberts (1983) favour a value, somewhere in the middle of this range, of $P/|\mathbf{B}|^2 = \mathcal{O}(10^{15})$ WT^{-2}. A field of 10 mT then requires $\mathcal{O}(10^{11})$ W. This gives a ball-park figure of the power requirement of the geodynamo. More accurate estimates can be expected from specific geodynamo models. For example, Glatzmaier & Roberts estimate that at least 2×10^{11} W is required to balance ohmic diffusion (the dominant loss mechanism) in models of their type (Glatzmaier & Roberts, 1995a,b, 1996a,b,c, 1997). Recent estimates have put the total heat transfer from the core to the mantle at 8 TW, with 6.8 TW of that due to conduction in the core (Anderson 2002) and 13 ± 4 TW (Lay *et al.*, 2006). Olson (2003) reviews the thermal interaction between the core and the mantle which has a vital controlling influence on the evolution of the whole of the Earth's deep interior.

The ohmic energy loss is made good by conversion [through the term $\nabla \times (\mathbf{u} \times \mathbf{B})$ in (4.5a)] from the kinetic energy of the flow $\mathbf{u}$. In turn this kinetic energy must be continually replenished. There are two possible means of driving the flow: internal, by buoyancy forces and external, through forcing by boundary motion. The main candidates are: thermal convection (T), compositional convection (C), and precessionally driven flows (P). Cooling and radiogenic heating can lead to (T). The latent heat and light constituent release at the *inner-core boundary* (ICB) associated with the freezing of the inner core (see Section 4.1.2) can lead to (T) and (C). Precessional driving of core flows is due to the gravitational torques exerted on the Earth by the Sun and the Moon (see for example Malkus, 1994).

Over the years, there has been considerable debate about the source of the core fluid motions driving the geodynamo. This has centered on two main issues: the power that can be extracted from a particular energy source, and the efficiency with which it can be converted into useful fluid motions (see for example Braginsky & Roberts, 1995 for a detailed discussion). In the late 1970s, precession had been discounted on efficiency grounds and there were doubts as to the amount of radiogenic heating in the core and the efficiency of its conversion into kinetic energy, see for example Verhoogen (1980). This led to the revival of the idea of a *gravitationally-powered dynamo* whose energy source is the gravitational potential energy stored in the outer core. The gravitational potential energy is released as the Earth cools and the dense (almost pure iron) inner core grows by the freezing of iron from the outer core. Verhoogen (1961) was the first to associate freezing in the core with the dynamo power source. He discounted the chemical segregation associated with freezing of a mixture, preferring convection driven by the latent heat released during the crystallization of the inner core and the specific heat given out by the cooling core. Braginsky (1963) was the first to recognise the contribution of compositional effects. Their inherent efficiency together with the estimated power available made this an attractive power source when the other candidates appeared wanting on efficiency grounds.

Considerable progress has been made in understanding the complex process of freezing of a mixture and applying the results to the Earth's core. Meanwhile, recent work on the other candidates means that precession should not be discounted [see for example Aldridge (2003)] and a question still remains about radiogenic heating [see for example Roberts *et al.* (2003)]. For a more detailed discussion, see for example Fearn (1998).

The debate about what is driving the geodynamo continues and is linked with models for the thermal history of the Earth and the age of the inner core, see for example Labrosse *et al.* (2001), Roberts *et al.* (2003) and Gubbins *et al.* (2003, 2004). Most dynamo models still adopt thermally driven convection as the basis for their driving force, using a combination of internal and differential heating.

4.2. GOVERNING EQUATIONS AND PARAMETERS

Our governing equations are the magnetic induction equation (1.14), the Navier-Stokes equations (1.60a,b) and the heat conduction equation (1.60c). It is most convenient to deal with these in non-dimensional form. Adopting the outer-core radius, r_o, as our length scale, the ohmic diffusion time, τ_η, defined in (4.1) as our timescale, r_o/τ_η as our velocity scale, $(\Omega\mu_0\rho_0\eta)^{1/2}$ (where $\Omega = 7.29\times10^{-5}\,\mathrm{s}^{-1}$ is the rotation frequency of the Earth and $\rho_0 = 1.1 \times 10^4\,\mathrm{kg\,m}^{-3}$ is the mean core density) as our scale for the magnetic field and $r_o(\mathrm{d}T/\mathrm{d}r)$ (where $\mathrm{d}T/\mathrm{d}r$ is a characteristic temperature gradient in the core) as the temperature scale, the governing equations become

$$\partial_t \mathbf{B} = \boldsymbol{\nabla} \times (\mathbf{u} \times \mathbf{B}) + \Delta \mathbf{B}\,, \tag{4.5a}$$

$$\boldsymbol{\nabla} \cdot \mathbf{B} = \boldsymbol{\nabla} \cdot \mathbf{u} = 0\,, \tag{4.5b}$$

$$\begin{aligned} \mathrm{E}_\eta\,[\partial_t \mathbf{u} + (\mathbf{u} \cdot \boldsymbol{\nabla})\mathbf{u}] + 2\mathbf{k} \times \mathbf{u} = -\boldsymbol{\nabla} p + \mathrm{q}\,\widetilde{\mathrm{Ra}}\,T\,\mathbf{r} + \mathrm{E}\,\Delta\mathbf{u} \\ + (\boldsymbol{\nabla} \times \mathbf{B}) \times \mathbf{B}\,, \end{aligned} \tag{4.5c}$$

$$\partial_t T + \mathbf{u} \cdot \boldsymbol{\nabla} T = q\Delta T + \mathcal{S}\,, \tag{4.5d}$$

where $\mathcal{S}$ represents a source of heat (the analog of the uniform density of heat sources in the sphere, $\widetilde{\beta}$, introduced in Section 3.4.3). In most models, this is taken to be uniform. The effects of compressibility are not believed to be of primary importance in the dynamics of the core so the Boussinesq approximation is usually adopted. In this context, when applied to the core, the temperature T should be interpreted as the deviation from the adiabatic temperature. To include the effects of compressibility the anelastic approximation can be used, see (1.48). Glatzmaier & Roberts use this for all but the earliest of their models, see Section 4.5.3.

The non-dimensional parameters appearing in (4.5a-d) are the modified Rayleigh number $\widetilde{\mathrm{Ra}}$ (sometimes called "the buoyancy number"), the magnetic Ekman number E_η, the Roberts number q and the previously defined Ekman number E:

$$\widetilde{\mathrm{Ra}} = \frac{g_0\,\alpha\,r_o^2}{\Omega\,\kappa}\frac{\mathrm{d}T}{\mathrm{d}r}, \quad \mathrm{E}_\eta = \frac{\eta}{\Omega\,r_o^2}, \quad \mathrm{E} = \frac{\nu}{\Omega\,r_o^2}, \quad \mathrm{q} = \frac{\kappa}{\eta}. \tag{4.6a,b,c,d}$$

In the above we have written the gravitational acceleration as $g_0\mathbf{g}$ and taken $\mathbf{g} = -\mathbf{r}$, the non-dimensional position vector, since, to a fair approximation, the strength of the gravitational acceleration increases linearly with radius in the core. The definition of $\widetilde{\mathrm{Ra}}$ is that most appropriate to a rotating magnetic system. The standard Rayleigh number, which is that normally used in non-rotating systems [see equation (1.59a), where $\Delta T L^{-1}$ is replaced by $\mathrm{d}T/\mathrm{d}r$] is

$$\mathrm{Ra} = \frac{g_0\,\alpha\,r_o^4}{\nu\kappa}\frac{\mathrm{d}T}{\mathrm{d}r} = \frac{\widetilde{\mathrm{Ra}}}{\mathrm{E}}. \tag{4.7}$$

Here we shall usually simply refer to (4.6a) as the "Rayleigh number" and only use the term "modified Rayleigh number" when it is necessary to contrast it with that given in (4.7).

The Ekman number E measures the strength of the viscous force (for length scales of the order of the core radius r_o) relative to the Coriolis force. The kinematic viscosity in the Earth's core is very poorly determined but most estimates are very much smaller than the magnetic diffusivity. A typical value for the Earth is $\nu \approx 10^{-6}\,\mathrm{m}^2\,\mathrm{s}^{-1}$ (De Wijs *et al.*, 1998), giving $\mathrm{E} = \mathcal{O}(10^{-15})$. It might therefore seem a very good approximation to neglect viscous effects altogether. In many fluid dynamical problems, provided the no-slip boundary conditions are also dropped, the *mainstream* flow (i.e. that away from narrow viscous boundary layers) is well approximated by an inviscid theory. (Then, the only role viscosity plays is to bring the tangential flow to zero at the boundaries. Such boundary layers are referred to as *passive*.) Unfortunately, in rotating systems, things are not so straightforward. The (Ekman) boundary layers are *active* or *controlling*; the mainstream solution cannot be completed without taking into consideration the flow in the Ekman layers, particularly the *Ekman pumping*, the flow out of the boundary layer into the mainstream (see Section 3.1.1). So far, two complementary approaches have been adopted to deal with the problem of very small E. The first is to retain viscous terms and, for reasons of numerical resolution, we have to accept very much larger values of E than that given above. The alternative is to neglect viscous terms but have to accept the complications associated with Taylor's (1963) constraint (see Section 4.3.1). This is an example of the controlling influence of the boundary layers on the mainstream.

The magnetic Ekman number (4.6b) acts here as a Rossby number, and is sometimes referred to as the Rossby number. In the fluid dynamics literature, the Rossby number is defined as the ratio of the fluid speed to the rotational speed. The fluid speed will only emerge as part of the solution to (4.5a-d) and the two definitions will only agree when the fluid speed is η/r_o. The magnetic Ekman number E_η is very much larger than E but is still small, $\mathcal{O}(10^{-9})$, so the inertial terms in (4.5c) are often neglected; an approximation that filters out inertial waves and torsional oscillations (see Section 4.3.4) .

For the Earth's core, molecular diffusivity values give $\kappa \approx 5 \times 10^{-6}\,\mathrm{m}^2\,\mathrm{s}^{-1}$ (Poirier, 2000), giving $\mathrm{q} = \mathcal{O}(10^{-6})$. Such a low value has important implications for the nature of convective flow, and for dynamo action. Current thought favours using $\mathrm{q} = \mathcal{O}(1)$ to avoid the various complications that arise when $\mathrm{q} \ll 1$. This is the sensible approach until the complex interaction between flow and field that maintains the field is better understood. We discuss the choice of parameter values and the various restrictions on these in detail in Section 4.4.

There are two other important non-dimensional parameters that do not appear in (4.5a-d). Had we chosen typical magnitudes $|\mathbf{B}|$ and $|\mathbf{u}|$ as our scales for the magnetic field and fluid velocity instead of those adopted above $[(\Omega\mu_0\rho_0\eta)^{1/2}$ and r_o/τ_η respectively], then $\mathbf{B}$ would be replaced by $\Lambda^{1/2}\mathbf{B}$ and $\mathbf{u}$ by $\mathrm{Rm}\,\mathbf{u}$, where the Elsasser number Λ and magnetic Reynolds number Rm are defined by

$$\Lambda = \frac{|\mathbf{B}|^2}{\Omega\mu_0\rho_0\eta} \left(= \frac{\sigma|\mathbf{B}|^2}{\Omega\rho_0} = \frac{\tau_\eta}{\tau_{MC}} \right), \qquad \mathrm{Rm} = \frac{|\mathbf{u}|r_o}{\eta}, \tag{4.8a,b}$$

where the slow MHD timescale identified by Soward in Section 3.3.3 is given by

$$\tau_{MC} = \frac{\Omega}{{\Omega_A}^2}, \qquad \text{where} \quad {\Omega_A}^2 = \frac{|\mathbf{B}|^2}{\mu_0\rho_0 r_o^2}. \tag{4.9a,b}$$

This is the timescale on which diffusionless magnetic waves evolve in a rapidly rotating system where $\Omega \gg \Omega_A$, the Alfvén frequency. For fully dynamic calculations, the scalings adopted here are the most appropriate since the amplitudes of $\mathbf{B}$ and $\mathbf{u}$ emerge as part of the solution. Hence Λ and Rm are not parameters that we can prescribe. In simpler model problems, though, the field and/or flow are often prescribed so that the alternative scalings based on $|\mathbf{B}|$ and $|\mathbf{u}|$ are often used. In either case, Λ and Rm are very useful non-dimensional measures of the field strength and flow speed respectively. Important alternative interpretations are in terms of the diffusivity η; both $\Lambda \to \infty$ and $\mathrm{Rm} \to \infty$ are associated with the perfectly conducting limit $\eta \to 0$.

For the Earth, a value of $\Lambda = 1$ roughly corresponds to a field strength of $1\,\mathrm{mT}$, while $\mathrm{Rm} = 1$ corresponds to a flow speed of about $4 \times 10^{-7}\,\mathrm{m\,s^{-1}}$. The latter, when compared with the flows inferred from the GSV suggest values of Rm of $\mathcal{O}(10^3)$ when based on the core radius as length scale. This exceeds the lower bounds that have been derived for Rm if there is to be net field generation by dynamo action [see, for example, Moffatt (1978), Roberts (1994), and Section 1.3.3] and is consistent with the values found in hydrodynamic dynamo models (see Section 4.5).

4.3. FUNDAMENTAL THEORETICAL RESULTS

4.3.1. TAYLOR'S CONSTRAINT

The smallness of the geophysical values of E and E_η suggests that both inertial and viscous terms be neglected in models of the core. This is the so-called *magnetostrophic approximation*. In this section we explore its fundamental consequences which clearly also have important implications for the behaviour of numerical solutions when E and E_η are small.

Setting $\mathrm{E}_\eta = 0$, $\mathrm{E} = 0$ in (4.5c) gives

$$2\,\mathbf{k} \times \mathbf{u} = -\boldsymbol{\nabla} p + \mathrm{q}\,\widetilde{\mathrm{Ra}}\,T\,\mathbf{r} + (\boldsymbol{\nabla} \times \mathbf{B}) \times \mathbf{B}\,. \tag{4.10}$$

Taking the ϕ-component gives

$$2\,u_s = -s^{-1}\,\partial_\phi p + [(\boldsymbol{\nabla} \times \mathbf{B}) \times \mathbf{B}]_\phi\,. \tag{4.11}$$

Integrating this over the cylinder $C(s)$, the cylinder of radius s coaxial with the rotation axis gives

$$2 \int_{C(s)} u_s\,\mathrm{d}S = \int_{C(s)} [(\boldsymbol{\nabla} \times \mathbf{B}) \times \mathbf{B}]_\phi\,\mathrm{d}S\,, \quad \forall\, s\,. \tag{4.12}$$

The term on the left hand side is twice the net flow of fluid out of the curved surface of the cylinder.

If viscosity is totally neglected, (4.11) applies throughout the core and the cylinder extends to the boundaries of the outer core. There can therefore be no flow of fluid into or out of the ends of the cylinder. Consequently, for an incompressible fluid, the left hand side of (4.12) must vanish, giving

$$\int_{C(s)} [(\boldsymbol{\nabla} \times \mathbf{B}) \times \mathbf{B}]_\phi\,\mathrm{d}S = 0, \qquad \forall\, s\,. \tag{4.13}$$

This condition was first derived by Taylor (1963) and is referred to as "Taylor's condition" or "Taylor's constraint". It can be interpreted as that the net magnetic torque on each cylinder must vanish. The generalisation of Taylor's constraint to regions with other boundary geometries, such as spherical surfaces with topography or parallel plane boundaries, is discussed in Appendix C.

If viscous effects are retained in the problem, (but are only important in thin Ekman layers at the boundaries of the outer core) then (4.11) is valid throughout the core except for the Ekman layers, and the cylinder $C(s)$ must be considered as extending, not to the boundaries of the outer core, but to the outer edges of the Ekman layers. The North-South flow in the Ekman layers leads to a net flow of fluid into the ends of the cylinder. This must be balanced by a net flow out of the curved surface of the cylinder, so the left hand side of (4.12) is in general non-zero. To evaluate $\int_{C(s)} u_s \, \mathrm{d}S$ we must calculate the flow in the Ekman layers.

The problem is analysed by splitting the core into three regions, a thin spherical shell that extends inward a short distance from the core-mantle boundary, a similar shell adjacent to the boundary with the inner core, and the *interior*, which is the remainder (and the bulk of) the outer core. In the interior, viscous effects are negligible, and (4.11) holds. In the two boundary regions, viscous effects are important. The short length scale in the radial direction permits a simplification to the governing equations and an analytical solution. This must then be matched to the solution in the interior. The spherical geometry is unimportant in the boundary layers and can locally be approximated by a plane layer. In general, we define, for any $f = f(r, \theta, \phi)$ its azimuthal mean

$$\overline{f}(r,\theta) \equiv \overline{f} \equiv \frac{1}{2\pi} \int_0^{2\pi} f \, \mathrm{d}\phi \, . \tag{4.14}$$

The analysis relates the mean azimuthal flow $\overline{u}_\phi$ at the edge of the boundary layer to the flow in the boundary layer in the θ–direction. The latter is related to the left hand side of (4.12) since, for an incompressible fluid, the flow into the top and bottom of the cylinder $C(s)$ must be matched by a flow out through its curved surface. The boundary-layer analysis then allows us to relate $\overline{u}_\phi$ to the right hand side of (4.12). Details can be found, for example in Fearn (1994) [but note that he uses an alternative definition of E that differs by a factor 2 and the factor 2 then does not appear in the Coriolis force term in (4.10)]. The boundary-layer analysis can be found in Chapter 3. We find, for the case where s is outside the tangent cylinder ($s > r_i$), and $\overline{u}_\phi(s, z_T)$ is assumed to take the same value as $\overline{u}_\phi(s, z_B)$ that

$$\overline{u}_\phi(s, z_T) = \frac{1}{2}\left(\frac{\cos\theta}{\mathrm{E}}\right)^{1/2} \mathcal{T} \, , \quad \text{where} \quad \mathcal{T}(s) = \int_{z_B}^{z_T} \overline{[(\mathbf{\nabla} \times \mathbf{B}) \times \mathbf{B}]_\phi} \, \mathrm{d}z \, , \tag{4.15a,b}$$

and $z_T = \sqrt{1 - s^2}$ and $z_B = -z_T$. This replaces (4.13) when the effects of Ekman boundary layers are included in the problem.

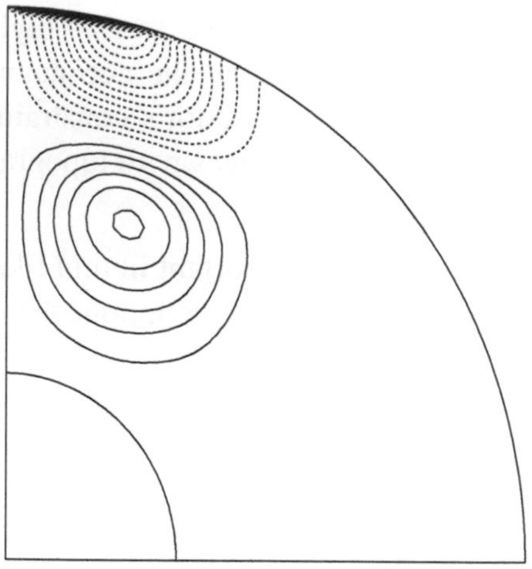

Figure 4.4 - An example of a solution satisfying Taylor's constraint. Shown are contour plots of $\overline{[(\nabla \times \mathbf{B}) \times \mathbf{B}]_\phi}$. Contour interval is 1. Solid lines represent positive and dashed lines represent negative contours.

We note here that (4.13) and (4.15a) are very different in character. The former is a constraint on $\mathbf{B}$ while the latter is a means of determining $\overline{u}_\phi$. In a typical small E solution we shall expect $\mathcal{T}$ also to be small. For $\mathcal{O}(1)$ values of $|\mathbf{B}|$ this is achieved by regions of positive $\overline{[(\nabla \times \mathbf{B}) \times \mathbf{B}]_\phi}$ cancelling with regions of negative $\overline{[(\nabla \times \mathbf{B}) \times \mathbf{B}]_\phi}$ in the integral, see Figure 4.4. The balance is delicate and can be expected to be difficult numerically.

4.3.2. THE "ARBITRARY" GEOSTROPHIC FLOW $u_G(s)$

If equation (4.13) is satisfied then we can solve equation (4.10) for $\mathbf{u}$, given $\mathbf{B}$ and T, as we see below. However, the solution is not unique. If we add any $u_G(s)\,\mathbf{e}_\phi$ to $\mathbf{u}$, then the additional Coriolis term can be written as

$$\mathbf{k} \times u_G(s)\,\mathbf{e}_\phi = -u_G(s)\,\mathbf{e}_s = -\nabla\left(\int u_G(s)\,\mathrm{d}s\right), \tag{4.16}$$

and hence can be absorbed into the pressure gradient term. Consequently if $\mathbf{u}$ is a solution of (4.10), then so is $\mathbf{u} + u_G(s)\,\mathbf{e}_\phi$ for arbitrary $u_G(s)$.

Taking the curl of (4.10), and using (4.5b) gives

$$-2\,\partial_z \mathbf{u} = \nabla \times [(\nabla \times \mathbf{B}) \times \mathbf{B}] + \mathrm{q}\,\widetilde{\mathrm{Ra}}\,(\nabla T \times \mathbf{r})\,. \tag{4.17}$$

Taking the axisymmetric part and integrating this with respect to z gives

$$2\overline{\mathbf{u}} = \int_z^{z_T} \overline{\nabla \times [(\nabla \times \mathbf{B}) \times \mathbf{B}]}\,\mathrm{d}z' + \mathrm{q}\widetilde{\mathrm{Ra}} \int_z^{z_T} \overline{(\nabla T \times \mathbf{r})}\,\mathrm{d}z' + \mathbf{F}(s)\,, \tag{4.18}$$

where $\mathbf{F}(s)$ is an arbitrary function of integration. The flow $\overline{\mathbf{u}}$ must satisfy the boundary condition that (to leading order) there is no normal flow at the top and bottom boundaries. These two boundary conditions give two expressions relating $\overline{u}_s$ and $\overline{u}_z$ which then determine F_s and F_z. In a non-axisymmetric system, the third component $\overline{u}_\phi$ of the flow would be determined from $\boldsymbol{\nabla} \cdot \mathbf{u} = 0$. However, in this axisymmetric system $\overline{u}_\phi$ is independent of ϕ and so does not appear in $\boldsymbol{\nabla} \cdot \mathbf{u}$. Consequently F_ϕ is undetermined and we call it the *"arbitrary" geostrophic flow* u_G. We then have

$$\overline{u}_\phi \;=\; u_M \;+\; u_T \;+\; u_G\,, \tag{4.19}$$

where

$$u_M \;=\; \tfrac{1}{2}\int_z^{z_T} \overline{\boldsymbol{\nabla}\times[(\boldsymbol{\nabla}\times\mathbf{B})\times\mathbf{B}]_\phi}\,\mathrm{d}z' \tag{4.20}$$

is the *magnetic wind*, and

$$u_T \;=\; \tfrac{1}{2}\mathrm{q}\,\widetilde{\mathrm{Ra}}\int_z^{z_T} \overline{(\boldsymbol{\nabla}T\times\mathbf{r})_\phi}\,\mathrm{d}z' \tag{4.21}$$

is the *thermal wind* familiar in the meteorological literature (see, for example, Roberts & Soward, 1978). Note that with our choice of the limits of integration in (4.20)-(4.21), $u_M = u_T = 0$ at $z = z_T$. Consequently the geostrophic flow $u_G = \overline{u}_\phi|_{z_T}$.

The apparent arbitrariness of the geostrophic flow u_G is a consequence of considering (4.10) in isolation; of considering the forcing terms on the right hand side as prescribed, rather than as determined through (4.5a) and (4.5d). In practice u_G is not arbitrary. The manner in which it is determined depends on the importance of the Ekman suction.

4.3.3. EKMAN STATES, TAYLOR STATES AND MODEL-Z: DETERMINATION OF THE GEOSTROPHIC FLOW u_G

If Taylor's condition (4.13) is satisfied, then (4.10) can be solved for $\mathbf{u}$ up to the unknown geostrophic flow. As Fearn & Proctor (1992) point out, it is the very existence of a "homogeneous solution" of the form $u_G(s)\mathbf{e}_\phi$ that makes a "solvability condition" of the form (4.13) necessary. Of course, $u_G(s)$ can only be considered as arbitrary in the context of solving (4.10) for a given right hand side when (4.13) is satisfied. In practice, u_G is determined in one of two ways. Either Ekman suction is important, so the left hand side of (4.12) is non-zero and Taylor's condition does not apply. Then we know $\overline{u}_\phi$ at the boundary through an expression like (4.15a). This extra piece of information determines u_G explicitly. Alternatively, Taylor's condition does apply. The system (4.5a-d) then must adjust the magnetic field so that Taylor's condition is satisfied. The mechanism used to achieve this is to adjust the differential rotation (the ω–effect discussed in Chapter 1) by varying u_G. The differential rotation stretches out poloidal field to generate toroidal field. By varying u_G, B_ϕ can be adjusted and perhaps (4.13) satisfied. This mechanism determines u_G in a very complicated implicit manner.

There is no guarantee that (4.13) can be satisfied. Fearn & Proctor (1987) tackled the problem of the determination of u_G through this mechanism, by choosing u_G to minimise the absolute value of the left hand side of (4.13) for a given poloidal magnetic field and a flow that is prescribed (apart from the geostrophic flow). Their method was very successful for certain choices of field and flow, but gave poor results for other choices.

Non-axisymmetric magnetoconvection models and kinematic α^2– and $\alpha\omega$–dynamo models and have been adapted to include a geostrophic flow determined by an expression similar to (4.15a). Without the feedback due to the geostrophic flow, both problems are linear and would show exponential growth of their solutions for a sufficiently large forcing. For forcing just above critical, the systems typically find themselves in an "Ekman state" where equilibration of the amplitude of the solution is achieved through the action of the geostrophic flow. In this respect, it is the condition (4.15a) that provides the dominant nonlinear effect, since it becomes important at much smaller amplitude of solution than all other nonlinear effects. The reason for this is the small value of the Ekman number, giving an equilibrated solution amplitude of $\mathcal{O}(\mathrm{E}^{1/4})$ (see Section 7.3 of Fearn, 1994). As the driving force is increased, the system usually evolves to a state where (4.13) is satisfied (a "Taylor state") and it is the other nonlinear effects that are responsible for equilibration, this time at higher [$\mathcal{O}(1)$] amplitude; viscous effects no longer have a major influence on the solution.

This picture of the nonlinear evolution of a dynamo has come to be known as the Malkus-Proctor scenario, see Malkus & Proctor (1975). It is not the only possible manner in which a dynamo can evolve. An (or the) alternative is where the Taylor state is replaced by a state in which the solution amplitude is $\mathcal{O}(1)$ but where viscous effects remain important, even in the limit $\mathrm{E} \to 0$. This is Braginsky's (1975) model–Z (see also Braginsky, 1994). Its fundamental difference from a Taylor state is the manner in which Taylor's constraint is satisfied or almost satisfied. In a Taylor state, with $|\mathbf{B}|$ of $\mathcal{O}(1)$, (4.13) is satisfied with $\overline{[(\nabla \times \mathbf{B}) \times \mathbf{B}]_\phi}$ [see (4.15b)] of $\mathcal{O}(1)$ everywhere and regions of positive $\overline{[(\nabla \times \mathbf{B}) \times \mathbf{B}]_\phi}$ cancelling with regions of negative $\overline{[(\nabla \times \mathbf{B}) \times \mathbf{B}]_\phi}$ when the integral over $C(s)$ is taken. (This cancellation effect is illustrated well in Fearn & Proctor, 1987.) There is strong coupling between adjacent cylinders [since $\overline{[(\nabla \times \mathbf{B}) \times \mathbf{B}]_\phi}$ is $\mathcal{O}(1)$], providing the mechanism for Taylor's condition (4.13) to be maintained as the system evolves. By contrast, in model–Z, (4.13) is almost satisfied, by $\overline{[(\nabla \times \mathbf{B}) \times \mathbf{B}]_\phi}$ being small everywhere. This is achieved by having B_s close to zero, while B_ϕ, B_z are $\mathcal{O}(1)$. This is because an alternative expression of Taylor's constraint (for an axisymmetric field and an insulating mantle) is

$$\int_{z_B}^{z_T} B_\phi B_s \,\mathrm{d}z = 0, \tag{4.22}$$

see, for example, Fearn (1994). The meridional field is then almost aligned with the z-axis, hence the name of the model. Since B_s is small, there is only small coupling between adjacent cylinders $C(s)$ and the system is unable to satisfy Taylor's condition exactly. Consequently, the geostrophic flow remains dependent on the strength of core-mantle coupling. Here, we have concentrated on viscous core-mantle coupling, so u_G remains dependent on E, and very large geostrophic flows are found in the limit of small E, see for example Braginsky & Roberts (1987).

Note that in the above discussion we have used the description "$\mathcal{O}(1)$" rather loosely. This has been to avoid too detailed a discussion of the appropriate scalings and to focus on the important distinction between the small amplitude Ekman state and the high amplitude Taylor and model–Z states. Model–Z is discussed in detail in Braginsky (1994), and the relationship between model–Z and Taylor states in Roberts (1989). The nonlinear role of the geostrophic flow on kinematic dynamo and magnetoconvection models is discussed in detail in Fearn (1994).

4.3.4. THE ROLE OF INERTIA

An arbitrary initial condition will not necessarily satisfy Taylor's constraint, in which case (4.10) has no solution. Of course, the full equation (4.5c) can quite happily be time-stepped for arbitrary initial conditions. Taylor (1963) comments that if his constraint is not satisfied then *"rapid torsional motion would be set up in which each concentric cylindrical annulus rotated as a rigid body. The adjacent annuli are coupled together, as if by elastic strings, through the magnetic field B_s. Because of this linkage, the torsional motion would modify the fields until a state was reached in which (4.13) was satisfied"*. Taylor (1963) envisaged that, in a short time, this adjustment would take place. Subsequently, the flow would continue to adjust in just the manner required to ensure that Taylor's constraint continued to be satisfied. Indeed this is the process that determines the geostrophic flow u_G in a Taylor state.

Should Taylor's constraint fail to be satisfied, the azimuthal torque (4.15b) on cylinders will be non-zero and the $\mathrm{E}^{-1/2}$ factor in (4.15a) indicates the generation of a strong geostrophic flow. Inertia must then inevitably play a role. If we retain inertia in our problem, but only for a geostrophic flow $u_G(s)\mathbf{e}_\phi$, we can repeat the analysis of Section 4.3.1 to obtain, in place of (4.15a),

$$2\, z_T\, \mathrm{E}_\eta\, \partial_t u_G + 2\left(\frac{\mathrm{E}}{\cos\theta}\right)^{1/2} u_G \;=\; \mathcal{T}\,. \tag{4.23}$$

The addition of the inertial term can be of assistance in finding steady solutions, playing a technical role, preventing numerical instabilities in the small E limit, see Jault (1995), in particular his equation (6). Glatzmaier & Roberts (1996a) introduced the axisymmetric azimuthal part of inertia into their 3D hydrodynamic dynamo model, having neglected inertia in their earlier work (Glatzmaier & Roberts, 1995a,b).

In the limit of small E, (4.23) becomes an equation for *torsional oscillations*

$$2\, z_T\, \mathrm{E}_\eta\, \partial_t u_G \;=\; \mathcal{T}\,. \tag{4.24}$$

An additional equation can be obtained by taking the expression (4.15b) and differentiating it with respect to t:

$$2\pi\, s\, \frac{\partial}{\partial t}\mathcal{T}(s,t) \;=\; \int_{C(s)} \left[(\boldsymbol{\nabla}\times\mathbf{B})\times\frac{\partial\mathbf{B}}{\partial t}\right]_\phi + \left[(\boldsymbol{\nabla}\times\frac{\partial\mathbf{B}}{\partial t})\times\mathbf{B}\right]_\phi \mathrm{d}S\,. \tag{4.25}$$

Substituting for $\partial\mathbf{B}/\partial t$ from the induction equation (4.5a) and neglecting the diffusion term, for an axisymmetric field it can be shown that

$$2\pi\, s\, \frac{\partial}{\partial t}\mathcal{T}(s,t) \;=\; a(s)\,\frac{\partial^2}{\partial s^2}\left(\frac{u_G}{s}\right) + b(s)\,\frac{\partial}{\partial s}\left(\frac{u_G}{s}\right) + c(s)\,, \tag{4.26a}$$

where

$$a(s) = \int_{C(s)} sB_s^2\,\mathrm{d}S\,, \qquad b(s) = \int_{C(s)} \left(s\,\mathbf{B}\cdot\boldsymbol{\nabla}B_s + 2B_s^2\right)\mathrm{d}S\,, \tag{4.26b,c}$$

and $c(s)$ contains all the other contributions that do not involve u_G. Note that Taylor (1963) derived a version of this equation, without the left-hand-side by differentiating (4.13) rather than (4.15b). It was Braginsky (1970) who first derived an equation describing torsional waves.

Equations (4.24) and (4.26a) form a hyperbolic system which may be expected to admit oscillatory solutions about the steady state for which $\mathcal{T} = 0$ (Moffatt, 1978). In these torsional oscillations, each cylinder $C(s)$ rotates about its axis. Recall that flows of the form $u_G(s)\mathbf{e}_\phi$ are unaffected by the Coriolis force, see (4.16). The oscillations are controlled by their magnetic linkage through B_s and the shearing of this field component by differential rotation; notice that it is only radial gradients of u_G/s that appear in (4.26a). The timescale of the torsional oscillations can be determined by combining equations (4.24) and (4.26a). We find that

$$\mathrm{E}_\eta \frac{\partial^2}{\partial t^2} \;\sim\; B_s^2\,. \tag{4.27}$$

Recalling that time has been non-dimensionalised using τ_η and $\mathbf{B}$ using $(\Omega\mu_0\rho_0\eta)^{1/2}$ we conclude that the characteristic time for torsional waves is

$$\frac{\mathrm{E}_\eta{}^{1/2}}{B_s}\,\tau_\eta = \frac{1}{B_s}\left(\tau_\eta\,\Omega^{-1}\right)^{1/2} = r_o\,\frac{\sqrt{\mu_0\,\rho_0}}{B_s^*}\,, \tag{4.28}$$

where B_s^* is the dimensional radial field. From this we can see that the timescale of torsional waves is essentially determined by the time it takes an Alfvén wave propagating on the radial field to travel a distance of the order of the radius of the core. If we choose an average radial field strength of $0.1\,\mathrm{mT}$ then the above gives a time of the order of 130 years. Stronger fields give shorter times. Section 3 of Roberts & Soward (1972) gives a more detailed analysis including a discussion of the decay of torsional oscillations.

Assuming a field strength of $0.186\,\mathrm{mT}$, Braginsky (1970) invoked torsional oscillations as a possible mechanism for explaining observations of variations in the length of the day with a period of about 60 years (see Roberts *et al.*, 2007). Jault & Le Mouël (1991) also investigated the problem, looking in detail at electromagnetic and topographic core-mantle coupling. Torsional oscillations continue to be of interest as mechanism for explaining observations. For example Pais & Hulot (2000) see evidence for torsional oscillations in their analysis of core flow at the CMB deduced from geomagnetic models, and Bloxham *et al.* (2002) in an analysis of geomagnetic data show that "geomagnetic jerks can be explained by the combination of a steady flow and a simple time-varying, axisymmetric, equatorially symmetric, toroidal zonal flow. Such a flow is consistent with torsional oscillations in the Earths core". Jault (2003) reviews this area.

4.4. PARAMETER CONSTRAINTS

The choice of the values for freely prescribable parameters in geodynamo models is a compromise between realistic values for planetary interiors and computationally tractable values. The balance will move to the former as our understanding of the problem, computational techniques and computational power advance, but initially a practical approach is wise; adopting values that produce well-resolved solutions. Here we summarise some results from model problems that illustrate the dependence of solutions on key parameters and discuss how these constrain what values of these parameters we can reasonably use in geodynamo models.

4.4.1. THE EKMAN NUMBER

As discussed in Section 4.2, using typical estimates of the molecular viscosity, and the radius of the Earth's outer core as our length scale, the Ekman number $\mathrm{E} = \mathcal{O}(10^{-15})$. It is this very small value that is the most fundamental source of difficulty in solving the geodynamo problem. The (Ekman) boundary layers at the CMB and ICB each have thickness of $\mathcal{O}(\mathrm{E}^{1/2})$ (equivalent to about $0.1\,\mathrm{m}$ for the Earth). In a non-magnetic system (where the leading order balance is geostrophic) both the convective length scale (see Section 3.4) and the shortest length scale of the Stewartson-layer structure (see Section 3.1.4) associated with the tangent cylinder are of $\mathcal{O}(\mathrm{E}^{1/3})$. Clearly, resolution of such structures in any numerical scheme is impossible for any realistic value of E. The Stewartson-layer structure is modified by the presence of a magnetic field (see Soward & Hollerbach, 2000) but any simulation has to be able to deal with situations where the field is, at least locally, weak. The stiffness associated with Taylor's constraint in the limit of small E (see Section 4.3.1) is a further difficulty. The only options are to work with much larger values of E or to adopt the magnetostrophic approximation $\mathrm{E} = 0$. Work by Walker *et al.* (1998) has identified a problem that affects both the $\mathrm{E} = 0$ and $\mathrm{E} \to 0$ cases; the development of small scale, high frequency waves, implying a time step of order $\mathrm{E}\tau_\eta$ to ensure numerical stability. This has proved an insurmountable problem in attempts to work in the magnetostrophic approximation.

Significant progress has been made using the finite E approach (see Section 4.5). Even with substantial supercomputer resources, it is not feasible to reduce E much below 10^{-5} (see for example Jones *et al.*, 1995; Kuang & Bloxham, 1997, 1999). To lower, what we might call the *headline Ekman number*, some authors have adopted so-called *hyperdiffusivities*. These enhance the diffusivity for short length scales in the θ– and ϕ–directions. Solutions are typically expressed through expansions in spherical harmonics, for example

$$T = \sum_{\ell=m}^{L} \sum_{m=0}^{M} T_\ell^m(r) P_\ell^m(\cos\theta) \mathrm{e}^{\mathrm{i}m\phi} \; + \; c.c., \tag{4.29}$$

where T is the toroidal part of the field, P_ℓ^m are associated Legendre functions (see for example Abramowitz & Stegun, 1965) and $c.c.$ denotes complex conjugate. Then, an example of hyperdiffusivity is defined by

$$\nu = \overline{\nu}\,(1 + h\,\ell^3), \tag{4.30}$$

where h is some constant and ℓ is the spherical harmonic degree. It is then $\overline{\nu}$ that appears in the definition of the Ekman number rather than ν. The effect of the hyperdiffusivity is to "damp small scales while allowing the large scales to experience much less diffusion" (Glatzmaier & Roberts, 1995b). Glatzmaier & Roberts (1995b) use $h = 0.075$, achieving a headline Ekman number of 2×10^{-6}. The other diffusivities κ and η are treated in the same manner. More recently, Glatzmaier & Roberts have used $h = 0.037$ for ν and κ and 0.016 for η (Glatzmaier and Roberts 1996b,c). The argument typically put forward to justify the use of hyperdiffusivities is that the diffusivities are assumed to be eddy (sub-grid) diffusivities and the small resolved eddies physically interact more strongly with the small unresolved eddies so they should have larger diffusivities. The use of hyperdiffusion does distort the dynamics of the core, resulting in viscosity retaining a controlling influence on the geostrophic flow, for example, see Zhang & Jones (1997).

An important issue that is relevant to the question of how low E must be in order to be characteristic of the geodynamo is highlighted by Jones (2000). He discusses whether the geodynamo is supercritical or subcritical, in the following sense. It is well known from linear studies (see for example Proctor, 1994) that, in a rapidly rotating system, the modified Rayleigh number for the onset of thermally driven convection $\widetilde{\mathrm{Ra}}_{c0} \equiv \widetilde{\mathrm{Ra}}_c(\mathbf{B} = \mathbf{0})$ is of $\mathcal{O}(\mathrm{E}^{-1/3})$ in the absence of any magnetic field (see Chapter 3). By contrast, in the presence of a prescribed magnetic field with $\Lambda = \mathcal{O}(1)$, $\widetilde{\mathrm{Ra}}_c$ reduces to $\mathcal{O}(1)$; the presence of the field facilitates convection. Numerical studies (see Section 4.5.3) show that $\widetilde{\mathrm{Ra}}$ may have to be well above its critical value in order to maintain a field. Nonetheless, when $\mathrm{E} \ll 1$, it is possible that a self-sustaining dynamo exists for $\widetilde{\mathrm{Ra}} < \widetilde{\mathrm{Ra}}_{c0}$. Jones (2000) calls this situation a *subcritical dynamo* and estimates that dynamos for $\mathrm{E} < 10^{-10}$ must be subcritical. They may exist at larger values of E but all numerical models so far have required $\widetilde{\mathrm{Ra}} > \widetilde{\mathrm{Ra}}_{c0}$ for dynamo action, see for example Busse *et al.* (2003). Gubbins (2001) argues that the Rayleigh number in the core is highly supercritical.

4.4.2. THE MAGNETIC REYNOLDS NUMBER

The magnetic Reynolds number Rm is a non-dimensional measure of the relative importance of the induction term $\boldsymbol{\nabla} \times (\mathbf{u} \times \mathbf{B})$ to the diffusive term $\Delta\mathbf{B}$. In the absence of the former, it can be shown that the field $\mathbf{B}$ inevitably decays. For field maintenance, the induction effect must be able to at least balance the diffusive losses; i.e. Rm must exceed some minimum value $[\geq \mathcal{O}(1)]$. Specific lower bounds on Rm for dynamo action can be derived, see for example Roberts (1994) and Section 1.3.3.

In a fully hydrodynamic dynamo model, the flow $\mathbf{u}$ emerges as part of the solution so Rm is not a parameter that can be freely chosen. The vigour of the flow depends on the forcing whose magnitude is determined by the modified Rayleigh number $\widetilde{\mathrm{Ra}}$ (see below). The level of the forcing must be sufficient that dynamo action is taking place, while at the same time ensuring that Rm (which measures the strength of the differential rotation) is not too large. Shear acts to inhibit convection (see for example Fearn & Proctor, 1983) and associated with this are short length scales typically $\mathcal{O}(\mathrm{Rm}^{-1/3})$. If Rm is too large, such length scales may be difficult to resolve.

4.4.3. THE ROBERTS NUMBER

The situation described above is worse when the Roberts number q is small. Differential rotation begins to have a significant effect when $\mathrm{Rm} = \mathcal{O}(\mathrm{q})$ with length scales $\mathcal{O}(\mathrm{Rm}/\mathrm{q})^{-1/3}$ having to be resolved. With the requirement for dynamo action that Rm be at least $\mathcal{O}(1)$ there is a clear problem when q is small, as is appropriate if we use molecular values of the diffusivities. The situation may be helped somewhat in that strong fields act to oppose shear (see for example Busse, 2002). Small values of q have also been problematical in attaining solutions satisfying Taylor's constraint (see Section 4.3.1 and Soward 1986; Skinner & Soward, 1988, 1990). Further, peculiar features are present in the $\widetilde{\mathrm{Ra}}$ versus Λ graph for the onset of thermally driven convection when $\mathrm{q} \ll 1$ (see Zhang & Jones 1996).

The source of many of these problems is the mismatch between the thermal and ohmic timescales [$\tau_\kappa = r_o^2/\kappa$ and τ_η, see (4.1)]. With molecular values of the diffusivities, $\tau_\eta = \mathcal{O}(10^5)$ years. This is an upper bound on the timescale on which the dynamo must regenerate magnetic field. With $\mathrm{q} = \mathcal{O}(10^{-5})$ (see Section 4.2), the natural thermal timescale is longer than the age of the Earth. This mismatch is probably also the source of the much higher values of $\widetilde{\mathrm{Ra}}$ found to be required for dynamo action when q is small, see for example the discussion in Jones (2000), Busse (2002, Figure 15) and below.

It is clear that, initially, to make progress, it is sensible to adopt $\mathcal{O}(1)$ values of q. Indeed, parameter surveys have shown no dynamo action for q less than some critical value q_c which decreases with decreasing E, see for example Christensen *et al.* (1999). They choose $\mathrm{Pr} = 1$ so $\mathrm{q} = \mathrm{Pm}$. They found dynamo action only for $\mathrm{Pm} > \mathrm{Pm}_c \sim 450\,\mathrm{E}^{3/4}$. For $\mathrm{E} = 10^{-3}$, $\mathrm{Pm}_c \sim 2$ and for $\mathrm{E} = 10^{-4}$ they found $\mathrm{Pm}_c \sim 0.5$.

4.4.4. THE RAYLEIGH NUMBER

The above discussion about disparate timescales suggests that very highly supercritical values of $\widetilde{\mathrm{Ra}}$ will be required for dynamo action if q is small. This is consistent with the findings of Glatzmaier and Roberts (1995a,b) who use $\mathrm{q} = 0.1$. Jones *et al.* (1995) find dynamo action at much smaller values of $\widetilde{\mathrm{Ra}} - \widetilde{\mathrm{Ra}}_{c0}$ for $\mathrm{q} = 10$ compared with $\mathrm{q} = 1$. The form of the buoyancy term in (4.5c) suggests that it is $\mathrm{q}\widetilde{\mathrm{Ra}}$ that is important for dynamo action.

4.4.5. THE MAGNETIC EKMAN NUMBER

Motivated by the smallness of E_η, some numerical models of the geodynamo neglect inertial terms altogether, a few choose the geophysical value, while many take the view that E_η should be no smaller than the Ekman number. The magnetic Prandtl number

$$\mathrm{Pm} = \mathrm{E}/\mathrm{E}_\eta = \nu/\eta \tag{4.31}$$

is small in the core but the numerical constraints on E mean that most numerical models take $\mathrm{Pm} = \mathcal{O}(1)$. Taking the geophysical value of E_η while accepting the numerical constraints on E would imply a large magnetic Prandtl number, while neglecting inertial effects altogether corresponds to the infinite magnetic Prandtl number limit. Whatever the choice, almost all studies choose a fixed value of E_η and focus on other aspects of the problem. Very little work has focussed in on the role of inertia in the dynamo problem (but see Fearn & Morrison, 2001; Fearn & Rahman, 2004b). Fearn & Morrison (2001) find that dynamo action shuts off as E_η is increased. This is consistent with the studies (see Section 4.4.3) that show no dynamo action if q or equivalently Pm is too small.

4.5. NUMERICAL MODELS

There has been considerable progress over the past 10 years in modelling the geodynamo, with many groups now actively involved. Reviews include those by Dormy *et al.* (2000), Jones (2000), Roberts & Glatzmaier (2000), Busse (2002), Glatzmaier (2002) and Kono & Roberts (2002). In making comparisons between different calculations, care should be taken to note that different authors use different definitions of the key non-dimensional parameters. The two main differences are:

- whether the factor 2 is retained, as here [see (4.5c)], in the Coriolis force or whether it is absorbed into the definitions of E, $\widetilde{\mathrm{Ra}}$, E_η and Λ,

- the choice of length scale. Here we have used the outer core radius r_o. Many papers use the core gap width $r_o - r_i$, and of course there are variations on the choice of r_i.

In the following discussion, where we give values of dimensionless parameters, we are simply quoting the values given by the authors; we have not attempted to normalise them to the definitions used here.

4.5.1. NONLINEAR α^2 AND $\alpha\omega$ MODELS

Proctor, in Chapter 1, has described linear mean-field α^2– and $\alpha\omega$–dynamos. For these, in the supercritical regime, a seed field will grow without bound. In the context of the geodynamo, the first nonlinear effect that becomes important as $\mathbf{B}$ grows in strength is the geostrophic flow discussed in Section 4.3.1. The $\mathrm{E}^{-1/2}$ factor in (4.15a) implies u_G is of $\mathcal{O}(1)$ when $|\mathbf{B}|$ is of $\mathcal{O}(\mathrm{E}^{1/4})$. The system is then in an Ekman state. A number of studies have investigated the role of the geostrophic flow in equilibrating mean-field dynamos. As the driving is increased (an increase in the strength of α and/or ω), the system typically evolves towards a state in which Taylor's constraint (4.13) is satisfied as envisaged by Malkus & Proctor (1975), though the manner in which this happens is model dependent, see for example Soward & Jones (1983a). If the only nonlinear effect that is included in a model is u_G, then, when (4.13) is satisfied, the solution is no longer viscously controlled and will grow without bound. Other (ageostrophic) nonlinear effects then must come into play to equilibrate the solution with $|\mathbf{B}| = \mathcal{O}(1)$ in a Taylor state. Studies using a spherical geometry include those by Proctor (1977c), Hollerbach & Ierley (1991), Barenghi (1992) and Hollerbach & Jones (1993, 1995). The latter solved

$$\partial_t \mathbf{B} = \boldsymbol{\nabla} \times (\mathbf{u} \times \mathbf{B} + \alpha \mathbf{B}) + \Delta \mathbf{B}\,, \tag{4.32a}$$

$$2\,\mathbf{k} \times \mathbf{u} = -\boldsymbol{\nabla} p + \vartheta\,\mathbf{r} + \mathrm{E}\,\Delta \mathbf{u} + (\boldsymbol{\nabla} \times \mathbf{B}) \times \mathbf{B}\,, \tag{4.32b}$$

together with (4.5b) for axisymmetric $\mathbf{B}$ and $\mathbf{u}$ and prescribed α and ϑ. The term $\vartheta\,\mathbf{r}$ models the buoyancy force, driving a meridional circulation as well as differential rotation [ϑ is a density perturbation as in Section 3.3.4 page 165, see equation (3.136e)].

This system retains the simplicity of the mean-field dynamo by parameterising non-axisymmetric effects by the α–effect while including the key nonlinear interactions between $\mathbf{B}$ and $\mathbf{u}$; the field drives a flow through the action of the Lorentz force in (4.32b) and this flow acts back on the field in (4.32a).

Hollerbach & Jones (1993, 1995) considered a system with a finitely conducting inner core and demonstrated the stabilising effect it could have on dynamo solutions. Fotheringham *et al.* (2002) and Fearn & Rahman (2004a) have extended the model to investigate the stability of axisymmetric fields to non-axisymmetric instabilities and found that such instabilities can significantly constrain the strength of field that can be generated.

4.5.2. 2.5D MODELS

The term "2.5D" is applied to models that solve the convectively-driven system (4.5a-d), resolving fully in radius r and colatitude θ but with only very restricted resolution in azimuth ϕ. The motivation for this is to produce a problem that is tractable with moderate computing resources. The justification is from Cowling's theorem (see Section 1.3.4). We know that an axisymmetric field cannot be maintained by fluid motions. The interaction between axisymmetric and non-axisymmetric parts of the system are therefore a key ingredient of the dynamo problem. So, simply considering the axisymmetric part and even one non-axisymmetric mode ensures that this key ingredient is present in the model.

Jones *et al.* (1995) reported the first results from a 2.5D model. The results were encouraging, producing fields of around the right magnitude and associated with flow strengths consistent with those deduced from the secular variation. Most of their calculations are for an Ekman number of 10^{-3} (note that their definitions of the Ekman and modified Rayleigh numbers have a factor 2 in the denominator and are based on the gap-width $r_o - r_i$ as length scale). Their single non-axisymmetric mode has azimuthal wavenumber $m = 2$. They consider both $\mathrm{q} = 10$ and $\mathrm{q} = 1$ and find that while for $\widetilde{\mathrm{Ra}}_J = 50$ (a subscript J denotes their definition of the parameter) is sufficient to maintain a dynamo for $\mathrm{q} = 10$, $\widetilde{\mathrm{Ra}}_J = 1600$ was required to sustain a field when $\mathrm{q} = 1$. This feature of increasing $\widetilde{\mathrm{Ra}}$ with decreasing q is one that is reinforced by 3D studies.

Subsequent studies have proceeded to use the model to investigate, for example, the effect of varying the Ekman number (Sarson & Jones, 1998), the effect of CMB heterogeneity (Sarson *et al.*, 1997) and reversals (Sarson & Jones, 1999, Sarson, 2000). The model has been used to good effect in elucidating the results from 3D models and in understanding dynamo behaviour in different parameter regimes (see below and Jones, 2000).

4.5.3. 3D MODELS

The first attempts at 3D calculations were by Zhang & Busse (1989, 1990). Their calculations used stress-free boundary conditions which reduces the problem of resolving viscous boundary layers. This was appropriate since computing resources were limited; allowing only modest space resolution. Furthermore, resolution in time was restricted to a single mode; they did not use a time-stepping method, instead following a bifurcation sequence from stationary fluid, to steadily drifting finite amplitude convection, to a finite amplitude convection driven dynamo. Solutions were sought proportional to $\exp[\mathrm{i}(\phi - ct)]$ so all components of the solution were forced to drift at the same wave speed c. They used $\mathrm{E} = 10^{-3}$ and found dynamo solutions. Unfortunately, attempts to follow these to lower E failed; the dynamo action found did not persist as E was reduced.

The fully 3D time-stepping calculation of Glatzmaier & Roberts (1995a,b) marked a major step forward. They used $\mathrm{q} = 0.1$ and found maintenance of a field of strength up to $56\,\mathrm{mT}$. Glatzmaier & Roberts found that very high values of their Rayleigh number[11] (defined as $g\,\alpha\,Q/2\,\Omega\,c_P\,\rho\,\kappa^2 \approx 6 \times 10^7$, where Q is the heat flux at the bottom of the core and c_P the specific heat at constant pressure defined on page 8) were required to give dynamo action. This high value at small q is consistent with the trend found by Jones *et al.* (1995) who speculate that the reason is that convective velocities scale with $\mathrm{q}\,\widetilde{\mathrm{Ra}}$ rather than $\widetilde{\mathrm{Ra}}$. The single integration of Glatzmaier & Roberts (1995a,b) required substantial supercomputer resources and simulated only 40,000 years. An important feature of the simulation was that it included a field reversal.

Following on from this pioneering work, several groups have produced their own geodynamo models. Increasing computing power has allowed longer integrations and modest exploration of parameter space. Most models use a similar (spectral) numerical approach (see for example Hollerbach, 2000) and give good agreement for a simple steady benchmark solution (see Christensen *et al.*, 2001). Jones (2000) has reviewed what has been learnt from the first 5 years of this work. He finds it useful to divide the calculations that have been done into two categories which he calls *Busse-Zhang* (BZ) and *Glatzmaier-Roberts* (GR) models. The distinction is made according to the choice of the key parameters q, $\widetilde{\mathrm{Ra}}$ and E. BZ calculations typically have $\widetilde{\mathrm{Ra}}$ a few times $\widetilde{\mathrm{Ra}}_{c0}$ (the critical value for the onset of convection in the rapidly rotating system in the absence of any magnetic field), $\mathrm{q} \sim 10$ and $10^{-4} \le \mathrm{E} \le 10^{-3}$. They are characterised by velocities having magnitude of $\mathcal{O}(10)$ on the thermal diffusion timescale [so the magnetic Reynolds number $\mathrm{Rm} = \mathcal{O}(100)$, sufficient for dynamo action]. The magnetic field has only a weak influence on convection which takes place mostly outside the tangent cylinder. There is no strong differential rotation, poloidal and toroidal fields are comparable in magnitude and the dynamo can be thought of as of α^2–type. GR calculations are more supercritical with $\widetilde{\mathrm{Ra}} \sim 100\,\widetilde{\mathrm{Ra}}_{c0}$ and therefore much more computationally intensive. (Consequently most published work is in the BZ regime.) The larger $\widetilde{\mathrm{Ra}}$ is permitted by going to lower E. (Increasing $\widetilde{\mathrm{Ra}}$ at fixed E can result in dynamo action shutting off.) Convective velocities are larger, permitting lower values of q. There is a stronger differential rotation and the dynamo is more of $\alpha\omega$–type, although the peak poloidal field strength remains comparable with the that of the toroidal field.

Most of the work described above is for what Jones (2000) refers to as the "zero-order model", that is a spherically symmetric Boussinesq basic state with only one buoyancy source. The main exception is the work by Glatzmaier & Roberts (1996a,b,c) (and their subsequent papers) which use the anelastic approximation

11 Which is (up to a factor $1/2$) formally equivalent to our modified Rayleigh number $\widetilde{\mathrm{Ra}}$, since their temperature is scaled with $Q\,/\,[c_P\,\rho\,\kappa\,(r_o - r_i)]$.

and both thermal and compositional buoyancy. Even within the zero-order model there is considerable scope for variation. As well as choice of the key governing parameters q, E, E_η and $\widetilde{Ra}$ models differ in:

- *Inner core.* Most models include an inner core of radius about one third of that of the outer core with an electrical conductivity comparable with (and usually the same as) that of the outer core. Some have no inner core while others choose an insulating or perfectly conducting inner core. Wicht (2002) has recently concluded that in his model, the inner core does not play an important role in Earth-like reversal sequences.

- *Buoyancy distribution.* Even with thermal buoyancy only, there is scope to choose differential heating, internal heating or some combination of the two.

- *Inertia.* Its size is determined by the choice of E_η. Some models neglect it altogether ($E_\eta = 0$) while others include it partially, for example only the axisymmetric azimuthal part important for the geostrophic flow.

- *Boundary conditions.* The values of E that we are forced to use in order to give a numerically tractable problem mean that the role of viscosity is significantly amplified compared with the real Earth. Kuang & Bloxham (1997, 1999) have applied stress-free boundary conditions, arguing that this reduces somewhat the influence of viscosity, in partial compensation for the effect of the larger Ekman number. Most calculations continue to apply rigid boundary conditions on $\mathbf{u}$.

Beyond the zero-order model, there have been several developments:

- *Density.* As mentioned above, Glatzmaier & Roberts (1996a,b,c, 1997, 1998) have championed the use of the anelastic approximation and of both thermal and compositional buoyancy. Their's is the only model that deals with the sources of buoyancy (mostly at the ICB) in a fully consistent manner, directly linking them to cooling at the CMB.

- *Heterogeneous boundary conditions.* The heat flow across the CMB is not spherically symmetric. Motivated by this, several groups have investigated the effect of heterogeneous thermal boundary conditions at the CMB. Glatzmaier *et al.* (1999) have shown the strong influence the choice of boundary condition has on reversal frequency. Olson & Christensen (2002) find that "When the amplitude of the boundary heat flow heterogeneity exceeds the average heat flow, the dynamos usually fail" and also find similarities between the present field and that produced by a model with boundary heat flow derived from lower-mantle seismic tomography, see also Christensen & Olson

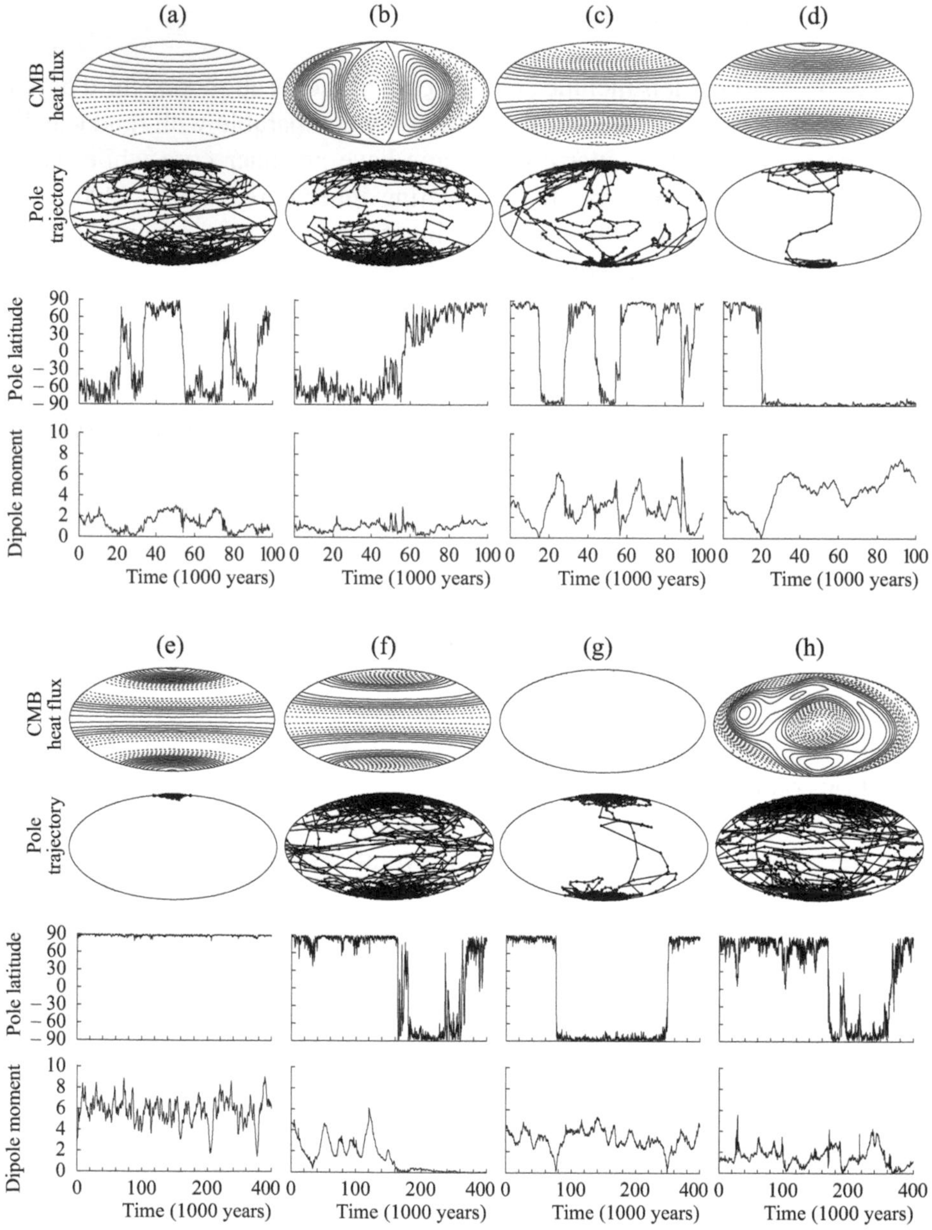

Figure 4.5 - An example of the influence heterogeneous heat flux boundary conditions on the reversal behaviour of a dynamo model. From Glatzmaier *et al.* (1999).

(2003). Bloxham (2000) compares the secular variation in the Kuang & Bloxham (1997, 1999) model with the paleomagnetic secular variation. He finds that, while there is a fair agreement for the meridional distribution, the amplitude in the numerical simulations is smaller by a factor of at least two. When he includes heterogeneous CMB heat flow in the model, he finds that he can match the amplitude of the paleomagnetic secular variation .

As computer power has increased and more groups have constructed their own models, a large number of model variations has been studied and several groups have undertaken parameter surveys (for example Christensen *et al.*, 1999, Kutzner & Christensen, 2002, Morrison & Fearn, 2000, Fearn & Morrison, 2001, Simitev & Busse 2002, 2003b).

4.6. TURBULENCE IN THE EARTH'S CORE: THE ENDS JUSTIFY THE MEANS?

During the past decade, a number of fully nonlinear, three-dimensional numerical simulations of the geodynamo have been published. As explained in the previous section, several of these have mimicked the geomagnetic field remarkably well. The success has been so great that some commentators have even gone so far as to say that the origin of the Earth's magnetism is a mystery no longer, even though the simulations cannot, by several orders of magnitude, employ estimated values for the material properties of the core, especially its diffusivities. Worse still, they inadequately resolve the flow and field. And yet the simulations are encouraging! This unreasonable success is called *the geodynamo paradox.* (See also Glatzmaier, 2002.)

The Earth's core is, so most geophysicists believe, driven into motion by buoyancy forces so strong that the flow and field are turbulent, fluctuating on every length and time scale. These irregularities, it is anticipated, are most pronounced for the small scale "eddies", but the occasional polarity reversals betray the stochastic nature of core flow on large length scales too. There is a lower limit, ℓ_F (say), on the length scales ℓ that can be resolved in numerical simulations. This limit is continually being reduced as computer technology advances and as high speed machines become increasingly available. Already the greater variability of small scale features has been confirmed, and polarity reversals have been mimicked; see, for example, Glatzmaier & Roberts (1995), Glatzmaier *et al.* (1999), Kageyama *et al.* (1999). Nevertheless, the physics of the unresolved scales, $\ell < \ell_F$, have been ignored. Disturbingly, the success of the simulations has sometimes been seen as sufficient justification for the drastically enhanced diffusivities and the neglect of the unresolved scales, the one magically compensating for the other; "the ends justify the means"! More generally, it is recognised that the basics of the simulations need to be better understood and improved. This review describes some of the attempts that have been made, and are being made, in this direction.

Two main approaches have been tried. The "traditional method" grew from techniques similar to those used with success in statistical mechanics. It seeks evolution equations governing variables averaged over an ensemble of identical systems. It is an approach that predates the advent of electronic computers and makes no reference to them. In contrast, the other approach, which we call the "engineering method" is geared to numerical computation. It recognises explicitly the existence of the *GS-SGS frontier*, $\ell = \ell_F$, separating the unresolved *sub-grid scales* (SGS) from the resolved *grid scales* (GS). It seeks evolution equations for GS variables that incorporate the SGS realistically. A numerical integration that makes use of these equations is called a *large eddy simulation* (LES). Somewhat arbitrarily, we shall assume that the GS-SGS frontier today is approximately at $\ell_F \approx 10^4\,\mathrm{m}$, so that, in terms of their wavenumbers, the SGS are defined by

$$k > k_F\,, \qquad \text{where} \qquad k_F = 1/\ell_F \approx 10^{-4}\,\mathrm{m}^{-1}\,. \qquad (4.33a,b)$$

"Why," the reader may ask, "should we concern ourselves with these matters? Surely, with time and patience, the geodynamo paradox will resolve itself? Isn't it certain that, as computer technology advances, all the important scales of core MHD will ultimately be numerically resolved?" While it is undoubtedly true that the GS-SGS frontier will be continually pushed back, there is no prospect in the foreseeable future of resolving all significant scales. Meanwhile, progress requires that the effect of the SGS on the LES is dealt with in a physically sensible way.

The notation in the sections on turbulence in the Earth's core departs slightly from that used elsewhere in the book. The main differences are:

- The fluid velocity is $\mathbf{U}$ and not $\mathbf{u}$. This is because large and small letters are used here in different ways for the same physical quantity;

- The fractional density excess, C, which is called the *codensity*, is used instead of temperature T or entropy S to describe buoyancy. Not only is this more general (since it recognises that compositional density differences might be a part of C) but also it avoids conflict with the use of large and small letters, since T and t (in place of C and c) would risk confusions with the time, t.

4.7. PRELIMINARY CONSIDERATIONS ON TURBULENCE

4.7.1. THE ENERGY SOURCE

As mentioned above, buoyancy is usually held responsible for core motions and fields. More specifically, it is argued that the inner core has been created by solidification during the general cooling of the Earth. As the inner core continues to advance upwards, freezing releases latent heat and light material from its surface, both of which provide buoyancy. This process must be potent enough to supply the heat lost by the core through the conduction of heat down the core adiabat, which is estimated to be on the order of $5\,\mathrm{TW}$ ($1\,\mathrm{TW} = 10^{12}\,\mathrm{W}$). It must also make good the energy dissipated by the magnetoconvection, which has been estimated to be $1 - 2\,\mathrm{TW}$ by Roberts *et al.* (2003) and $0.2 - 0.5\,\mathrm{TW}$ by Christensen & Tilgner (2004). Although this energy is fed back into the core and helps to maintain the convection, considerations of the thermodynamic efficiency of the core, regarded as a heat engine, require a convective heat flow from the core, over and above the adiabatic heat flow, of perhaps as much as $5\,\mathrm{TW}$, making a total of $10\,\mathrm{TW}$.

A crude estimate of the radial heat flux is $\overline{\rho} C_p \langle \theta U_r \rangle$, where C_p ($\approx 800\,\mathrm{J\,kg^{-1}\,K^{-1}}$) is the specific heat, and θ is the temperature excess above the adiabat; this is positive where hot fluid rises and negative where it descends. The horizontal average, $\langle \theta\, U_r \rangle$, is therefore positive. We may crudely estimate $\langle \theta\, U_r \rangle$ as $\frac{1}{3}\, V\, \Delta T$, where ΔT is the temperature difference in K between the rising and falling material; the factor $1/3$ is included to allow for the fact that only the vertical component of $\mathbf{U}$ is relevant. This leads to a radial heat flux of approximately $300\,\Delta T\,\mathrm{W\,m^{-2}}$. Assuming that this is approximately the same everywhere beneath the boundary layer at the core surface, the convective flow of heat is $4 \times 10^{16} \Delta T\,\mathrm{W}$. The postulated 5 TW requires that $\Delta T \approx 10^{-4}\,\mathrm{K}$. The corresponding density difference is approximately $\Delta\overline{\rho} = \overline{\rho}\overline{\alpha}\Delta T \approx 10^{-5}\,\mathrm{kg\,m^{-3}}$, where $\overline{\alpha} \approx 10^{-5}\,\mathrm{K^{-1}}$ is the coefficient of thermal expansion. Braginsky & Roberts (1995, 2002), who included compositional buoyancy also, estimated that $\Delta C = \Delta\overline{\rho}/\overline{\rho} \approx 10^{-8}$.

The light material created at the surface of the inner core is mixed throughout the fluid core by its own buoyancy. Many laboratory experiments and numerical simulations of convection show the light material rising in "plumes", mixing with its surroundings as it does so. It should be emphasised however that the buoyancy in the core, as quantified by the appropriate dimensionless parameter (the Rayleigh number, Ra), is much stronger than in the experiments and simulations. For example, the simulations of Chen & Glatzmaier (2005) show plumes but their Ra is only 10^9; in contrast, Ra is probably closer to 10^{24} in the core! Under these circumstances, the coherence required for a plume to exist is absent: turbulent convection spreads the light material almost evenly throughout the fluid core. The fractional density excess, $\Delta C = \Delta\overline{\rho}/\overline{\rho} \approx 10^{-8}$, creates a uniform top-heavy gradient, β, of order $5 \times 10^{-15}\,\mathrm{m}^{-1}$. The destabilizing effect of this gradient is represented by the (imaginary) Brunt frequency, ω_α, defined by

$$\omega_\alpha^2 = g\beta\,. \tag{4.34}$$

[We use this notation in order to ease comparison between the analysis that follows and that of Braginsky & Meytlis (1990). It should be understood that the suffix α relates to the thermal expansion coefficient and not to the α–effect of mean field electrodynamics.] Taking $g \approx 5\,\mathrm{m\,s^{-2}}$, we find that $\omega_\alpha \approx 10^{-7}\,\mathrm{s^{-1}}$, the value we use below.

4.7.2. ORDERS OF MAGNITUDE

Let $\mathbf{\Omega} = \Omega\,\mathbf{e}_z$ be the angular velocity of the Earth, and $\mathbf{U}$ the fluid velocity relative to the reference frame that rotates with the mantle; $\mathbf{e}_a$ is the unit vector in the direction of coordinate a and ∂_a will be the partial derivative with respect to a. The density of the core, $\overline{\rho}$, is assumed to be uniform except when evaluating the buoyancy force created by variations in the codensity C, i.e. for simplicity we adopt again the Boussinesq approximation. We shall consider thermally-created C only, so that $C = -\overline{\alpha}\theta$. The thermal diffusivity is denoted by κ, the kinematic viscosity by ν and the magnetic diffusivity by η. The constancy of $\overline{\rho}$ makes it convenient to use what are sometimes called *Alfvén units*: the magnetic field, $\mathbf{B}$, is divided by $(\mu_0\overline{\rho})^{1/2}$, and therefore has the dimensions of (Alfvén) velocity, $1\,\mathrm{cm\,s^{-1}}$ being approximately equivalent to $11\,\mathrm{G} = 1.1\,\mathrm{mT}$. The same factor is removed from the electric field $\mathbf{E}$. The current density, $\mathbf{J}$, is divided by $(\overline{\rho}/\mu_0)^{1/2}$ so that $\mathbf{J} = \boldsymbol{\nabla} \times \mathbf{B}$; here μ_0 is the magnetic permeability.

One reason why the MHD of the core differs so much from the MHD of plasmas (discussed in Chapter 6 and Chapter 7) lies in the dissimilar physical properties of the working fluids. The Earth's fluid core is a liquid metal alloy, and it is characteristic of such materials that $\mathrm{Pm} \ll 1$, $\mathrm{q} \ll 1$, and $\mathrm{Pr} = \mathcal{O}(1)$ (see Table II). For discussions of plasma turbulence, see for example Biskamp (2003) and Yoshizawa *et al.* (2003).

As we have seen Section 4.2, the relative importance of viscous and Coriolis forces is quantified by the Ekman number $\mathrm{E} = \nu/\Omega L^2$. If we take $L \approx 2 \times 10^6\,\mathrm{m}$ to be characteristic of the large length scales of the core, we obtain $\mathrm{E} = \mathcal{O}(10^{-15})$, from which we may infer that the viscous forces are much the smaller on the large scales L. If E is redefined using a smaller scale ℓ instead of L, E is increased, showing that viscous forces are always significant on small enough length scales (such as those of boundary layers).

It is generally agreed that the fluid core is in a *strong field state*, meaning that $\mathbf{J} \times \mathbf{B}$, which is the Lorentz force (per unit mass, i.e. acceleration), is comparable with (or somewhat greater than) the Coriolis force, $2\boldsymbol{\Omega} \times \mathbf{U}$. This statement is to be understood in a general sense, for special circumstances exist in which these forces are small or ineffective. For example, in a nonmagnetic system, rotation favours flows that are two-dimensional with respect to the direction of $\boldsymbol{\Omega}$. In such a "geostrophic motion", the Coriolis force is irrotational and cannot drive convective circulations, a point that is made transparent by absorbing it into the pressure gradient. When the motion is slightly ageostrophic, the Coriolis force can mostly be absorbed into the pressure gradient, the remnant being small compared with $2\Omega U$, which then becomes a poor estimate of the efficacy of the Coriolis force, $2\boldsymbol{\Omega} \times \mathbf{U}$. Similarly, a nonrotating MHD system favours $\mathbf{U}$ and $\mathbf{J}$ that are nearly two-dimensional with respect to the direction of $\mathbf{B}$; then JB becomes a poor estimate of the efficacy of the Lorentz force, $\mathbf{J} \times \mathbf{B}$. More generally, in a rotating MHD system, the Coriolis and Lorentz forces favour states that change slowly in the directions of $\boldsymbol{\Omega}$ and $\mathbf{B}$ compared with their variation in the direction of $\boldsymbol{\Omega} \times \mathbf{B}$, perpendicular to both $\boldsymbol{\Omega}$ and $\mathbf{B}$. Although the "constraints" imposed by the Coriolis and Lorentz forces are important in determining the state of a rotating MHD system such as the Earth's fluid core, we ignore them in the order-of-magnitude estimates of this subsection.

The westward drift of large scale magnetic features across the Earth's surface corresponds to a motion at the core surface of order $10^{-4}\,\mathrm{m\,s^{-1}}$ and this is usually assumed to be the characteristic speed $U = |\mathbf{U}|$ of core flow. The relative magnitude of the Coriolis force and inertial force $\mathbf{U} \cdot \boldsymbol{\nabla}\mathbf{U}$ is quantified by the Rossby number, $\mathrm{Ro} = U/\Omega L = \mathcal{O}(10^{-6})$, from which we may infer that the inertial forces are much the smaller on the large scales L. If Ro is redefined using the smaller scale ℓ instead of L, Ro is increased, showing that inertial forces are always significant on small enough length scales.

In nonrotating systems, it is often argued that there is an approximate equipartition between the kinetic and magnetic energy densities, $\mathcal{K}$ and $\mathcal{M}$, per unit volume. This is not true in a strong field system. The large scale velocity U and field B contribute most to $\mathcal{K} \sim \frac{1}{2}\overline{\rho}|\mathbf{U}|^2$ and $\mathcal{M} \sim \frac{1}{2}\overline{\rho}|\mathbf{B}|^2$. Since the Coriolis and Lorentz forces are comparable and both greatly exceed the inertial force, $\mathcal{K}$ is smaller than $\mathcal{M}$ by a factor of order Ro.

For the length scale ℓ, on which the typical flow and field are of order u and b (say), the viscous and ohmic dissipation rates are $\sigma^\nu \sim \overline{\rho}\nu(v/\ell)^2$ and $\sigma^\eta \sim \overline{\rho}\eta(b/\ell)^2$, respectively. Even though U and B may be larger than u and b respectively, it obviously does not necessarily follow that $U/L > v/\ell$ and $B/L > b/\ell$. In other words, when σ^ν and σ^η are summed for all length scales ℓ, the large scales may not make the greatest contributions to the total viscous and ohmic dissipation rates, $\mathcal{Q}^\nu$ and $\mathcal{Q}^\eta$, per unit volume. Consider an extreme case: suppose the scales for which

$\mathrm{Ro} = u/\Omega\ell$ is $\mathcal{O}(1)$ or larger are the scales that make the greatest contributions to $\mathcal{Q}^\nu$ and $\mathcal{Q}^\eta$. On these scales, the Coriolis force is relatively unimportant and equipartition, $u \sim b$, is plausible. Then $\mathcal{Q}^\eta/\mathcal{Q}^\nu \sim \eta/\nu \gg 1$. This disparity is only increased if larger scales, for which $\mathrm{Ro} < \mathcal{O}(1)$, make the greatest contributions to $\mathcal{Q}^\nu$ and $\mathcal{Q}^\eta$. We conclude that the energy received by the motions and fields from buoyancy is mainly dissipated by Joule heating.

4.7.3. BASIC EQUATIONS AND THEIR AVERAGES

Let us rewrite Boussinesq equations (see Chapter 1) governing magnetoconvection in the core using the codensity C (note that we use here U to denote the velocity field, as u will be used for an expansion below)

$$\partial_t \mathbf{U} + \mathbf{U}\cdot\nabla\mathbf{U} + 2\boldsymbol{\Omega}\times\mathbf{U} - \nu\,\Delta\mathbf{U} = -\nabla P + \mathbf{B}\cdot\nabla\mathbf{B} + C\mathbf{g}, \tag{4.35a}$$

$$\partial_t \mathbf{B} + \mathbf{U}\cdot\nabla\mathbf{B} - \eta\,\Delta\mathbf{B} = \mathbf{B}\cdot\nabla\mathbf{U}, \tag{4.35b}$$

$$\partial_t C + \mathbf{U}\cdot\nabla C - \kappa\,\Delta C = S_C, \tag{4.35c}$$

$$\nabla\cdot\mathbf{U} = \nabla\cdot\mathbf{B} = 0. \tag{4.35d}$$

As earlier in the chapter, P is a reduced pressure including the centrifugal and magnetic pressure terms and S_C is the source of C, supposed given.

The traditional approach of turbulence theory envisages averages over an ensemble of identical systems. The idea of an ensemble average is an old one, long predating electronic computation and the difficulties of numerical simulation. We shall therefore initially refrain from using the GS, SGS, LES terminology. When the statistical properties of the turbulence at a point $\mathbf{x}$ and time t are independent of $\mathbf{x}$, the turbulence is "homogeneous"; when they are also independent of direction, it is "isotropic"; if they are independent of t, it is (statistically) "steady". The statistical properties may be different when viewed in a mirror (i.e. under parity inversion; see below). Since the core is inhomogeneous, we shall not be dealing with homogeneous or isotropic turbulence but, in an approximate sense described below, it may be regarded as "almost" homogeneous or isotropic.

Ensemble averages obey *Reynolds rules*:

$$\overline{F+G} = \overline{F} + \overline{G}, \qquad \overline{\overline{F}\,\overline{G}} = \overline{F}\,\overline{G}, \qquad \overline{\overline{F}\,g} = 0, \tag{4.36a,b,c}$$

where $g = G - \overline{G}$ is the "fluctuating part" of G. It is useful, but not essential, to represent all variables in this way

$$\mathbf{U} = \overline{\mathbf{U}} + \mathbf{u}, \quad \mathbf{B} = \overline{\mathbf{B}} + \mathbf{b}, \quad C = \overline{C} + c, \quad P = \overline{P} + p. \tag{4.37a,b,c,d}$$

The primary aim is to determine the evolution of the mean fields, $\overline{\mathbf{U}}$, $\overline{\mathbf{B}}$ and $\overline{C}$. On substituting (4.37a–d) into (4.35a–d), and averaging, the equations governing the mean fields result:

$$\begin{aligned}
\partial_t\overline{\mathbf{U}} &+ \overline{\mathbf{U}}\cdot\nabla\overline{\mathbf{U}} + 2\boldsymbol{\Omega}\times\overline{\mathbf{U}} - \nu\Delta\overline{\mathbf{U}} \\
&= -\nabla\overline{P} + \overline{\mathbf{B}}\cdot\nabla\overline{\mathbf{B}} + \overline{C}\mathbf{g} + \overline{\mathbf{M}}^u, && (4.38a)\\
\partial_t\overline{\mathbf{B}} &+ \overline{\mathbf{U}}\cdot\nabla\overline{\mathbf{B}} = \overline{\mathbf{B}}\cdot\nabla\overline{\mathbf{U}} + \eta\Delta\overline{\mathbf{B}} + \overline{\mathbf{M}}^b, && (4.38b)\\
\partial_t\overline{C} &+ \overline{\mathbf{U}}\cdot\nabla\overline{C} = S_C + \kappa\Delta\overline{C} + \overline{M}^c, && (4.38c)\\
\nabla\cdot\overline{\mathbf{U}} &= \nabla\cdot\overline{\mathbf{B}} = 0, && (4.38d)
\end{aligned}$$

where $\overline{\mathbf{M}}^u = -\nabla\cdot\mathbf{R}, \qquad \overline{\mathbf{M}}^b = \nabla\times\boldsymbol{\mathcal{E}}, \qquad \overline{M}^c = -\nabla\cdot\mathbf{I},$ (4.39a,b,c)

Here $\mathbf{R}$ is the *Reynolds stress tensor*, $\boldsymbol{\mathcal{E}}$ is the mean *electromotive force* (EMF), and $\mathbf{I}$ is the *turbulent codensity flux*:

$$\mathbf{R} = \overline{\mathbf{uu} - \mathbf{bb}}, \qquad \boldsymbol{\mathcal{E}} = \overline{\mathbf{u}\times\mathbf{b}}, \qquad \mathbf{I} = \overline{\mathbf{u}c}. \qquad (4.40a,b,c)$$

Despite its name, $\mathbf{R}$ is dimensionally a stress tensor only after multiplication by $\overline{\rho}$.

Clearly, one cannot make use of (4.38a–d) if $\overline{\mathbf{M}}^u$, $\overline{\mathbf{M}}^b$ and $\overline{M}^c$ are unknown. If they cannot be computed, they must be specified. An *ad hoc* specification has the best chance of success if it is simple, and is based on a physically plausible picture of how the fluctuating fields influence the mean fields.

4.7.4. QUALITATIVE DESCRIPTIONS OF TURBULENCE

In the spirit of kinetic theory, the ensemble average, $\overline{Q}$, of a quantity Q at $\mathbf{x}$ and t is sometimes visualised as a *local average*, taken over a small length scale ℓ and a small timescale τ surrounding $(\mathbf{x}, t)$. Because turbulence exists on many length and time scales, it is hard to make this idea precise; the average generally depends on the choice of ℓ and τ. One may however imagine that, amongst the many length and time scale, there exists one length scale ℓ_D and one time scale τ_D that defines the "dominant" mode, in the sense that only length and time scales of order ℓ_D and τ_D and smaller influence the large scales significantly. Plausibly, when the local average of Q is computed using any ℓ and τ in the intervals $\ell_D \ll \ell \ll L$ and $\tau_D \ll \tau \ll T = L/U$, the result is insensitive to the chosen ℓ and τ, and is a good approximation to the ensemble average, $\overline{Q}$. Systems in which such ranges of ℓ and τ exist are called *two-scale systems*, L and T forming the "macroscale" and ℓ_D and τ_D defining the other, dominating, "microscale".

The two-scale approximation provides heuristic support for *local turbulence theory*, a form of words intended to signify that the effect of the turbulent eddies on the mean flow, at point $\mathbf{x}$ and time t, depends only on the properties of the mean motion at the same $(\mathbf{x}, t)$. The dominant microscales exist in an environment controlled by the macroscale fields $\overline{\mathbf{U}}$, $\overline{\mathbf{B}}$ and $\overline{C}$. The statistical properties of the turbulence on this scale, together with the values at $(\mathbf{x}, t)$ of averages such as (4.40a–c), depend only on $\overline{\mathbf{U}}$, $\overline{\mathbf{B}}$, $\overline{C}$ and their gradients, at the same values $(\mathbf{x}, t)$.

Although the two-scale approximation confers a modicum of respectability on the idea of a local average and on the concept of local turbulence, the question of whether naturally-occurring turbulence is well described by the approximation is an entirely separate matter. Another serious issue is also apparent: how is the local turbulence kept supplied with energy to offset its dissipative losses?

Consider *classical turbulence*, by which we mean turbulence of shear flows in a non-rotating non-magnetic systems. Turbulence arises when the macroscale Reynolds number, $\mathrm{Re} = UL/\nu$ is sufficiently great, where U and L are characteristic of the shear flow. Although all the energy is injected at large scales L, much of it is dissipated by the small-scale turbulence. This happens through an *inertial cascade* of energy created by the $\mathbf{U} \cdot \boldsymbol{\nabla}\mathbf{U}$ term in (4.35a). This progressively transfers energy from the large scale shear to the turbulent eddies, the latter becoming increasingly isotropic as their size diminishes until their motion resembles the Brownian motion of molecules that diffuse momentum isotropically. Let us suppose that these are the dominant microscales. Then, if molecular diffusion is represented by $\nu\,\Delta\mathbf{U}$, what can be more natural than to represent the action of the small isotropic eddies on the mean motion by $\nu_{\mathrm{T}}\,\Delta\overline{\mathbf{U}}$? This *Boussinesq-Reynolds ansatz* (BRA) originated in the nineteenth century and is still alive today; see, for example, Frisch (1995). It is the oldest and simplest way of specifying $\overline{\mathbf{M}}^u$. The diffusion of $\overline{\mathbf{U}}$ is controlled by the sum, $\nu^S = \nu + \nu_{\mathrm{T}}$, of the molecular viscosity and a "turbulent viscosity", ν_{T}. This enhancement in the effective viscosity is the physical expression of the energy that cascaded to the turbulent scales and was dissipated there.

One can take the BRA further. According to molecular transport theory, $\nu \approx \frac{1}{3}\overline{u}\ell$ in a dilute gas, where $\overline{u}$ is the root-mean-square molecular velocity and l is the mean-free-path between molecular collisions. Similarly, we may introduce a root-mean-square turbulent velocity $u = (\overline{|\mathbf{u}|^2})^{1/2}$ and a typical scale ℓ for the eddies. If the turbulent Reynolds number, $\mathrm{Re}_\ell = u\ell/\nu$, is large, the diffusion of large scale momentum by the turbulent eddies is a dynamic process, negligibly affected by viscous forces, so that ν_{T} is independent of ν. On dimensional grounds, $\nu_{\mathrm{T}} \sim u\ell$ is the only possibility, although it is difficult to make this rough estimative more quantitative. Because $\mathrm{Re} \gg 1$, it follows that $\nu_{\mathrm{T}} \gg \nu$, so that $\nu^S \approx \nu_{\mathrm{T}}$.

The replacement of $\overline{\mathbf{M}}^u$ by $\nu_{\mathrm{T}}\,\Delta\overline{\mathbf{U}}$ is (for constant ν_{T}) effectively the same as approximating $\mathbf{R}$ by $-\nu_{\mathrm{T}}\boldsymbol{\nabla}\overline{\mathbf{U}}$. More precisely,

$$R_{ij} = \tfrac{1}{3}R_{kk}\delta_{ij} - 2\nu_{\mathrm{T}}\overline{S}_{ij}, \tag{4.41}$$

where S is the rate of strain tensor, defined in (1.27f), we recall:

$$S_{ij} = \tfrac{1}{2}\left(\partial_i U_j + \partial_j U_i\right) . \tag{4.42}$$

The form (4.41) has the required symmetry, $R_{ji} = R_{ij}$. The first term on the right-hand side is required because $S_{kk} = 0$ by (4.35d) but $R_{kk} = \overline{|\mathbf{u}|^2}$ is nonzero. This term is, however, ineffective because its divergence is a gradient that can be absorbed into $\boldsymbol{\nabla}\overline{P}$; it is usually omitted without comment. In the simplest form of the theory, ν_{T} is a constant but more sophisticated choices have been made. For example, Smagorinsky (1963) specified $\nu_{\mathrm{T}}(\mathbf{x},t)$ as a function of the $\overline{S}_{ij}$ at $\mathbf{x}$ and t, the idea being that the larger the mean shear the stronger the local turbulence and the greater the ν_{T}.

When applied to magnetoconvection, the simplest generalisation of the BRA is

$$\overline{\mathbf{M}}^u = \boldsymbol{\nabla}\cdot(\nu_{\mathrm{T}}\boldsymbol{\nabla}\overline{\mathbf{U}}) , \qquad \overline{\mathbf{M}}^b = -\boldsymbol{\nabla}\times(\eta_{\mathrm{T}}\boldsymbol{\nabla}\times\overline{\mathbf{B}}) , \qquad \overline{M}^c = \boldsymbol{\nabla}\cdot(\kappa_{\mathrm{T}}\boldsymbol{\nabla}\overline{C}) , \tag{4.43a,b,c}$$

where ν_{T}, η_{T} and κ_{T} are turbulent diffusivities. When these are constant, (4.38a–d) have the same form as (4.35a–d), but with ν, η and κ replaced by $\nu^S = \nu + \nu_{\mathrm{T}}$, etc. Although magnetic field and heat are diffused by the small-scale eddies in somewhat different ways from momentum, it is commonly assumed (as for ν_{T}) that $\eta_{\mathrm{T}} \sim v\ell$ and $\kappa_{\mathrm{T}} \sim v\ell$. Because κ is so small (of much the same size as ν), the turbulent diffusion of heat in the core dominates its molecular diffusion: $\kappa^S \approx \kappa_{\mathrm{T}}$.

The generalised BRA (4.43a–c) raises a number of issues. One salient difference between the inertial cascades of classical turbulence theory and of convectively-driven turbulence should be stressed. The former envisages an inertial range of eddy sizes in which eddies of any one scale receive energy from the larger eddies and pass it on to the smaller eddies. In convectively-driven turbulence however, an inertial range of this type cannot occur, since buoyancy injects energy on every scale. Since buoyancy has a preferred direction, the local turbulence may be significantly anisotropic with respect to the direction, $\mathbf{e}_g$, of gravity. It may therefore be necessary to replace (4.43a–c) by

$$\mathbf{M}^u = \boldsymbol{\nabla}\cdot(\boldsymbol{\nu}_{\mathrm{T}}\cdot\boldsymbol{\nabla}\mathbf{U}) , \qquad \mathbf{M}^b = \boldsymbol{\nabla}\cdot(\boldsymbol{\eta}_{\mathrm{T}}\cdot\boldsymbol{\nabla}\mathbf{B}) , \qquad M^c = \boldsymbol{\nabla}\cdot(\boldsymbol{\kappa}_{\mathrm{T}}\cdot\boldsymbol{\nabla}C) , \tag{4.44a,b,c}$$

where $\boldsymbol{\nu}_{\mathrm{T}}$, $\boldsymbol{\eta}_{\mathrm{T}}$ and $\boldsymbol{\kappa}_{\mathrm{T}}$ are tensor diffusivities, symmetric with respect to $\mathbf{e}_g$. Such anisotropies have been introduced in descriptions of turbulent convection in stars; see, for example, Rüdiger (1989). As we shall recognise in Section 4.8, turbulent convection in the Earth's core is strongly affected by Coriolis and Lorentz forces, so that the turbulence and the tensor diffusivities are highly anisotropic, the directions of $\boldsymbol{\Omega}$ and $\overline{\mathbf{B}}$ being preferred.

The physical basis of (4.43) and (4.44a–c) is the idea that the action of the turbulent eddies on the macroscale fields is entirely diffusive. This, however, is not necessarily the case; there may also be "up-the-spectrum" processes or "back-scatteer". For instance, rotating, stratified flows generally lack mirror symmetry. As a result, $\mathbf{R}$ contains a term proportional to $\overline{\mathbf{U}}$ rather than to its gradient. This is called the (anisotropic) *kinetic α–effect*; see Frisch (1995). This name is a reminder of the α–effect in *mean field electrodynamics* (MFE). Not only does turbulence in an electrically conducting fluid create a mean EMF, $\boldsymbol{\mathcal{E}} = \overline{\mathbf{u} \times \mathbf{b}}$ proportional to the gradient of $\overline{\mathbf{B}}$ and represented by $-\eta_{\mathrm{T}}\boldsymbol{\nabla} \times \overline{\mathbf{B}}$ in (4.43); as explained in Chapter 1, when the flow is non-mirror symmetric, it also contains a term proportional to $\overline{\mathbf{B}}$ rather than to its gradient. In the simplest case, this EMF is simply $\alpha\overline{\mathbf{B}}$, parallel to $\overline{\mathbf{B}}$. Through this accident of notation, the phenomenon is called *the α–effect.*

When the turbulent magnetic Reynolds number, $\mathrm{Rm}_\ell = u\ell/\eta$, is large, as is the case in some astrophysical contexts, the α–effect and turbulent diffusion are very important contributors to the total mean EMF, $\overline{\mathbf{U}} \times \overline{\mathbf{B}} + \boldsymbol{\mathcal{E}}$. This is unlikely to be the case in the Earth's core. Because the fluid core generates the geomagnetic field, we can be sure that the magnetic Reynolds number of the large scale motions,

$$\mathrm{Rm} = UL/\eta^S\,, \tag{4.45}$$

is not small, but neither is it very large; numerical models suggest that $\mathrm{Rm} \approx 100$. Taking $U \approx 10^{-4}\,\mathrm{m\,s^{-1}}$ and $L \approx 2\times 10^6\,\mathrm{m}$ as before, we see that $\mathrm{Rm} \approx 100$ implies that $\eta^S \approx 2\,\mathrm{m^2\,s^{-1}}$. Taken literally, $\eta^S \approx 2\,\mathrm{m^2\,s^{-1}}$ and $\eta \approx 2\,\mathrm{m^2\,s^{-1}}$ (see above) imply $\eta_{\mathrm{T}} \approx 0$. Given the uncertainties of geophysical estimation, one should not make such a bold statement, but it seems probable that η_{T} is, at most, comparable with η. Only when $\mathrm{Rm}_\ell \equiv u\ell/\eta \gg 1$ is $\eta_{\mathrm{T}}/\eta \gg 1$ and (even assuming that u is as large as U which is unlikely) $\mathrm{Rm}_\ell < 1$ for all scales, ℓ, less than $L/100 \approx 2\times 10^4\,\mathrm{m}$.

There is an observational clue also indicating that turbulent induction does not dominate the electrodynamics of the core. The minimum time necessary for the polarity of the dipole component of the geomagnetic field to reverse is of order of the longest free decay time, $\tau_d = L^2/\pi^2\eta^S$ [see (4.2)]. This is reduced if η_{T}/η is increased but interpretations of paleomagnetic data are consistent with $\tau_d = L^2/\pi^2\eta$. It therefore seems unlikely that the theory of core MHD would be seriously compromised by setting $\eta_{\mathrm{T}} = \alpha = 0$ and more generally $\overline{\mathbf{M}}^b = \mathbf{0}$.

Even if $\nu_{\mathrm{T}} \approx \eta_{\mathrm{T}} \approx 1\,\mathrm{m^2\,s^{-1}}$, the macroscale Ekman number $\mathrm{E} = \nu_{\mathrm{T}}/\Omega L^2$ is only about 4×10^{-9}. The turbulent Ekman number $\mathrm{E_T} = \nu_{\mathrm{T}}/\Omega\ell^2$ based on a length scale ℓ exceeds 1 only for $\ell < 0.3\,\mathrm{km}$. It seems that, even when enhanced by the turbulence, viscous forces do not influence the macroscale motions significantly (except in boundary layers). This suggests that (if numerical algorithms allowed it), one could simplify core MHD by assuming that $\overline{\mathbf{M}}^u = \mathbf{0}$. This step is, however, questionable. According to (4.35a), there is, in addition to the inertial cascade from $\mathbf{U} \cdot \boldsymbol{\nabla}\mathbf{U}$, a *magnetic cascade* of kinetic energy created by $\mathbf{B} \cdot \boldsymbol{\nabla}\mathbf{B}$ and represented by $-\overline{\mathbf{bb}}$ in (4.40a). This suggests that, in the isotropic case (4.43),

$$R_{ij} = \tfrac{1}{3}R_{kk}\delta_{ij} - 2\nu_{\mathrm{T}\,U}\overline{S}_{ij} + 2\nu_{\mathrm{T}\,B}\overline{T}_{ij}, \quad \text{where} \quad T_{ij} = \tfrac{1}{2}(\partial_i B_j + \partial_j B_i) \tag{4.46}$$

is a more appropriate BRA than (4.41). On the face of it, the magnetic cascade is stronger for scales on which $\mathrm{Ro} = u/\Omega\ell$ is less than $\mathcal{O}(1)$. The last term in the expression (4.46) for R_{ij} recognises that mean motion gains kinetic energy from the turbulent magnetic field at a rate approximated here by $\nu_{B\,\mathrm{T}}\overline{\mathbf{J}} \cdot \nabla \times \overline{\mathbf{U}}$. [This provides a motivation for including a corresponding term $-\nu_{B\,\mathrm{T}}\nabla \times \overline{\mathbf{U}}$ in $\boldsymbol{\mathcal{E}}$; see Yoshizawa *et al.*(2003).]

4.8. THE TRADITIONAL APPROACH TO TURBULENCE

4.8.1. A THREE-STEP PROGRAM

Evidently, a wide gulf exists between the physical basis of the BRA and its mathematical expression (4.43). Can it be bridged? A deductive theory would require precise knowledge of $\mathbf{u}$, $\mathbf{b}$ and c. A three-step program has been attempted. In step 1, the averaged equations (4.38a–d) are subtracted from (4.35a–d) to obtain equations governing the turbulent variables:

$$(\mathrm{D}_t - \nu\,\Delta)\mathbf{u} + 2\boldsymbol{\Omega} \times \mathbf{u} = -\nabla p + \mathbf{B}\cdot\nabla\mathbf{b} + c\mathbf{g} + \mathbf{L}^u + \mathbf{m}^u, \tag{4.47a}$$
$$(\mathrm{D}_t - \eta\,\Delta)\mathbf{b} = \mathbf{B}\cdot\nabla\mathbf{u} + \mathbf{L}^b + \mathbf{m}^b, \tag{4.47b}$$
$$(\mathrm{D}_t - \kappa\,\Delta)c = -\mathbf{u}\cdot\nabla C + m^c, \tag{4.47c}$$
$$\nabla\cdot\mathbf{u} = 0, \qquad \nabla\cdot\mathbf{b} = 0, \tag{4.47d}$$

where $\mathrm{D}_t = \partial_t + \mathbf{U}\cdot\nabla$ and

$$\mathbf{L}^u = \mathbf{b}\cdot\nabla\mathbf{B} - \mathbf{u}\cdot\nabla\mathbf{U}, \qquad \mathbf{L}^b = \mathbf{b}\cdot\nabla\mathbf{U} - \mathbf{u}\cdot\nabla\mathbf{B}, \tag{4.47e}$$
$$\mathbf{m}^u = \nabla\cdot(\mathbf{bb} - \mathbf{uu}) - \mathbf{M}^u, \quad \mathbf{m}^b = \nabla\times(\mathbf{u}\times\mathbf{b}) - \mathbf{M}^b, \tag{4.47f}$$
$$m^c = -\nabla\cdot(\mathbf{u}c) + M^c. \tag{4.47g}$$

The overbars on $\overline{\mathbf{U}}$ etc have been omitted, but will be restored later if clarity demands it. In principle, these equations (and appropriate boundary conditions) determine $\mathbf{u}$, $\mathbf{b}$ and c as functionals of $\mathbf{U}$, $\mathbf{B}$ and C. Step 2 consists in determining these functionals, a truly daunting task. In step 3, the functionals are inserted into (4.40a–c) so that $\mathbf{M}^u$, $\mathbf{M}^b$ and M^c can be evaluated, also as functionals of $\mathbf{U}$, $\mathbf{B}$ and C. These might, but most likely would not, be of the form (4.44a–c).

An equally challenging alternative to steps 2 and 3 is to multiply the $i^{\rm th}$–component of (4.47a) by u_j and add the result to what is obtained by multiplying the $j^{\rm th}$–component of (4.47a) by u_i; after averaging, the evolution equation for R_{ij} is obtained. Similar operations with (4.47b) and (4.47c) generate analogous equations governing $\overline{\mathbf{bb}}$, $\overline{\mathbf{ub}}$ and $\overline{\mathbf{u}c}$. These evolution equations involve triple moments, such as $\overline{\mathbf{uuu}}$, $\overline{\mathbf{ubb}}$, etc. One is forced again either to make *ad hoc* choices for these tensors or to derive evolution equations to govern them, and these will involve quartic moments, and so on. This is the celebrated closure problem of turbulence theory. No matter how the hierarchy of equations is closed, the equations governing $\overline{\mathbf{uu}}$, $\overline{\mathbf{bb}}$, $\overline{\mathbf{ub}}$ and $\overline{\mathbf{u}c}$ must be solved. The required $\mathbf{M}^u$, $\mathbf{M}^b$ and M^c are then obtained by differentiation. We shall not adopt this alternative procedure here. For further details, see for example Yoshizawa *et al.* (2003).

Returning to the direct solution of (4.47a–d), one must recognise that step 2 is so challenging that approximation is inevitable. One possibility is to linearise (4.47a–d) by setting

$$\mathbf{m}^u = \mathbf{0}\,, \qquad \mathbf{m}^b = \mathbf{0}\,, \qquad m^c = 0\,. \tag{4.48a,b,c}$$

This is often called *first order smoothing* but can seldom be justified in the contexts in which it is applied. Nevertheless, linear equations are tractable and contain all the essential physics apart from the nonlinear cascades. They are therefore well worth investigating, as we do in the following subsections.

4.8.2. LINEARISED MODES OF A SIMPLE MODEL

We consider a simple magnetoconvective system consisting of a fluid "box",

$$0 \le x \le \ell_0\,, \qquad 0 \le y \le \ell_0\,, \qquad 0 \le z \le \ell_0\,, \tag{4.49a,b,c}$$

through which a uniform field passes; we assume the orientations

$$\mathbf{g} = -g\,\mathbf{e}_z, \quad \boldsymbol{\beta} = -\boldsymbol{\nabla} C = -\beta\,\mathbf{e}_z, \quad \mathbf{B} = B\,\mathbf{e}_y, \quad \mathbf{U} = \mathbf{0}\,. \tag{4.50a,b,c,d}$$

This defines the unperturbed state. This model and variants have frequently been studied; see below.

Boundary conditions, such as periodicity, must be satisfied by $\mathbf{u}$, $\mathbf{b}$ and c, so it is convenient to expand these in Fourier series. The available wavenumbers are discrete

$$k_x = (2\pi/\ell_0)\,n_x\,, \qquad k_y = (2\pi/\ell_0)\,n_y\,, \qquad k_z = (2\pi/\ell_0)\,n_z\,, \tag{4.51a,b,c}$$

where n_x, n_y and n_z are integers. The discreteness of $\mathbf{k}$ is an unnecessary complication in what follows, and we lose nothing essential by supposing that the Fourier representations are continuous functions of $\mathbf{k}$. It will become important however to recognise that the finite size of the box imposes a lower bound k_0 on the available wavenumbers

$$k_x \geq k_0\,, \quad k_y \geq k_0\,, \quad k_z \geq k_0\,, \tag{4.52a,b,c}$$

so that $k \geq \sqrt{3}k_0$, where $k = (k_x^2 + k_y^2 + k_z^2)^{1/2}$ is the total wavenumber. We shall sometimes replace k_x etc, by $1/\ell_x$ etc. We shall take $k_0 \approx 10^{-6}\,\mathrm{m}^{-1}$, so that $\ell_0 = 1/k_0$ is typical of the physical dimensions of the fluid core.

At first we ignore the inertial and viscous terms in (4.47a). The linearised equations (4.47a–d) are then

$$2\boldsymbol{\Omega} \times \mathbf{u} = -\boldsymbol{\nabla} p + \mathbf{B}\cdot\boldsymbol{\nabla}\mathbf{b} + c\mathbf{g}\,, \tag{4.53a}$$

$$(\partial_t - \eta\,\Delta)\mathbf{b} = \mathbf{B}\cdot\boldsymbol{\nabla}\mathbf{u}\,, \tag{4.53b}$$

$$(\partial_t - \kappa\,\Delta)c = -\mathbf{u}\cdot\boldsymbol{\nabla} C\,, \tag{4.53c}$$

$$\boldsymbol{\nabla}\cdot\mathbf{u} = 0\,, \qquad \boldsymbol{\nabla}\cdot\mathbf{b} = 0\,. \tag{4.53d}$$

A solution can be found in the form of a tessalated pattern of convective cells. For example, we may write

$$c = \widetilde{c}(t)\sin x' \sin y' \sin z'\,, \tag{4.54a}$$

where

$$x' = k_x x + \delta_x, \qquad y' = k_y y + \delta_y\,, \qquad z' = k_z z + \delta_z\,, \tag{4.54b,c,d}$$

and δ_x, δ_y and δ_z are arbitrary phases. In linear theory, $\widetilde{c}$ is proportional to $\exp(\gamma t)$ and ∂_t is equivalent to multiplication by γ; $-\Delta$ is equivalent to k^2.

On substituting (4.54a) into (4.53a), (4.53b) and (4.53d), cumbersome expressions for $\mathbf{u}$ and $\mathbf{b}$ emerge. To simplify these, we follow Braginsky & Meytlis (1990) by introducing the abbreviations

$$\gamma_\eta = \eta k^2\,, \quad \gamma_\kappa = \kappa k^2\,, \quad \Omega_* = 2\Omega k_z/k\,, \quad \gamma_* = \gamma_B k_y^2/k^2\,, \tag{4.55a,b,c,d}$$

where $\gamma_B = |\mathbf{B}|^2/\eta \approx 2\times 10^{-4}\,\mathrm{s}^{-1}$ and is scale-independent; Ω_* and γ_* are also independent of the absolute scale ℓ but depend on the relative scales, ℓ_z/ℓ and ℓ_y/ℓ, of the assumed periodicities. The relation

$$\mathbf{b} = B\partial_y\mathbf{u}/(\gamma + \gamma_\eta)\,, \tag{4.56}$$

obtained from (4.53b), can be used to eliminate $\mathbf{b}$ from (4.53a) in favour of $\mathbf{u}$. On removing p by taking the curl of the resulting equation, and using (4.53c) also, one obtains the dispersion relationship that determines the growth rate γ:

$$(\gamma + \gamma_\kappa)[\Omega_*^2(\gamma + \gamma_\eta)^2 + \gamma_*^2\gamma_\eta^2] = \omega_\alpha^2(\gamma + \gamma_\eta)\gamma_*\gamma_\eta(k^2 - k_z^2)/k^2\,. \tag{4.57}$$

This implies that

$$0 \leq \gamma + \gamma_\kappa \leq \frac{\omega_\alpha^2}{4\Omega}\frac{k^2 - k_z^2}{kk_z}\,. \tag{4.58}$$

4.8.3. THE MOST EASILY EXCITED MODE

It is of interest to determine the maximum growth rate, $\gamma_{\max}$, and the general behaviour of γ as a function of $\mathbf{k}$. The fact that $\mathrm{q} = \kappa/\eta$ is so small means that the system is not prone to overstability, i.e. the interesting root of the cubic (4.57) is real. If we ignored (4.52a–c), we would find that the maximising $\mathbf{k}$ is zero, and the corresponding γ is infinite. By imposing (4.52a–c), we improve physical realism, make $\gamma_{\max}$ finite and, in a small way, add *geometry-dependence* to the solutions.

It is convenient to maximise γ first over k_y. The equation $\partial\gamma/\partial\gamma^* = 0$ gives

$$k_y^2 = \frac{\omega_\alpha^2}{2|\mathbf{B}|^2}\frac{(k^2 - k_z^2)}{k^2} + \frac{2\Omega}{|\mathbf{B}|^2}(\eta - \kappa)kk_z\,, \tag{4.59a}$$

$$\gamma = \frac{\omega_\alpha^2}{4\Omega}\frac{(k^2 - k_z^2)}{kk_z} - \kappa k^2\,. \tag{4.59b}$$

This is "optimal" according to (4.58), but it is not the only relevant case. The *Elsasser number* (which is scale independent),

$$\Lambda = |\mathbf{B}|^2 / \Omega\,\eta\,, \quad (= \gamma_B/\Omega) \tag{4.60}$$

is moderately large in the core: $\Lambda \approx 50$, if we take $B \approx 0.1\,\mathrm{m\,s^{-1}}$ in Alfvén units, corresponding to a field of 0.01 T (or 100 G). It is therefore possible that the k_y given by (4.59a) lies outside the admissible range (4.52a–c). If (4.59a) gives $k_y < k_0$ or $k_x \equiv (k^2 - k_z^2 - k_y^2)^{1/2} < k_0$, the optimal case must be abandoned and, by (4.58), γ is reduced. The maximum growth rate is then obtained from (4.57) after the substitutions

$$k_y = k_0\,, \quad \text{and} \quad \gamma_* = |\mathbf{B}|^2\,k_0^2/2\,\gamma_\eta\,, \tag{4.61a,b}$$

have been made. Calling this "the non-optimal case", we show in Figure 4.6 a dashed curve, labelled "SO" (standing for "switchover"). This marks where the optimal case (above and to the right of the dashed curve) switches to or from the non-optimal case. No admissible region arises in which (4.59a) implies that $k_x < k_0$.

Figure 4.6 is a regime diagram that summarises the properties of the dispersion relationship (4.57) or, more precisely (since a 2D plot cannot show the dependence of γ on all three variables k_x, k_y and k_z), it displays information about $\gamma(k, k_z)$ for the maximising k_y. This is shown in a log-log plot of k/k_* versus k_z/k_*, where k_* is defined by

$$k_*^2 = \omega_\alpha^2/2\Omega\kappa \tag{4.62}$$

and is approximately $3.7 \times 10^{-3}\,\mathrm{m}^{-1}$ so that $\ell_* \approx 1\,\mathrm{km}$. The geometric constraints (4.52a–c) exclude everything outside the "wedge" defined by $k_z = k_0$ and $k_z = (k^2 - 2\,k_0^2)^{1/2}$.

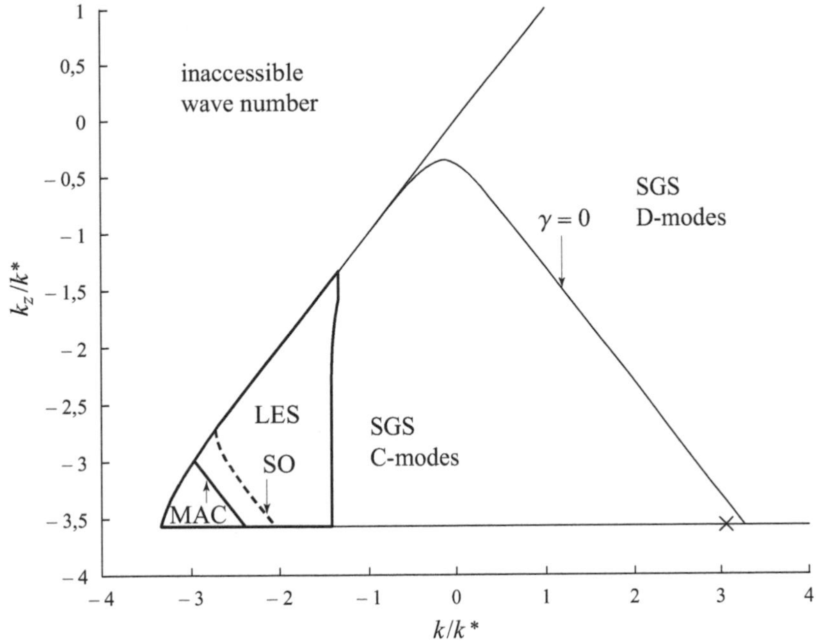

Figure 4.6 - Regime diagram for the linear dispersion relationship. This log-log plot displays information about the growth rate γ, maximised over k_y. The abscissa is k/k^*; the ordinate is k_z/k^*, where k^* is defined by (4.62). Solutions exist only in the "wedge" shown; other wavenumbers are inaccessible. The most unstable mode is marked with an "×". For the convective C-modes, $\gamma > 0$; for the damped D-modes, $\gamma < 0$. The "triangle" with a bold perimeter indicates modes that are accessible in large eddy simulations. For further explanation, see text.

Figure 4.6 has implications for LES. In numerical work, the integers in (4.51a–c) are bounded above by N_F (say). For continuous **k**, this corresponds to k_F in (4.33a), and the SGS are defined by

$$k_x > k_F\,, \qquad k_y > k_F\,, \qquad \text{or} \qquad k_z > k_F\,. \tag{4.63a,b,c}$$

The "triangle" highlighted by thick lines and marked "LES" contains only the geometrically accessible modes that are also numerically accessible according to (4.33a).

The curve labelled "$\gamma = 0$" defines the neutral modes, separating the unstable solutions $\gamma > 0$ labelled "C–modes" (standing for "convective modes") from the stable solutions $\gamma < 0$ labelled "D–modes" (damped modes). The curve labelled "MAC" will be discussed later. Because $k_* \gg k_0$, the overall maximum growth rate, $\gamma_{\max}$, belongs to one of the SGS, namely $(k, k_z, k_y) = (k_{\max}, k_{z,\max}, k_{y,\max})$ where

$$(k_{\max}, k_{z,\max}, k_{y,\max}) \approx ({k_*}^2/4\,k_0,\ k_0,\ \tfrac{1}{2}k_*/\Lambda^{1/2})\,, \quad \gamma_{\max} \approx \kappa k_*^4/16\,k_0^2\,. \tag{4.64a,b}$$

This maximising wave vector is marked with a cross in Figure 4.6. The growth rate is large: $\gamma_{\max} \approx 6 \times 10^{-4}\ \mathrm{s}^{-1}$, corresponding to a time of less than a day. This is because the convection is highly supercritical. We may restate this in terms of the (modified) *Rayleigh number*:

$$\widetilde{Ra} = \omega_\alpha^2\, L^2/2\,\Omega\,\kappa = (k_* L)^2\,. \tag{4.65}$$

This nondimensional measure of the density stratification is often used in convection studies. Here it is about 10^8, even though β is so tiny. According to (4.64a,b), the maximising wavenumbers are $k_x \approx k \approx 3.4\,\mathrm{m}^{-1}$, $k_y \approx 4.3 \times 10^{-3}\,\mathrm{m}^{-1}$, $k_z \approx 10^{-6}\,\mathrm{m}^{-1}$, corresponding to

$$\ell_x \approx 0.3\,\mathrm{m}\,, \qquad \ell_y \approx 1/4\,\mathrm{km}\,, \qquad \ell_z \approx 1000\,\mathrm{km}\,. \tag{4.66a,b,c}$$

These large disparities in the dimensions of the convection cells, and particularly their thinness in the x–direction, led Braginsky & Meytlis (1990) to call them "plate-like". This feature is not totally unexpected. A nonrotating, non-magnetic system extracts buoyant energy from unit horizontal area most effectively by up-and-down motions in as close a proximity as possible, i.e. having the smallest possible horizontal scales ℓ_x and ℓ_y. In fact $\ell_x = \ell_y = 0$ is preferred if viscosity is ignored! In the present case, the rotational and magnetic constraints suppress modes of small ℓ_y, but they leave the x–direction relatively unfettered. Cells of small ℓ_x "stack plates" close together and draw on the buoyancy source efficiently. See also St. Pierre (1996) and Siso-Nadal & Davidson (2004).

The smallness of ℓ_x reopens the question of the roles of inertia and viscosity, a topic we examine in the next subsection.

4.8.4. MORE COMPLICATED AND LESS COMPLICATED MODELS

To explore the wisdom of excluding the inertial and viscous forces from (4.53a), we may restore $(\partial_t - \nu\,\Delta)\mathbf{u}$ to its left-hand side. The dispersion relation replacing (4.57) becomes

$$\begin{aligned}(\gamma + \gamma_\kappa)[\Omega_*^2(\gamma + \gamma_\eta)^2 + \{\gamma_*\gamma_\eta + (\gamma + \gamma_\nu)(\gamma + \gamma_\eta)\}^2] \\ = \omega_\alpha^2\,(\gamma + \gamma_\eta)\,\{\gamma_*\gamma_\eta + (\gamma + \gamma_\nu)(\gamma + \gamma_\eta)\}(k^2 - k_z^2)/k^2\,. \end{aligned} \tag{4.67}$$

Maximisation of γ over k_y again gives (4.59b) but this optimal case applies over a smaller domain than in Section 4.8.3. The maximising k_y is now given by

$$k_y^2 = \left[\frac{\omega_\alpha^2}{2|\mathbf{B}|^2}\frac{(k^2 - k_z^2)}{k^2} + \frac{2\Omega}{|\mathbf{B}|^2}(\eta - \kappa)kk_z\right](1 - \mu^2 F)\,, \tag{4.68a}$$

where

$$F = \frac{(k^2 - k_z^2)}{k_z^2} + \frac{k^3}{k_z k_*^2}\left(\frac{\nu}{\kappa} - 1\right), \qquad \mu = \frac{\omega_\alpha}{2\Omega}. \tag{4.68b,c}$$

The switchover between the optimal and non-optimal cases now occurs near the curve $\gamma = 0$ in Figure 4.6. The inertial force is responsible for this; it reduces the growth rate at the vertex, $(k, k_z) = (\sqrt{3}k_0, k_0)$, of the wedge in Figure 4.6 by about 80%, but the location and magnitude of $\gamma_{\max}$ are little changed.

One of the enduring themes of geodynamo theory is that field generation is mainly the responsibility of finite amplitude MAC waves (see Section 3.3.4 and Braginsky, 1964b, 1967). The acronym "MAC" is conspicuous for what it leaves out: the inertial force, $\partial_t \mathbf{U} + \mathbf{U}\cdot\boldsymbol{\nabla}\mathbf{U}$, in (4.35a). The evolution of the waves is determined by the time derivatives in (4.35b) and (4.35c) alone; inertia is unimportant. In the present plane layer model, for which only the real root of (4.57) is relevant, the MAC "waves" do not propagate. In the curved geometry of the Earth however, they move longitudinally and have been associated with the observed westward drift of the geomagnetic field.

Small amplitude MAC waves are governed by (4.47a–d); they also regenerate field but, if the magnetic Reynolds number $\mathrm{Rm}_\ell^* = \gamma/\eta k^2$ is small, electromagnetic induction is insignificant and $\partial_t \mathbf{b}$ in (4.53a) is negligible compared with $\gamma_\eta \mathbf{b}$. The line on which $\mathrm{Rm}_\ell^* = 1$ is shown in Figure 4.6 and is labelled "MAC". As it falls in the LES triangle, it appears that field regeneration by the SGS is relatively unimportant, confirming the suggestion of Section 4.7.4 that the turbulent α–effect and the turbulent enhancement of the magnetic diffusivity can be safely ignored.

For the SGS of small Rm_ℓ^*, the neglect of $\partial_t \mathbf{b}$ transforms (4.47b) into the *low conductivity induction equation* which is a consequence of Ohm's law in the form

$$\eta \mathbf{j} = -\boldsymbol{\nabla}\phi + \mathbf{u}\times\mathbf{B}, \tag{4.69}$$

where ϕ is the (scaled) electric potential. This *low conductivity approximation* is commonly used in laboratory MHD. The last term in (4.69) creates a Lorentz force, $\mathbf{j}\times\mathbf{B}$, of $-\mathbf{u}_\perp/\tau_B$, where $\mathbf{u}_\perp$ is the component of $\mathbf{u}$ perpendicular to $\mathbf{B}$ and $\tau_B = 1/\gamma_B$ is variously called the *magnetic damping time* and the *Joule damping time*. In brief, motions perpendicular to $\mathbf{B}$, unless maintained, are damped out in a characteristic time of order τ_B that, like the Elsasser number, is scale-independent; see Davidson (2001).

The laboratory approximation (4.69) is good for the SGS of small Rm_ℓ^* but the rotation of the core adds a further anisotropic force, $2\boldsymbol{\Omega}\times\mathbf{u}$, that opposes motions perpendicular to $\boldsymbol{\Omega}$ and also "releases the magnetic constraint" for motions parallel to $\mathbf{B}$. The only component of $\mathbf{u}$ that is strongly suppressed is perpendicular to both $\boldsymbol{\Omega}$ and $\mathbf{B}$, i.e. parallel to $\mathbf{e}_x$ in the present model, and parallel to $\boldsymbol{\Omega}\times\mathbf{B}$ for orientations more general than (4.50a–d).

Braginsky & Meytlis (1990) developed the theory of the SGS of small Rm_ℓ^*. In place of (4.57) they obtained

$$(\gamma + \gamma_\kappa)(\Omega_*^2 + \gamma_*^2) = \omega_\alpha^2 \gamma_* (k^2 - k_z^2)/k^2 \,. \tag{4.70}$$

Maximisation over k_y again gives (4.59b), the maximising k_y now being given by

$$\Omega_* = \gamma_* \,, \qquad \text{i.e.} \qquad k_y^2 = k\, k_z/\Lambda \,. \tag{4.71a,b}$$

Because $k_y/k_O \leq \sqrt{3}$ and $k_z/k_O \leq 1$, (4.71b) shows that $k_y/k_O \leq 3^{1/4}/\Lambda^{1/2}$. Since $\Lambda > \sqrt{3}$, it follows that (4.59b) and (4.71b) hold for all admissible **k**, i.e. throughout the wedge of Figure 4.6. The most easily excited mode is given by (4.54a) and

$$\begin{aligned}
u_x &= \widetilde{u}(k_z \cos x' \sin y' \cos z' + k_y \sin x' \cos y' \cos z')/k, & (4.72a)\\
u_y &= -\widetilde{u} \cos x' \sin y' \cos z', \qquad u_z = \widetilde{u} \sin x' \sin y' \sin z', & (4.72b)\\
b_x &= \widetilde{b}(k_z \cos x' \cos y' \cos z' - k_y \sin x' \sin y' \cos z')/k, & (4.72c)\\
b_y &= -\widetilde{b} \cos x' \cos y' \cos z', \qquad b_z = \widetilde{b} \sin x' \cos y' \sin z', & (4.72d)
\end{aligned}$$

where $\widetilde{b} = (Bk_y/\gamma_\eta)\widetilde{u}$; also $\widetilde{u} = (g/2\gamma_*)\widetilde{c}$ which expresses the equality of the rate of Joule heating and the rate of working of the buoyancy force. The expressions (4.72a–d) have been simplified by using the thin plate approximation, $k_x \approx k$.

4.8.5. Finite amplitudes

Growth rates as large as $\gamma_{\max}$ in (4.64a,b) are misleading consequences of (4.48a–c). In reality, small $\mathbf{u}$, $\mathbf{b}$ and c grow in amplitude only until $\mathbf{m}^u$, $\mathbf{m}^b$ and m^c are large enough to establish equilibrium, or more precisely (since large values of $\widetilde{Ra}$ imply turbulent convection) until statistical equilibrium is established; the modes (4.64a,b) are then statistically steady but their amplitudes should be large because they are the most easily excited according to linear theory. Also, Figure 4.6 has a fresh interpretation: the nonlinearities restore the inertial and magnetic cascades, and these maintain motions even in the domain labelled “D-modes”, which should now be renamed “driven-modes”.

A method often used in MFE and mean field MHD to compute $\mathbf{R}$, $\boldsymbol{\mathcal{E}}$ and $\mathbf{I}$ is based on the idea that the turbulence is “nearly” homogeneous and isotropic, so that $\boldsymbol{\nabla}\mathbf{U}$, $\boldsymbol{\nabla}\mathbf{B}$ and $\boldsymbol{\nabla}C$ are small perturbations of a turbulent flow all of whose statistical properties are known; see, for example Krause & Rädler (1980) and Rüdiger (1989). This is a poor description of core turbulence. A more persuasive scenario is that of Braginsky & Meytlis (1990). They argued that, on the background of the basic unstable stratification, perturbations consisting of independent plate-like “parcels”, resembling the cells of the linear theory but not part of the tessalated pattern (4.54a), grow exponentially. When the amplitude of the parcels becomes large enough for strongly nonlinear effects to come into play, the parcel is destroyed and subsides into the mean state, i.e. it is "smoothed out". In its place a new parcel is born, grows and dies, in a never ending sequence. The system is filled by these independent parcels, all in different stages of development.

In this picture, a parcel is destroyed before it can overturn completely; its lifetime is of order $\tau = \ell_z/u_z = 1/k_z\widetilde{u}$. Thermal diffusion has little time to act in these rapid processes. In the statistically-steady state, the source m^c ($\sim k_z\widetilde{uc}$) in (4.47c) is balanced by $\mathbf{u}\cdot\boldsymbol{\nabla}C$ ($\sim \beta\widetilde{u}$) so that

$$\widetilde{c} \sim \frac{\beta}{k_z}, \qquad \widetilde{u} \sim \frac{g\widetilde{c}}{2\gamma_*} \sim \frac{\omega_\alpha^2 k}{4\Omega\, k_z^2}. \tag{4.73a,b}$$

It follows that $\tau \sim 1/\gamma$, where $\gamma = \omega_\alpha^2 k/4\Omega k_z$ is the small–κ form of the growth rate (4.59b) of the linear theory.

When $\mathrm{Rm}_\ell \ll 1$, the Braginsky-Meytlis scenario can be modelled by (4.72a–d), which show, after some reductions, that $\mathbf{m}^u$ contains the contribution $\mathbf{m}_1^u$, where

$$\mathbf{m}_1^u = \tfrac{1}{4}k_z\mathbf{e}_y(\widetilde{b}^2 - \widetilde{u}^2)\sin 2x'. \tag{4.74}$$

A little consideration of (4.47a) shows that neither the Coriolis nor the Lorentz force can balance $\mathbf{m}_1^u$. The viscous force can do so only through a part of u_y that is proportional to $1/\gamma_\nu$. This is large *unless*

$$\widetilde{b} = \widetilde{u}, \qquad \text{i.e.} \qquad \Omega_* = \gamma_* = \gamma_\eta. \tag{4.75a,b}$$

Because $\mathrm{Ro} = u\,k/\Omega$ increases with k, the energy in mode k is increasingly kinetic (see Section 4.7.2). It is now seen from (4.75a,b) that there is equipartition for the SGS of small Rm_ℓ (Braginsky & Meytlis, 1990).

The result (4.75a,b) also suggests that the inertial and magnetic cascades for the SGS of small Rm_ℓ cancel each other out. This is consistent with the argument given in Appendix C of Braginsky & Roberts (1990). The rate of ohmic dissipation by the SGS is given by $Q^{\eta,t} = \eta\,\overline{j^2} = \overline{\mathbf{B}}\cdot\overline{\mathbf{j}\times\mathbf{u}} + \overline{\mathbf{j}\cdot\mathbf{e}}$. After ignoring inertial and viscous forces, we deduce from (4.47a) that $\overline{\mathbf{B}}\cdot\overline{\mathbf{j}\times\mathbf{u}} = \overline{c\mathbf{u}}\cdot\mathbf{g} - \boldsymbol{\nabla}\cdot(\overline{p\mathbf{u}})$, so that

$$Q^{\eta,t} = \overline{c\mathbf{u}}\cdot\mathbf{g} - \boldsymbol{\nabla}\cdot(\overline{p\mathbf{u}}) + \overline{\mathbf{j}\cdot\mathbf{e}}. \tag{4.76}$$

Divergences such as $\boldsymbol{\nabla}\cdot(\overline{p\mathbf{u}})$ vanish when the overbar is interpreted as a local average (Section 4.7.3). If $\mathrm{Rm}_\ell \ll 1$, (4.69) holds and $\overline{\mathbf{j}\cdot\mathbf{e}} = -\boldsymbol{\nabla}\cdot(\overline{\phi\mathbf{j}})$; this also vanishes on averaging. Equation (4.76) then states that these modes draw the energy they need to pay their ohmic expenses from, and only from, buoyancy; cascade is not involved. If $\mathrm{Rm}_\ell = \mathcal{O}(1)$ however, $\overline{\mathbf{j}\cdot\mathbf{e}} = -\boldsymbol{\nabla}\cdot(\overline{\mathbf{e}\times\mathbf{b}}) + \frac{1}{2}\partial_t\overline{b^2}$, which does not vanish. The lost ohmic energy is partially derived from a cascade from other modes.

Because of the enormous disparity in the relative dimensions of the parcels, their diffusion of large-scale momentum, field and energy is extremely anisotropic. Clearly (4.44a–c) is far more likely to be successful than (4.43) but, when diagonalised, the tensor diffusivities $\boldsymbol{\nu}_\mathrm{T}$, $\boldsymbol{\eta}_\mathrm{T}$ and $\boldsymbol{\kappa}_\mathrm{T}$ have much smaller components in the direction of $\boldsymbol{\Omega} \times \mathbf{B}$ than in the $\boldsymbol{\Omega}\mathbf{B}$–plane. Estimates of the relative magnitudes of their component can be derived from the relative dimensions of the parcels, which follow from (4.75a,b):

$$\frac{k_y}{k} = \frac{\eta k}{B}, \qquad \frac{k_z}{k} = \frac{\eta k^2}{2\Omega}, \qquad \text{so that} \qquad \frac{\ell_y}{\ell_z} = \frac{kB}{2\Omega}. \tag{4.77a,b,c}$$

Nothing has singled out one particular band of SGS as being dominant, in the sense of the two-scale approximation of Section 4.7.4. Braginsky & Meytlis proposed a heuristic principle according to which the dominant parcels have an approximately square yz–cross–section. Then, by (4.77a–c),

$$k_D = \frac{2\Omega}{B}, \qquad \text{so that} \qquad k_{Dy} = k_{Dz} = \frac{4\,\Omega^2\,\eta}{B^3}, \tag{4.78a,b}$$

corresponding to $\ell_{Dx} \approx 0.7\,\mathrm{km}$ and $\ell_{Dy} = \ell_{Dz} \approx 25\,\mathrm{km}$. Braginsky & Meytlis estimated that, for the all important $\kappa_{\mathrm{T}\,zz}$ determining the turbulent codensity flux,

$$\kappa_{\mathrm{T}\,zz} \sim u_{Dz}\ell_{Dz} \sim \frac{\widetilde{u}}{k_{Dz}} \sim \frac{\omega_\alpha^2\,B^8}{128\,\Omega^6\,\eta^3}. \tag{4.79}$$

This sensitive dependence on B is a consequence of the B–dependence of k_D. For the values we have used earlier, $\kappa_{\mathrm{T}\,zz} \approx 0.8\,\mathrm{m}^2\,\mathrm{s}^{-1}$.

The Braginsky–Meytlis scenario is similar to one used successfully in the theory of stellar structure, *mixing length theory* (MLT); see Chapter 5 of Hansen & Kawaler (1994) or Chapter 14 of Cox (1968), and Section 5.6.1. Loper *et al.* (2003) envisage core processes that are even closer to MLT than in the Braginsky–Meytlis picture; see also Chulliat *et al.* (2005).

4.8.6. AN ALTERNATIVE APPLICATION: DNS

Before leaving the present model, we should point out that it can be re-invented and put to a different use. The "box" (4.49a–c) was presented earlier as an ultra-simplistic model of the entire core. Obvious shortcomings such as (4.50a–d) were ignored. An alternative is to consider a similar, but much smaller, "box" that represents a small volume v of the core. It is reasonable to suppose that, even though their orientations may differ from (4.50a–d), g, β, B, and U, are nearly uniform across the box. Although U would generally be nonzero, the reference frame can be chosen to move with the box (Braginsky & Meytlis, 1990).

Many models of this general type have been studied. Several of these have been "direct numerical simulations" (DNS) in which the full nonlinear equations have been integrated. See for example Matsushima *et al.* (1999), Matsushima (2001, 2004) and Buffett (2003) and Setion 5.6. Averaging over the box can provide estimates of the diffusivity tensors as functions of position $\mathbf{x}$; these may be regarded as local space averages at $\mathbf{x}$. One should not forget however, that the wavenumbers accessible to a small box do not include much of the area depicted in Figure 4.6, and in particular may not include significant wavenumbers such as (4.64a,b), corresponding to buoyant parcels some of whose dimensions exceed those of the box.

4.9. THE ENGINEERING APPROACH TO TURBULENCE

4.9.1. FILTERING

We now describe an approach that differs substantially from the traditional one described above, and is specifically oriented towards numerical work. Let us focus on $C(\mathbf{x}, t)$. This might be the highly resolved codensity obtained by DNS and used to test different methods of SGS modelling; see Meneveau & Katz (2000), Buffett (2003) and Chen & Glatzmaier (2005). Alternatively, it might be the unavailable, "infinitely well resolved", codensity that actually exists in the core. This will be our choice. As we have already mentioned, an LES that employs spectral methods excludes all the spectral components of the SGS part of this C. These are "filtered out" completely; see (4.33a) and (4.63a–c). We seek to modify the LES so as to minimise this loss.

Considering as before $\mathbf{k}$ to be a continuous variable, we focus on the Fourier transform $\breve{C}(\mathbf{k}, t)$ of $C(\mathbf{x}, t)$. The filtering just described consists of the Draconian elimination of all wavenumbers greater than k_F. Equivalently, $\breve{C}(\mathbf{k}, t)$ has been replaced by $\breve{G}_F(\mathbf{k})\breve{C}(\mathbf{k}, t)$, where $\breve{G}_F(\mathbf{k})$ is a *filter* that vanishes for $k > k_F$. A product of Fourier transforms corresponds in physical space to a convolution:

$$\overline{C}(\mathbf{x}, t) = \int G_F(\mathbf{x}')\, C(\mathbf{x} - \mathbf{x}', t)\, \mathrm{d}\mathbf{x}' . \qquad (4.80)$$

When grid point methods are used to simulate core MHD, the operation (4.80) corresponds to a different filter. But in either case the filter has a finite "width", Δ, which is the "range" over which the smoothing operation (4.80) acts in physical space; it is conventionally defined by

$$\int G_F(\mathbf{x})\,\mathrm{d}\mathbf{x} = 1\,, \qquad \int x^2 G_F(\mathbf{x})\,\mathrm{d}\mathbf{x} = (\tfrac{1}{2}\Delta)^2\,, \tag{4.81a,b}$$

the first of which is demanded of all filters.

Equation (4.80) gives the overbar operation a completely fresh interpretation, but it should be particularly noticed that, with this interpretation, *only the first of Reynolds rules (4.36a) is valid.* The application of (4.80) to (4.35a–d) generates *filtered equations* that have exactly the same form as (4.38a–d). These are usually written differently; the decomposition (4.37a–d) is not made explicitly. Instead, (4.40a–c) is written as

$$\mathrm{R} = (\overline{\mathbf{UU}} - \overline{\mathbf{U}}\,\overline{\mathbf{U}}) - (\overline{\mathbf{BB}} - \overline{\mathbf{B}}\,\overline{\mathbf{B}})\,, \tag{4.82a}$$

$$\boldsymbol{\mathcal{E}} = \overline{\mathbf{U}\times\mathbf{B}} - \overline{\mathbf{U}}\times\overline{\mathbf{B}}\,, \qquad \mathbf{I} = \overline{\mathbf{U}C} - \overline{\mathbf{U}}\,\overline{C}\,. \tag{4.82b}$$

Also, different words are used, e.g. $\mathbf{R}$ is called "the sub-grid scale stress", $\mathbf{I}$ is "the SGS codensity flux", etc.

The action of another filter, $G_\Delta(\mathbf{x})$, on $\overline{C}$ is analogous to (4.80). It creates a doubly filtered quantity, $\widetilde{\overline{C}}$, where

$$\widetilde{\overline{C}}(\mathbf{x},t) = \int G_\Delta(\mathbf{x}')\,\overline{C}(\mathbf{x}-\mathbf{x}',t)\,\mathrm{d}\mathbf{x}'\,. \tag{4.83}$$

The function $G_\Delta(\mathbf{x})$ may be so small for $|\mathbf{x}| > \Delta$ that the integral in (4.83) can be taken over all space although, when $\mathbf{x}$ lies within a distance Δ of a boundary, the integration limits require careful handling. There are a number of ways G_Δ can be selected. One popular choice is the Gaussian filter:

$$G_\Delta(\mathbf{x}) = (6/\pi\,\Delta^2)^{3/2}\exp(-6\,|\mathbf{x}|^2/\Delta^2)\,. \tag{4.84}$$

In what follows, the first filter, denoted by the overbar, is always the natural filter, $G_F(\mathbf{x})$, intrinsic to the LES. The doubly-filtered variables, $\widetilde{\overline{\mathbf{U}}}$, $\widetilde{\overline{\mathbf{B}}}$ and $\widetilde{\overline{C}}$ satisfy (4.38a–d) with $\mathbf{R}$, $\boldsymbol{\mathcal{E}}$ and $\mathbf{I}$ appropriately redefined, as suggested by (4.82a) and (4.82b).

4.9.2. SIMILARITY AND DYNAMICAL SIMILARITY

The question now arises of how to approximate the unknown $\mathbf{R}$, $\boldsymbol{\mathcal{E}}$ and $\mathbf{I}$. The similarity method is based on the idea that the solution for the unresolved scales smaller than Δ has a similar structure to the solution for the resolved scales slightly larger than Δ. In the case of the codensity, this leads to

$$\mathbf{I} = A_I(\widetilde{\overline{\mathbf{U}}\,\overline{C}} - \widetilde{\overline{\mathbf{U}}}\,\widetilde{\overline{C}}), \tag{4.85}$$

where A_I is a constant and $\widetilde{C}$ etc denote the result of applying a second, coarser filter to C etc, e.g. one of width 2Δ [so that 2Δ replaces Δ in (4.83)].

To make the approximation (4.85) seem plausible, observe that, if we substitute $\mathbf{U}$ and C for $\overline{\mathbf{U}}$ and $\overline{C}$ in (4.85), we would obtain (for $A_I = 1$) the actual SGS codensity flux, as evaluated on the coarser grid, i.e. $\widetilde{\mathbf{U}C} - \widetilde{\mathbf{U}}\,\widetilde{C}$. This is just as inaccessible as $\overline{\mathbf{U}C} - \overline{\mathbf{U}}\,\overline{C}$ (= $\mathbf{I}$) because it too involves contributions from the SGS smaller than Δ. But when these contributions are small compared with the contributions from scales between Δ and 2Δ, (4.85) should be a good approximation to $\mathbf{I}$ from the coarser grid. Constants such as A_I are often replaced by 1 in turbulence modelling, but their inclusion adds flexibility, the possibility of compensating for the fact that the estimate (4.85) ignores scales smaller than Δ. The application of (4.85) runs into difficulties near boundaries, which require special treatment and increased numerical resolution.

The approximation (4.85), and its companions

$$\mathbf{R} = A_R^u(\widetilde{\overline{\mathbf{U}}\,\overline{\mathbf{U}}} - \widetilde{\overline{\mathbf{U}}}\,\widetilde{\overline{\mathbf{U}}}) - A_R^b(\widetilde{\overline{\mathbf{B}}\,\overline{\mathbf{B}}} - \widetilde{\overline{\mathbf{B}}}\,\widetilde{\overline{\mathbf{B}}}), \quad \boldsymbol{\mathcal{E}} = A_E(\widetilde{\overline{\mathbf{U}} \times \overline{\mathbf{B}}} - \widetilde{\overline{\mathbf{U}}} \times \widetilde{\overline{\mathbf{B}}}), \tag{4.86a,b}$$

constitute the *similarity approximation*. For given values of A_I, A_R^u, A_R^b and A_E, it is easily applied since the required fields $\overline{\mathbf{U}}$, $\overline{\mathbf{B}}$ and $\overline{C}$ are the LES fields available at each time step. If spectral methods are employed, the computational overhead is about 80%, but the effective k_F is at least doubled (Glatzmaier, private communication).

Instead of assigning the constants A_R^u, A_R^b, A_E and A_I in an *ad hoc* way, they can be computed by a method called *dynamical similarity*; see Im *et al.* (1997). One way of implementing this is to apply a third filter even coarser than the other two, e.g. one of width 4Δ; its operation on C is denoted by $\widehat{C}$. In exactly the same way as expression (4.85) is equivalent to the statement

$$\overline{\mathbf{U}C} - \overline{\mathbf{U}}\,\overline{C} \approx A_I(\widetilde{\overline{\mathbf{U}}\,\overline{C}} - \widetilde{\overline{\mathbf{U}}}\,\widetilde{\overline{C}}), \tag{4.87}$$

we have

$$\widetilde{\overline{\mathbf{U}C}} - \widetilde{\overline{\mathbf{U}}}\,\widetilde{\overline{C}} \approx A_I(\widehat{\widetilde{\overline{\mathbf{U}}}\,\widetilde{\overline{C}}} - \widehat{\widetilde{\overline{\mathbf{U}}}}\,\widehat{\widetilde{\overline{C}}}). \tag{4.88}$$

The left-hand sides of these equations involve the SGS and are unknown, but we may use Germano's identity,

$$(\widetilde{\overline{\mathbf{U}C}} - \widetilde{\overline{\mathbf{U}}}\,\widetilde{\overline{C}}) - (\widetilde{\overline{\mathbf{U}C}} - \widetilde{\overline{\mathbf{U}}\,\overline{C}}) = \widetilde{\overline{\mathbf{U}}\,\overline{C}} - \widetilde{\overline{\mathbf{U}}}\,\widetilde{\overline{C}}, \tag{4.89}$$

to obtain an equation for A_I, every coefficient in which can be computed:

$$A_I[(\widehat{\tilde{\overline{\mathbf{U}}}\,\tilde{\overline{C}}} - \hat{\tilde{\overline{\mathbf{U}}}}\,\hat{\tilde{\overline{C}}}) - (\widetilde{\widetilde{\overline{\mathbf{U}}\,\overline{C}}} - \widetilde{\tilde{\overline{\mathbf{U}}}\,\tilde{\overline{C}}})] = \widetilde{\overline{\mathbf{U}}\,\overline{C}} - \tilde{\overline{\mathbf{U}}}\,\tilde{\overline{C}}\,. \tag{4.90}$$

Equation (4.90) is too much good news. We sought an equation that would determine A_I and have obtained instead a myriad of equations, one for each grid point and for each vector component. These will give different, though hopefully not very different, values for A_I. One recourse is to apply (4.90) in some average sense, e.g. by taking the scalar product of (4.90) with $\nabla\overline{C}$ and summing over the grid; see Chen & Glatzmaier (2005). One might also consider employing *all* of (4.90), by using them to eliminate A_I from (4.85) on a point-by-point component-by-component basis.

Similar techniques can be applied to $\mathbf{R}$ and $\boldsymbol{\mathcal{E}}$ too.

4.9.3. Related methods

We have focussed on two possible methods, similarity and dynamic similarity, but there are others of a similar nature, e.g. the gradient model, the dynamic Smagorinsky model, and various "mixed models". Meneveau & Katz (2000) review these and confront them with the results of DNS and laboratory experiments. Buffett (2003) does likewise, his test model being close to that of Section 4.8.6.

Another method that has received much attention is the Navier-Stokes-alpha model, where α is a length, best chosen to be about 2Δ (this length has nothing to do with either the α of Section 4.7.1 or the α of Section 4.7.4). The filtered and unfiltered variables are related by

$$C = (1 - \alpha^2\,\Delta)\,\overline{C}, \tag{4.91}$$

which, in an infinite domain, is equivalent to the action of the filter

$$G_\alpha(\mathbf{x}) = \frac{1}{4\pi\alpha^2|\mathbf{x}|}\exp\left(-\frac{|\mathbf{x}|}{\alpha}\right). \tag{4.92}$$

A full explanation of this method is beyond the scope of this review. Suffice it to say here that it generates equations not unlike (4.38a–d) but ones that are closed and that preserve integral properties of (4.35a–d); see Holm (2002). It is most easily applied when the Boussinesq approximation is used or when the fluid is incompressible. See also Jones & Roberts (2005).

4.10. WHERE ARE WE NOW, AND THE FUTURE

4.10.1. THE GEODYNAMO

Our knowledge of the geomagnetic field comes from a number of distinct sources:

- *Paleomagnetic measurements.* These give the long-time behaviour; showing that the field is maintained on times very much longer than τ_η and give information about reversals, excursions, and the long-term secular variation. For example, in analyses of paleomagnetic data, Love (2000a) finds an inverse correlation between angular secular variation and field strength and Love (2000b) confirms the statistical significance of the paths taken by the virtual geomagnetic pole (VGP) during reversals having preferred locations.

- *Surface observations.* These are available for about the past 400 years. Early measurements giving global coverage were largely made from ships. More recently a network of land-based observatories has provided good quality data.

- *Satellite observations.* The high quality data from MAGSAT (1979-1980) is now being complemented by ØRSTED allowing detailed models of the field and the secular variation over the past 20 years, see for example Hulot *et al.* (2002) and Jackson (2003).

In addition, there are other sources of information relevant to geodynamo simulations:

- *Seismological measurements.* In addition to giving vital information on the structure and composition of the core, recent work has used the anisotropy of the inner core to determine its rotation rate, see for example Tromp (2001).

There are two clear distinct aims in geodynamo modelling:

(i) to understand the key physical processes of convection-driven hydrodynamic dynamos in parameter regimes characteristic of the Earth, and

(ii) to try to explain specific features of the observed geomagnetic field.

An example of point (i) above is to demonstrate the maintenance of a magnetic field of strength comparable with that of the Earth over times long compared with τ_η. An example of point (ii) is to explain the observed variation in the reversal frequency. We can expect that simpler models such as the zero-order models to be adequate for point (i) while features specific to the Earth such as its heterogeneous CMB heat flow and its thermal history resulting in its inner-core growth to be necessary for point (ii). Ultimately, we may hope to learn new facts about the interior of the Earth by matching the results of sophisticated modelling to observations.

Numerical simulations give reasonable results for the morphology and strength of the field at the CMB, and the models are also capable of giving reversals and excursions which can be compared with palaeomagnetic observations. They also predict differential rotation between the inner core and the mantle. Given the parameter values we are able to use, particularly for E and q, the success of our models is better than we might expect. Jones (2000) comments "The parameter regime in which the current generation of numerical models can be run is very far from the regime of geophysical parameter values; so far, indeed, that the strong similarity between the model outputs and the geodynamo is quite surprising."

We can expect progress in a number of directions in the coming years. Increasing computing power and improved numerical methods such as the inclusion of subgrid-scale models, discussed in this chapter, should benefit all classes of models. Improved data and its analysis will identify generic and specific features of the Earth's field requiring explanation, motivating further developments away from the zero-order model.

4.10.2. A CRITICAL SUMMARY OF TURBULENCE

Section 4.8 has shone some light onto the physical nature of the SGS. Their ineffectiveness in field generation has been exposed; the structure of the turbulent eddies has been clarified. This has led to a substantial insight: the turbulent eddies are highly anisotropic and consequently tensor forms (4.44a–c) of the Boussinesq-Reynolds ansatz are much more realistic than the original scalar forms (4.43). Nevertheless, the difficulties in completing the three-step program have not been properly overcome. The program required that $\mathbf{u}$, $\mathbf{b}$ and c be derived as functionals of $\overline{\mathbf{U}}$, $\overline{\mathbf{B}}$ and $\overline{C}$ for general $\mathbf{U}$, $\mathbf{B}$ and C. This objective has not been reached, and it is hard to see how numerical computation can help. Consequently, a different gulf exists, one that separates (4.44a–c) from its physical basis. Order of magnitude estimates such as (4.79) are suggestive but not completely convincing.

Although the idea of using turbulent transport coefficients still lacks proper theoretical backing, it has some physical content and is therefore not devoid of merit. But there is a basic objection to the 3-step program: the turbulence has been studied using molecular transport coefficients. This means that a conceptually troubling discontinuity exists at the GS/SGS frontier. To remove this by adopting turbulent diffusivities when studying the turbulence would conflict with the requirement that only molecular diffusion acts at sufficiently large k. One might consider introducing a sliding scale on which the turbulent coefficients gradually give way to molecular coefficients as k increases, in a manner similar to that envisaged long ago by Heisenberg (1945); see also Chandrasekhar (1955). This would introduce further complications into an analysis that is already too difficult.

One virtue of the simple Boussinesq-Reynolds ansatz (4.43) is that it is the easiest of all techniques to apply. It has, however, not been endorsed by DNS and laboratory experiments; see Meneveau & Katz (2000) and Buffett (2003). Difficulties are also faced in selecting numerical values for the turbulent diffusivities. It is less than satisfactory to choose the values that secure the best agreement between the numerical results and the phenomena they are meant to describe! The same applies to the tensor diffusivities for use in (4.44a–c) and also, when hyperdiffusion is invoked, to the coefficients quantifying that.

The generalised Boussinesq-Reynolds ansatz (4.43) is *not* easy to apply. The direction $\boldsymbol{\Omega} \times \overline{\mathbf{B}}$ of weak diffusion constantly varies with the direction of $\overline{\mathbf{B}}$; probably an adaptive grid technique would be required to overcome the concomitant numerical difficulties. This obstacle does not arise when $\overline{\mathbf{B}}$ is zonally dominant; Donald & Roberts (2004) explored this case and ominously discovered that the anisotropy of κ_{T} strongly influenced their dynamo.

These difficulties do not bode well for the future of the traditional approach, but criticisms can be levelled at the methods described in Section 4.9 too. While it is true that the similarity method has a common sense basis, namely the continuity of physical behaviours across the man-made GS/SGS frontier, it is devoid of physical content. This is sometimes seen as one of its advantages: since it assumes nothing about the physics of the SGS, there is no risk of distorting that physics! But, on the general principle that inaction has as many consequences as action, this is unconvincing. The similarity method assumes that the solution near the frontier is typical of all SGS. This belief has no theoretical basis, and the analyses of Section 4.8 indicate that it is untrue. The Navier-Stokes-alpha model may be in a happier situation, but that too involves approximations that are hard to justify.

Engineering methods were devised for the computer age, and not surprisingly they are numerically convenient to apply. Their value in studying core convection and the geodynamo has yet to be proven. Meanwhile the traditional approach, though physically illuminating, seems to be at an unsatisfactory dead end. Perhaps the insights needed to guide theory and to resolve the geodynamo paradox will emerge first from laboratory experiments (see Chapter 8).

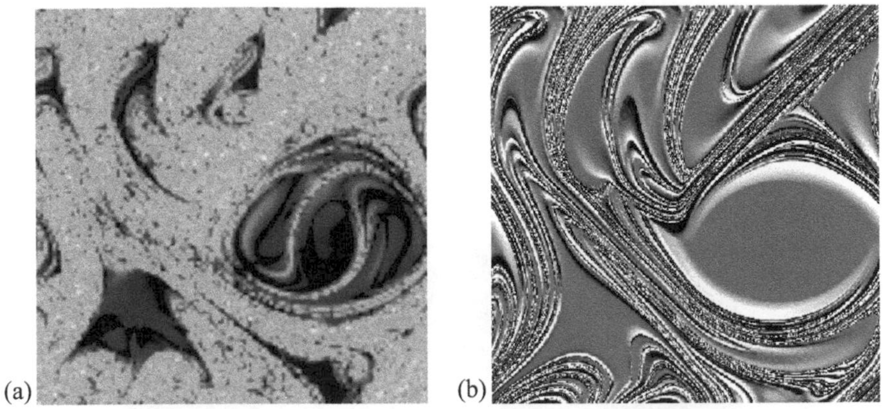

Figure 1.5 - Chaos in the GP-flow at time $t \approx 20$. (a) Finite-time Liapunov exponents (after Cattaneo *et al.* 1996) for $\omega = 1, \varepsilon = 1$, showing there is exponential stretching almost everywhere. (b) Normal field B_z (courtesy of F. Cattaneo). Note the large regions of multiply folded field.

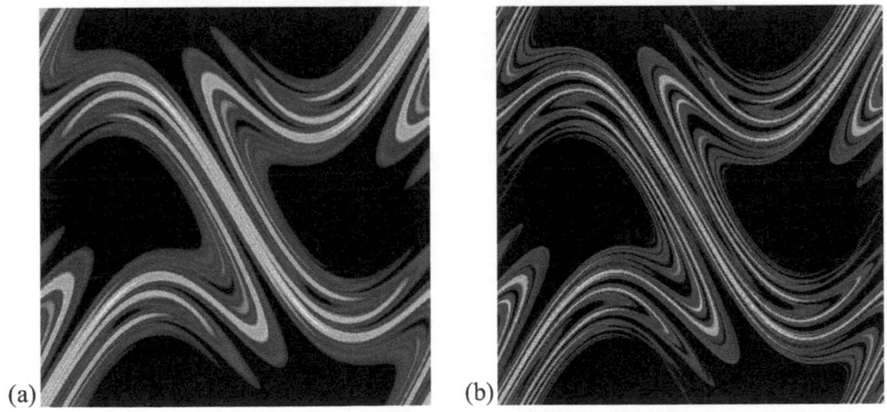

Figure 1.14 - Eigenfunctions of Otani's flow for $k = 0.8$ and (a) $\varepsilon = 5 \times 10^{-4}$ and (b) $\varepsilon = 5 \times 10^{-5}$. The magnitude of the magnetic field is shown, with black indicating zero field.

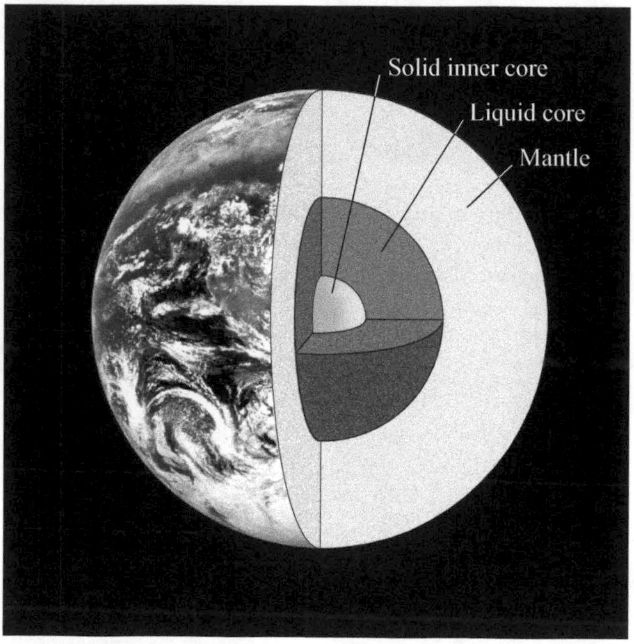

Figure 4.1 - The interior of the Earth. The outer region, the mantle, is composed mainly of silicates, it is a good electrical insulator. Some 2885 km below the surface lies the Earth core, composed mainly of liquid iron. It occupies a little more than half the radius of the Earth. At the center the pressure increase is such that the iron solidifies, this is the solid inner core, occupying 1215 km in radius.

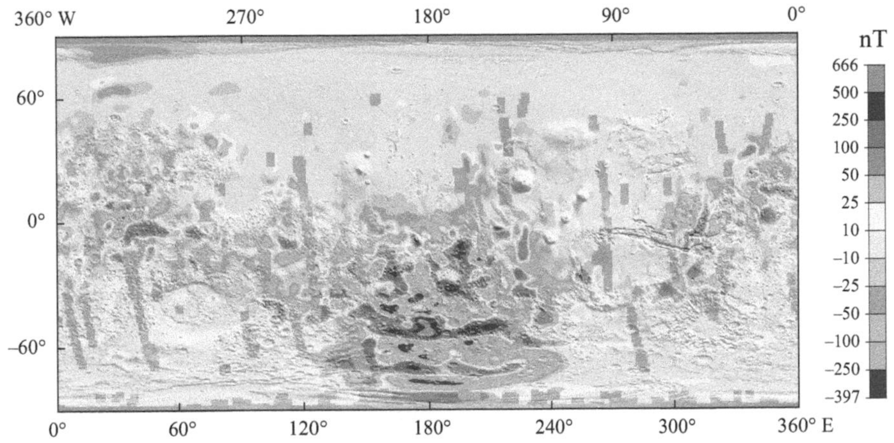

Figure 5.1 - Radial magnetic field at 200 km altitude in colour, overlain on a gray-shaded topographic gradient map of Mars (MOLA data). The dark grey bands show regions of inadequate data coverage (courtesy of M. Purucker).

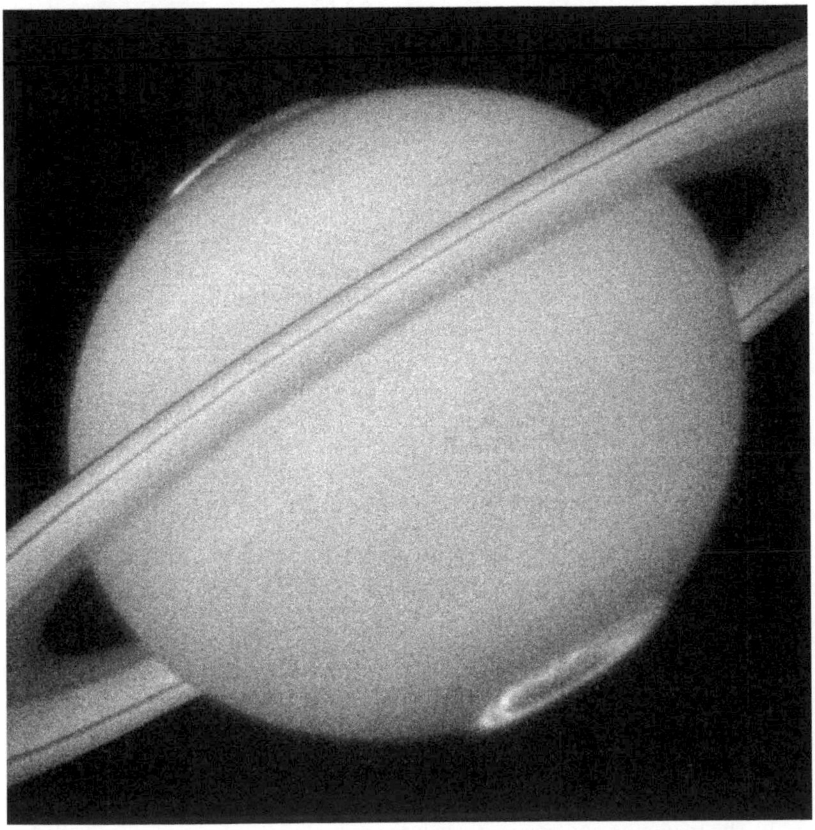

Figure 5.2 - The aurorae of Saturn observed by the Hubble space telescope (courtesy of NASA).

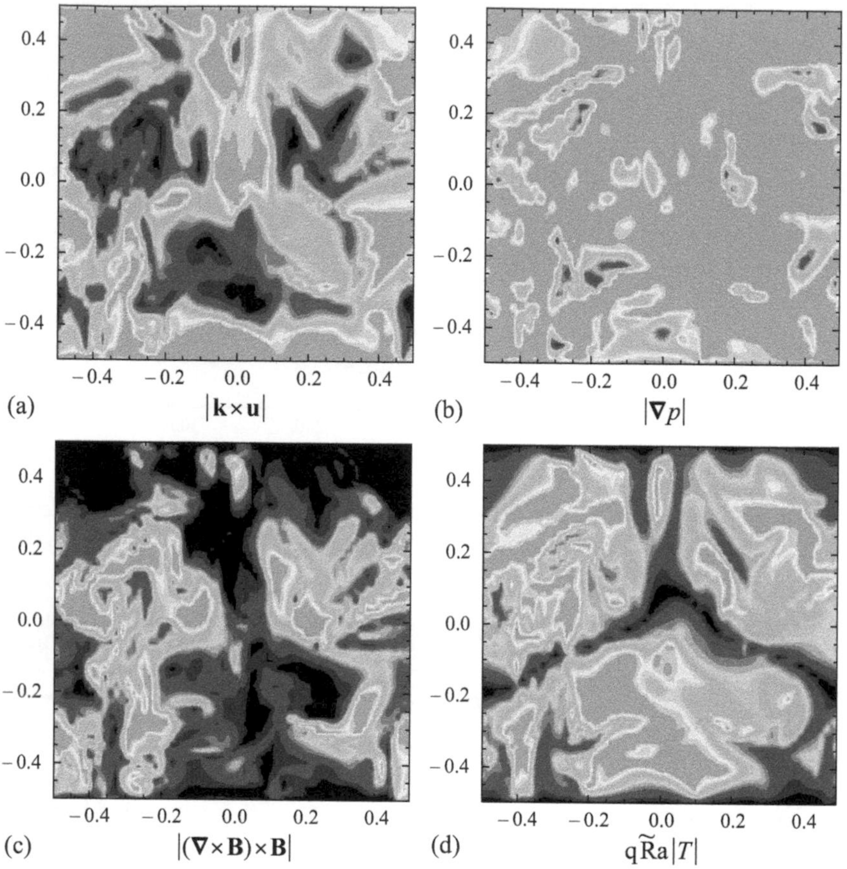

Figure 5.3 - Snapshot of the various forces in the $y = 0$ plane for a plane layer dynamo at $\mathrm{E} = 10^{-5}$. (a) Coriolis force, (b) pressure force, (c) Lorentz force, (d) Buoyancy force.

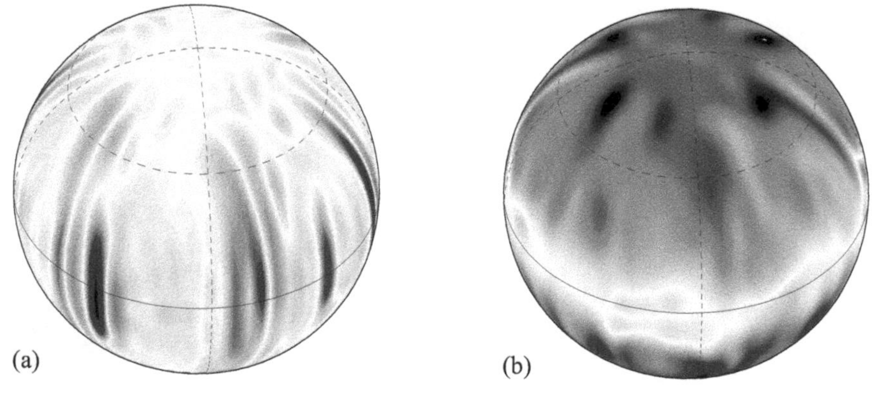

Figure 5.4 - (a) Shaded contour plots of the radial velocity, u_r at an arbitrary radius $r = 0.8\,r_o$. $\mathrm{Pr} = \mathrm{q_T} = 1$. The minimum and maximum values are $[-295.35,\ 135.55]$. (b) Shaded contour plots of the radial magnetic field, b_r at $r = r_o$. $\mathrm{Pr} = \mathrm{q_T} = 1$. The minimum and maximum values are $[-0.9379, 0.9007]$.

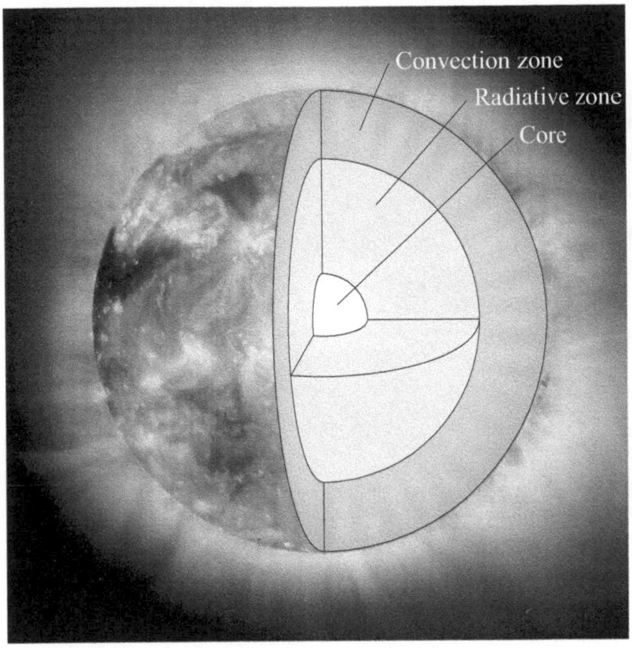

Figure 6.1 - Internal structure of the Sun. The cutaway image shows the visible surface (the photosphere, with a radius $R_{\odot} \approx 700\,\mathrm{Mm}$), together with an outer region where energy is carried mainly by convection, and an inner region where energy is transported by radiation. The narrow interface between the convection zone and the radiative zone, at a radius of approximately $0.7R_{\odot}$, has a thickness of only $0.02R_{\odot}$ and is the site of the tachocline, where there is a strong radial gradient in angular velocity. The temperature rises from $6000\,\mathrm{K}$ at the surface to about $2 \times 10^{6}\,\mathrm{K}$ at the base of the convection zone and then to $1.5 \times 10^{7}\,\mathrm{K}$ in the central core, where energy is generated by thermonuclear fusion.

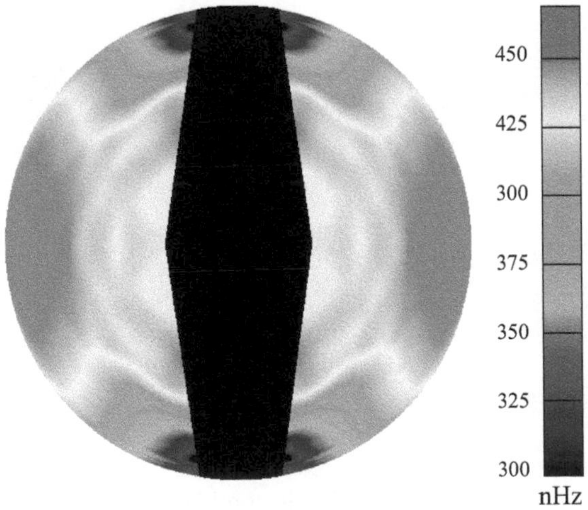

Figure 6.4 - Differential rotation in the solar interior. The rotation rate is approximately constant along radii in the convection zone, whose base is indicated by the dashed line. A frequency of 450 nHz corresponds to a period of about 26 days (courtesy of J. Christensen-Dalsgaard).

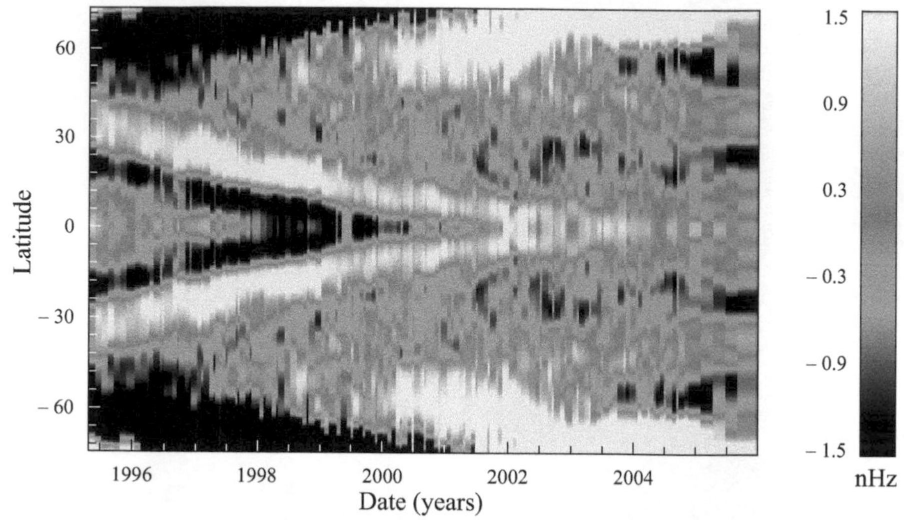

Figure 6.6 - Torsional oscillations with an 11–year period near the solar surface, derived from p–mode splitting measured with the GONG network (GONG RLS, $r = 0.99R_{\odot}$). Zones of slightly more rapid rotation progress towards the equator and towards the poles (courtesy of R. Howe).

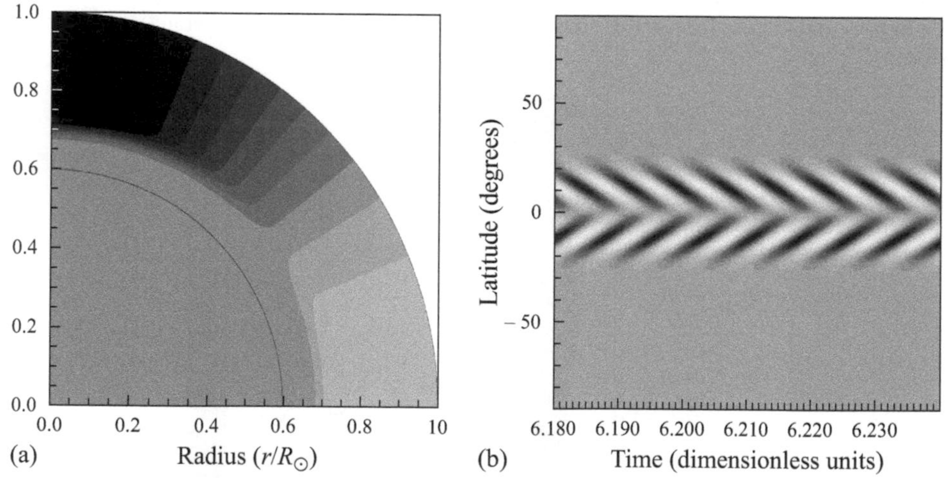

Figure 6.7 - (a) An analytical fit to the measured solar differential rotation, showing abrupt variation at the tachocline, near a radius of $0.7R_{\odot}$. (b) Cyclic toroidal fields at the base of the convection zone: a butterfly diagram for a nonlinear interface dynamo with the field limited by α–quenching, and an α–effect that is concentrated towards the equator (courtesy of P.J. Bushby).

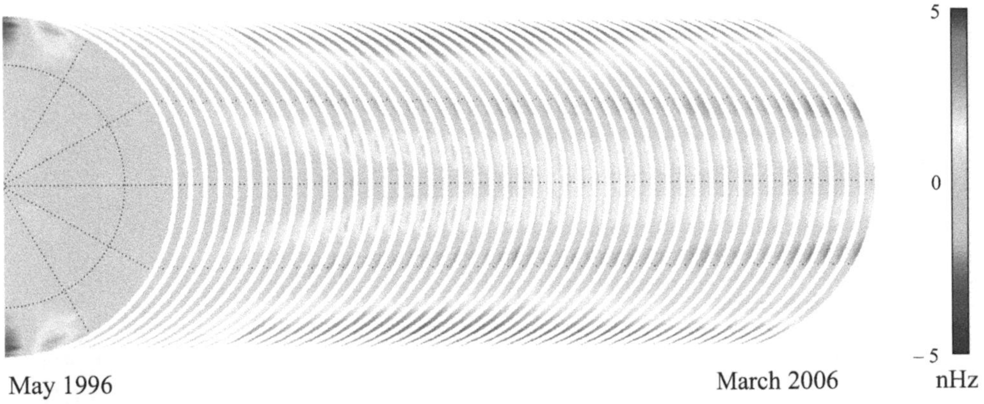

Figure 6.8 - Zonal shear flows extending throughout the convection zone. Results from the inversion of SOHO MDI data by Vorontsov *et al.* (2002), showing bands of slightly more rapid motion diverging from mid-latitudes as the activity cycle progresses (courtesy of S.V. Vorontsov).

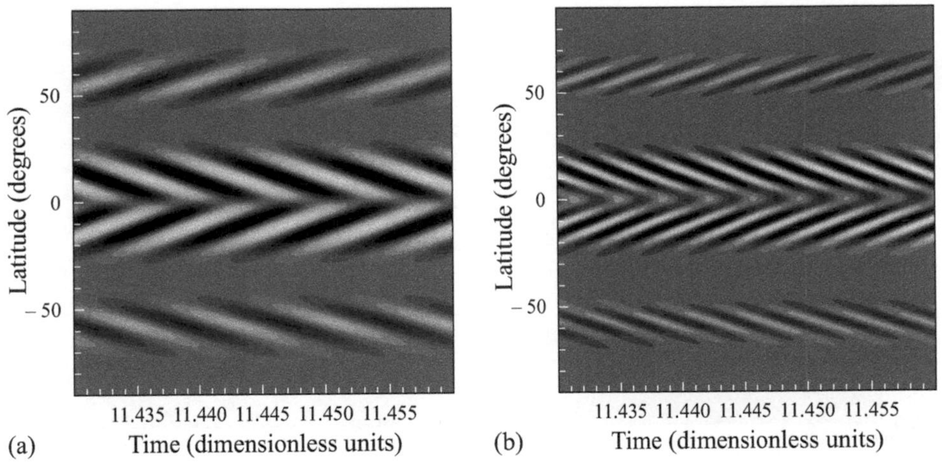

Figure 6.9 - A nonlinear interface dynamo with activity limited by the macrodynamic Malkus–Proctor effect. (a) Butterfly diagram, showing poleward and equatorward branches. (b) Zonal shear flows at the base of the convection zone, again showing both branches (courtesy of P.J. Bushby).

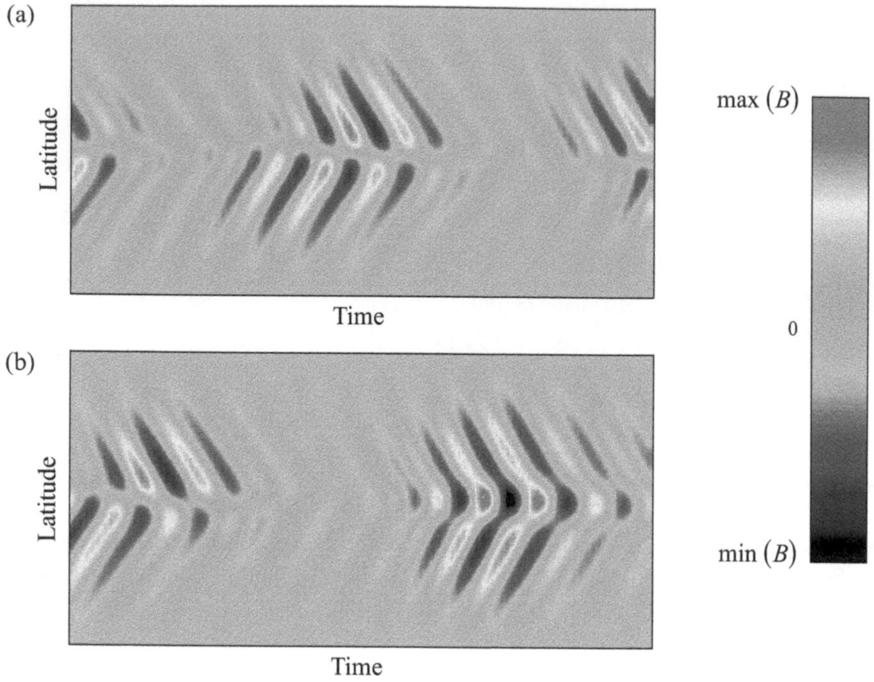

Figure 6.12 - Butterfly diagrams with grand minima, for the same parameters as Figure 6.11. (a) The interval covered in Figure 6.11, showing loss of dipole symmetry as the cycles emerge from grand minima. (b) A later sequence, in which the solution flips from dipole to quadrupole symmetry during a grand minimum (after Beer *et al.*, 1998).

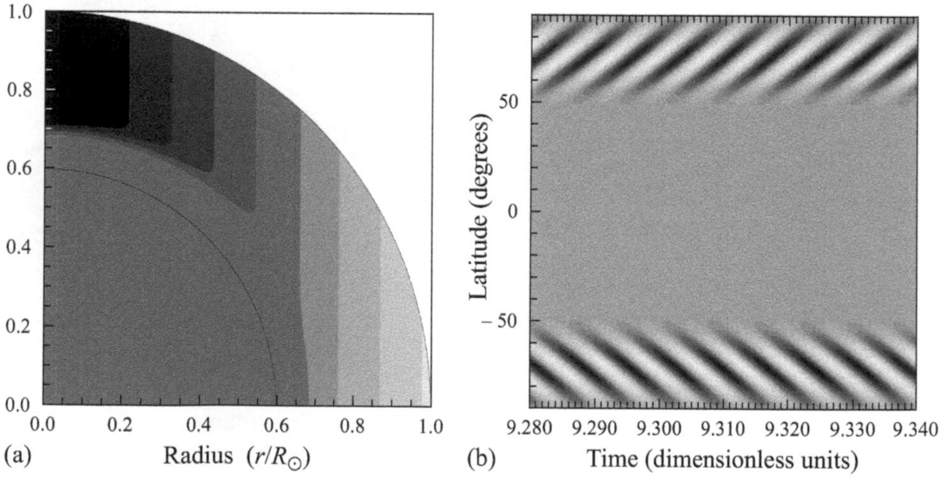

Figure 6.16 - (a) Conjectured variation of angular velocity in the convection zone of a rapidly rotating star, with behaviour dominated by the Proudman–Taylor theorem. Outside the tangent cylinder ω is constant on cylindrical surfaces but there is still a tachocline near the poles. (b) Corresponding butterfly diagram, with strong cyclic activity at high latitudes that could give rise to polar spots (courtesy of P.J. Bushby).

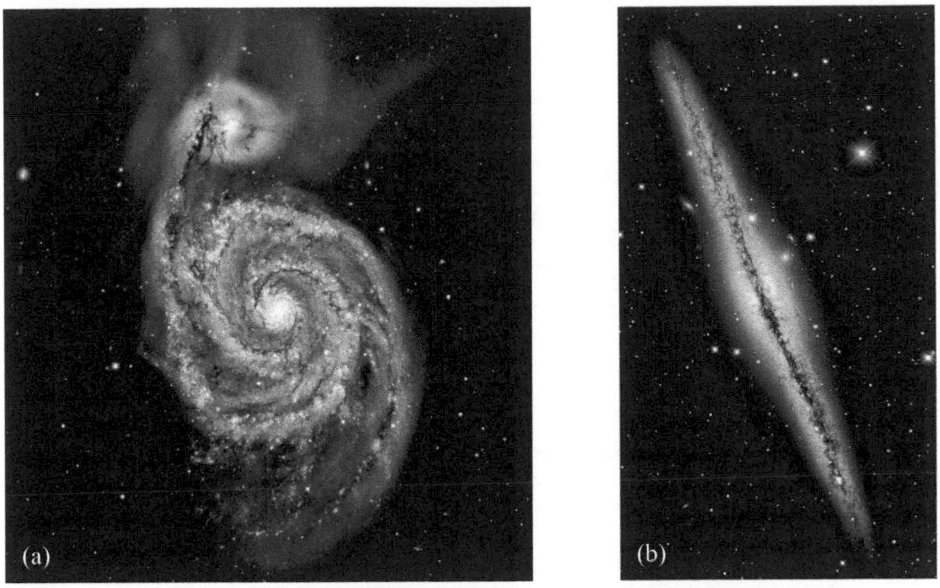

Figure 7.1 - Optical images of two nearby spiral galaxies. (a) M51, the Whirlpool galaxy (with a satellite galaxy at the top). (b) NGC 891 (Both courtesy of the Canada–France–Hawaii Telescope/J.-C. Cuillandre/Coelum). M51 is one of nearby galaxies (distance 9.6 Mpc) notable for its prominent spiral pattern. M51 is the first external galaxy where a well ordered, large-scale magnetic field was detected (Segalovitz *et al.*, 1976) and studied in fine detail. NGC 891 is at about the same distance as M51, but seen nearly edge-on, so the thinness of the galactic disc is evident. The dark strip along the galactic disc and filaments extended away from the galactic plane are due to obscuration by interstellar dust. The filaments trace gas outflow from the disc into the halo.

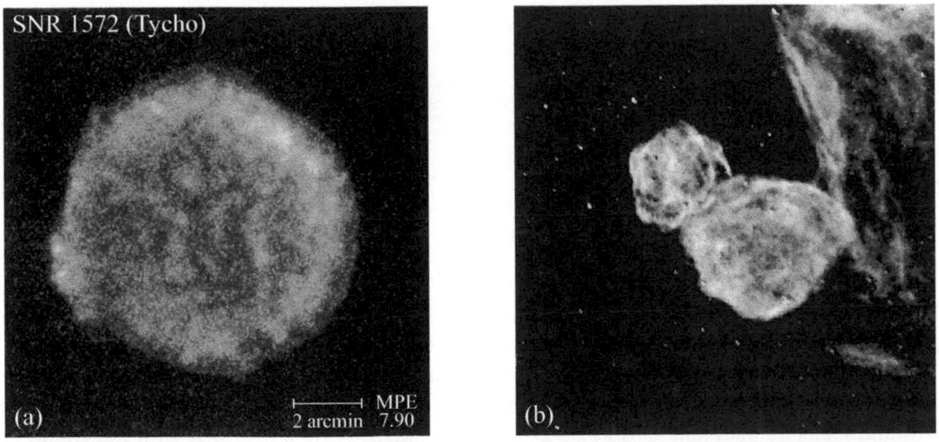

Figure 7.2 - SN remnants are expanding bubbles of hot gas that emits thermal X-rays. (a) This is illustrated by the X-ray image of Tycho's supernova remnant (courtesy of the ROSAT Mission and the Max-Planck-Institut für extraterrestrische Physik) whose parent star's explosion in 1572 was recorded by the famous Danish astronomer Tycho Brahe. The hot gas cools only slowly, and SN remnants often merge. (b) False-colour optical (Hα) image of two SN remnants DEM L316 in the Large Magellanic Cloud which appear to be colliding (Williams *et al.*, 1997; image produced by the Magellanic Cloud Emission-Line Survey, reprinted with permission).

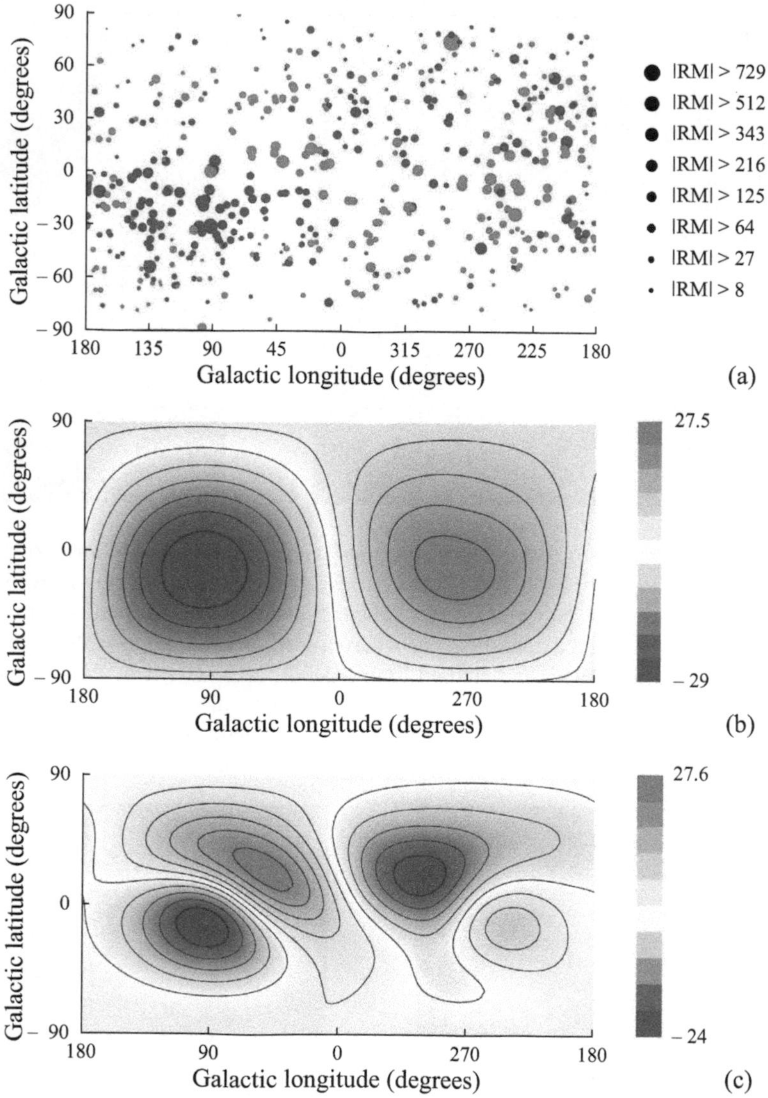

Figure 7.6 - (a) Faraday rotation measures of 551 extragalactic radio sources from the catalogue of Simard-Normandin & Kronberg (1980) shown in the (l, b)–plane, where (l, b) are the Galactic longitude and latitude in a reference frame centred at the Sun with the Galactic center in the direction $l = 0$ and Galactic midplane at $b = 0$. Positive (negative) RMs are shown with red (blue) circles whose radius indicates $|\mathrm{RM}|$ ($\mathrm{rad\,m^{-2}}$) as shown to the right of the panel. The lower two panels (b) & (c) show the wavelet transform of these data at scales 76° (b) and 35° (c) (Frick *et al.*, 2001). The transform at 76° has been obtained with the region of the Radio Loop I removed (this radio feature is a nearby supernova remnant). The wavelet transform at the scale 35° is dominated by local magneto-ionic features.

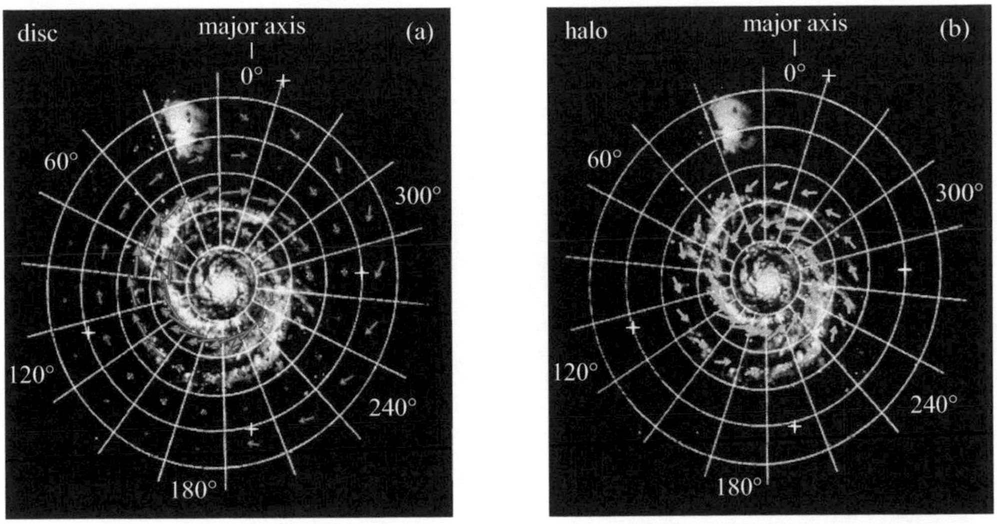

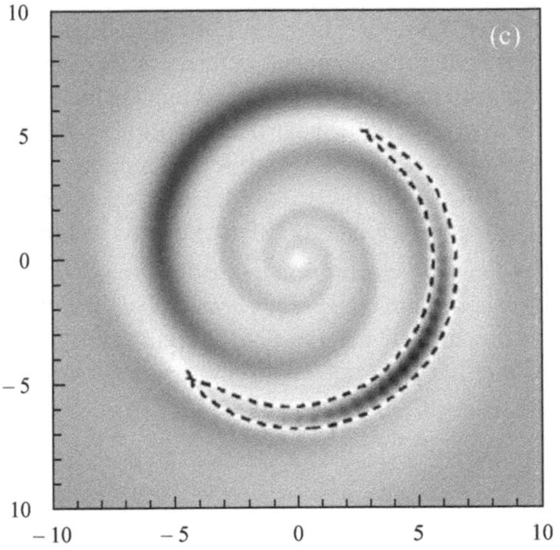

Figure 7.7 - The global magnetic structure of the galaxy M51, in the disc (a) and halo (b). Arrows show the direction and strength of the regular magnetic field on a polar grid shown superimposed on the optical image (Berkhuijsen *et al.*, 1997). The grid radii are 3, 6, 9, 12 and 15 kpc. (c) Magnetic field strength from the dynamo model for the disc of M51 (Bykov *et al.*, 1997) is shown with shades of grey (darker shade means stronger field). Magnetic field is reversed within the zero–level contour shown dashed; scale is given in kpc. The magnetic structure rotates rigidly together with the spiral pattern visible in the shades of grey.

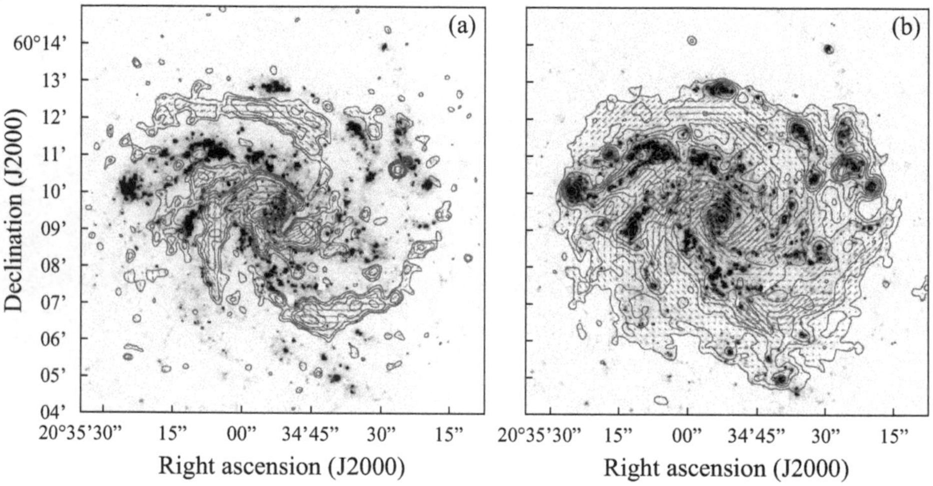

Figure 7.8 - (a) Magnetic arms in the galaxy NGC 6946: polarized intensity at the wavelength $\lambda = 6\,\mathrm{cm}$ (blue contours), a tracer of the large-scale magnetic field $\overline{B}$, superimposed on the galactic image in the Hα spectral line of ionized hydrogen (grey scale). Red dashes indicate the orientation of the B–vector of the polarized emission (parallel to the direction of intrinsic magnetic field if Faraday rotation is negligible), with length proportional to the fractional polarization – see (7.5). The spiral arms visualized by Hα are the sites where gas density is maximum. The large-scale magnetic field is evidently stronger between the arms where gas density is lower. (b) As in the left panel, but now for the total synchrotron intensity, a tracer of the total magnetic field $B^2 = \overline{B}^2 + \overline{b^2}$. The total field is enhanced in the gaseous arms. Given that the large-scale field concentrates between the arms, this means that the random field is significantly stronger in the arms, a distribution very different from that of the large-scale field (images courtesy of R. Beck, MPIfR, Bonn).

CHAPTER 5

PLANETARY DYNAMOS

Christopher Jones

This chapter is devoted to planetary dynamos. After reviewing observational facts (Section 5.1), we identify some outstanding problems in planetary dynamo theory (Section 5.2). In the light of the observational evidence, we address the conditions needed for dynamo action (Section 5.3), the possible energy sources for planetary dynamos (Section 5.4), the internal structure of the planets (Section 5.5), and the dynamics of planetary interiors (Section 5.6). We finally review, in Section 5.7, the existing numerical planetary dynamo models.

5.1. OBSERVATIONS OF PLANETARY MAGNETIC FIELDS

The solar system is conventionally divided into the Sun, the planets, their satellites and the asteroids, comets, gas and dust which are gravitationally bound to the Sun. For our purposes, planets means any planet or satellite with a radius greater than 1000 km. Bodies smaller than this are unlikely to have dynamos, because their interiors are not electrically conducting. The largest satellites are comparable in magnitude to the smaller planets, and the structure of their cores is not likely to be fundamentally different. The bodies in the solar system that have radii over 1000 km are the giant planets Jupiter and Saturn, the ice giants Uranus and Neptune and the terrestrial planets which have rocky mantles and iron cores. These terrestrial planets comprise the four inner planets, Mercury, Venus, Earth and Mars, our Moon, the four

Galilean satellites of Jupiter, Io, Europa, Ganymede and Callisto, the largest satellites of Saturn and Neptune, Titan and Triton, and the planet Pluto. A useful source of basic information about the planets and the space missions which explored them is the *Encyclopaedia of Planetary Sciences*, (1997). At present, we do not have the means to detect magnetic fields in the recently discovered extra-solar planets, but it is very likely that strong magnetic fields exist there. Recent reviews of planetary magnetic fields, including dynamo aspects, have been given by Stevenson, (2003) and Jones (2003). Older reviews, but still containing much useful information, are Stevenson (1982a, 1983), Levy (1995).

Most of our knowledge of planetary magnetic fields has come from the space probes launched over the past thirty years. The Earth's magnetic field was discovered in ancient times, but only Jupiter was known to have a magnetic field before space exploration began. Jupiter's field was first discovered from its radio emission (Burke and Franklin, 1955), although subsequent missions have given far more detail about the nature of the jovian field. Mercury's magnetic field was explored by Mariner 10 in 1974/5, while the two Voyager probes explored the fields of the outer planets. Voyager 1 and 2 both reached Jupiter in 1979, going on to Saturn, while Voyager 2 flew by Uranus and arrived at Neptune in 1989. The Galileo mission explored Jupiter and its satellites from 1995. Our understanding of planetary dynamos really dates from these misions. It was not possible to successfully predict which planets had fields or the form and strength of those fields before the planets were visited. At the time of writing, this exploration of planetary magnetic fields is still being actively undertaken. The Cassini mission will reach Saturn this year (2004) and will explore both Titan and Saturn. Several missions to revisit Mercury are in the advanced planning stage, and it is likely that the martian field will be better mapped soon.

We can at the outset divide planetary magnetic fields into those which are being maintained by current systems in the planetary interior and those where the field comes from only from remanent magnetism. It was established by C.F. Gauss early in the 19th century that the main field of the Earth is of internal origin, though the much smaller external components which originate from the interaction of the ionosphere with the solar wind can also be detected. There are also significant contributions from remanent crustal magnetic fields to the geomagnetic field; the discovery of permanently magnetised rocks goes back to antiquity. In the case of the Earth, we know that the field evolves slowly but significantly with time (the secular variation of the geomagnetic field) and also that the field can undergo complete reversals over a time-scale of around a few thousand years (Merrill *et al.*, 1996). Permanent magnetism is therefore not an adequate explanation of the Earth's main field. Of course, we have very limited information about the history of other planetary magnetic fields, so the same arguments cannot be used. However, the Earth's crustal magnetic fields are typically rather weak (less than 0.1% of the main field), and it seems unlikely that crustal magnetic fields on other planets would be enormously stronger. Furthermore, crustal fields cannot occur at high temperature, since rocks rapidly lose their magnetism if heated more than a few hundred degrees. Since the temperature of the deep interior of the larger planets is believed to be several thousand degrees, remanent magnetism would be confined to a relatively thin surface shell. It is then rather unlikely that the field would be coherent over the whole surface of the planet, required to explain the existence of a large-scale dipolar field. It is these difficulties which have suggested to most scientists that the magnetic fields on planets are mainly generated by dynamo processes going on in the interior . To describe the form of a planetary magnetic field we need a mathematical model of that field. This is done by expansion in spherical harmonics, a technique that also goes back to C.F. Gauss (see e.g. Merrill *et al.*, 1996). The natural system of coordinates for this problem is spherical polar coordinates (r, θ, ϕ) based on the rotation axis. Outside the conducting core of the planet, where no currents can exist, the potential Φ, where $\nabla\Phi = -\mathbf{B}$, satisfies $\Delta\Phi = 0$ [see (1.62a,b)]. It follows that Φ can be expanded as

$$\Phi = a \sum_{\ell=1}^{\infty} \sum_{m=0}^{\ell} \left(\frac{a}{r}\right)^{\ell+1} P_\ell^m(\cos\theta)(g_\ell^m \cos m\phi + h_\ell^m \sin m\phi)\,, \qquad (5.1)$$

where a is the planetary radius. The coefficients g_ℓ^m and h_ℓ^m are known as the Gauss coefficients. For the Earth, these coefficients have been measured with increasing accuracy since the initiative of Gauss led to the setting up of magnetic observatories in the 19th century. For many of the planets, these coefficients have been reconstructed from one or two fly-bys of a space probe. The largest component for the geomagnetic field is the g_0^1 component which corresponds to a dipole field. This field is antisymmetric about the equator, but it is perfectly possible for the g_0^2 component quadrupole fields, which are symmetric about the equator, to be generated by the dynamo process. Part of the dynamo problem is to explain which type of field is generated in the core.

Of the planets listed above, Pluto and Triton are as yet unexplored, as is Titan, though this will be visited soon. In Table III we list some of the observed properties of planetary magnetic fields. The sources for this data are Lodders & Fegley (1998), Russell (1993), and Connerney (1993). The dipole moment M is given in units of $\mathrm{A\,m^2}$, and is related to the Gauss coefficients by the formula $M = 4\pi a^3 g_1^0/\mu_0$. Note that in some older papers a different definition of the dipole moment, without the factor $4\pi/\mu_0$, is used.

We start by considering the four inner planets. Of these, only the Earth definitely has a dynamo. Far more is known about the form and history of the Earth's magnetic field, than about other planets and we discuss this in more detail below. Mercury is

Table III - Planetary magnetic fields.

	Dipole moment $A\,m^2$	Planetary radius 10^6 m	Core radius 10^6 m	Max Field at CMB 10^{-4} T	Dipole inc. degrees	Rotation rate Ω $10^{-5}\,s^{-1}$
Mercury	4.3×10^{19}	2.438	1.9	0.014	< 10	0.124
Earth	8×10^{22}	6.371	3.48	7.6	11.5	7.3
Ganymede	1.3×10^{20}	2.634	0.48	2.5	≈ 10	1.02
Jupiter	1.5×10^{27}	69.95	56	17	9.6	17.6
Saturn	4.2×10^{25}	58.30	32	2.5	0.8	16.4
Uranus	3.8×10^{24}	25.36	18	1.3	58.6	10.1
Neptune	2.0×10^{24}	24.62	20	0.52	47.0	10.8

a difficult case, because its magnetic field seems a bit too strong to be explained by crustal magnetism, but rather weak in comparison with the other dynamo generated fields. Another difficulty is that Mercury has not been revisited since 1974/5, and the Mariner 10 fly-by gave only very limited data. In particular, we do not know whether the field is really a dipole dominated field or is dominated by higher order spherical harmonics. We now come to two absentees in Table III, Venus and Mars. The question of why some planets do not have magnetic fields is almost as interesting as why others do have such fields, and the case of Venus is particularly intriguing, as the overall structure of Venus is apparently not that dissimilar to the Earth. Venus has, however, no measurable large-scale magnetic field. Mars has only a rather small magnetic field, which appears to be mainly remanent magnetism. Observations of the martian field from the Mars Global Surveyor are shown in Figure 5.1, by courtesy of M. Purucker (see Purucker *et al.*, 2000 for details). This gives the radial field just above the surface of Mars, and raises the question of how and when the martian rocks got magnetised. The most popular view is that Mars had a dynamo in the past, but it has now ceased to operate (Ruzmaikin, 1991; Stevenson, 2001). Our moon also has magetised rocks but no current dynamo, so it too may have had a dynamo in the past.

Of the explored outer planets, Jupiter, Saturn, Uranus and Neptune all have quite strong magnetic fields, and so probably have dynamos acting in their interior. Jupiter, the largest planet, has the strongest magnetic field. It is dipole dominated, but as with the Earth, the nonaxisymmetric components are sufficient to incline the dipole axis about 10° from the rotation axis. Saturn's field is somewhat smaller than Jupiter's and is remarkably axisymmetric, so its magnetic axis is almost perfectly aligned with its rotation axis. This was quite a surprise when the field was measured by the

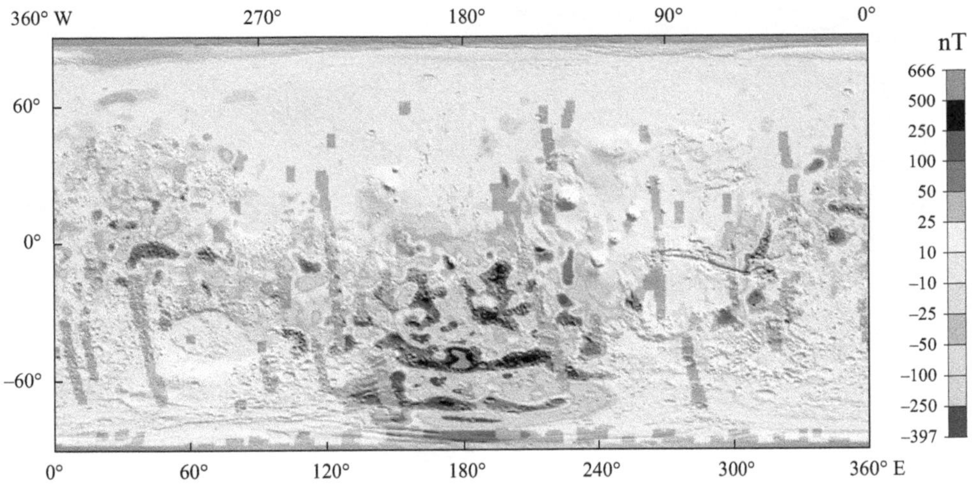

Figure 5.1 - Radial magnetic field at 200 km altitude in colour, overlain on a gray-shaded topographic gradient map of Mars (MOLA data). The dark grey bands show regions of inadequate data coverage (courtesy of M. Purucker). (**See colour insert**.)

Voyager 2 mission, because all other known planetary fields at that time suggested that a dipole inclined by around 10° was the standard behaviour. In Figure 5.2, a rather beautiful picture taken from the Hubble space telecope shows Saturn's aurorae. These are produced by particles from the solar wind being trapped by Saturn's magnetic field near its magnetic poles, just as our own northern and southern lights are produced on Earth. It is clear from alignment of the aurorae and the ring system that the magnetic poles of Saturn are closely coincident with its rotation axis.

The magnetic fields of Uranus and Neptune created yet more surprise at the time of the Voyager 2 fly-by. Up to that time, it was believed that all planetary magnetic fields were dipole dominated, but the fields of these planets are more quadrupolar than dipolar. The fields are also highly non-axisymmetric, with no obvious alignment with the rotation axis. An equatorial dipole, that is a dipole whose axis points along the equator rather than along the axis, is a better model of the field, but a detailed analysis of the observations (Holme & Bloxham, 1996) shows that the equatorial dipole model is too simple, and quadrupolar components are essential for describing the field.

The very successful Galileo mission made many observations of the magnetic fields of the four Galilean moons of Jupiter. A new problem emerges here, because all the moons are quite close to Jupiter, and are therefore bathed in the magnetic field of Jupiter itself. Note that this problem does not arise with the solar magnetic field; the solar wind does carry some solar field out to the planets, but it is much weaker

Figure 5.2 - The aurorae of Saturn observed by the Hubble space telescope (courtesy of NASA). (**See colour insert**.)

than the internal field. This problem is particularly severe with Io, the innermost moon of Jupiter. The magnetic field of Jupiter is measurably perturbed by Io, but is this because Io is generating its own magnetic field or is this signal just an inductive response from an electrically conducting core to the time-varying jovian magnetic field? (Kivelson *et al.*, 2001). The problem of dynamo generation in an ambient magnetic field is significantly different from the problem of dynamo generation in an isolated system.

The third moon of Jupiter, Ganymede, has a very distinct magnetic signal, and theoretical considerations (Sarson *et al.*, 1997, 1999) suggest that this cannot be simply a response to Jupiter's field. The signal is too strong and Jupiter's field not strong enough at that distance. The remaining moons, Europa and Callisto, appear not to have internal dynamos. The inductive response of Europa is of particular scientific interest, though, because it suggests that Europa has liquid water below its icy surface (Spohn & Schubert, 2003), raising the possibility that it might support life. Ganymede is quite small to have a dynamo, and the question arises as to why its core has not cooled down and frozen out (Showman *et al.*, 1997), particularly as Mars has apparently lost its dynamo. It is clear that the thermal history of a planet has a very significant bearing on whether or not it has a magnetic field, and so this is an issue we consider below.

5.2. SOME OUTSTANDING PROBLEMS IN PLANETARY DYNAMO THEORY

As we shall see below, there are formidable difficulties in solving the mathematical equations governing dynamo theory, and many of the basic physical quantities in planetary cores are only very poorly known. Nevertheless, a consensus view is beginning to emerge that planetary magnetic fields are generated by convection driven dynamo action in their cores. It seems likely that this mechanism can produce magnetic fields of the required strength and form without the necessity of adopting wildly unrealistic values of the physical parameters. However, there are a number of serious unresolved problems that face this theory, and general acceptance of the theory will only come when reasonable solutions of these issues can be provided. As we discuss the theory below, we refer to these problems and suggest possible explanations. Working outwards from the Sun the issues are

(i) Why is Mercury's field so weak?

It is evident from Table III alone that Mercury has a rather weak field compared to the other planets. Let us recall that dimensionless parameter that theory suggests should be used to measure field strength is the Elsasser number [see Section 4.2, (4.8a)],

$$\Lambda = \frac{|\mathbf{B}|^2}{\rho\Omega\eta\mu}. \tag{5.2}$$

These quantities can all be estimated, and if the surface field strength is used, the Elsasser number is just less than unity for all the planets except Mercury, where it is very much less than unity.

(ii) Why does Venus not have a magnetic field?

Some authors (e.g. Levy, 1995) have suggested this may be connected with the slow rotation rate of Venus ($0.30 \times 10^{-6}\mathrm{s}^{-1}$), but others disagree (e.g. Stevenson, 2003; Nimmo, 2002). The possibility that the lack of a venusian dynamo might be connected with the different type of mantle convection occurring there is considered in Section 5.5. Although the rotation rate is low, in dimensionless terms the relevant parameter is the Rossby number

$$\mathrm{Ro} = \frac{U_*}{\mathcal{L}\Omega}, \tag{5.3}$$

where U_* is the typical velocity of the convective flow, and $\mathcal{L}$ is the typical size of the convecting core. For the dynamics to be strongly affected by rotation, Ro must be small. Even though the rotation rate of Venus is small, the convective velocities in planetary cores are also very small, so that any plausible estimate gives Venus a small Rossby number. So it is not sufficient just to say that the rotation of Venus is too small for a dynamo to be expected.

(iii) How did the Earth's dynamo work before the inner core formed?

Studies of the thermal history of the Earth suggest that the solid inner core of the Earth only started forming 1 to 2 billion years ago (Labrosse *et al.*, 2001). Paleomagnetic evidence indicates that the magnetic field of the Earth existed at least 3.5×10^9 years ago. It is generally believed that compositional convection, based on the deposition of almost pure iron on the inner core and the stirring produced by the release of buoyant light material at the inner core boundary, is currently driving the geodynamo. Compositional convection of this type requires an inner core, so what was the energy source for the geodynamo when there was no inner core?

(iv) Why did the dynamo of Mars fail?

The strong coherent remanent magnetism found in martian rocks suggests that there must have been a strong field on Mars in the past (Stevenson, 2001). There is no dynamo operating now, so something must have happened to shut down the dynamo on Mars. Since Mars is smaller than the Earth, a natural suggestion is that the core has completely frozen and now consists of solid iron incapable of sustaining the fluid motion necessary for dynamo action. Unfortunately, this simple idea is almost certainly wrong, as measurements of the solar tides from the Mars global surveyor indicates that Mars still has a substantial fluid core (Yoder, 2003).

(v) How does Ganymede maintain a dynamo when its core is so small?

The magnetic Reynolds number, defined in Section 5.3 below, is believed to be a few hundred for the Earth's dynamo. Since Ganymede's core is about ten times smaller, if the electrical conductivity and flow speeds are similar to the those in the Earth's core, the magnetic Reynolds number for Ganymede's dynamo is only 20–30. No convection driven dynamo models have been found that sustain a magnetic field at such low magnetic Reynolds numbers.

(vi) Why is Saturn's field so axisymmetric?

This actually presents an interesting theoretical puzzle, because of Cowling's famous anti-dynamo theorem, which states that it is not possible to maintain a purely axisymmetric magnetic field by fluid motion, see Section 1.3. Possible ways out of this paradox are discussed below, and are based on the idea that the actual field of Saturn in the core is non-axisymmetric, but we only see a filtered version of the field, with these non-axisymmetric components removed (Stevenson, 1982b).

(vii) Why are the fields of Uranus and Neptune so unusual?

While all the other planets have fields which are basically dipolar, these two ice giant planets have large scale fields which are not even approximately aligned with the rotation axis, and contain substantial non-dipolar components. Clearly, some rather different process is going on in these planets, but why should this be so if the basic driving mechanism is similar to that of the other planets? There has been some recent work on this problem discussed in Section 5.7 below.

This is a formidable list of challenges, and it might be concluded that there must be other fundamental mechanisms at work generating planetary magnetic fields beside convection driven fluid dynamos. This may of course be the case, but we do not think we can draw this conclusion at the present time. It seems more likely that these problems reflect our ignorance of the physical and chemical conditions inside planetary cores, our lack of knowledge of the behaviour of matter at very high pressure, and our inability to solve the dynamo equations other than in a highly restricted region of parameter space. Progress towards removing these theoretical obstacles is being made. If it turns out that these seven challenges still cannot be addressed, then the basic picture will have to be changed, but we have not yet reached that position.

5.3. CONDITIONS NEEDED FOR DYNAMO ACTION IN PLANETS

The magnetic induction equation (4.5a) can be thought of as a competition between the induction term $\nabla \times (\mathbf{u} \times \mathbf{B})$ through which the fluid motion creates magnetic field, and the ohmic diffusion term $\eta \Delta \mathrm{B}$ which dissipates field through electrical resistance. If the typical velocity is U_* and the size of the electrically conducting region is r_o, the ratio of these terms is the magnetic Reynolds number

$$\mathrm{Rm} = \frac{U_* r_o}{\eta} . \tag{5.4}$$

For a dynamo to work, Rm must be at least around 10, and numerical dynamo simulations usually fail to sustain a dynamo unless Rm is at least close to 100.

More precise conditions on Rm for dynamo action were presented in Chapter 1.

So to discover whether a dynamo is operating, we need to know the magnetic diffusivity (or equivalently the electrical conductivity), the typical velocity and the size of the dynamo region. Electrical conductivity is discussed in detail in Section 5.5 below. For the Earth, there are two independent ways of estimating the typical velocity from observations, through the secular variation and through decadal length of day variations. Secular variation is the rate of change of the geomagnetic field. It is not possible to rigorously derive the flow velocity from secular variation data, but estimates of the typical velocity of $U_* \approx 2 \times 10^{-4}\,\mathrm{m\,s^{-1}}$ are suggested by Bloxham & Jackson (1991). If the decadal variations in the length of the day are ascribed to core motions, and total angular momentum is conserved, an independent estimate of core velocity also around this value is obtained (Jault *et al.*, 1988; Jackson, 1997). The size of the Earth's fluid core is accurately known from seismology ($3485\,\mathrm{km}$), and using $\eta \approx 2\,\mathrm{m^2\,s^{-1}}$, we obtain $\mathrm{Rm} \approx 350$, large enough for dynamo action.

When we come to the other planets, it is much more difficult to estimate Rm reliably. We can make plausible estimates of the size of the electrically conducting region for most of the planets, though Uranus and Neptune are difficult. More data is now available about electrical conductivity (see Section 5.5 below). It is the estimation of the typical velocity that is most difficult. In the case of Jupiter, some observational evidence is available (Russell *et al.*, 2001) because we can estimate the change in the magnetic field over the 25 year interval between the Voyager and Galileo missions. This suggests a typical velocity of around ten times the Earth's value $U_* \approx 2 \times 10^{-3}\,\mathrm{m\,s^{-1}}$, which leads to a value of Rm substantially larger than for the Earth. For all other planets we can only make estimates from theoretical considerations (Starchenko & Jones, 2002), discussed in Section 5.6 below.

The fluid velocity expected in the core depends on the energy source that is maintaining the field. The magnetic energy equation is formed by taking the scalar product of (4.5a) with $\mathbf{B}$ and integrating over the volume of the core,

$$\frac{1}{2}\int_{\mathcal{D}} \frac{\partial}{\partial t}|\mathbf{B}|^2\,\mathrm{d}\mathbf{x} = \int_{\partial\mathcal{D}} \mathbf{E} \times \frac{\mathbf{B}}{\mu_o}\,\mathrm{d}S - \int_{\mathcal{D}} [\mathbf{u}\cdot(\mathbf{j}\times\mathbf{B}) - Q_j]\,\mathrm{d}\mathbf{x}\,, \qquad Q_j = \eta\mu_0|\mathbf{j}|^2\,. \tag{5.5a,b}$$

The surface integral is the Poynting flux of magnetic energy out of the core, and Q_j is the ohmic dissipation. The remaining term is the work done by the Lorentz force, and this must be balanced by the work done by the driving forces in the equation of motion to maintain the dynamo.

In summary, for a dynamo to operate inside a planet we must have

(a) A large electrically conducting region in the interior.

(b) This region must be in a liquid or gaseous state, and not frozen out.

(c) Some energy source must be present to stir the fluid at a velocity corresponding to a magnetic Reynolds number of around 100 or more.

(d) The form of the flow must be such as to generate a large scale dynamo.

Issue (a) and (b) are discussed in Section 5.5 on the internal structure of the planets. Issue (c) is discussed in Sections 5.4 & 5.6, and issue (d) can be addressed by numerical dynamo simulations, Section 5.7 below. All of these areas of planetary science are in a state of active development, and many fundamental uncertainties remain.

5.4. ENERGY SOURCES FOR PLANETARY DYNAMOS

A number of different mechanisms have been suggested to drive the Earth's magnetic field other than a dynamo, and these could in principle power planetary magnetic fields. They are listed by Stevenson (1983) and Merrill *et al.* (1996), but none has gained any widespread acceptance, because they are incapable of maintaining the field strength found in planetary dynamos. The only exception is the thermoelectric effect, which Stevenson (1987) suggested as a possible power source for the rather weak field of Mercury.

If the dynamo origin is accepted, there remains the question of the driving mechanism stirring the fluid. This is still controversial. The four main contenders are thermal convection, compositional convection, precession and tidal forcing. The front runners are thermal and compositional convection, and we focus on these here, although precession driven dynamos are still being actively studied (Aldridge, 2003). Tides and precession draw their energy from the Earth's kinetic energy of rotation (Malkus, 1994). The thermal energy of a planet comes from the gravitational energy liberated at the time of formation, 4.5×10^9 years ago. Compositional convection in the Earth (Braginsky, 1963) arises from the gradual freezing of the inner core of the Earth (see Section 4.1.2). The density of the solid iron inner core is somewhat greater than that of the liquid outer core, because the outer core has a significant amount of light elements in addition to the molten iron. As the Earth cools, the inner core is growing because solid iron freezes onto it. As the iron freezes, the lighter components rise up through the core, stirring it by this compositional convection. Since heavier material is accumulating at the bottom of the outer core, gravitational energy is liberated by this process. Ultimately, this process must stop when all the iron in the outer core has frozen out, but this may take many billions of years.

Whatever the energy source, the work done must be sufficient to maintain the field against Ohmic diffusion [equation (5.5a,b)], so it is of interest to try to estimate how much energy is lost by this process. This has recently been discussed in detail in Roberts *et al.* (2003); see also Christensen & Tilgner (2004). It turns out that Ohmic loss depends largely on the small scale components of the field, and on whether the observed external field is all there is in a planetary core. If the internal field in the Earth's core is as smooth as possible, and there is no hidden toroidal component of the field, the ohmic dissipation could be as small as $43\,\mathrm{MW}$. This can be compared with $44\,\mathrm{TW}$ of heat coming out of the interior of the Earth, and around $0.5\,\mathrm{TW}$ liberated by compositional convection. However, the estimate of $43\,\mathrm{MW}$ is very much a lower bound. Dynamo simulations typically show only a rather small fraction of their magnetic field emerging through the core-mantle boundary. The typical magnetic field wound up in the core can be ten times the size of the escaping field given in Table III. Also, at magnetic Reynolds numbers large enough to sustain a dynamo, there are substantial small-scale components to the field which enhance the dissipation. In the Glatzmaier-Roberts (1997) simulations the dissipation was about $0.3\,\mathrm{TW}$, but this could be a significant underestimate due to the use of enhanced diffusion coefficients, which makes the computed field smoother than the real thing.

The current rotational energy of the Earth is 2×10^{29} J, so if it was originally twice this value, rotational energy is liberated at a rate of around $1.5\,\mathrm{TW}$. Most of this energy is going into driving oceanic and solid Earth tides, so that the energy budget for a precessionally driven dynamo is quite tight. The other difficulty is that the forcing is on the short rotational timescale, so the modes directly excited are the high frequency inertial modes of oscillation. To drive a dynamo, these have to be converted into flows varying on the very long (thousand year) dynamo timescale without a big loss of efficiency. Nevertheless theoretical progress is being made on precession driven flow (e.g. Aldridge, 2003) and with our present state of knowledge, precessionally and tidally driven dynamos cannot be excluded.

Since the amount of heat coming out of the planets is large compared to the amount of ohmic dissipation, it is natural to invoke this as the primary energy source for the dynamo. However, it is not necessarily true that the heat coming through the surface of the planet passes through the dynamo region. In the Earth, less than half of the $44\,\mathrm{TW}$ coming through the surface comes out of the core. Indeed estimates for the heat flux through the CMB vary from about $3\,\mathrm{TW}$ to $15\,\mathrm{TW}$. Furthermore, by no means all of the heat flux coming out of the core is available to drive the dynamo. There is a "Carnot" type efficiency factor involved (Gubbins, 1977, Roberts *et al.*, 2003) which means that the useful heat flux is no more than a fraction $\delta T/T$. Here δT is the temperature difference across the conducting core, probably somewhat over $1,000\,\mathrm{K}$ for the Earth, and T is the mean core temperature, probably around $4,500\,\mathrm{K}$ for the Earth. The actual efficiency factor may of course be significantly less than this, depending on where in the core the bulk of the field is generated. In consequence, it is believed that although compositional convection is estimated as generating only 0.5 TW, because of its high efficiency factor, it could be making a similar or greater contribution to driving the geodynamo at present.

5.5. INTERNAL STRUCTURE OF THE PLANETS

The planets can conveniently be grouped into three main categories, the gas giants Jupiter and Saturn, the ice giants Uranus and Neptune, and the rest, which are called the terrestrial planets. The structure of these different categories are quite distinct, so we discuss them separately, though when we come to dynamo modelling, we see that it may be possible to use the same basic methods to treat all the planets.

Terrestrial planets

As a first step, we assume that all the terrestrial planets have a composition similar to that of the Earth, that is an iron core, which may be divided into a liquid outer core and a solid inner core, surrounded by a rocky mantle. This picture seems to be consistent with the gravity fields of the terrestrial planets. The size of the metallic core in these planets is inferred from its radius, mass and moment of inertia. The moment of inertia is related to the gravity field, and for all planets visited by space probes, this is quite well-known. The moment of inertia gives the degree of central condensation, 0.4 times the mass times the radius squared being the value for a uniform density planet. Assuming the core has the density of iron (at the appropriate pressure) we can estimate the core size, to arrive at the data in Table III. It might be wondered why we are confident that the liquid iron core starts to freeze from the centre which is the hottest part of the core. The melting point rises with pressure, and the centre of the planet has the highest pressure, so although the temperature is higher there, the liquid first freezes at the centre. We therefore expect that during the hot formation phase the temperature will be hot enough for the core to be entirely liquid, but as the planet cools a solid inner core starts to form at the centre, gradually expanding as time goes on. For the Earth, the inner core is believed to have started forming between 1 and 2 billion years ago, and its present radius is 1220 km, the liquid outer core having a radius of 3485 km. The freezing point of iron is strongly depressed by the presence of impurities. Since the solid inner core is mainly pure iron, whereas the outer core contains impurities, this means the impurity content of the liquid outer core is continually rising as the outer core expands, making it very difficult for the outer core to freeze completely.

It is not sufficient for the planet to have a liquid electrically conducting core; the core must also be convecting heat out. For this to happen, the amount of heat coming out of the core must be greater than that which can be carried down the adiabatic gradient (see Sections 1.1.3 & 4.2). If this is not the case, no convection occurs, as the required heat flux is entirely carried by conduction. An alternative is possible if the core is stirred by compositional convection; then a negative convective heat flux can occur (Loper, 1978).

For liquid metal cores, we can give plausible estimates of the adiabatic temperature gradient in the core. Let us recall the adiabatic temperature gradient

$$\frac{1}{T_{\mathrm{ad}}}\left(\frac{\mathrm{d}T}{\mathrm{d}r}\right)_{\mathrm{ad}} = -\frac{\alpha\, g}{c_P}\,, \tag{5.6}$$

where g is the magnitude of the local gravity, α is the coefficient of thermal expansion and c_P is the specific heat at constant pressure (see page 8). The pressure is determined by

$$\frac{\mathrm{d}p}{\mathrm{d}r} = -g\rho\,, \tag{5.7}$$

together with an equation of state $\rho = \rho(p, T)$. When solving (5.6), some reference temperature is needed at one point in the core. This should come from the Core-Mantle boundary temperature. For a complete understanding of the temperature structure of the planet, we should solve a mantle convection model to find the temperature difference between the surface of the planet and the Core-Mantle boundary, which is then the boundary condition for (5.6). Integrating inward, we would then reach the pressure and temperature at which the solid inner core is in equilibrium with the liquid outer core, the liquidus temperature. This then tells us the size of the inner core, and we have the temperature and pressure structure of the fluid outer core. This procedure is described in Schubert *et al.* (2001). In practice, there are considerable uncertainties because of lack of knowledge of mantle convection, in particularly how to treat tectonic plates satisfactorily. In the Earth, geophysicists use seismic evidence about the location of the CMB and ICB, and use theory to predict the melting temperature of the ICB. (5.6) is then used to find the CMB temperature. Unfortunately, it is not particularly easy to estimate the liquidus temperature as a function of pressure, because it is strongly affected by the amount (and chemical composition) of the light element component of the outer core. However, this is an area where the recently developed *ab initio* quantum calculations (Vočadlo et al. 2003) can help, and it seems that the ICB temperature is around $5,400\,\mathrm{K}$ and the CMB is near $4,000\,\mathrm{K}$. For other planets, we have no direct evidence as to whether an inner core exists or not, and no information as to the depth of the ICB, although this may become available through data on planetary nutation (Dehant *et al.*, 2003). We therefore have to do the best we can using simple mantle convection models (Schubert *et al.*, 2001) to predict the CMB temperature, and then use this as the boundary condition for (5.6). In principle, the solution of (5.6) tells us whether an inner core exists, and hence whether compositional convection is occurring.

The heat flux carried down this adiabatic gradient by conduction and radiation is

$$F_{\rm ad} = -\kappa\,\rho\,c_P \left(\frac{{\rm d}T}{{\rm d}r}\right)_{\rm ad}. \tag{5.8}$$

If the actual heat flux is F, then the conduction gradient is defined by

$$F = -\kappa\,\rho\,c_P \left(\frac{{\rm d}T}{{\rm d}r}\right)_{\rm cond}. \tag{5.9}$$

In the Boussinesq approximation, convection occurs if

$$\mathrm{Ra} = \frac{g\,\alpha\,d^4}{\kappa\,\nu}\left[\left(\frac{{\rm d}T}{{\rm d}r}\right)_{\rm cond} - \left(\frac{dT}{dr}\right)_{\rm ad}\right] > \mathrm{Ra}_{\rm crit}, \tag{5.10}$$

where $\mathrm{Ra}_{\rm crit}$ is some number depending on the geometry, the boundary conditions and whether rotation and magnetic fields are important. In practice $\kappa\,\nu\,\mathrm{Ra}_{\rm crit}/g\,\alpha\,d^4$ is a very small temperature gradient, so $F > F_{\rm ad}$, the Schwarzschild criterion, governs whether convection occurs. In the Earth, and indeed all terrestrial planets, F and $F_{\rm ad}$ are of similar magnitude, so it is non-trivial to establish whether convection is occurring.

The thermal conductivity, k is required for this calculation. It can be estimated in terms of the magnetic diffusivity using the Wiedemann-Franz law for electronic conduction,

$$k = \kappa\,\rho\,c_P = 0.02\,T/\eta\,, \tag{5.11}$$

where T is the temperature and η is the magnetic diffusivity, which is $\approx 2\,\mathrm{m}^2\,\mathrm{s}^{-1}$ in the Earth's core, and is probably not that different in the other terrestrial planets. Stevenson (2003) notes that if η were reduced, which would help to increase the magnetic Reynolds number and hence enhance dynamo action, k is increased so the conducted heat flux down the adiabat is increased. This makes it more difficult to for the fluid to convect, as then a larger F is needed to ensure the Schwarschild criterion for convection is satisfied. This means that terrestrial planet dynamo action is always going to be a marginal affair. This may, however, help to explain some of the mysteries mentioned in Section 5.2 above. If the Earth has F just above $F_{\rm ad}$, convection occurs and a dynamo is possible. Possibly the rather different nature of mantle convection on Venus means that on that planet F is just below $F_{\rm ad}$ and so there is no convection and so no dynamo (Stevenson, 2003, Schubert *et al.*, 2001). Similarly, Mars might have had sufficient heat flux F in the past for it to be greater than $F_{\rm ad}$ but in the course of time it fell below, stopped convecting and so lost its dynamo.

These are attractive ideas, but there are still many uncertainties. Even the heat balance in the Earth's core is still controversial. Recent estimates (Roberts *et al.*, 2003) suggest that the core is losing just over $2\,\mathrm{TW}$ due to secular cooling, and just over

4 TW due to latent heat release at the inner core boundary. The heat conducted down the adiabatic is around 6 TW, so with these numbers the core is just convective everywhere. However, it is also possible there is radioactive heating from potassium in the core (discussed in Roberts *et al.*, 2003), in which case there could be more vigorous convection throughout the core. This might help to explain how the Earth maintained a dynamo before the inner core formed; if there was then vigorous thermal convection driven by radioactivity this might be sufficient to power the dynamo without the need for compositional convection. Another source of uncertainty is the amount of light element (possibly sulphur) in planetary cores. With a high proportion of sulphur, cores stay liquid longer, and this may explain why Mars has a substantial liquid core at present (Yoder, 2003).

5.5.1. GIANT PLANETS

There has been significant progress in our understanding of the interiors of giant planets, fuelled partly by the search for extra-solar giant planets (Hubbard *et al.*, 2002). The first step is to construct models based on solutions of (5.6) and (5.7), but the equation of state is now considerably more complex because the pressure in giant planets can reach values at which quantum effects become important, turning hydrogen into a metal with free electrons, pressure ionization. This makes the equation of state a complicated issue (Hubbard *et al.*, 2002) which is studied by theoretical calculations and high pressure shock experiments. The important issues for dynamo theory are whether the interior is convective throughout or has stably stratified zones, whether there is a phase change boundary and how the electrical conductivity varies with depth.

It is generally believed that the interiors of the giant planets are fully convective, the heat flux being blocked by the high opacity of the very outermost layers in the atmosphere (Hubbard, 2002). However, uncertainties remain and a possible stably stratified region may exist at a pressure of around $100\,\mathrm{GPa}$ ($1\,\mathrm{Mbar}$) where metallization occurs. The current view (Guillot, 1999) is that there is no sharp jump between the molecular hydrogen-helium atmosphere and the metallic core, so the electrical conductivity rises smoothly as we go into the interior, reaching a plateau of $2 \times 10^5\,\mathrm{S\,m^{-1}}$ (the Earth's value is usually taken as $5 \times 10^5\,\mathrm{S\,m^{-1}}$) at a pressure of $140\,\mathrm{GPa}$ (Nellis, 2000) which corresponds to a radius of about $0.9\,R_J$, where the temperature is about $5,000\,\mathrm{K}$. In Saturn, the critical pressure of $140\,\mathrm{GPa}$ is only reached at about $0.5\,R_S$, so the conducting core of Saturn is a much smaller fraction of the planet than is the case for Jupiter. It is important to note that the magnetic Reynolds numbers of the giant planets will be much larger than for the Earth. If the typical velocities are roughly ten times those in the Earth's core, and the size of Jupiter's core is roughly ten times that of the Earth, Rm will be at least 50 times the Earth's value. It is therefore possible that dynamo action could be occurring at lower pressures than the 140 GPa where full metallization occurs. In this zone, for pressures between 50 GPa and 140 GPa, Nellis (2000) suggests $\log_{10} \sigma \approx 3 \times 10^{-11} p + 1.1$ where σ is in Sm^{-1} and p is the pressure in Pa.

5.5.2. ICE GIANTS

The theory of the internal structure of Uranus and Neptune is much less developed than that of the giant planets (Hubbard *et al.*, 1996, Miner, 1998). Both planets are believed to have three zones, a rocky core with a radius of about one third the planetary radius, an "ice" layer consisting of a mixture of water and ammonia extending to about 0.75 of the planetary radius, and a deep outer hydrogen/helium atmosphere extending to the surface. The electrically conducting zone is believed to be the ionic ice zone. This zone is believed to be fluid on the basis of *ab initio* quantum calculations (Cavazzoni *et al.*, 1999), and these calculations also support the shock experiment results of Nellis *et al.* (1997) for the electrical conductivity. This is due entirely to the movement of ions in the liquid, and rises to a value of around $3 \times 10^3\,\mathrm{S\,m^{-1}}$ at a pressure of 40 GPa. At higher pressures, Cavazzoni *et al.* (1999) suggest that the ionic conductivity rises slowly to reach $10^4\,\mathrm{S\,m^{-1}}$ at 300 GPa. Above this, they suggest that metallization may occur, that is electron conductivity, which could lead to conductivities of the order of $2 \times 10^5\,\mathrm{S\,m^{-1}}$ in the deepest part of the ice layer. The pressure at the bottom of the ice layer (at 0.3 of the planetary radius) is around 600 GPa, the temperature there being about 7,000 K. Note that ionic electrical conductivity is typically a hundred times less than electronic conductivity, so that unless velocities are much higher than in the Earth's core, the magnetic Reynolds number will be marginal despite these planets being ten times as big as the Earth.

5.6. DYNAMICS OF PLANETARY INTERIORS

The structure of planetary interiors is mostly calculated on the assumption that the temperature gradient is close to its adiabatic value. Our treatment is based on the geodynamo, but the same principles can be applied to all planetary cores. Dynamical models start by assuming that the departure from the adiabatic state is small, so we write

$$T = T_{\mathrm{ad}} + \Theta, \quad p = p_{\mathrm{ad}} + p, \quad \text{etc.} \tag{5.12}$$

As discussed in the previous Sections, both thermal and viscous diffusion are small in the Earth's core and to obtain sensible models it is necessary to introduce turbulent values, κ_{T} and ν_{T}. However, we need to retain both the molecular κ as well as the turbulent κ_{T} in our description. In the Earth, the adiabatic temperature varies by over $1000\,\mathrm{K}$ over the core, but the convective temperature fluctuations driving the motion are typically less than $10^{-3}\,\mathrm{K}$. The temperature gradients arising from the adiabatic static profile are therefore at least six orders of magnitude greater than the convective temperature gradients. So $\kappa\,\boldsymbol{\nabla} T_{\mathrm{ad}}$ can be of the same order as $\kappa_{\mathrm{T}}\,\boldsymbol{\nabla}\Theta$ even though κ_{T} may be six orders of magnitude larger than κ. The turbulence does not directly affect the transport of heat down the adiabat, so no terms involving the product of κ_{T} and $\boldsymbol{\nabla} T_{\mathrm{ad}}$ appear.

The Boussinesq equations (4.5a-d) are often used to discuss the dynamics of planetary interiors, though the anelastic equations are more suited to dynamo models of the giant planets, since they can take into account the substantial density variations occurring in their cores. In the anelastic approximation, the temperature equation is replaced by an entropy equation, but in planets it is often the case that $\alpha T_* \ll 1$, α being the coefficient of expansion and T_* the typical temperature. Then the temperature fluctuation Θ dominates the pressure fluctuation in the entropy, (the anelastic liquid approximation) and equations rather similar to the Boussinesq equations emerge, see Anufriev *et al.* (2005) for a recent discussion of the anelastic equations and their relation to the Boussinesq equations. Note that the typical value of Θ is much smaller than T in planetary cores; typical values of Θ are in the range $10^{-3} - 10^{-4}\,\mathrm{K}$ only.

The magnetic Ekman number E_η can be written as $[U_* r_o/\eta]^{-1}[U_*/r_o\Omega]$, i.e. $\mathrm{Rm}^{-1}\mathrm{Ro}$ (Ro being the Rossby number). This is very small in the core, as is the Ekman number, so we expect the inertial and viscous terms to be unimportant in the planetary cores, except on very small length and time–scales. The heat source term $\mathcal{S}$ in equation (4.5c) requires some explanation (Braginsky & Roberts, 1995, Anufriev *et al.*, 2005). The heat sources in the Earth's core are the latent heat release at the inner core boundary and (possibly) radioactive heating. The latent heat is more conveniently taken into account through the boundary condition, but we must not forget that the core is cooling and the adiabatic gradient is not in general divergence free. Any heat that is not conducted down the adiabat must be convected, so this is an effective source term. So we obtain

$$\mathcal{S} = \Delta T_{\mathrm{ad}} - \dot{T}_{\mathrm{ad}} + \frac{Q}{\delta T\, c_P}, \tag{5.13}$$

where the adiabatic temperature is measured in units of δT, the temperature difference across the core, and Q is the radioactive heating in units of Wkg^{-1} (for details see Anufriev *et al.*, 2005). In practice, dynamo codes are often written without the heat source term, but replacing it with a term $\beta(r)\,u_r$ to represent advection down a mean temperature gradient. This can be justified by writing $\Theta = \widehat{\Theta} + \widehat{T}(r)$ where $\widehat{T}$ is a solution of $\Delta\widehat{T} + \mathcal{S} = 0$ satisfying appropriate boundary conditions at the core boundaries. Then $\widehat{\Theta}$ replaces Θ throughout, and the term $\beta(r)\,u_r$ is introduced into the heat equation. No adjustment to the momentum equation is required, because a radially symmetric term can be absorbed in the pressure gradient. It is however important to realise that this $\widehat{T}$ is not the adiabatic temperature, but is a much smaller temperature.

Even when they are enhanced by turbulence, the thermal and viscous diffusion terms are small in (4.5c) and (4.5d). They are needed in numerical schemes for stability reasons, but we expect that they are not part of the primary balance of terms. This balance is the MAC balance, defined in Section 3.3.4 Besides the magnetic force, Archimedean force and Coriolis force, pressure forces are also required, though (4.5c) is usually solved by taking its curl to form the vorticity equation. We therefore expect the viscous term, $\mathrm{E}\,\Delta\mathbf{u}$ in (4.5c), to be small. Unfortunately, in spherical geometry codes it has not yet proved possible to reduce E to very low values, though this has been achieved in plane layer geometry at low Rossby number. The plane layer dynamo equations have the same terms as the spherical equations (Jones & Roberts, 2000, Rotvig & Jones, 2002), and so exhibit similar dynamics, but the geometry makes the equations much more tractable numerically.

Rotvig & Jones (2002) solved the equations (4.5a-d) – (4.6a-d) in the limit of low inertia, that is assuming $\mathrm{E}_\eta = 0$, and they used a low Ekman number $\mathrm{E} = 10^{-5}$; further details are given in their paper. They adopted a plane layer geometry, in which the z–direction corresponds to the radial direction in which gravity acts, and the rotation vector points in the (y, z)–plane, at $45°$ to the direction of gravity. This therefore models conditions in a piece of a planetary core at latitude $45°$. To explore the relative strength of the various forces, a snapshot was taken in the course of a numerical simulation of the dynamo equations, shown in Figures 5.3 (a-d). The magnitudes are shown in the plane $y = 0$, though the pictures do not depend critically on which plane cross-section is taken. Figure 5.3 (a) gives the Coriolis force $|2\boldsymbol{\Omega} \times \mathbf{u}|$ in the dimensionless units. The maximum value, corresponding to the darkest shade of red, is 1389. Fig 5.3 (b) gives the equivalent picture for the pressure gradient $|\boldsymbol{\nabla} p|$, and the maximum value here was 1949, slightly greater than the maximum Coriolis force. The Lorentz force $|\mathbf{j} \times \mathbf{B}|$ is shown in Figure 5.3c, with a maximum value of 2256. The Lorentz force therefore is larger than the Coriolis and pressure forces at some places, but note that the Lorentz force is somewhat more localised than the other forces. The buoyancy force $\widetilde{\mathrm{Ra}}\, g\, \Theta$, which has a maximum of 576, is shown in Figure 5.3d. The fluid motion in dynamo simulations is continually changing, so no particular significance attaches to the exact magnitudes of the forces at any particular time. The important point is that all the forces are of comparable strength, so all play a major role in the dynamics of planetary cores. The viscous forces can also be constructed, but they are typically very small in the interior of the fluid, though they are significant in the boundary layers.

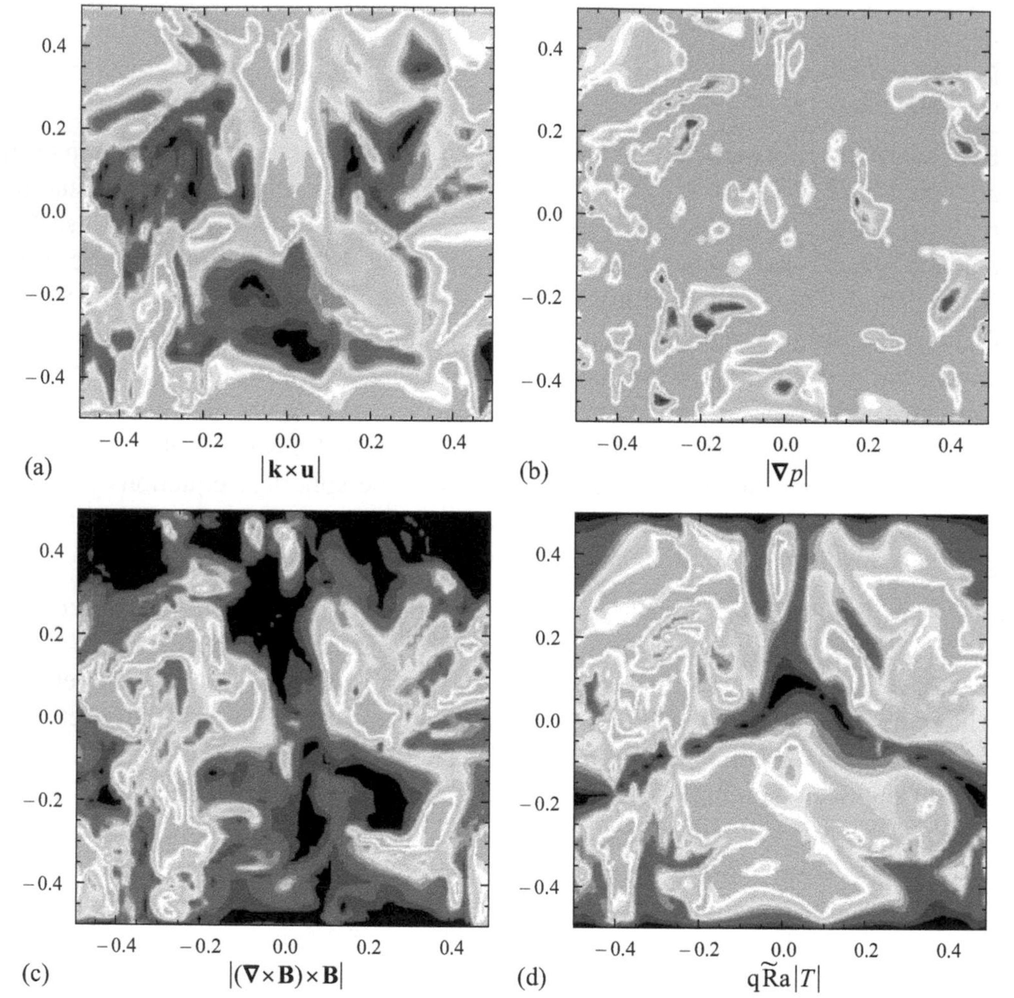

Figure 5.3 - Snapshot of the various forces in the $y = 0$ plane for a plane layer dynamo at $\mathrm{E} = 10^{-5}$. (a) Coriolis force, (b) pressure force, (c) Lorentz force, (d) Buoyancy force. (**See colour insert**.)

5.6.1. TYPICAL VELOCITY AND FIELD ESTIMATES

When investigating the MAC balance, it is helpful to consider the convected heat flux

$$F_{\mathrm{conv}} = \frac{1}{4\pi r^2} \int_S \rho\, c_P\, u_r\, \Theta\, \mathrm{d}S\,, \tag{5.14}$$

which is better constrained observationally than the temperature fluctuation. In the giant planets, F_{conv} will be close to the total heat flux coming out of the planet, while for terrestrial planets F_{conv} is typically a substantial fraction of the heat flux coming out of the core. If $\langle\cdot\rangle$ denotes the volume average, we can define the correlation between u_r and Θ as

$$\chi = \frac{\langle \int_S u_r \Theta \, dS \rangle}{4\pi r^2 \langle u_r \rangle \langle \Theta \rangle} . \tag{5.15}$$

An important question is whether this correlation is close to unity or much less. If it is close to unity, then

$$\langle \frac{1}{4\pi r^2} \int_S u_r \Theta \, \mathrm{d}S \rangle \approx \langle u_r \rangle \langle \Theta \rangle . \tag{5.16}$$

We might expect there to be a high correlation, because hot fluid will rise and cold fluid sink in general, but in a rapidly rotating magnetic fluid it is perhaps not so certain. Experiments on rapidly rotating fluids without magnetic field (Aubert *et al.* 2001) suggest that the correlation remains of order unity. We write

$$F_{\mathrm{conv}} \approx \chi \, \rho \, c_P \, U_* \, \Theta_* , \tag{5.17}$$

where U_* is a typical value of velocity and Θ_* is a typical value of temperature. The results of Rotvig & Jones (2002) suggest that the Coriolis force and the buoyancy force are comparable, so

$$2\Omega U_* \approx g\alpha\Theta_* \tag{5.18}$$

is expected. If we take the curl of the equation of motion and look at the component parallel to the rotation, we obtain

$$2\Omega \, \partial_z u_z \approx \mathbf{e}_z \cdot \boldsymbol{\nabla} \times g \, \alpha \, \Theta \, \mathbf{e}_r . \tag{5.19}$$

In a rapidly rotating fluid the length scale of variation perpendicular to the rotation axis L_H is smaller than the length scale parallel to the rotation axis L_z, by a factor $\mathrm{E}^{1/3}$ (see e.g. Jones *et al.*, 2000). This would give

$$2\Omega U_* \approx (L_z/L_H) g\alpha\Theta_* \approx \mathrm{E}^{-1/3} g\alpha\Theta_* , \tag{5.20}$$

rather than (5.18). Since E is very small in planetary cores, this is a very substantial difference. However, in the presence of magnetic field, L_z/L_H becomes smaller, and ultimately is $\mathcal{O}(1)$ for strong fields. Retaining the factor L_z/L_H and eliminating Θ_* using we obtain

$$U_* \approx \left[\frac{g\alpha F_{\mathrm{conv}} L_z}{2\chi \, L_H \, \rho \, c_P \, \Omega} \right]^{1/2} , \tag{5.21}$$

and Starchenko & Jones (2002) used

$$U_{\mathrm{MAC}} \approx \left[\frac{g\alpha F_{\mathrm{conv}}}{\rho c_p \Omega} \right]^{1/2} , \tag{5.22}$$

taking $L_z/(2L_H\chi) = 1$.

This MAC balance regime is by no means the only possible one for planetary dynamos. Stevenson (1979, 2003) has suggested that mixing length theory provides an alternative. Here the dominant balance is between buoyancy and inertia:

$$\frac{U_*^2}{\ell} \approx g\,\alpha\,\Theta_*\,, \tag{5.23}$$

where ℓ is the mixing length, typically the core size or the density scale height. As before, the heat flux equation (5.14) gives

$$F_{\rm conv} \approx \rho\, c_P\, U_*\, \Theta_*\,. \tag{5.24}$$

Eliminating Θ_* between (5.23) and (5.24) yields

$$U_{ml} \approx \left[\frac{g\,\alpha\,F_{\rm conv}\,\ell}{\rho\,c_P}\right]^{1/3} = U_{\rm MAC}^{2/3}(\Omega\,\ell)^{1/3}\,. \tag{5.25}$$

Stevenson (2003) has a coefficient 0.3 in this formula. The mixing length estimate gives somewhat larger values for the velocity than the MAC value.

5.7. NUMERICAL DYNAMO MODELS FOR THE PLANETS

It is possible to solve the full three-dimensional system of equations (4.5a-d) using direct numerical simulation. The usual geometry is to consider a spherical shell for the dynamo region, and to include the effects of rotation. The equations are very demanding computationally, and many hundreds of processor hours are needed to obtain solutions even when moderate values of the parameters are used. A dynamo benchmark (Christensen *et al.*, 2001) has been established to allow researchers to check the very complex numerical codes used to integrate the equations. In all the planets, E and $q_{\rm T}$ are extremely small, but in simulations it is difficult to get $q_{\rm T}$ significantly below unity, and Ekman numbers below 10^{-5}, without inducing numerical instability. For comparison, even if a turbulent value of the Earth's viscosity is used, E works out at around 10^{-9}. Results from simulations must therefore be treated with caution. A further difficulty is that most codes use the Boussinesq equations, whereas the giant planets have large density variations. Despite all these difficulties, the simulations produce interesting results, and many of the features of the geomagnetic field can be reproduced, even including field reversals. This is a very active field of research.

One result emerging from the simulations is that the distribution of the energy source has an important effect on the type of field generated. Christensen *et al.* (1999) assumed that the heat source lies inside the dynamo region, so that the dynamo is

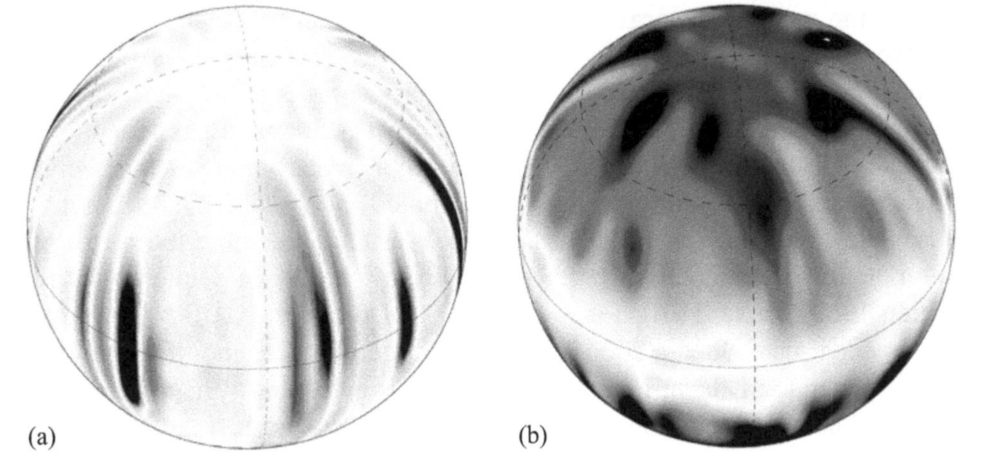

Figure 5.4 - (a) Shaded contour plots of the radial velocity, u_r at an arbitrary radius $r = 0.8\,r_o$. $\mathrm{Pr} = \mathrm{q_T} = 1$. The minimum and maximum values are $[-295.35,\ 135.55]$. (b) Shaded contour plots of the radial magnetic field, b_r at $r = r_o$. $\mathrm{Pr} = \mathrm{q_T} = 1$. The minimum and maximum values are $[-0.9379, 0.9007]$. **(See colour insert.)**

powered by a flux of heat (or buoyancy from compositional convection) entering the spherical shell from below. They found mostly, but not exclusively, axial dipolar fields. This is the form of field which approximately models that of the Earth, Jupiter, Saturn and Ganymede. On the other hand, Busse *et al.* (2003), using models with uniform heating, found that axial dipoles are not typical. They find quadrupolar modes, and even hemispherical dynamos in which the field is generated primarily in one hemisphere. Another factor that can affect the form of the field is the strength of the heating as measured by the Rayleigh number. Aubert & Wicht (2004) find that for $\widetilde{\mathrm{Ra}}$ not far above critical, equatorial dipole modes can be found. This means that the dipole is aligned with its axis in the equatorial plane, not parallel to the axis of rotation. Such a field might be a first approximation to the fields of Neptune and Uranus.

Another factor which can influence the form and strength of the generated field is the Prandtl number, $\mathrm{Pr} = \nu/\kappa$. In liquid metals this number is small, though it is often argued that since the diffusion processes in planetary cores are probably turbulent, a value of $\mathrm{Pr_T} = \nu_\mathrm{T}/\kappa_\mathrm{T} = 1$ is appropriate. At low Prandtl numbers, inertial effects are important and the generated fields have a rather complicated morphology, but at higher Prandtl numbers strong axial dipoles become common (Sreenivasan & Jones, 2006). At low Pr, the kinetic energy is usually at least as large as the magnetic energy in numerical dynamo models. Unless the velocity is much larger in the Earth's core than the velocity indicated by the secular variation, which seems unlikely, the magnetic energy in the core is much larger than the kinetic energy. With the constraints imposed by numerical stability, the only way to get the ratio of kinetic to magnetic energy correct in planetary cores is to assume a rather large Prandtl number. A description of the dynamo process in numerical dynamo models is given by Olson *et al.* (1999), but it is not a simple matter to understand why some models are axial dipole dominated when other models have much more complicated fields. The flow patterns found in the models usually resemble the cartridge belt pattern of convection columns predicted by linear theory (see Section 3.4, and also Roberts, 1968; Busse, 1970; Jones *et al.*, 2000; Dormy *et al.*, 2004). An example from Sreenivasan & Jones (2006) is shown in Figure 5.4(a), and the corresponding dynamo generated field in Figure 5.4(b). These figures are a snapshot from a numerical soluxtion of equations (4.5a-d). The parameters are $\mathrm{E} = 10^{-4}$, $\mathrm{Pr} = \mathrm{q} = 1$ and $\widetilde{\mathrm{Ra}} = 750$, which is about ten times critical. There are no internal heat sources, and the radius ratio is 0.35, appropriate to the Earth's liquid outer core. The velocities given in the figure caption are in units corresponding to the magnetic Reynolds number, so the velocities are slightly lower than those expected in the Earth's core (where an estimate of 300 – 500 is most usual). The columnar structure expected in a rapidly rotating fluid is clearly seen in Figure 5.4(a). The magnetic field is shown at the core-mantle boundary in Figure 5.4(b), and the strong dipole dominance is immediately apparent.

Yet another factor that may influence the form and strength of the generated fields is the radius ratio of the dynamo region. It has been suggested (Ruzmaikin & Starchenko, 1991) that the irregular fields of Uranus and Neptune could be connected with the dynamo region in these planets being comparatively thin. This tends to generate magnetic fields which are more irregular, with the dynamo process occurring in local patches which are only weakly coupled together, in contrast to the model in Figures 5.4(a) and 5.4(b), where the convection columns strongly couple all parts of the dynamo process in the core.

It is clear that the study of numerical dynamo models is still in its infancy, and there is still much that is not understood even in these mathematically clean idealised models. Nevertheless, our understanding has improved significantly in the last decade, and it is likely that dynamo theory will become an important part of planetary science in the years to come.

CHAPTER 6

STELLAR DYNAMOS

Steven Tobias & Nigel Weiss

6.1. STELLAR MAGNETIC ACTIVITY

Stars that are magnetically active owe this activity to a combination of turbulent convection and rotation. In this review we shall focus on stars like the Sun, which lie on the main sequence and are sufficiently cool that hydrogen becomes ionised below their surfaces, resulting in the presence of a deep outer convection zone. Their magnetic fields can be measured directly through the Zeeman broadening of spectral lines, or inferred from proxy evidence. This is provided by coronal X–ray emission, by H and K emission from singly ionised Ca^+, by photometric variability (associated with starspots) or by optical and radio flares – all of which are known to be associated with magnetic activity on the Sun (Tayler, 1997). The Sun is unique, however, in that we can observe detailed magnetic structures on its surface, and we have records of its activity extending back through many centuries; its internal structure is also well-established (see Figure 6.1). On the other hand, the Sun is a single star whose large-scale properties evolve extremely slowly. So it is only through exploiting the solar-stellar connection and examining the magnetic properties of other stars that we can understand how magnetic activity depends on such key parameters as rotation (Wilson, 1994; Mestel, 1999; Schrijver & Zwaan, 2000).

Chromospheric Ca^+ emission has been measured for a large number of nearby stars, revealing a wide range of activity (Vaughan & Preston, 1980; Soderblom, 1985; Henry *et al.*, 1996). Comparison of middle-aged stars like the Sun with similar stars in young clusters shows that magnetic activity declines with age. Moreover, there

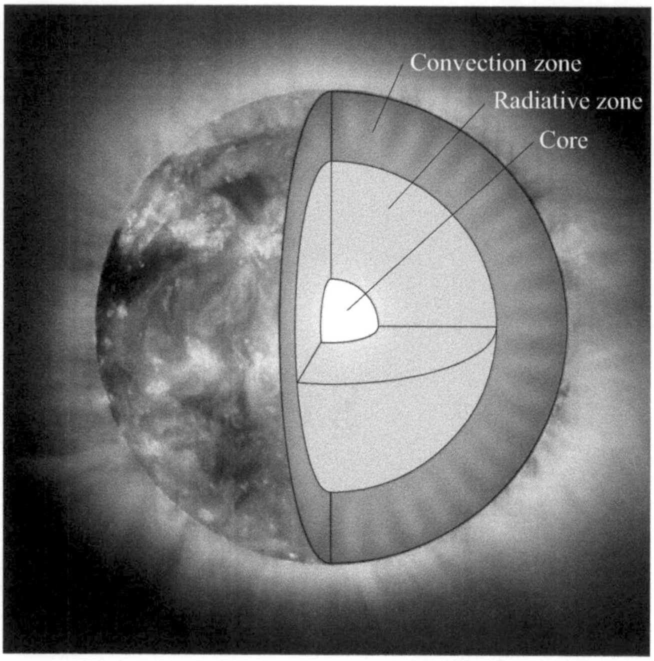

Figure 6.1 - Internal structure of the Sun. The cutaway image shows the visible surface (the photosphere, with a radius $R_\odot \approx 700\,\mathrm{Mm}$), together with an outer region where energy is carried mainly by convection, and an inner region where energy is transported by radiation. The narrow interface between the convection zone and the radiative zone, at a radius of approximately $0.7R_\odot$, has a thickness of only $0.02R_\odot$ and is the site of the tachocline, where there is a strong radial gradient in angular velocity. The temperature rises from $6000\,\mathrm{K}$ at the surface to about $2 \times 10^6\,\mathrm{K}$ at the base of the convection zone and then to $1.5 \times 10^7\,\mathrm{K}$ in the central core, where energy is generated by thermonuclear fusion. (**See colour insert**.)

is a strong correlation between activity and rotation (Noyes *et al.*, 1984; Baliunas & Vaughan, 1985; Saar & Brandenburg, 1999). When stars first arrive on the main sequence and begin to burn hydrogen they are spinning rapidly (Soderblom, Jones & Fischer, 2001), with rotation periods of order a day, but they gradually lose angular momentum to magnetic braking owing (Mestel, 1999) and spin down. It is only in slowly rotating middle-aged stars like the Sun (with rotation periods of order a month) that cyclic activity is found (Baliunas *et al.*, 1995). The cycle periods are all around 10 years: Figure 6.2 shows the time-dependent Ca^+ emission in a solar-type star, exhibiting cyclic variation with a period of 8.2 yr.

The Sun's own magnetic activity varies cyclically, with an average period of about 11 years (Stix, 2002). The most dramatic manifestation of this activity is in sunspots, which are dark because they are the sites of strong magnetic fields that locally inhibit

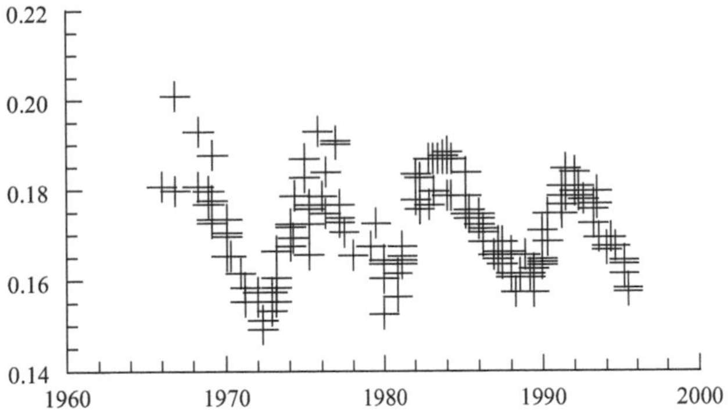

Figure 6.2 - Cyclic activity in a star. Chromospheric Ca^+ emission as a function of time for the K0 star HD 81809 (Mount Wilson Observatory H-K Project).

convection. The cyclic variation of the incidence of sunspots is demonstrated by the well-known butterfly diagram in Figure 6.3. Spots typically appear in pairs with opposite polarity, oriented nearly parallel to the equator. The spots are contained within active regions, which are formed by the emergence of almost azimuthal (or toroidal) magnetic flux, whose orientation obeys Hale's laws. The polarities of leading and following spots are consistent in each hemisphere but antisymmetric about the equator; and these polarities reverse from one activity cycle to the next. Hence the *magnetic* cycle has a period of 22 years. The axis of a sunspot group or active region is actually inclined at a small angle so that leading spots are closer to the equator, and this angle increases systematically with latitude (Joy's law). This result, with the large horizontal scale of active regions, suggests that the emerging flux is deep-seated and not a merely superficial phenomenon.

The arguments for ascribing the origin of these magnetic fields to a dynamo are different from those for planetary dynamos. Whereas the Earth's magnetic field has to be maintained by a geodynamo, since it has been present for billions of years despite an Ohmic decay time of only 10^4 yr, the solar problem is to explain how the field reverses every 11 years when the decay time is 10^9–10^{10} yr. It has been claimed that cyclic behaviour could be driven by an oscillator, with a steady poloidal field and alternating shears in differential rotation, though no mechanism for producing such shears has been suggested. In fact, the Sun possesses a large-scale poloidal field that is most prominent in polar regions and has dipole symmetry, and this field reverses near sunspot maximum (i.e. 90° out of phase with the activity cycle). Furthermore, the only observed fluctuations in angular velocity have a period of 11 years, not 22 years, and an 11-yr periodicity is precisely what is expected from a nonlinear dynamo, since the Lorentz force is quadratic in the magnetic field. We may therefore assume that this cyclic solar activity is maintained by a large-scale homogeneous dynamo, which generates systematic fields (magnetic *climate*) as opposed to the

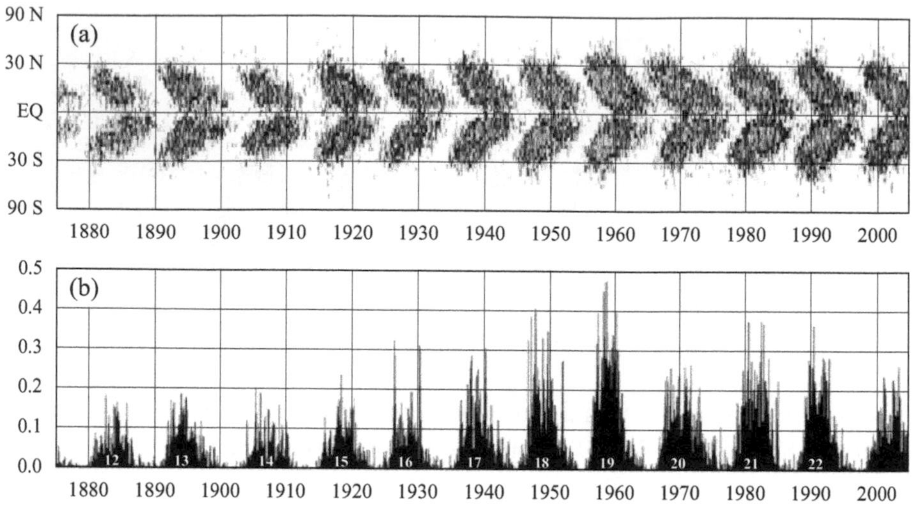

Figure 6.3 - Cyclic activity in the Sun (1874–2004). (a) butterfly diagram, showing the incidence of sunspots as a function of latitude and time; new spots appear at $\pm 30°$ as the old cycle dies away at the equator. (b) area covered by sunspots as a function of time (courtesy of D.H. Hathaway).

small-scale disordered fields (magnetic *weather*) which could be produced by local dynamo action near the photosphere.

The current state of solar dynamo theory forces most of our discussion to be physical rather than mathematical, backed up by numerical rather than by analytical results. In the next two sections we introduce mean field ($\alpha\omega$) dynamos for the solar cycle. Then, in Section 6.4, we focus on dynamos located at the interface between the convective and radiative zones, where the radial shear is greatest. Long-term modulation of cyclic activity is the subject of Section 6.5. Next, in Section 6.6, we consider the enhanced activity in rapidly rotating stars and go on to comment briefly on dynamos in protostellar accretion discs. Finally, we summarise future prospects for stellar dynamo theory. Many of these issues have already been discussed in various recent reviews (e.g. Stix, 1991; Weiss, 1994; Rosner, 2000; Tobias, 2002a; Choudhuri, 2003; Ossendrijver, 2003, Rüdiger & Arlt, 2003).

6.2. LINEAR $\alpha\omega$–DYNAMOS FOR THE SOLAR CYCLE

A proper treatment of the solar dynamo would require an accurate simulation of the nonlinear interactions between rotation, convection and magnetic fields. Direct numerical simulation of these processes in a regime where the magnetic Reynolds number $\mathrm{Rm} \approx 10^9$ remains beyond the capacity of the largest computers and, in any case, the key physical mechanisms of differential rotation, helicity and magnetic buoyancy are not adequately understood. Apart from some early brave attempts (Gilman, 1983; Glatzmaier, 1985), stellar dynamo theory has had to rely on the mean field approximation (discussed in Section 1.5).

Since differential rotation is so effective at creating toroidal fields, nearly all stellar models are axisymmetric $\alpha\omega$–dynamos. Then the poloidal field $\mathbf{B}_P = \boldsymbol{\nabla} \times (A\mathbf{e}_\phi)$ and the toroidal field $\mathbf{B}_T = B_\phi \mathbf{e}_\phi$ satisfy the linear equations

$$\partial_t A = \alpha B_\phi + \eta \mathcal{D}^2 A\,, \qquad \partial_t B_\phi = r \sin\theta\, \mathbf{B}_P \cdot \boldsymbol{\nabla}\omega + \eta \mathcal{D}^2 B_\phi\,, \qquad \text{(6.1a,b)}$$

referred to spherical polar co-ordinates, where ω is the local angular velocity, η here denotes the total (laminar plus turbulent) diffusivity and $\mathcal{D}^2 = \Delta - 1/r^2 \sin^2\theta$.

6.2.1. DYNAMO WAVES

Parker (1955, 1979) provided the simplest (and earliest) example of a mean field dynamo. He considered a Cartesian model with $A,\ B \propto \exp(\mathrm{i}k\,x)$, where the x–direction corresponds to increasing θ and $U(z)$ represents the sheared zonal velocity with z corresponding to a local radial co-ordinate. He showed that there was exponential growth when the dynamo number (see Section 1.5.3), $\mathrm{D} = \alpha U'/(2\eta^2 k^3)$, was greater in magnitude than unity (prime is used to note a derivative, i.e. $U' = \mathrm{d}U(z)/\mathrm{d}z$). The waves travel "equatorward" if $\mathrm{D} < 0$. This result from a relatively simple model has had a profound effect on stellar dynamo theory; it is now widely claimed that dynamo waves always travel poleward if $\mathrm{D} > 0$. However this is not always the case (although often true) and some solar dynamo models *are* able to reproduce equator-propagating magnetic fields even for $\mathrm{D} > 0$. This result can readily be extended to other geometries and, more generally, the waves travel along surfaces of constant ω.

It is important to realise that the local behaviour of travelling waves with periodic boundary conditions may differ qualitatively from the global behaviour of solutions that are spatially confined, whether in Cartesian or in spherical geometry. For linear theory, this corresponds to the difference between convective and absolute instability (Tobias *et al.*, 1998b). For waves of frequency ω and wavenumber k, governed by the dispersion relation

$$\omega(k; \mathrm{D}) \equiv 0\,, \qquad (6.2)$$

instability in an infinite (or periodic) domain occurs at the smallest value of D that satisfies the dispersion relation for some real k. It is possible therefore to generate a marginal curve of $\mathrm{D}_{\mathrm{crit}}$ versus k (Worledge *et al.*, 1997). In a finite domain of length

Table IV - Some orders of magnitude for the solar dynamo.

Solar radius	$R_\odot$	6.96×10^8	m
Solar mass	$M_\odot$	1.99×10^{30}	kg
Surface temperature		5780	K
Central temperature		15.6×10^6	K
Surface density		2.0×10^{-4}	$\mathrm{kg\,m^{-3}}$
Central density		1.5×10^5	$\mathrm{kg\,m^{-3}}$
Solar age		4.57×10^9	yr
Large scale magnetic field strength		5×10^4	nT (0.5 G)
Diameter of surface granulation	L	10^6	m
Sunspot magnetic field strength		0.3	T (3000 G)
Rotation time at equator		25	d
Rotation time at 60° latitude		29	d
Ekman number, base of the c.z.	E	$\nu/\Omega R_\odot^2$	$\mathcal{O}(10^{-15})$
photosphere			$\mathcal{O}(10^{-17})$
Rossby number, base of the c.z.	Ro	$\lvert\mathbf{u}\rvert/\Omega R_\odot$	$\mathcal{O}(10^{-2})$
photosphere		$\lvert\mathbf{u}\rvert/\Omega L$	$\mathcal{O}(10^{3})$
Prandtl number, base of the c.z.	Pr	ν/κ	$\mathcal{O}(10^{-6})$
photosphere			$\mathcal{O}(10^{-13})$
Magnetic Prandtl number, base of the c.z.	Pm	ν/η	$\mathcal{O}(10^{-1})$
photosphere			$\mathcal{O}(10^{-7})$
Reynolds number, base of the c.z.	Re	$\lvert\mathbf{u}\rvert R_\odot/\nu$	$\mathcal{O}(10^{13})$
photosphere		$\lvert\mathbf{u}\rvert L/\nu$	$\mathcal{O}(10^{13})$
Magnetic Reynolds number, base of the c.z.	Rm	$\lvert\mathbf{u}\rvert R_\odot/\eta$	$\mathcal{O}(10^{11})$
photosphere		$\lvert\mathbf{u}\rvert L/\eta$	$\mathcal{O}(10^{6})$

L, instability of a global mode is governed by a more stringent condition. In the limit $L \to \infty$, this condition approaches that for absolute instability of a periodic wavetrain; that is both (6.2) and the condition $\partial\omega/\partial k = 0$ must be satisfied simultaneously. This can only be achieved if k is allowed to be complex (see Kuzanyan & Sokoloff, 1995). It has its origins in the development of the Maximally–Efficient–Generation Approach (MEGA, see Ruzmaikin *et al.*, 1990), as explained in detail in the Appendix of Bassom *et al.* (2005). The two criteria for periodic and finite domains yield very different critical dynamo numbers and frequencies. Moreover, nonlinear behaviour can be qualitatively affected (Tobias 1998), with the frequency in the nonlinear regime being determined by the interaction of the global mode with the boundaries, and the possible presence of secondary absolute instabilities.

6.2.2. SPHERICAL MODELS

The spherical problem possesses two important symmetries with respect to reflection about the equator. The governing equations (6.1a,b), with appropriate boundary conditions and suitable constraints on α and ω, are invariant under the transformations

$$d: \quad (\theta,\, t) \rightarrow (\pi - \theta,\, t), \;\; (A,\, B) \rightarrow (A,\, -B) \tag{6.3a}$$

and

$$q: \quad (\theta,\, t) \rightarrow (\pi - \theta,\, t), \;\; (A,\, B) \rightarrow (-A,\, B)\,. \tag{6.3b}$$

These symmetries generate an abelian group (D_2) with four elements, including $i = dq: \quad (\theta,\, t) \rightarrow (\theta,\, t), \;\; (A,\, B) \rightarrow (-A,\, -B)$ and the identity (Jennings & Weiss, 1991). The trivial solution $A = B = 0$ possesses the full D_2 symmetry, which is broken at the initial Hopf or pitchfork bifurcation. The linear problem then allows two distinct families of eigenfunctions, with different symmetries about the equator. For *dipole* solutions, with the symmetry d, the toroidal field B is antisymmetric about the equator, while A is symmetric; for *quadrupole* solutions, with the symmetry q, A is antisymmetric and B is symmetric. If an appropriate dynamo number is defined by setting $\mathrm{D} = \alpha\omega' R_\odot^4/\eta^2$, where $R_\odot$ is the solar radius, then the critical values of D at which dipolar and quadrupolar modes become unstable differ only slightly. Provided that $\mathrm{D} < 0$ in the northern hemisphere, oscillatory dipole modes are marginally favoured and the pattern drifts equatorward. Thus it is easy to construct butterfly diagrams that are qualitatively similar to that in Figure 6.3 (Steenbeck & Krause, 1969; Stix, 1976, 2002). Note that the symmetry of one or other of these solutions can only be broken at a subsequent bifurcation in the nonlinear domain. If this happens at a pitchfork bifurcation, further symmetries of periodic solutions can be classified (Jennings & Weiss, 1991) but symmetry–breaking more commonly involves a Hopf bifurcation that leads to quasiperiodic behaviour.

6.2.3. THE ω–EFFECT

It has long been known that the angular velocity varies with latitude at the surface of the Sun: the equatorial regions rotate distinctly more rapidly (with a sidereal period of 25 days) than the poles (with a period of about 35 days). More recently, one of the triumphs of helioseismology has been the determination of the Sun's internal rotation (Thompson *et al.*, 2003). Measurements of p–mode frequencies have revealed that there is very little radial shear in the convection zone, where $\omega \approx \omega(\theta)$, while ω is nearly uniform in the radiative interior. Between the two is a thin layer (with thickness around $0.02R_\odot$) with a very strong radial shear, the *tachocline*. This observed pattern of differential rotation is displayed in Figure 6.4; since the radiative core rotates at an intermediate rate, $\partial\omega/\partial r$ changes sign at a latitude around $30°$.

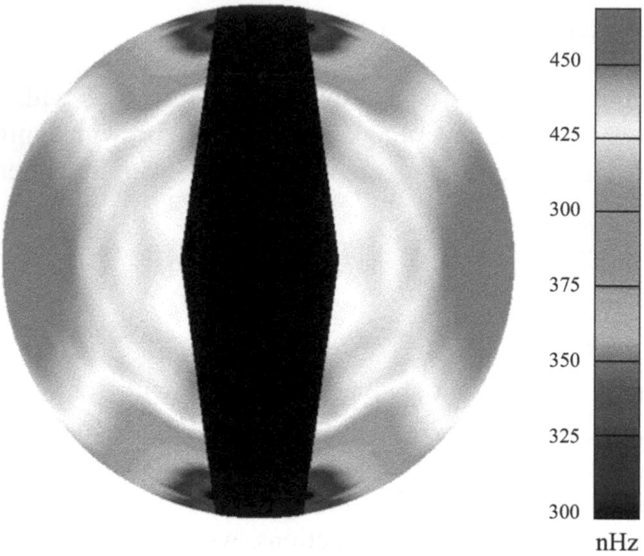

Figure 6.4 - Differential rotation in the solar interior. The rotation rate is approximately constant along radii in the convection zone, whose base is indicated by the dashed line. A frequency of 450 nHz corresponds to a period of about 26 days (courtesy of J. Christensen-Dalsgaard). (**See colour insert**.)

The dynamics within the tachocline is not yet understood (Tobias, 2004) but it is generally accepted that strong toroidal fields are generated and stored within this region of shear.

6.2.4. THE α–EFFECT

The source of the α–effect is much less clear. The earliest treatments assumed that poloidal fields were regenerated by cyclonic eddies that were distributed throughout the convection zone and that α (which is antisymmetric about the equator) reverses its sign in such a way that $D < 0$ at the base of the convection zone in the northern hemisphere (Parker, 1979; Krause & Rädler, 1980). Some recent authors have revived a surface flux-transport model due originally to Babcock (1961) and to Leighton (1967), in which the α–effect is ascribed to the decay, through turbulent diffusion, of active regions whose orientation is determined by Joy's law. The opposing fields of leading spots cancel out as they approach the equator, while the trailing fields of following spots spread polewards and eventually reverse the polar fields at sunspot maximum. In that case, the amplitude of the activity cycle should determine the strength of the high-latitude poloidal field at the next sunspot minimum. This can be checked by studying the incidence of recurrent geomagnetic activity, caused by high-speed streams emerging from coronal holes. Detailed investigations show that the toroidal fields at sunspot maximum are more closely related to the mid-latitude poloidal fields that *precede* them than to those that follow afterwards (Simon & Legrand 1986; Hathaway, Wilson & Reichmann 1999; Ruzmaikin & Feynman 2001). This evidence implies that flux transport is only a superficial process.

The dynamo is obviously more efficient if the α–effect is located near the base of the convection zone, where the ω–effect is strong. Indeed, Mason, Hughes & Tobias (2002) have shown that the influence of a surface source in generating dynamo waves is swamped by that of a much weaker α near the tachocline. There are several buoyancy-driven mechanisms that might provide the latter. These include magnetostrophic waves (Moffatt, 1978; Schmitt, 1987), instabilities of flux tubes (Ferriz-Mas, Schmitt & Schüssler, 1994; Ossendrijver, 2000b) and instabilities driven by magnetic buoyancy (Brandenburg & Schmitt, 1998; Thelen, 2000a, b), These last have been studied in considerable detail (see Hughes & Proctor, 1988; Tobias, 2004) in both the linear and nonlinear (Matthews, Hughes & Proctor, 1995; Wissink *et al.*, 2000) regimes, and their interactions with rotational shear have also been explored (Cally, 2000; Hughes & Tobias, 2001; Cline, Brummell & Cattaneo, 2003; Tobias & Hughes, 2004). Another possible source of kinetic helicity arises from MHD instabilities associated with differential rotation within the tachocline, which have been studied and classified (e.g. Gilman & Fox 1999; Cally, 2001, 2003; Gilman & Dikpati, 2002; see Tobias, 2004 for a review).

These instabilities are joint instabilities of the strong toroidal field and latitudinal differential rotation just below the base of the solar convection zone. The global mode associated with the instability is known to possess non-zero kinetic helicity which may be related to the α–effect. However, this connection can only be reliably achieved in a small Rm analysis and for high Rm a straightforward association between kinetic helicity and α–effect is not possible (see Courvoisier, Hughes & Tobias, 2006).

One feature of these instabilities is that they are triggered by a finite-amplitude toroidal field, which itself has to be built up by dynamo action following a supercritical bifurcation; once the buoyancy driven instabilities set in the dynamo can become much more efficient. Thus the branch of nonlinear dynamo solutions may have two turning points, with an intermediate segment of unstable solutions, leading to subcritical behaviour and hysteresis (Ossendrijver, 2000b; *cf.* Figure 3 of Weiss & Tobias, 2000).

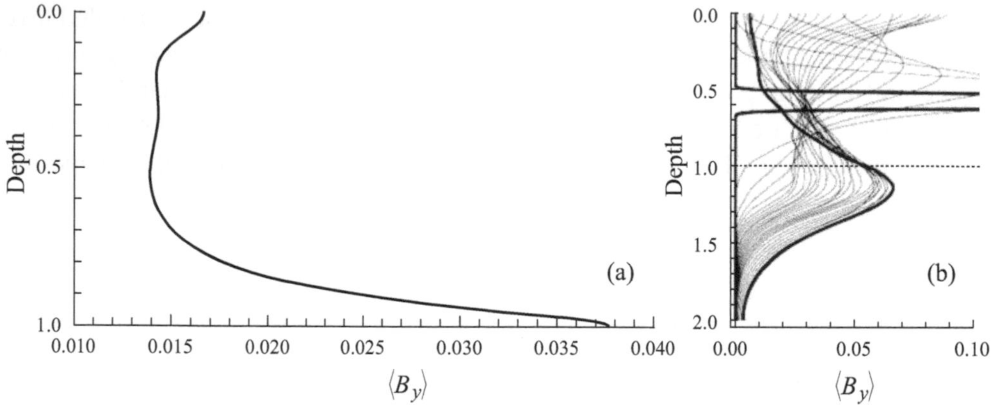

Figure 6.5 - Downward pumping of magnetic flux. Results for a horizontal field, initially in the y–direction and confined to a thin sheet, in a strongly stratified layer. (a) The field $\langle B_y \rangle$, averaged horizontally and over time, as a function of depth in a vigorously convecting layer (courtesy of N.H. Brummell). (b) Evolution with time of $\langle B_y \rangle$ when the convecting layer lies above a layer that is strongly stably stratified; magnetic flux is expelled into the stable region (after Tobias *et al.*, 2001).

6.2.5. MAGNETIC PUMPING

In addition to producing turbulent diffusion and regenerating large-scale fields by the α–effect, turbulent motion can also lead to net transport of magnetic fields. In mean field electrodynamics this is represented by the antisymmetric part of the α–effect ($\alpha^a_{ij} = \varepsilon_{ijk}\gamma_k$). Physically, this corresponds to flux expulsion down the gradient of turbulent intensity and γ can be calculated and interpreted as a pumping velocity (Krause & Rädler 1980; Zeldovich, Ruzmaikin & Sokoloff, 1983; Moffatt, 1983). In Boussinesq convection, with up-down symmetry flux is expelled equally towards the top and bottom of the convecting layer. In a stratified layer, however, there is a preferred direction which leads to a net downward transport of magnetic flux. Two distinct mechanisms are involved. For mildly nonlinear convection there are isolated gentle upflows enclosed by a coherent network of downflows and this pattern can give rise to topological pumping (Drobyshevski & Yuferev, 1974). In turbulent convection the sinking network is focused into rapidly descending plumes and this pattern leads to a net downward transport (Weiss, Thomas, Brummell & Tobias, 2004), as illustrated in Figure 6.5a. This process becomes much more effective when there is a stably stratified region beneath the convectively unstable layer as shown in Figure 6.5b (Tobias *et al.*, 1998a, 2001; Dorch & Nordlund 2001). It follows that any large-scale fields within the convection zone will tend to be pumped downwards and into the stably stratified tachocline, where they can accumulate within an even thinner shell that is penetrated by overshooting convection.

6.2.6. MERIDIONAL CIRCULATION

There is a mean meridional flow at the solar surface, moving poleward in each hemisphere with a typical speed of about $10\,\mathrm{m\,s^{-1}}$ (Hathaway, 1996), though this pattern is erratic. Helioseismic inversions indicate that this flow extends downwards through at least the upper half of the convection zone (Braun & Fan, 1998) and mass conservation then requires that there should be an equatorward counterflow at its base, where the density is much higher and the velocity correspondingly less. Nevertheless, a flow speed of $1\,\mathrm{m\,s^{-1}}$ (if attainable) would be enough to traverse the sunspot zone within 11 years, and such a counterflow has been invoked as a conveyor belt to explain the equatorward drift of activity as the cycle progresses (see Choudhuri, 2003). [This idea actually goes back to Bjerknes (1926) and Bullard (1955), who postulated a flow towards the equator at the photosphere.] It has also been suggested that this conveyor belt transports poloidal flux generated by a surface α–effect all the way down to the tachocline, where it can be stretched out to form a strong toroidal field (Dikpati & Charbonneau, 1999). Later, more elaborate calculations show, however, that any realistic dynamo model requires a powerful source of poloidal flux at the base of the convection zone (Dikpati *et al.*, 2004).

6.3. NONLINEAR QUENCHING MECHANISMS

Kinematic growth of the oscillatory field is eventually limited by the nonlinear action of the Lorentz force. Within the mean field approximation this is most easily represented by quenching the α–effect, in the simplest case by setting [see also (2.51), page 82]

$$\alpha = \frac{\alpha_0}{1 + |\mathbf{B}|^2/\mathcal{B}^2}, \tag{6.4}$$

where $\mathcal{B}$ is, say, the equipartition field strength (such that $\mathcal{B}^2 = \mu_o\langle\rho|\mathbf{u}|^2\rangle$, where $\mathbf{u}$ is the turbulent velocity). This formalism is extremely convenient and has been widely used, although there are strong reasons for believing that α–quenching is much more drastic, and that (6.4) should be replaced, for instance, by [see also (2.74), page 95]

$$\alpha = \frac{\alpha_0}{1 + \mathrm{Rm}^\gamma\,|\mathbf{B}|^2/\mathcal{B}^2}, \tag{6.5}$$

with $0 < \gamma \leq 2$ (see Section 2.7 and Vainshtein & Cattaneo, 1992; Diamond, Hughes & Kim, 2004). In the Sun, this would imply that α is quenched when the mean field B is less than 1 G, which would be disastrous for a dynamo. Analogous (and equally contentious) arguments suggest that the turbulent diffusivity η should be similarly quenched. A proper understanding of the transport coefficients of mean-field theory in both the kinematic and nonlinear regimes is essential (see

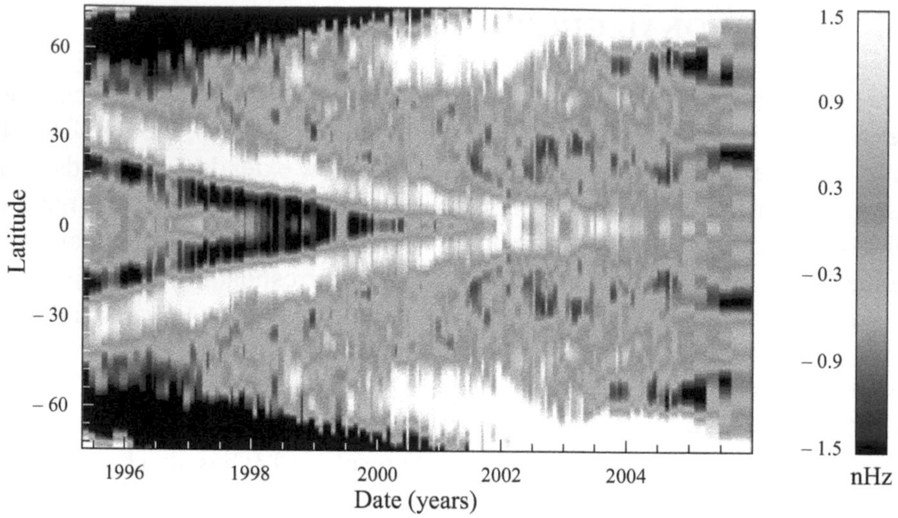

Figure 6.6 - Torsional oscillations with an 11–year period near the solar surface, derived from p–mode splitting measured with the GONG network (GONG RLS, $r = 0.99R_{\odot}$). Zones of slightly more rapid rotation progress towards the equator and towards the poles (courtesy of R. Howe). (**See colour insert.**)

Section 2.7 for an in-depth discussion). This remains a controversial issue, though one of fundamental importance for stellar dynamos.

An alternative is to let the Lorentz force limit differential rotation. Here there are two possibilities, depending on whether microdynamic or macrodynamic effects are more important. If we regard the overall differential rotation as being driven by the impact of Coriolis forces on small-scale turbulence – the Λ–effect (see page 89 and Rüdiger, 1989; Rüdiger & Arlt, 2003) – then Λ–quenching depends on the *microdynamic* balance between Maxwell and Reynolds stresses (see page 89 and Rüdiger & Kitchatinov, 1990; Kitchatinov, Rüdiger & Küker, 1994). If, on the other hand, the shear is concentrated in the tachocline (or maintained by large-scale motion in the convection zone) it is appropriate to consider the the rotational effect of the couple exerted by the *macrodynamic* Lorentz force (Malkus & Proctor, 1975). As we have already pointed out, this Malkus–Proctor effect generates a secondary flow with *twice* the frequency of the magnetic cycle. Moreover, that is just what is observed, as so-called "torsional oscillations" with an 11–year periodicity, at the solar surface – see Figure 6.6 (Howe *et al.*, 2000a). That suggests that the dynamo equations (6.1a,b) should be augmented by adding a third equation for the evolution of ω (the large-scale angular velocity), in which the magnetic torque competes with turbulent "viscous" damping.

The remaining possibility is that kinematic growth is saturated by flux loss from the dynamo domain; magnetic flux simply rises, owing to magnetic buoyancy, and escapes through the surface of the Sun. This is not straightforward, since plasma is tied to the field lines: for instance, axisymmetric toroidal flux can only be lost if a flux ring kinks and reconnects to form a band of loops that can decay diffusively. The role of magnetic helicity (whether large-scale at the tachocline, or small-scale in the convection zone) also becomes important in this context (see Ossendrijver, 2003).

6.4. INTERFACE DYNAMOS

Solar magnetic fields can only be measured at the photosphere and in the atmosphere above it; the magnetism of the Sun's interior has to be inferred from theory. Hence it is important to distinguish primary properties of the solar dynamo (notably the cyclic eruption of toroidal fields) from secondary properties of the surface fields. The latter – termed epiphenomena by Cowling (1975) – include the decay of active regions and perhaps even the observed reversals of the polar fields. As already stated, the scale and systematic properties of the toroidal fields that emerge in sunspots indicate that they must be formed deep down in the interior. Any such fields within the convection zone would be magnetically buoyant and would rise to the surface within a month. So it seems clear that the toroidal field must be stored at the interface between the convective and radiative zones, where it can be pumped downwards into the tachocline. Moreover, that is just where azimuthal fields are most readily generated by differential rotation.

It is not difficult to construct plausible mean field models with suitably chosen distributions of ω and α. Parker (1993) realised that catastrophic α–quenching could be avoided if the α– and ω–effects were spatially separated. Thus he introduced an *interface* dynamo model, with the shear confined to the tachocline, where $\alpha = 0$, and an α–effect in the lower part of the convection zone, where the toroidal field is small. Moreover the turbulent diffusivity responsible for allowing the spatially separated regions to communicate had a similar radial dependence to the α–effect, being large in the convection zone and small in the tachocline. His linear Cartesian model showed that dynamo waves could be maintained and he noted that the toroidal field was confined to a thin layer just below the interface. Parker also realised that the toroidal field strength was related to the ratio of the diffusivities in the tachocline and the convection zone. He proposed that the interface scenario would be consistent in the nonlinear regime if both the transport coefficients (α and η) were strongly quenched by the magnetic field.

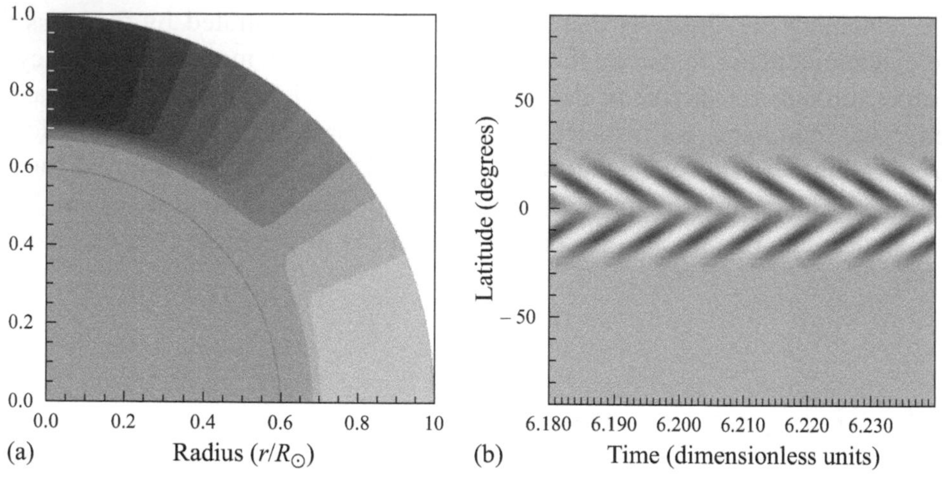

Figure 6.7 - (a) An analytical fit to the measured solar differential rotation, showing abrupt variation at the tachocline, near a radius of $0.7R_\odot$. (b) Cyclic toroidal fields at the base of the convection zone: a butterfly diagram for a nonlinear interface dynamo with the field limited by α–quenching, and an α–effect that is concentrated towards the equator (courtesy of P.J.Bushby). (**See colour insert**.)

Nonlinear solutions were later computed by Tobias (1997a) for a range of quenching mechanisms. The model was subsequently extended to describe global behaviour, subject to lateral boundary conditions corresponding to the poles and the equator – see Section 6.4.2 below (Tobias, 1996, 1997b).

6.4.1. SPHERICAL INTERFACE MODELS

The most straightforward way of representing the solar dynamo is to construct an analytical fit to the pattern of differential rotation determined by helioseismology, coupled with a suitably chosen form for $\alpha(r,\theta)$, which has to be antisymmetric about the equator, and then to adopt the α–quenching expression (6.4). This was done by Charbonneau & MacGregor (1996), who also demonstrated that it was possible to attain toroidal fields of equipartition strength in an interface configuration even if α was strongly quenched, using the expression (6.5). The same procedure has been followed by Markiel & Thomas (1999) and by Bushby (2004, 2005). Figure 6.7a shows the simplified form adopted for $\omega(r,\theta)$ in the latter calculation. The strong radial shear changes sign at mid-latitudes and butterfly diagrams show a poleward branch as a result. It follows that α has to be concentrated near the equator in order to produce a butterfly diagram with a dominant equatorward branch that resembles what is observed; Figure 6.7b shows an example calculated with $\alpha \propto \cos\theta \sin^4\theta$ (Bushby, 2004). This enhancement of the α–effect at low latitudes argues strongly

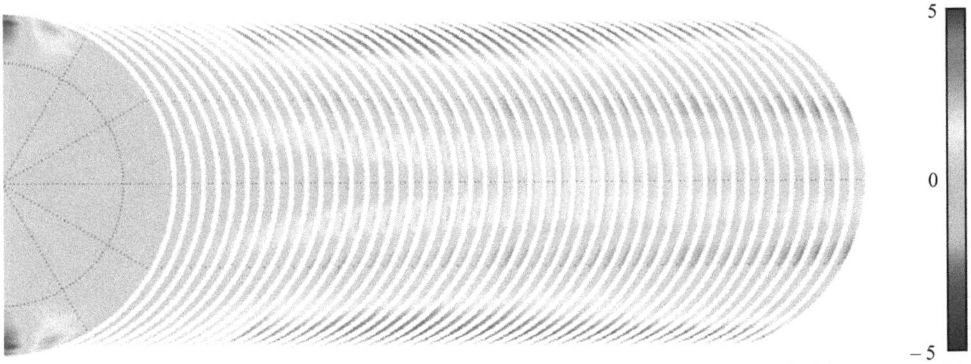

Figure 6.8 - Zonal shear flows extending throughout the convection zone. Results from the inversion of SOHO MDI data by Vorontsov *et al.* (2002), showing bands of slightly more rapid motion diverging from mid-latitudes as the activity cycle progresses (courtesy of S.V. Vorontsov). (**See colour insert.**)

in favour of magnetic buoyancy instabilities as its source. (The alternative of an equatorward meridional flow would damage the weaker poleward branch.)

Other nonlinear quenching mechanisms are able to yield qualitatively similar results. Microdynamic Λ–quenching has been successfully modelled both by Küker, Arlt & Rüdiger (1999) and by Pipin (1999). Similarly, the macrodynamic Malkus–Proctor effect has been used e.g. by Covas *et al.* (2000). The advantage of this last approach is that it automatically generates zonal flows that can be compared with those that are observed.

6.4.2. ZONAL SHEAR FLOWS

The "torsional oscillations" that have been directly measured at the solar surface (Ulrich *et al.*, 1988; Howe *et al.*, 2000a) can be followed deep into the convection zone by measuring the rotational splitting of p–mode frequencies (Vorontsov *et al.*, 2002; Thompson *et al.*, 2003). Figure 6.8 shows a ten–year dataset (Thompson *et al.*, 2003) with twin bands of marginally more rapid rotation accompanying the sunspot zones towards the equator. (The effect is small, around 0.01ω, but definitely present.) There is an even more prominent pair of bands that migrate simultaneously towards the poles. It should be emphasised that these zonal shear flows (Kosovichev & Schou, 1997) are a robust deduction from the helioseismic data: moreover, the zonal flows certainly extend through at least the outer third of the convection zone and probably penetrate to its base (Vorontsov *et al.*, 2002).

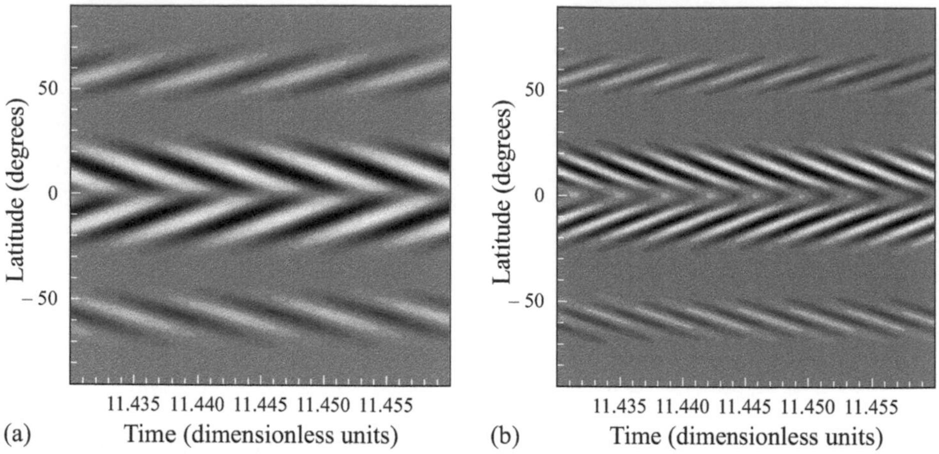

(a) (b)

Figure 6.9 - A nonlinear interface dynamo with activity limited by the macrodynamic Malkus–Proctor effect. (a) Butterfly diagram, showing poleward and equatorward branches. (b) Zonal shear flows at the base of the convection zone, again showing both branches (courtesy of P.J. Bushby). (**See colour insert**.)

The cyclic behaviour of the zonal shear flows indicates that they must be driven by the quadratic Lorentz force (Schüssler, 1981; Yoshimura, 1981). This can be modelled as a consequence of the macrodynamic Malkus–Proctor effect (Belvedere, Pidatella & Proctor, 1990; Covas *et al.*, 2000; Bushby, 2004). Figure 6.9a shows an idealised butterfly diagram with both a strong equatorial branch and a weaker polar branch. The corresponding zonal shear flows are displayed in Figure 6.9b (Bushby, 2004, 2005). Some similar calculations (Covas, Tavakol & Moss, 2001b; Tavakol *et al.*, 2002) have a richer structure, with phase shifts near the base of the convection zone ("spatiotemporal fragmentation") that have been tentatively linked to possible variations in ω with a shorter period ($1.3\,\mathrm{yr}$) at the base of the convection zone (Howe *et al.*, 2000b).

The obvious location of the magnetic couple that generates fluctuations in zonal angular momentum is at the tachocline, where strong toroidal fields are created. Yet the variations in ω are largest at the photosphere, where the "torsional oscillations" were originally detected. This can be explained as resulting from the large decrease in density with radius across the convection zone. Nonlinear dynamo models that include strong stratification (Covas, Moss & Tavakol, 2004) do indeed have zonal shear flows whose magnitude increases outwards towards the surface of the Sun (Bushby, 2004, 2005). Moreover, the zonal flows derived from helioseismic measurements show a definite lag in phase that *increases* with increasing radius (Vorontsov *et al.*, 2002). This confirms that these fluctuating flows are driven from below rather than from above.

6.5. MODULATION OF CYCLIC ACTIVITY

The butterfly diagram in Figure 6.3 is nearly symmetric about the equator but the sunspot record is definitely not periodic. The extended timeseries in Figure 6.10a (lower curve) shows the group sunspot number, a measure of solar activity established by Hoyt & Schatten (1998), running back from the present day to the first telescopic observations by Galileo and Scheiner in 1610. The most striking feature is the dearth of sunspots between 1645 and 1715 – the Maunder Minimum (Eddy, 1976). That this effect was real and not due to inadequate observations is amply confirmed by the contemporary records at the Paris Observatory (Ribes & Nesme-Ribes, 1993). Since then, the amplitude of the activity cycles has varied irregularly, with another marked dip around 1800 (the Dalton Minimum).

Fortunately, there are proxy data that extend the record back for many thousands of years. Galactic cosmic rays are deflected by magnetic fields in the solar wind and there is therefore a well-established anti-correlation between solar activity and the incidence of cosmic rays on the Earth's atmosphere. The cosmic rays lead to the production of cosmogenic isotopes such as ^{14}C (which is stored in tree rings after circulating for about 30 years in the atmosphere) and ^{10}Be (which descends in rain or snow and is preserved in polar icecaps). Because of its importance in age determination, the ^{14}C record has been closely studied and variations in abundance provide a smoothed record of solar activity extending for about 10,000 years into the past (Stuiver & Braziunas 1993; Stuiver *et al.*, 1998). The 11–year activity cycle shows up clearly in ^{10}Be abundances, which are anti-correlated with solar activity, as can be seen from Figure 6.10a; note that weaker cycles persisted throughout the Maunder Minimum (Beer, Tobias & Weiss, 1998). Longer term variability has been measured from abundances in polar ice cores dating back to 50,000 years BP, yielding the production rates in Figure 6.10b. The corresponding power spectrum, in Figure 6.10c, shows a sharp peak at a period of 205 yr (also present in the ^{14}C data) which almost certainly has a solar origin, as well as hints of longer periodicities (Wagner *et al.*, 2001). These records demonstrate that grand minima, of which the Maunder Minimum is the most recent example, are a regular feature of solar activity, associated with a characteristic timescale of about 200 years. Nor is the Sun unique in this respect: there are examples of almost identical stars, of which one is active while the other is quiescent, and it has been estimated that between 10% and 30% of all stars are undergoing a grand minimum at any time (Baliunas & Jastrow, 1990; Henry *et al.*, 1996).

Any theoretical explanation of this modulation must explain not only the occurrence of grand minima but also the persistence of the 205–year periodicity in the data (Tobias, 2002b). The simplest assumption is that this behaviour has a stochastic origin. For instance, the effect of large fluctuating fields may be to add stochastic

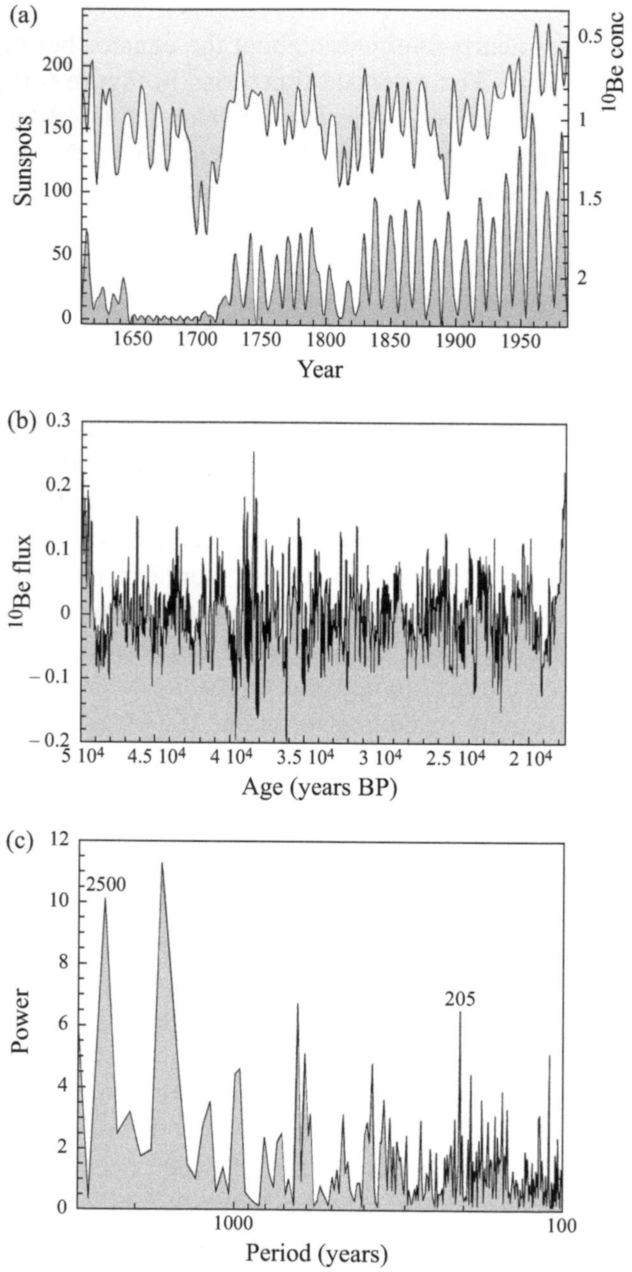

(a) The sunspot group number (lower curve) since the earliest telescopic observations, showing the Maunder Minimum (1645–1715), when scarcely any sunspots appeared, and the drop in activity around 1800 (the Dalton Minimum). The modulation appears to be chaotic. This is compared with the variation in the production of the cosmogenic isotope ^{10}Be: the upper curve shows a filtered record of ^{10}Be abundance in the Dye 3 ice core from Greenland (in units of $10,000\,\mathrm{atoms\,g^{-1}}$). The two records are anticorrelated and the 11–year activity cycle is apparent in both.

(b) The ^{10}Be flux data from the GRIP ice core for the interval from $50,000$ to $25,000\,\mathrm{yr}$ BP, band-pass filtered with a $100 - 3000\,\mathrm{yr}$ filter.

(c) Power spectrum (after Lomb–Scargle) for the unfiltered GRIP data (after Wagner *et al.*, 2001). The 205–year peak is strongly significant (courtesy of J. Beer).

Figure 6.10 - Modulation of the solar activity cycle.

fluctuations, which are indubitably present, to the α–effect (Hoyng, 1988; Ossendrijver & Hoyng, 1996) and these could reduce the dynamo number below its critical value so that magnetic activity is switched off. This presumes, of course, that the dynamo number is never far above the critical value, which seems unlikely for the Sun and even less likely for other stars. In another version, with a surface α–effect and a meridional flow, the conveyor belt is stochastically disturbed in order to produce modulation (Charbonneau & Dikpati, 2000). Schmitt, Schüssler & Ferriz-Mas (1996; see also Ossendrijver, 2000a) introduced a more sophisticated process: in their dynamo model, there are two contributions to the α–effect. The *dynamic* α–effect, which leads to cyclic activity, is produced by instabilities of toroidal flux tubes at the base of the convection zone, which only set in after the field exceeds some critical strength. If a combination of nonlinear and stochastic effects reduces the field strength below this critical value, the dynamo is switched off, and it remains quiescent until the *stochastic* α–effect lifts the poloidal and toroidal fields above the critical value once again. More generally, stochastic fluctuations are bound to become significant if the systematic fields fade away. Nevertheless, the weakness of these stochastic mechanisms is that they cannot readily explain the persistent 205–year frequency of modulation that is observed in the Sun.

6.5.1. DETERMINISTIC MODULATION

The alternative is that temporal modulation of cyclic activity is a natural feature of nonlinear dynamos. The claim here is that as the dynamo number is increased there are transitions, in a wide variety of nonlinear models, first to cyclic activity, then to periodically modulated (quasiperiodic) cycles and finally to chaotically modulated activity, and that this pattern is generic (Tobias, Weiss & Kirk, 1995). This bifurcation sequence is most conveniently modelled by including a dynamic coupling of the magnetic fields to the velocity through the macrodynamic Malkus–Proctor effect, thereby introducing an additional viscous timescale and increasing the order of the system. (This can be achieved in various other ways: for instance, Yoshimura (1978) introduced an explicit time-delay in order to obtain periodic and aperiodic modulation.) The simplest demonstration (Tobias 1996, 1997b) is in two-dimensional Cartesian geometry, with a shear velocity

$$v = V(z)\sin(\pi x/2L) + u(x,z,t)\,, \tag{6.6}$$

when the linear equations (6.1a,b) are replaced by the nonlinear system

$$\frac{\partial A}{\partial t} = \alpha(z)\cos\left(\frac{\pi x}{2L}\right) + \eta\Delta A\,, \tag{6.7a}$$

$$\frac{\partial B}{\partial t} = \mathrm{D}\left\{\left[V'\sin\left(\frac{\pi x}{2L}\right) + \frac{\partial u}{\partial z}\right]\frac{\partial A}{\partial x} - \left[\frac{\pi}{2L}V\cos\left(\frac{\pi x}{2L}\right) + \frac{\partial u}{\partial x}\right]\frac{\partial A}{\partial z}\right\} + \eta\Delta B\,, \tag{6.7b}$$

and
$$\frac{\partial u}{\partial t} = \mathrm{sgnD}\left(\frac{\partial B}{\partial z}\frac{\partial A}{\partial x} - \frac{\partial A}{\partial z}\frac{\partial B}{\partial x}\right) + \nu\Delta u\,. \tag{6.7c}$$

These equations are solved subject to boundary conditions $A = B = u = 0$ at $z = -1$, $\partial_z A = B = \partial_z u = 0$ at $z = 1$, while $A = B = u = 0$ at $x = 0,\ 2L$ (the poles). The variables A, B, u can then be expanded in Fourier sine series, which greatly facilitates computation. For an interface model, α drops smoothly from unity to zero as $z \downarrow 0$, while $V \to 0$ as $z \uparrow 0$.

Bifurcations from the trivial solution then give rise to two distinct families of solutions with different symmetries about the equator ($x = L$). For the dipole (antisymmetric) solutions, $\partial_x A = B = \partial_x u = 0$ at $x = L$; for the quadrupole (symmetric) solutions, $A = \partial_x B = \partial_x u = 0$ at $x = L$. For $\mathrm{D} < 0$, so that dynamo waves travel towards the equator, the initial bifurcation is to a dipole mode, closely followed by a quadrupole. A nonlinear solution will be confined to the dipole or quadrupole subspace unless its symmetry is broken at a bifurcation.

Tobias (1996) imposed dipole symmetry and showed that, as $|\mathrm{D}|$ was increased with $\mathrm{Pm} = \nu/\eta \ll 1$, the initial Hopf bifurcation (giving rise to periodic oscillations and trajectories that are attracted to limit cycles in the phase space of the system) was followed by a secondary Hopf bifurcation, leading to quasiperiodic behaviour (with periodically modulated cycles and trajectories that lie on a two-torus in the phase space). The modulation period apparently varies as $\mathrm{Pm}^{-1/2}$. If $|\mathrm{D}|$ is further increased the two-torus eventually gives way to a chaotic attractor. Frequency analysis of such a chaotic system can nevertheless pull out periodicities that correspond to the periods of the unstable periodic and quasiperiodic orbits that are embedded in the chaotic attractor. We conjecture that the 205–year periodicity in the ^{14}C and ^{10}Be records, which seem to be aperiodic, corresponds to a period of such a "ghost attractor".

Relaxing the symmetry constraint naturally allows a much richer variety of behaviour (Tobias, 1997b; Brooke, Moss & Phillips, 2002; Phillips, Brooke & Moss, 2002). It is convenient, following Brandenburg *et al.* (1989), to introduce the total magnetic energies E_d and E_q associated, respectively, with the dipole and quadrupole components of the toroidal field, and to define the parity $\mathcal{P} = (E_q - E_d)/(E_q + E_d)$. Thus $\mathcal{P} = 1$ for a pure quadrupole field and $\mathcal{P} = -1$ for a pure dipole. Tobias (1997b) once again explored behaviour as $|\mathrm{D}|$ was increased, with $\mathrm{Pm} = 0.1$; it is instructive to follow the bifurcation sequence, although the detailed pattern of transitions is clearly model-dependent. The initial Hopf bifurcation leads to periodic cycles with $\mathcal{P} = -1$ but stability is transferred to quadrupole cycles (with $\mathcal{P} = +1$) via an intermediate branch of quasiperiodic mixed-mode solutions (with $-1 < \mathcal{P} < 1$). The periodic quadrupole solutions then undergo a Hopf bifurcation giving rise to weakly modulated cycles; following a symmetry–breaking pitchfork bifurcation, stability is gained by a branch of mixed-mode quasiperiodic solutions and then transferred to quasiperiodic dipoles. The modulation becomes increasingly

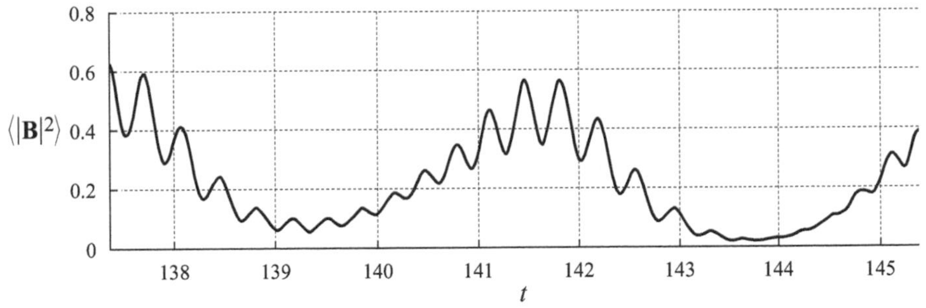

Figure 6.11 - Grand minima in a nonlinear $\alpha\omega$–dynamo model in Cartesian geometry. Variation of the toroidal magnetic energy $\langle|\mathbf{B}|^2\rangle$ with time, showing modulated cyclic activity, from a calculation with the macrodynamic Malkus–Proctor effect $(\mathrm{Pm} = 0.025,\ \mathrm{D} = -1100)$ and mild α–quenching (after Beer *et al.*, 1998).

prominent as $|\mathrm{D}|$ is further increased, until a new effect appears. Trajectories spend most of the time very close to the dipole subspace but in a grand minimum, when the field is very weak and E_d is extremely small, a small quadrupolar component causes $\mathcal{P}$ to deviate perceptibly from -1. This symmetry change becomes much more noticeable for yet higher values of $|\mathrm{D}|$.

Figure 6.11 shows the variation of the magnetic energy, $\langle|\mathbf{B}|^2\rangle = (E_q + E_d)$, for a short stretch of time including two grand minima, with parameters chosen to match the behaviour of the Sun, as revealed by the ^{10}Be measurements in Figure 6.10 (Beer *et al.*, 1998). The corresponding butterfly diagram is illustrated in Figure 6.12a: note that the field shows dipole symmetry when the cycles are most active but that this symmetry is distinctly broken as the dynamo emerges from a grand minimum, when all the activity is confined to one "hemisphere". That is precisely what happened at the end of the Maunder Minimum (Ribes & Nesme-Ribes, 1993). None of the dozen or so spots observed between 1680 and 1700 was in the northern hemisphere; when the first sunspot cycle reappeared in 1702, all the spots were in the southern hemisphere and it was only after 1714 that activity spread to both hemispheres as usual. This preference for hemispheric behaviour has been analysed by Bushby (2003a). A similar symmetry breaking is seen during reversals in geodynamo simulations (Glatzmaier & Roberts, 1995): when the field reverses the dipole field passes through zero and higher multipoles predominate.

Even more striking is the butterfly diagram, for the same parameter values, in Figure 6.12b. Here the cycles enter a deep grand minimum with parity $\mathcal{P} = -1$ but emerge from it with $\mathcal{P} = +1$. Apparently the dynamo can flip from dipole to quadrupole symmetry (and back again) during grand minima. Figure 6.13a illustrates this process by projecting the trajectory in phase space onto the three-dimensional space spanned by E_d, E_q and the mean square velocity $\langle|\mathbf{u}|^2\rangle$. Cycles,

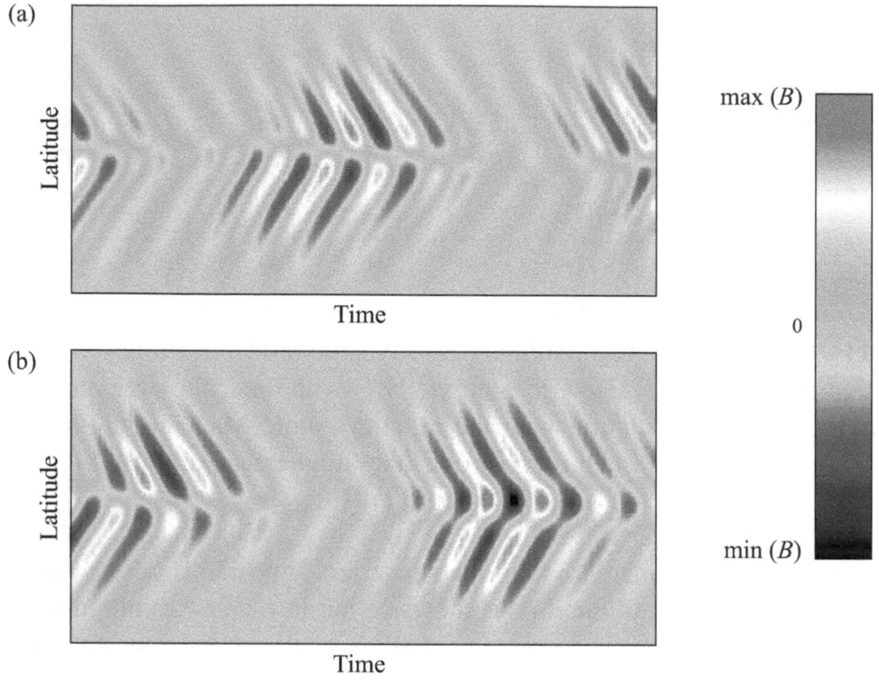

Figure 6.12 - Butterfly diagrams with grand minima, for the same parameters as Figure 6.11. (a) The interval covered in Figure 6.11, showing loss of dipole symmetry as the cycles emerge from grand minima. (b) A later sequence, in which the solution flips from dipole to quadrupole symmetry during a grand minimum (after Beer *et al.*, 1998). (**See colour insert.**)

modulation and flipping are all apparent in this phase portrait. What these results clearly show is that the parity of a stellar magnetic field is likely to change as the star evolves and spins down: the fact that the Sun has exhibited dipole symmetry for the past 300 years does not preclude its having had quadrupole or mixed symmetry in the past, or flipping symmetry in the future.

Although the bifurcation sequences appear most clearly in these Cartesian models, grand minima have also been found in various spherical dynamo calculations, with nonlinear growth limited by microdynamic (Λ–quenching) or macrodynamic (Malkus–Proctor) processes (Küker, *et al.*, 1999; Pipin, 1999; Kitchatinov *et al.*, 1999; Bushby, 2004). A qualitatively different type of modulation appears in models with algebraic or dynamic α–quenching, as well as in some with Λ–quenching, where there are major fluctuations in parity without grand minima in the magnetic energy $\langle|\mathbf{B}|^2\rangle$ (Brandenburg *et al.*, 1989; Kitchatinov, Rüdiger & Küker, 1994; Brooke *et al.*, 1998; Tworkowski *et al.*, 1998). Thus it is possible to distinguish two extreme forms of modulation: in Type 1 modulation there are parity changes

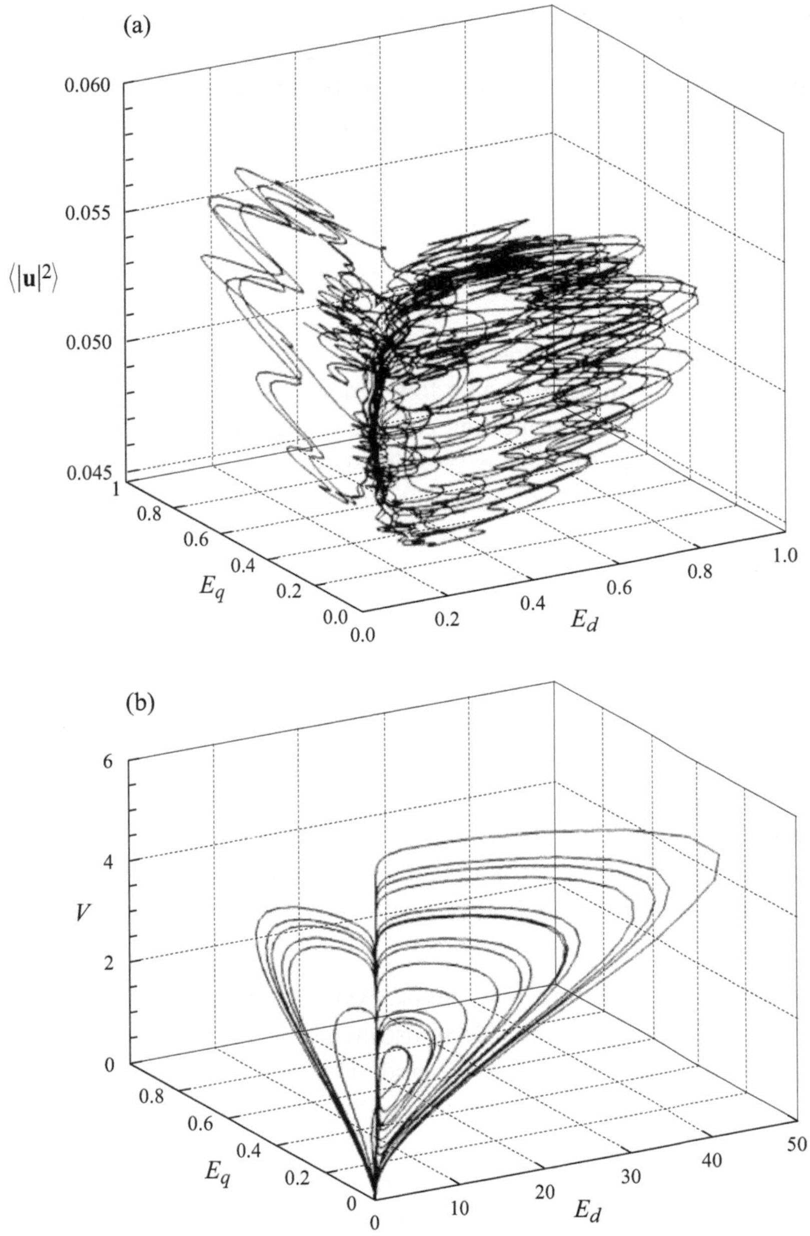

Figure 6.13 - Phase portraits showing flipping of symmetry during grand minima, with trajectories projected onto the three-dimensional space spanned by E_d, E_q and $\langle|\mathbf{u}|^2\rangle$. (a) For the PDEs, with the parameters used in Figures 6.11 and 6.12, showing cyclic behaviour, modulation and occasional flipping from the dipole to the quadrupole subspace. (b) A similar plot for the sixth-order ODE model. Here the basic cycle is filtered out, so modulation and flipping are more clearly shown (from Knobloch *et al.*, 1998).

without great changes in amplitude, and only modest variations in the mean square velocity perturbation $\langle|\mathbf{u}|^2\rangle$; by contrast, in Type 2 there are amplitude changes without great changes in parity, and correspondingly significant variations in $\langle|\mathbf{u}|^2\rangle$ (Tobias, 1997b; Knobloch *et al.*, 1998). This distinction is best clarified by considering low-order models.

6.5.2. LOW-ORDER MODELS

The bifurcation structures underlying Type 1 and Type 2 modulation can be established by constructing appropriate low-order systems of nonlinear ordinary differential equations, whose properties can then be explored in detail. In Cartesian geometry this is easily done by expressing solutions as truncated Fourier series. Thus Schmalz & Stix (1991) studied a simple one-dimensional model with dynamic α–quenching, where the lowest-order truncation yielded the familiar Lorenz system (see also Zeldovich *et al.*, 1983) while higher truncations introduced oscillatory, quasiperiodic and chaotic behaviour. The nonlinear development of one-dimensional dynamo waves, limited by the macrodynamic Malkus–Proctor effect, can be described by a complex generalisation of the Lorenz equations (Weiss *et al.*, 1984; Jones, Weiss & Cattaneo 1985; Feudel *et al.*, 1993). Solutions of this truncated model exhibit Type 2 modulation; moreover the same sequence of transitions, to periodic, quasiperiodic and chaotic modulation, also appears for solutions of the partial differential equations (PDEs) governing the corresponding two-dimensional problem (Tobias, 1997a). Nevertheless, these severely truncated models share a fundamental weakness, for the qualitative results are sensitive to the level of truncation (Covas *et al.*, 1997).

A more satisfactory alternative is to devise systems whose behaviour is generic and robust. Thus Type 2 modulation is demonstrated by a third-order model governed by the normal form equations for a saddle-node/Hopf bifurcation (Tobias *et al.*, 1995). Here all the hydrodynamics (convection and rotation) is collapsed onto the z–axis of cylindrical polar co-ordinates (s, ϕ, z), while the poloidal and toroidal magnetic fields are represented by x, y, where $s^2 = x^2 + y^2$. Two fixed points (one stable and the other unstable) appear on the invariant z–axis in a saddle-node bifurcation. The hydrodynamically stable fixed point undergoes a Hopf bifurcation, shedding a limit cycle corresponding to periodic magnetic activity, while the other fixed point remains magnetically stable. After removing a degeneracy (Guckenheimer & Holmes, 1986; Kirk, 1991, 1993), the normal form equations are

$$\dot{s} = \lambda s + a z s , \qquad \dot{\phi} = \omega , \tag{6.8a,b}$$

and

$$\dot{z} = \mu - z^2 - s^2 + b z^3 , \tag{6.8c}$$

where λ, μ are control parameters and ω, a, b are real constants ($a > 0$). Since this system is still axisymmetric it cannot exhibit chaotic behaviour. To break the axial symmetry, Tobias *et al.* (1995) replaced (6.8a,b) by

$$\dot{s} = \lambda s + a z s + c s^2 z \cos\phi, \qquad \text{and} \qquad \dot{\phi} = \omega - c s z \sin\phi. \qquad (6.9a,b)$$

As pointed out by Ashwin, Rucklidge & Sturman (2004), equations (6.9a,b) do not respect the symmetry $(x, y) \to (-x, -y)$ and it would be preferable to choose cubic terms that do; this omission does not, however, have any qualitative effect on the results (Wilmot-Smith *et al.*, 2005). Increasing the dynamo number corresponds to following an appropriate path in the (λ, μ)–plane. If $c = 0$ the primary Hopf bifurcation is followed by a secondary Hopf bifurcation leading to periodically modulated cycles, with quasiperiodic trajectories lying on a two-torus that encloses the unstable limit cycle. The torus swells until it is destroyed in a heteroclinic bifurcation, with an orbit linking the two fixed points. Setting $c \neq 0$ destroys the axial symmetry and allows chaotic modulation, associated with resonant tongues, horseshoes and a heteroclinic tangle. Figure 6.14 shows examples of trajectories in the quasiperiodic and chaotic regimes. Because these are normal form equations, we can expect such behaviour to be robust.

Knobloch & Landsberg (1996) adopted a similar approach for Type 1 modulation. Since the dipole and quadrupole fields appear in rapid succession and with similar frequencies, their interaction can be described by a normal form for the Hopf bifurcation with 1:1 resonance. If z_1, z_2 are the complex amplitudes of the two fields the relevant equations are

$$\dot{z}_1 = (\mu + \sigma + \mathrm{i}\,\omega_1)\, z_1 + a\,|z_1|^2\, z_1 + b\,|z_2|^2\, z_1 + c\, z_2^2\, \overline{z}_1, \qquad (6.10a)$$

$$\dot{z}_2 = (\mu + \mathrm{i}\,\omega_2) z_2 + a'\,|z_2|^2\, z_2 + b'\,|z_1|^2\, z_2 + c'\, z_1^2\, \overline{z}_2, \qquad (6.10b)$$

where the parameters μ, σ represent the dynamo number and the splitting, respectively, and $\omega_{1,2}$, a, b, c etc. are real constants. This system possesses solutions corresponding to pure dipole and quadrupole osillations, as well as periodic and quasiperiodic mixed-mode solutions. Knobloch *et al.* (1998) [see also Ashwin *et al.*(2004)] combined these equations with a reduced form of the saddle-node/Hopf equations, to obtain a sixth-order system that reproduces the key features of the PDEs, including both Type 1 and Type 2 modulation and also the interaction between the two, as well as flipping from dipole to quadrupole symmetry and vice versa. Figure 6.13b shows a phase portrait for this system that corresponds to that for the PDEs in Figure 6.13a. Since these forms of modulation appear in normal form equations we should expect to find them not only in PDEs but in real stars as well.

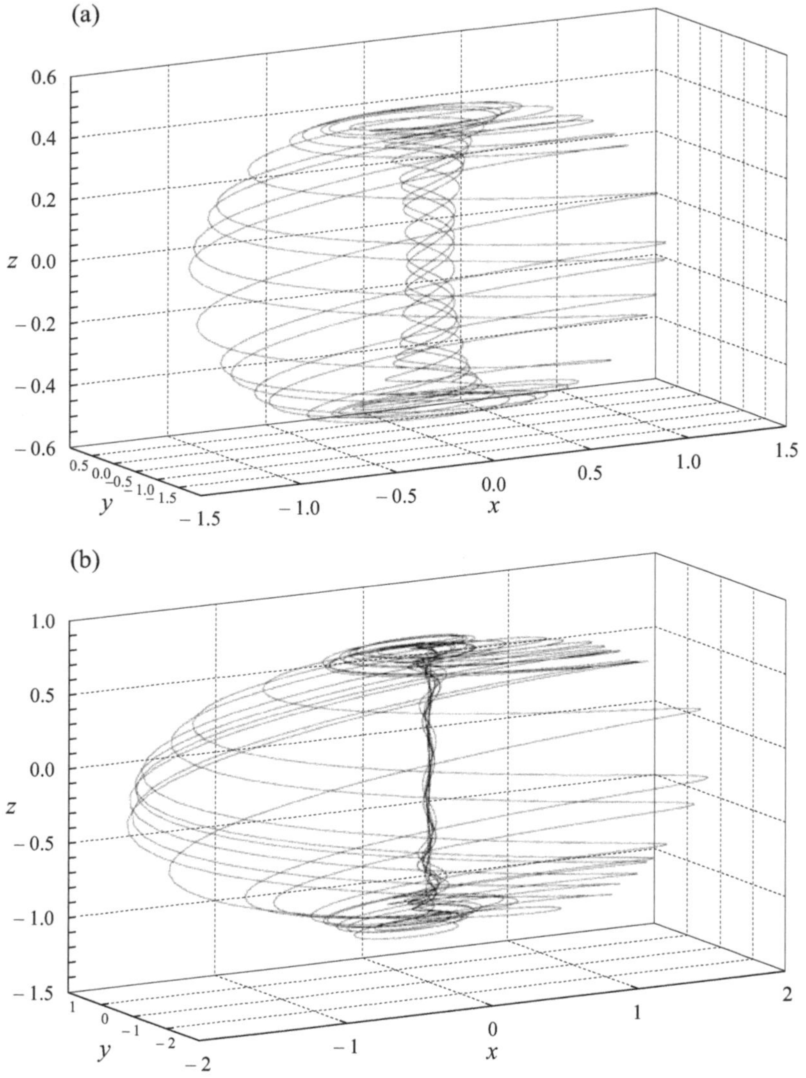

Figure 6.14 - Phase portraits for the third-order system (6.8c), (6.9a,b), which is the normal form for a saddle-node/Hopf bifurcation. (a) Quasiperiodic motion on a two-torus, corresponding to periodic modulation of cyclic behaviour: trajectories spiral out from the upper fixed point on the z–axis and move downwards till they converge towards the lower fixed point, before spiralling upwards around the invariant z–axis. (b) Chaotic modulation after the torus has been destroyed (from Weiss & Tobias, 2000).

6.5.3. ON–OFF AND IN–OUT INTERMITTENCY

From a more mathematical viewpoint, these different types of modulation could be regarded as examples of intermittency associated with the existence of invariant subspaces in the phase space of the system. If we restrict attention to idealised large-scale dynamo action (Tobias *et al.*, 1995) governed by nonlinear axisymmetric mean field dynamo equations, then the system possesses three invariant subspaces. These are the purely hydrodynamic subspace $\mathcal{M}_h$, the MHD dipole subspace $\mathcal{M}_d$ and the MHD quadrupole subspace $\mathcal{M}_q$.

For modulation of type 2 (as observed in the Sun) it suffices to restrict attention to orbits that are confined to $\mathcal{M}_h \cup \mathcal{M}_d$. One might for simplicity postulate that the dynamics in $\mathcal{M}_h$ is independent of the control parameters (such as the dynamo number) that determine behaviour in $\mathcal{M}_d$; in the simplest case dynamics in $\mathcal{M}_h$ is also unaffected by the magnetic field and we then say that the system has normal parameters and skew product structure. Suppose that behaviour in $\mathcal{M}_h$ is chaotic and that a trajectory within this subspace may sometimes be transversely stable and sometimes be transversely unstable. This gives rise to *on–off* intermittency deterministic rather than stochastic switching. Platt *et al.* (1993a,b) considered a simple example where dynamics in $\mathcal{M}_h$ was governed by the Lorenz equations in a chaotic regime, while dynamics in $\mathcal{M}_d$ was described by the normal form equation for a supercritical Hopf bifurcation with a control parameter that depended on the chaotic output from the Lorenz system. The resulting blowout bifurcations (Ott & Sommerer, 1994) led to bursts of cyclic activity in $\mathcal{M}_d$, interspersed with quiescent episodes as the trajectory approached closer to $\mathcal{M}_h$.

More generally, we should expect hydrodynamic behaviour to be influenced by the magnetic field and ω (and hence D) so that the system no longer has normal parameters or skew product structure. If there are two invariant sets in $\mathcal{M}_h$, one transversely stable but non-attracting in $\mathcal{M}_h$ and the other transversely unstable but attracting in $\mathcal{M}_h$, we speak of *in–out* intermittency (Ashwin, Covas & Tavakol, 1999; Covas *et al.*, 2001a). The normal form equations (6.10a,b) provide a simple example. Note, however, that the interesting dynamics of the modulated cycles discussed in the preceding subsections is determined by behaviour in $\mathcal{M}_d$; for suitably chosen parameter values, this can of course lead to intermittent bursts of activity.

Type 1 modulation corresponds to interactions between $\mathcal{M}_d$ and $\mathcal{M}_q$ only. The system then has non-normal parameters and non-skew symmetric structure, and behaviour in either subspace can be chaotic. In–out intermittency manifests itself as chaotic changes in parity as trajectories move towards and away from one or other subspace. With suitably chosen parameters, behaviour can become strikingly intermittent (Brooke *et al.*, 1998; Covas & Tavakol, 1999; Covas *et al.*, 2001a). However, the situation illustrated in Figure 6.13, where solutions flip from the neighbourhood of $\mathcal{M}_d$ to that of $\mathcal{M}_q$ as the trajectory approaches $\mathcal{M}_h$, clearly involves all three invariant subspaces.

6.6. RAPIDLY ROTATING STARS

Of around 100 nearby late-type stars whose Ca^+ H and K emission has been monitored for over 30 years (Baliunas *et al.*, 1995) there are about a dozen slow rotators that exhibit cycles similar to those observed in the Sun. The measured cycle frequency $\omega_{\rm cyc}$ increases with increasing rotation rate in a manner consistent with a power law of the form $\omega_{\rm cyc} \propto \omega^b$ for a star of given structure, where $1 < b \leq 2$ (Ossendrijver, 1997; Saar & Brandenburg, 1999). This is consistent with a variety of nonlinear dynamo models but unfortunately the observations cannot be used to discriminate between different quenching mechanisms (Tobias, 1998).

When a measure of chromospheric Ca^+ emission is plotted as a function of spectral type there appears to be a gap separating old, inactive stars from stars that are younger, rapidly rotating and more active (Vaughan & Preston, 1980; Henry *et al.*, 1996). Those stars above this Vaughan-Preston gap whose cycle periods can be measured again show a power law relationship, with a similar exponent but with $\omega_{\rm cyc}$ about ten times lower at a given rotation rate than it would be for less active stars below the gap (Saar & Brandenburg, 1999). This indicates that there is a fundamental difference between the dynamo processes in rapidly rotating stars and in the Sun. It is therefore dangerous to extrapolate to other stars from models that have been tuned to match solar observations.

Such a difference is indeed to be expected (Knobloch *et al.*, 1981). We know that in the solar convection zone the angular velocity ω varies with latitude rather than with radius but in a sufficiently rapid rotator Coriolis forces must predominate and the Proudman–Taylor theorem will lead to elongated convection cells and an angular velocity that tends to be constant on cylindrical surfaces (see Chapter 3). This change in the convection pattern has been demonstrated experimentally for electrostatically driven convection in a zero-gravity environment in space (Hart *et al.*, 1986) and the altered rotation pattern is a regular feature in numerical experiments (e.g. Rüdiger *et al.*, 1998; Brun & Toomre, 2002; Thompson *et al.*, 2003).

The most rapid rotators, with periods of a few days or less, have strong fields over a large fraction of their surfaces, produce vigorous X–ray emission from their coronae and exhibit photometric variability, associated with starspots, as they rotate. Over a longer interval they show aperiodic variability with characteristic timescales of several years. By means of Doppler imaging (Rice, 2002; Strassmeier, 2002) it is possible to determine the positions of starspots and to measure magnetic fields on these stars. Figure 6.15 shows the brightness distribution on the surface of one of the

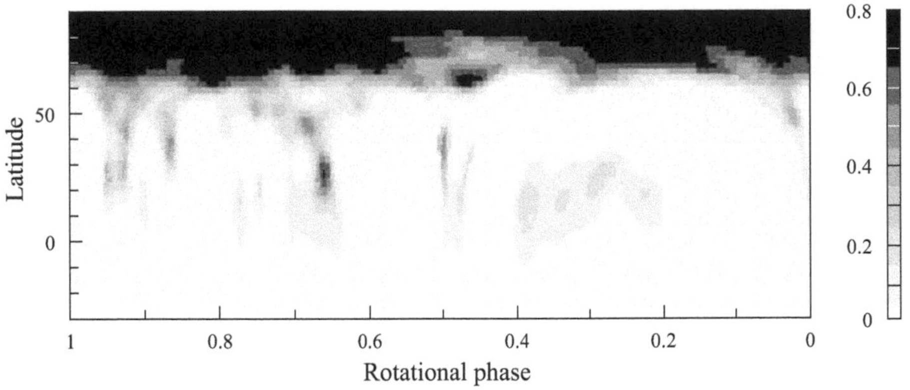

Figure 6.15 - Doppler imaging of the rapidly rotating star AB Doradus: the shading denotes the fractional spot occupancy at different positions on the stellar surface and shows a dark polar spot as well as non-axisymmetric patterns at lower latitudes (from Donati *et al.*, 1999).

best observed examples, AB Doradus, whose rotation period is only 0.5 days (Donati *et al.*, 1999). Like many similar stars it has a prominent polar spot. (The other pole is not visible but is expected also to be spotted.) There are also non-axisymmetric magnetic features at low latitudes. Differential rotation can be measured too: the absolute range of variation in ω is similar to that in the Sun but proportionately much less (so that $\Delta\omega/\omega$ is a decreasing function of ω).

This magnetic pattern is quite different from solar activity and several explanations have been proposed. Schüssler *et al.* (1996) considered the effect of a strong Coriolis force on the non-axisymmetric instability of an isolated flux tube in the deep convection zone, and argued that it would be deflected towards the poles. Schrijver & Title (2001) drew an analogy with the surface flux-transport model of the solar dynamo and suggested that in a much more active star magnetic flux would still be swept towards the poles, where it could accumulate to form polar spots. Bushby (2003b), on the other hand, focused on the stellar interior and on the changed pattern of differential rotation. He proposed the rotation profile illustrated in Figure 6.16a. Outside the tangent cylinder that encloses the radiative zone ω is constant on cylindrical surfaces, while the tachocline survives within the tangent cylinder and towards the poles. (This is reminiscent of rotation in the Earth's core.) With this profile fields can be generated by a distributed α–effect at low latitudes, while a vigorous interface dynamo operates at high latitudes, producing dynamo waves that propagate towards the poles, as illustrated in Figure 6.16b.

Models of lower main-sequence stars show that as the stellar mass decreases the radiative core shrinks until stars eventually become fully convective. Although there is no scope for an interface dynamo in these low-mass M stars, they still remain

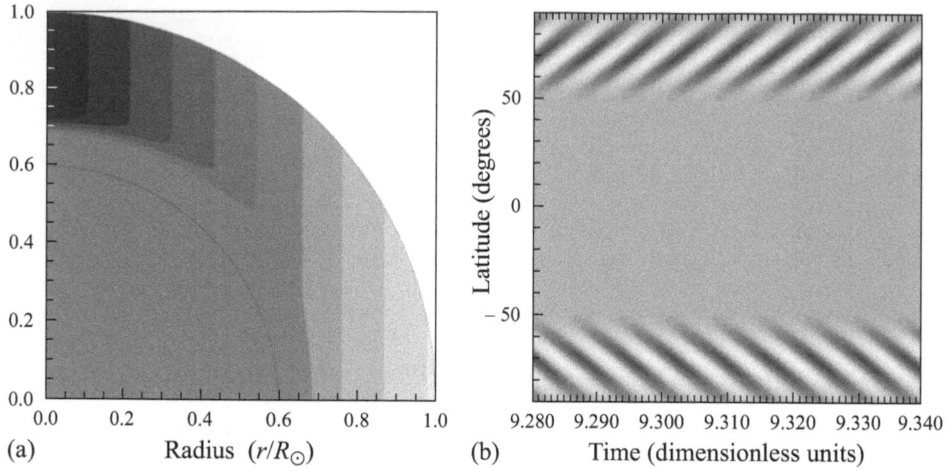

Figure 6.16 - (a) Conjectured variation of angular velocity in the convection zone of a rapidly rotating star, with behaviour dominated by the Proudman–Taylor theorem. Outside the tangent cylinder ω is constant on cylindrical surfaces but there is still a tachocline near the poles. (b) Corresponding butterfly diagram, with strong cyclic activity at high latitudes that could give rise to polar spots (courtesy of P.J. Bushby). (**See colour insert**.)

extremely active. Doppler imaging of rapidly rotating M dwarfs reveals dark patterns at low and moderate latitudes but no polar spots (Barnes *et al.*, 2001). The corresponding magnetic field is apparently generated by a vigorous $\alpha^2\omega$–dynamo operating within the convection zone. The dynamo process in these stars is clearly quite different from that which leads to cyclic activity on the Sun.

Starspots have also been observed on T Tauri stars, which are fully convective but have not yet evolved onto the main sequence. These stars may be surrounded by Keplerian accretion discs, within which planets can be formed. A field-free disc is dynamically stable and incapable of acting as a kinematic dynamo. However, a finite initial field (whether axial or azimuthal) leads to the development of a magneto–rotational instability (Balbus & Hawley, 1998). This allows a novel type of bootstrapping dynamo, for the nonlinear development of the magneto–rotational instability can act as a turbulent dynamo to maintain the field that facilitates the instability (Brandenburg *et al.*, 1995; Hawley, Gammie & Balbus, 1996). Since this process requires the presence of an initial field its bifurcation structure may resemble that for Poiseuille flow in a pipe, perhaps with a subcritical bifurcation at infinity.

6.7. THE FUTURE

We have perforce had to concentrate our discussion on mean field dynamo models. Although their limitations are extremely obvious, they do seem able to explain the broad features of magnetic activity in stars like the Sun. Two different approaches can be distinguished. The first, which we have emphasised here, aims to construct *illustrative* models, designed to probe particular aspects of the overall problem and to demonstrate specific types of qualitative behaviour. The Cartesian and spherical models described in Sections 6.4 and 6.5, as well as the low-order systems of Section 6.5.2, all belong to this category. It is also possible to produce *imitative* models, where the free parameters are tuned so as to reproduce the observed behaviour of the solar cycle (as currently perceived). Such models inevitably require frequent revision. There is no clear procedure for calculating the free input parameters and their values will not be constrained until MHD turbulence is properly understood. Unless these parameters are predetermined it will always be possible to produce models that imitate the solar cycle – although it is by no means clear that such an enterprise can provide a deeper understanding of the underlying physics. It must also be stressed that none of these models, whether illustrative or imitative, have any long-term predictive power, for they are all too primitive – and, moreover, the solar cycle appears to be chaotic.

The time has now come for mean field dynamos to be superseded by properly self-consistent nonlinear computations. Looking ahead, the first need is to isolate the key problems that need to be tackled before embarking on a full calculation. So far as the Sun is concerned, there are three issues to be settled. First of all, what is the origin of the tachocline and how is it maintained? Are magnetic fields inextricably bound up with the tachocline's internal circulation, as suggested by Gough & McIntyre (1998)? Next, what is the structure of convection below the visible surface and deep down in the convection zone? And, finally, how does this convection drive the pattern of differential rotation that helioseismology has revealed? Once these questions have been satisfactorily answered, we shall be ready to tackle the magnetohydrodynamic dynamo itself.

CHAPTER 7

GALACTIC DYNAMOS

Anvar Shukurov

7.1. INTRODUCTION

Galaxies are attractive objects to study. The magnetism of their natural beauty adds to the fascinating diversity of physical processes that occur over an enormous range of scales from the global dimension of order $10\,\mathrm{kpc}$[12] down to the viscous turbulent scales of 1000 km and less. The visual image of a galaxy (see Figure 7.1) is dominated by the optical light mostly produced by stars that contribute most of the visible galactic mass ($2 \times 10^{11} M_\odot$ for the Milky Way, where $M_\odot = 2 \times 10^{30}$ kg is the mass of the Sun). A few percent of the galactic mass is due to the interstellar gas that resides in the gravitational field produced by stars and dark matter. Spiral galaxies are flat (Figure 7.1) because the stars and gas rapidly rotate. The gas is ionised by the UV and X-ray radiation and by cosmic rays; the degree of ionization of diffuse gas ranges from 30% to 100% in various phases – see Section 7.2.1. Interstellar gas is involved in turbulent motions that can be detected because the associated Doppler shifts broaden spectral lines emitted by the gas beyond their width expected from thermal motions alone. The effective mean free path of interstellar gas particles is small enough to justify a fluid description under a broad range of conditions. Altogether, interstellar gas can be reasonably described as an electrically conducting, rotating, stratified turbulent fluid – and thus a site of MHD processes discussed elsewhere in this volume, including various types of dynamo action.

12 A length unit appropriate to galaxies is $1\,\mathrm{kpc} \approx 3.1 \times 10^{19}\,\mathrm{m} \approx 3262$ light years. The distance of the Sun from the centre of the Milky Way is $s_\odot \approx 8.5\,\mathrm{kpc}$.

Figure 7.1 - Optical images of two nearby spiral galaxies. (a) M51, the Whirlpool galaxy (with a satellite galaxy at the top). (b) NGC 891 (Both courtesy of the Canada–France–Hawaii Telescope/J.-C. Cuillandre/Coelum). M51 is one of nearby galaxies (distance 9.6 Mpc) notable for its prominent spiral pattern. M51 is the first external galaxy where a well ordered, large-scale magnetic field was detected (Segalovitz *et al.*, 1976) and studied in fine detail. NGC 891 is at about the same distance as M51, but seen nearly edge-on, so the thinness of the galactic disc is evident. The dark strip along the galactic disc and filaments extended away from the galactic plane are due to obscuration by interstellar dust. The filaments trace gas outflow from the disc into the halo. (**See colour insert**.)

The energy density of interstellar magnetic fields is observed to be comparable to the kinetic energy density of interstellar turbulence and cosmic ray energy density, and apparently exceeds the thermal energy density of interstellar gas (Cox, 1990). Therefore, interstellar gas, magnetic field and cosmic rays form a complex, nonlinear physical systems whose behaviour is equally affected by each of the three components. The system is so complex that magnetic fields and cosmic rays – the components that are more difficult to observe and model – are often neglected. Such a simplification is perhaps justifiable at very large scales of order $10\,\mathrm{kpc}$, where the motions of interstellar gas (mainly the overall rotation) are governed by gravity: systematic motions at a speed in excess of $10\text{–}30\,\mathrm{km\,s^{-1}}$ are too strong to be affected by interstellar magnetic fields. However, motions at smaller scales (comparable to and less than the turbulent scale, $\ell \approx 0.1\,\mathrm{kpc}$) are strongly influenced by magnetic fields. In particular, interstellar turbulence is in fact an MHD turbulence. In this respect, the interstellar environment does not differ much from stellar and planetary interiors.

Until recently, interstellar magnetic fields had been a rather isolated area of galactic astrophysics. The reason for that was twofold. Firstly, magnetic fields are difficult to observe and model. Secondly, they were understood too poorly to provide useful insight into the physics of interstellar gas and galaxies in general. The widespread attitude of galactic astrophysicists to interstellar magnetic fields was succinctly described by Woltjer (1967):

> The argument in the past has frequently been a process of elimination: one observed certain phenomena, and one investigated what part of the phenomena could be explained; then the unexplained part was taken to show the effects of the magnetic field. It is clear in this case that, the larger one's ignorance, the stronger the magnetic field.

The attitude hardly changed in 20 subsequent years, when Cox (1990) observed that

> As usual in astrophysics, the way out of a difficulty is to invoke the poorly understood magnetic field. ...One tends to ignore the field so long as one can get away with it.

The situation has changed dramatically over the last 10–15 years. Theory and observations of galactic magnetic fields are now advanced enough to provide useful constraints on the kinematics and dynamics of interstellar gas, and the importance and role of galactic magnetic fields are better appreciated.

In this chapter, we review in Section 7.2 those aspects of galactic astrophysics that are relevant to magnetic fields, and briefly summarise in Section 7.3 our observational knowledge of magnetic fields in spiral galaxies. Section 7.4 is an exposition of the current ideas on the origin of galactic magnetic fields, including the dynamo theory. The confrontation of theory with observations is the subject of Section 7.5 where we summarise the advantages and difficulties of various theories and argue that the mean-field dynamo theory remains the best contender. Magnetic fields in elliptical galaxies are briefly discussed in Section 7.6.

7.2. INTERSTELLAR MEDIUM IN SPIRAL GALAXIES

7.2.1. TURBULENCE AND MULTI-PHASE STRUCTURE

The interstellar medium (ISM) is much more inhomogeneous and active than stellar and planetary interiors. The reason for that is ongoing star formation where massive young stars evolve rapidly (in about 10^6 yr) and then explode as supernova stars (SN) releasing large amounts of energy ($E_{\rm SN} \approx 10^{51}$ erg $\approx 10^{44}$ J per event). These explosions control the structure of the ISM.

SN remnants are filled with hot, overpressured gas and first expand supersonically; at this stage the gas surrounding the blast wave is not perturbed. However, a pressure disturbance starts propagating faster than the SN shell as soon as the expansion velocity becomes comparable to or lower than the speed of sound in the surrounding gas – at this stage the expanding SN remnant drives motions in the surrounding gas, and its energy is partially converted into the kinetic energy of the ISM. When pressure inside an SN remnant reduces to values comparable to that in the surrounding gas, the remnant disintegrates and merges with the ISM. Since SN occur at (almost) random times and positions, the result is a random force that drives random motions in the ISM that eventually become turbulent. The size of an SN remnant when it has reached pressure balance determines the energy-range turbulent scale,

$$\ell \approx 0.05\text{–}0.1\,\mathrm{kpc}\,.$$

A useful review of supernova dynamics can be found, e.g. in Lozinskaya (1992), and the spectral properties of interstellar turbulence are discussed by Armstrong *et al.* (1995). Among numerous reviews of the multi-phase ISM we mention that of Cox (1990) and a recent text of Dopita & Sutherland (2003).

About $f = 0.07$ of the SN energy is converted into the ISM's kinetic energy. With the SN frequency of $\nu_{\rm SN} \approx (30\,\mathrm{yr})^{-1}$ in the Milky Way (i.e. one SN per 30 yr), the kinetic energy supply rate per unit mass is $\dot{e}_{\rm SN} = f\nu_{\rm SM}E_{\rm SN}M_{\rm gas}^{-1} \approx 10^{-2}\,\mathrm{erg\,g^{-1}\,s^{-1}} \approx 10^{-6}\,\mathrm{J\,kg^{-1}\,s^{-1}}$, where $M_{\rm gas} \approx 4\times10^{9}\,M_\odot \approx 8\times10^{39}$ kg is the total mass of gas in the galaxy. This energy supply can drive turbulent motions at a speed u_0 such that $2u_0^3/\ell = \dot{e}_{\rm SN}$ (where the factor 2 allows for equal contributions of kinetic and magnetic turbulent energies), which yields

$$u_0 \approx 10\text{–}30\,\mathrm{km\,s^{-1}}\,,$$

a value similar to the speed of sound at a temperature $T = 10^4$ K or higher. The corresponding turbulent diffusivity follows as

$$\eta_{\rm T} \approx \tfrac{1}{3}\ell u_0 \approx (0.5\text{–}3)\times10^{22}\,\mathrm{m^2\,s^{-1}}\,. \qquad (7.1)$$

Table V - The multi-phase ISM. The origin and parameters of the most important phases of interstellar gas: n, the mid-plane number density in hydrogen atoms per cm^3; T, the temperature in K; c_s, the speed of sound in $\mathrm{km\,s^{-1}}$; h, the scale height in kpc; and f_V, the volume filling factor in the disc of the Milky Way, in per cent.

Phase	Origin	n	T	c_s	h	f_V
Warm		0.1	10^4	10	0.5	60 – 80
Hot	Supernovae	10^{-3}	10^6	100	3	20 – 40
Hydrogen clouds	Compression	20	10^2	1	0.1	2
Molecular clouds	Self-gravity, thermal instability	10^3	10	0.3	0.075	0.1

Supernovae are the main source of turbulence in the ISM. Stellar winds is another significant source, contributing about 25% of the total energy supply (e.g. Section VI.3 in Ruzmaikin *et al.*, 1988).

The time interval between supernova shocks passing through a given point is about (McKee & Ostriker, 1977; Cox, 1990)

$$\tau = (0.5\text{–}5) \times 10^6\,\mathrm{yr}\,.$$

After this period of time, the velocity field at a given position completely renovates to become independent of its previous form. Therefore, this time can be identified with the correlation time of interstellar turbulence. The renovation time is 2–20 times shorter than the "eddy turnover" time $\ell/u_0 \approx 10^7\,\mathrm{yr}$. This means that the short-correlated (or δ–correlated) approximation, so important in turbulence and dynamo theory (e.g. Zeldovich *et al.*, 1990; Brandenburg & Subramanian, 2005), can be quite accurate in application to the ISM – this is a unique feature of interstellar turbulence. Note that the standard estimate (7.1) is valid if the correlation time is ℓ/u_0. If the renovation time was used instead, the result would be $\eta_\mathrm{T} \approx \ell^2/\tau \approx 10^{23}\,\mathrm{m^2\,s^{-1}}$, a value an order of magnitude larger than the standard estimate.

Another important result of supernova activity is a large amount of gas heated to $T = 10^6\,\mathrm{K}$ (Figure 7.2). The gas is so tenuous that the collision rate of the gas particles is low, and so its radiative cooling time is very long and exceeds τ: the hot bubbles produced by supernovae can merge before they cool (Figure 7.2). A result is a network of hot tunnels that form the hot component of the ISM. Altogether, the interstellar gas is found in several distinct states, known as "phases" (this usage may be misleading as most of them are not proper thermodynamic phases) whose parameters are presented in Table V. Some of the parameters (especially the volume filling factors) are not known confidently, so estimates of Table V should be approached

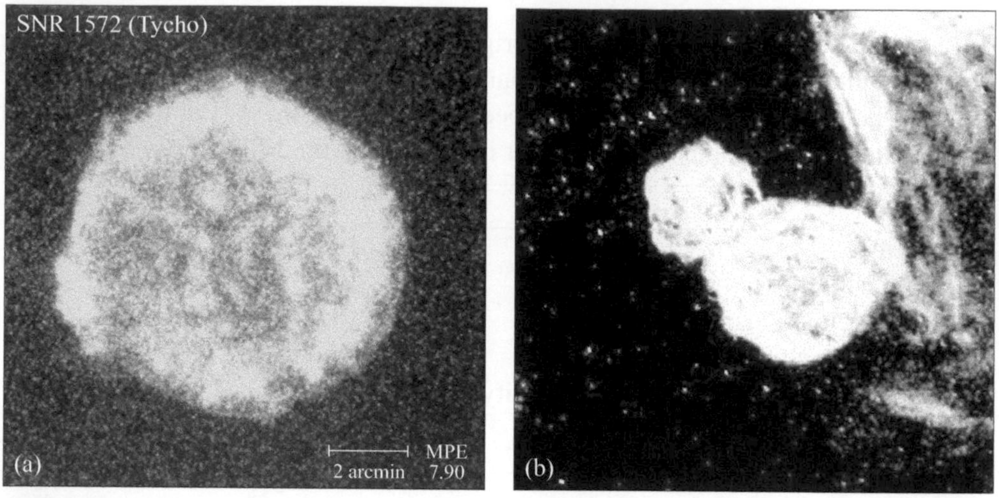

Figure 7.2 - SN remnants are expanding bubbles of hot gas that emits thermal X-rays. (a) This is illustrated by the X-ray image of Tycho's supernova remnant (courtesy of the ROSAT Mission and the Max-Planck-Institut für extraterrestrische Physik) whose parent star's explosion in 1572 was recorded by the famous Danish astronomer Tycho Brahe. The hot gas cools only slowly, and SN remnants often merge. (b) False-colour optical (Hα) image of two SN remnants DEM L316 in the Large Magellanic Cloud which appear to be colliding (Williams *et al.*, 1997; image produced by the Magellanic Cloud Emission-Line Survey, reprinted with permission). (**See colour insert**.)

with healthy skepticism. The warm diffuse gas can be considered as a background against which the ISM dynamics evolves; this is the primary phase that occupies a connected (percolating) region in the disc, whereas the hot gas may or may not fill a connected region. The warm gas is ionised by the stellar ultraviolet radiation and cosmic rays; its degree of ionization is about 30% at the Galactic midplane. The hot gas is so hot that it is fully ionised by gas particle collisions.

The locations of SN stars are not entirely random: 70% of them cluster in regions of intense star formation (known as OB associations as they contain large numbers of young, bright stars of spectral classes O and B) where gas density is larger than on average in the galaxy. Collective energy input from a few tens (typically, 50) SN within a region about $0.5–1\,\mathrm{kpc}$ in size produces a superbubble that can break through the galactic disc (Tenorio-Tagle & Bodenheimer, 1988). This removes the hot gas into the galactic halo and significantly reduces its filling factor in the disc (from about 70% to 10–20%). This also gives rise to a systematic outflow of the hot gas to large heights where the gas eventually cools, condenses and returns to the disc after about $10^9\,\mathrm{yr}$ in the form of cold, dense clouds of neutral hydrogen (Wakker & van Woerden, 1997). This convection-type flow is known as the galactic fountain

(Shapiro & Field, 1976), and it can plausibly support a mean-field dynamo of its own (Sokoloff & Shukurov, 1990). Another aspect of its role in galactic dynamos is discussed in Section 7.4.3. The vertical velocity of the hot gas at the base of the fountain flow is 100–200 $\mathrm{km\,s^{-1}}$ (e.g. Kahn & Brett, 1993; Korpi *et al.*, 1999a,b).

7.2.2. GALACTIC ROTATION

Spiral galaxies have conspicuous flat components because they rotate rapidly enough. The Sun moves in the Milky Way at a velocity of about $u_\odot = s_\odot \Omega_\odot = 220\,\mathrm{km\,s^{-1}}$, to complete one orbit of a radius $s_\odot \approx 8.5\,\mathrm{kpc}$ in $2\pi/\Omega_\odot = 2.4 \times 10^8\,\mathrm{yr}$. These values are representative for spiral galaxies in general. The Rossby number is estimated as

$$\mathrm{Ro} = \frac{u_0}{\ell \Omega_\odot} \approx 4\,.$$

The vertical distribution of the gas is controlled, to the first approximation, by hydrostatic equilibrium in the gravity field produced by stars and dark matter, with pressure comprising thermal, turbulent, magnetic and cosmic ray components in roughly equal proportion (e.g. Boulares & Cox, 1990; Fletcher & Shukurov, 2001). The semi-thickness of the warm gas layer is about $h = 0.5\,\mathrm{kpc}$, i.e. the aspect ratio of the gas disc is

$$\varepsilon = \frac{h}{s_\odot} \approx 0.06\,. \tag{7.2}$$

Since the gravity force decreases with radius s together with the stellar mass density, h grows with s at $s \gtrsim 10\,\mathrm{kpc}$ (see Section VI.2 in Ruzmaikin *et al.*, 1988, for a review).

However, the hot gas has larger speed of sound and turbulent velocity, and its Rossby number can be as large as 10 given that its turbulent scale is about $0.3\,\mathrm{kpc}$ (see Poezd *et al.*, 1993). Hence, the hot gas fills a quasi-spherical volume, where its pressure scale height of order $5\,\mathrm{kpc}$ is comparable to the disc radius.

$\mathrm{Ro} = 1$ at a scale $0.4\,\mathrm{kpc}$ in the warm gas, which is similar to the scale height of the gas layer. This implies that rotation significantly affects turbulent gas motions, making them helical on average. A convenient estimate of the associated α–effect can be obtained from F. Krause's formula,

$$\alpha_0 \approx \frac{\ell^2 \Omega}{h} \approx 0.5\,\mathrm{km\,s^{-1}}\,, \tag{7.3}$$

where Ω is the angular velocity, and the numerical estimate refers to the Solar neighbourhood of the Milky Way. Thus, $\alpha_0 \approx 0.05\,u_0$ near the Sun and increases in the inner Galaxy together with Ω.

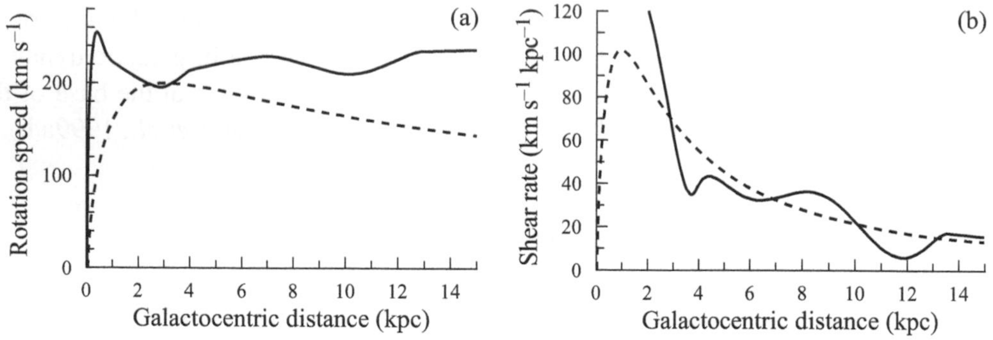

Figure 7.3 - (a) The rotation speed $s\Omega(s)$ in the galactic midplane versus galactocentric radius s in the Milky Way (solid) (Clemens, 1985), and the generic Schmidt's rotation curve with $U_0 = 200\,\mathrm{km\,s^{-1}}$, $s_0 = 3\,\mathrm{kpc}$ and $n = 1$ (dashed). (b) The corresponding rotation shear rates (taken with minus sign), $-s\partial\Omega/\partial s$.

The spatial distribution of galactic rotation is known for thousands galaxies (Sofue & Rubin, 2001) from systematic Doppler shifts of various spectral lines emitted by stars and gas. In this respect, galaxies are much better explored than any star or planet (including the Sun and the Earth) where reliable data on the angular velocity in the interior are much less detailed and reliable or even unavailable. The radial profile of the galactic rotational velocity is called the rotation curve. Rotation curves of most galaxies are flat beyond a certain distance from the axis, so $\Omega \propto s^{-1}$ is a good approximation for $s \gtrsim 5\,\mathrm{kpc}$. The rotation curve of a generic galaxy, known as the Schmidt rotation curve and shown in Figure 7.3, has the form

$$s\Omega(s) = U_0 \frac{s}{s_0} \left[\frac{1}{3} + \frac{2}{3} \left(\frac{s}{s_0} \right)^n \right]^{-3/2n},$$

where the parameters vary between various galaxies in the range $s_0 \approx 5$–$20\,\mathrm{kpc}$, $U_0 \approx 200\,\mathrm{km\,s^{-1}}$ and $n \approx 0.7$–1. This rotation curve is not flat at large radii, but it provides an acceptable approximation at moderate distances from galactic centre where magnetic field generation is most intense. Some galaxies have more complicated rotation curves. Notably, the Milky Way and M31 are among them – see Figures 7.3 and 7.5. The complexity of the rotation curves is explained by a complicated distribution of the gravitating (stellar and dark) mass in those galaxies. It is evident from Figure 7.3b that the rotation shear is strong at all radii even for the Schmidt rotation curve, and so the rotation in the inner part of a spiral galaxy cannot be approximated by the solid-body law, even if the shape of some rotation curves tempts to do so.

The vertical variation of the rotation velocity is only poorly known. In a uniform gravitating disc of infinite radial extent the angular velocity of rotation would be constant in z. Then it is natural to expect that Ω should decrease along z at a scale comparable to the radial scale length of the gravitating mass in the disc, typically $s_* = 3$–$5\,\mathrm{kpc}$. Recent observations of gas motions in galactic halos have confirmed such a decrease (Fraternali *et al.*, 2003). In the absence of detailed models, an approximation $\Omega \propto \exp(-z/s_*)$ seems to be appropriate.

7.3. MAGNETIC FIELDS OBSERVED IN GALAXIES

Estimates of magnetic field strength in the diffuse interstellar medium of the Milky Way and other galaxies are most efficiently obtained from the intensity and Faraday rotation of synchrotron emission. Other methods are only sensitive to relatively strong magnetic fields that occur in dense clouds (Zeeman splitting) or are difficult to quantify (optical polarisation of star light by dust grains). The total I and polarised P synchrotron intensities and the Faraday rotation measure RM are weighted integrals of magnetic field over the path length L from the source to the observer, so they provide a measure of the average magnetic field in the emitting or magneto–active volume:

$$I = K\int_L n_{\mathrm{cr}} B_\perp^2\,\mathrm{d}s\,, \qquad P = K\int_L n_{\mathrm{cr}} \overline{B}_\perp^2\,\mathrm{d}s\,, \tag{7.4a,b}$$

$$\mathrm{RM} = K_1\int_L n_{\mathrm{e}} B_\parallel\,\mathrm{d}s\,, \tag{7.4c}$$

where n_{cr} and n_{e} are the number densities of relativistic and thermal electrons, $\mathbf{B}$ is the total magnetic field comprising a regular $\overline{\mathbf{B}}$ and random $\mathbf{b}$ parts, $\mathbf{B} = \overline{\mathbf{B}} + \mathbf{b}$ with $\langle \mathbf{B} \rangle = \overline{\mathbf{B}}$, $\langle \mathbf{b} \rangle = 0$ and $\langle B^2 \rangle = B^2 + \langle b^2 \rangle$, angular brackets denote averaging, subscripts $\perp$ and $\parallel$ refer to magnetic field components perpendicular and parallel to the line of sight, and K and $K_1 = e^3/(2\pi\,\mathrm{m}_{\mathrm{e}}^2 c^4) = 0.81\,\mathrm{rad\,m^{-2}\,cm^3\,\mu G^{-1}\,pc^{-1}}$ are certain dimensional constants (with e amd m_{e} the electron charge and mass and c the speed of light). The power of $B_\perp$ and $\overline{B}_\perp$ in (7.4a,b) in fact depends on the energy spectral index $-q$ of cosmic ray electrons, being equal to $(q+1)/2$. With the observed value $q \simeq 3$, we have $(q+1)/2 \simeq 2$. The degree of polarisation p is related to the ratio $\langle b^2 \rangle / \overline{B}^2$:

$$p \equiv \frac{P}{I} \approx p_0 \frac{\overline{B}_\perp^2}{B_\perp^2} = p_0 \frac{\overline{B}_\perp^2}{\overline{B}_\perp^2 + \frac{2}{3}\langle b^2 \rangle}\,, \tag{7.5}$$

where the random field $\mathbf{b}$ has been assumed to be isotropic in the last equality, n_{cr} is assumed to be a constant, and $p_0 \approx 0.75$ weakly depends on the spectral index of the emission. This widely used relation is only approximate. In particular, it does not allow for any anisotropy of the random magnetic field, for the dependence of n_{cr}

on B, and for depolarisation effects; some generalisations are discussed by Sokoloff *et al.* (1998).

The orientation of the apparent large-scale magnetic field in the sky plane is given by the observed B-vector of the polarised synchrotron emission. Due to Faraday rotation, the true orientation can differ by an angle of $\mathrm{RM}\lambda^2$, which amounts to 10°–20° at a wavelength $\lambda = 6\,\mathrm{cm}$. The special importance of the Faraday rotation measure, RM, is that this observable is sensitive to the direction of $\mathbf{B}$ (the sign of $\overline{B}_\parallel$) and this allows one to determine not only the orientation of $\overline{\mathbf{B}}$ but also its direction. Thus, analysis of Faraday rotation measures can reveal the three-dimensional structure of the magnetic vector field (Berkhuijsen *et al.*, 1997; Beck *et al.*, 1996).

Since n_{cr} is difficult to measure, it is often assumed that magnetic field and cosmic rays are in pressure equilibrium or energy equipartition; this allows to express n_{cr} in terms of B. The physical basis of this assumption is the fact that cosmic rays (charged particles of relativistic energies) are confined by magnetic fields. An additional assumption involved is that the energy density of relativistic electrons responsible for synchrotron emission (energy of several GeV per particle) is one percent of the proton energy density in the same energy interval, as measured near the Earth.

The cosmic ray number density n_{cr} in the Milky Way can be determined independently from γ–ray emission produced when cosmic ray particles interact with the interstellar gas. Then magnetic field strength can be obtained without assuming equipartition (Strong *et al.*, 2000); the results are generally consistent with the equipartition values. However, equation (7.5) is not consistent with the equipartition or pressure balance between cosmic rays and magnetic fields as it assumes that $n_{\mathrm{cr}} = \mathrm{const}$. Therefore, $\overline{B}$ obtained from (7.5) can be inaccurate (Beck *et al.*, 2003).

The mean thermal electron density n_{e} in the ISM can be obtained from the emission measure of the interstellar gas, an observable defined as $\mathrm{EM} = \int_L n_{\mathrm{e}}^2\,\mathrm{d}s$, but this involves the poorly known filling factor of interstellar clouds. In the Milky Way, the dispersion measures of pulsars, $\mathrm{DM} = \int_L n_{\mathrm{e}}\,\mathrm{d}s$ provide information about the mean thermal electron density, but the accuracy is limited by our uncertain knowledge of distances to pulsars. Estimates of the strength of the regular magnetic field in the Milky Way are often obtained from the Faraday rotation measures of pulsars simply as

$$\overline{B}_\parallel = \frac{\mathrm{RM}}{K_1\,\mathrm{DM}}\,. \tag{7.6}$$

This estimate is meaningful if magnetic field and thermal electron density are statistically uncorrelated. If the fluctuations in magnetic field and thermal electron density are correlated with each other, they will contribute positively to RM and (7.6) will yield overestimated $\overline{B}_\parallel$. In the case of anticorrelated fluctuations, their contribution is negative and (7.6) is an underestimate. As shown by Beck *et al.* (2003), physically

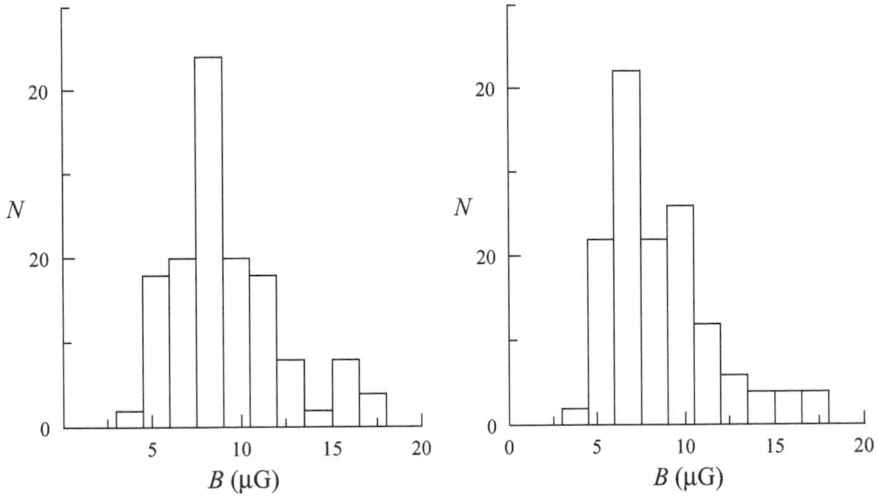

Figure 7.4 - The distributions of the strength of the total magnetic field in a sample of spiral galaxies obtained from the observed synchrotron intensity I using energy equipartition between magnetic fields and cosmic rays (p. 109 in Niklas, 1995) under slightly different assumptions. The estimates of the left-hand panel were derived from integrating the observed synchrotron intensity in the range corresponding to the relativistic electron energies from 300 MeV to infinity, and in the right-hand panel the integration was over a frequency range 10 MHz–10 GHz. Results presented in the left-hand panel are better justified physically (Section 2.1 in Beck *et al.*, 1996; Section III.A.1 in Widrow, 2002).

reasonable assumptions about the statistical relation between magnetic field strength and electron density can lead to (7.6) being in error by a factor of 2–3.

The observable quantities (7.4a–c) have provided extensive data on magnetic field strengths in both the Milky Way and external galaxies (Ruzmaikin *et al.*., 1988; Beck *et al.*, 1996; Beck, 2000, 2001). The average total field strengths in nearby spiral galaxies obtained from total synchrotron intensity I range from $B \approx 4\,\mu$G in the galaxy M31 to about $15\,\mu$G in M51, with the mean for the sample of 74 galaxies of $B = 9\,\mu$G (Beck, 2000). Figure 7.4 shows the distribution of magnetic field strength in a sample of spiral galaxies. The typical degree of polarisation of synchrotron emission from galaxies at short radio wavelengths is $p = 10$–20%, so (7.5) gives $\overline{B}/B = 0.4$–0.5; these are always lower limits due to the limited resolution of the observations, and $\overline{B}/B = 0.6$–0.7 is a more plausible estimate. Most existing polarisation surveys of synchrotron emission from the Milky Way, having much better spatial resolution, suffer from Faraday depolarisation effects and missing large-scale emission and cannot provide reliable values for p. The total equipartition magnetic field in the Solar neighbourhood is estimated as $B = 6 \pm 2\,\mu$G from the synchrotron intensity of the diffuse Galactic radio background (E. M. Berkhuijsen, in Beck, 2001). Combined with $\overline{B}/B = 0.65$, this yields a strength of the local regular field of $\overline{B} = 4 \pm 1\,\mu$G. Hence, the typical strength of the local Galactic random magnetic fields, $b = (B^2 - \overline{B}^2)^{1/2} = 5 \pm 2\,\mu$G, exceeds that of the regular field by a factor $b/\overline{B} = 1.3 \pm 0.6$.

Meanwhile, the values of $\overline{B}$ in the Milky Way obtained from Faraday rotation measures seem to be systematically lower than the above values (see Beck *et al.*, 2003, and references therein). RM of pulsars and extragalactic radio sources yield $\overline{B} = 1$–$2\,\mu$G in the Solar vicinity, a value about twice smaller than that inferred from the synchrotron intensity and polarisation. There can be several reasons for the discrepancy between the estimates of the regular magnetic field strength from Faraday rotation and synchrotron intensity. Both methods suffer from systematic errors due to our uncertain knowledge of thermal and relativistic electron densities, so one cannot be sure if the difference is significant. Nevertheless, the discrepancy seems to be worrying enough to consider carefully its possible reasons.

The discrepancy can be explained, at least in part, if the methods described above sample different volumes. The observation depth of total synchrotron emission, starlight polarisation and of Faraday rotation measures are all of the order of a few kpc. Polarised emission, however, may emerge from more nearby regions. However, a more fundamental reason for the discrepancy can be partial correlation between fluctuations in magnetic field and electron density. Such a correlation can arise from statistical pressure balance where regions with larger gas density have weaker magnetic field, and vice versa. As discussed by Beck *et al.* (2003), the term $\langle b_{\|} n_{\mathrm{e}} \rangle$ then differs from zero and contributes to the observed RM leading to underestimated $\overline{B}$. In a similar manner, correlation between B and the cosmic ray number density biases the estimates of magnetic field from synchrotron intensity and polarisation (see also Sokoloff *et al.*, 1998). Altogether, $\overline{B} = 4\,\mu$G and $b = 5\,\mu$G seem to be acceptable estimates of magnetic field strengths near the Sun. The geometry and three-dimensional structure of the magnetic fields observed in spiral galaxies are further discussed in Section 7.5.

7.4. THE ORIGIN OF GALACTIC MAGNETIC FIELDS

There are two basic approaches to the origin of global magnetic structures in spiral galaxies – one of them asserts that the observed structures represent a primordial magnetic field twisted by differential rotation, and the other that they are due to ongoing dynamo action within the galaxy. The simplicity of the former theory is appealing, but it fails to explain the strength, geometry and apparent lifetime of galactic magnetic fields (Ruzmaikin *et al.*, 1988; Beck *et al.*, 1996; Kulsrud, 1999; Widrow,

2002; see Section 7.5 below). Furthermore, there are no mechanisms known to produce cosmological magnetic fields of required strength and scale (Beck *et al.*, 1996), although Kulsrud *et al.* (1997) argue that suitable magnetic field can be produced in protogalaxies. Dynamo models appear to be much better consistent with the observational and theoretical knowledge of interstellar gas, and all models of magnetic fields in specific galaxies, known to the author, have been formulated in terms of dynamo theory. It seems to be very plausible that galactic magnetic fields are generated by some kind of dynamo action, i.e. that they are produced *in situ.* The most promising is the mean-field turbulent dynamo.

7.4.1. MEAN-FIELD MODELS OF THE GALACTIC DYNAMO

As discussed in Section 7.2.2, the discs of spiral galaxies are thin. This provides a natural small parameter, the disc aspect ratio, equation (7.2). This greatly facilitates modelling of many global phenomena in galaxies, including large-scale magnetic fields. Parker (1971) and Vainshtein & Ruzmaikin (1971, 1972) were the first to suggest mean-field dynamo models for spiral galaxies. These were local models discussed in Section 7.4.1, where only derivatives across the disc (in z) are retained. The theory has been extended to two and more dimensions and applied to specific galaxies (see Ruzmaikin *et al.*, 1988; Beck *et al.*, 1996; Widrow, 2002 and references therein). Rigorous asymptotic solutions for the $\alpha\omega$–dynamo in a thin disc were developed by Soward (1978, 1992a,b) and further discussed by Priklonsky *et al.* (2000) and Willis *et al.* (2003). Reviews of these results can be found in Ruzmaikin *et al.* (1988), Beck *et al.* (1996), Kulsrud (1999) and Soward (2003).

In this section we present asymptotic solutions of the mean-field dynamo equations (1.103) in a thin disc surrounded by vacuum. We first consider axially symmetric solutions of the kinematic problem, and then discuss generalisations to non-axisymmetric modes and to nonlinear regimes. Cylindrical coordinates (s, ϕ, z) with the origin at the galactic centre and the z–axis parallel to the galactic angular velocity are used throughout this chapter. In this section we use dimensionless variables, with s and z measured in the units of the characteristic disc radius and disc half-thickness (e.g. $s_0 = s_\odot \approx 8.5\,\mathrm{kpc}$ and $h_0 = 0.5\,\mathrm{kpc}$), respectively. Then the dimensionless radial and axial distances are both of order unity within the disc as they are measured in different units in order to make the disc thinness explicit. The corresponding time unit is the turbulent magnetic diffusion time across the disc, $h_0^2/\eta_\mathrm{T} \approx 7.5 \times 10^8\,\mathrm{yr}$.

It is convenient to introduce a unit rotational shear rate G_0:

$$G = s\frac{\partial\Omega}{\partial s} \equiv G_0 g(s, z)\,, \tag{7.7}$$

with $g(s,z)$ its dimensionless value and $G_0 = -\Omega_\odot$ for a flat rotation curve, $\Omega \propto s^{-1}$, and adopt the characteristic magnitude of the α–coefficient near the Sun as given by equation (7.3).

KINEMATIC, AXIALLY SYMMETRIC SOLUTIONS

The three components of an axially symmetric magnetic field can be expressed in terms of the azimuthal components of the large-scale magnetic field $\overline{B}_\phi$ and vector potential $\overline{A}_\phi$:

$$\overline{\mathbf{B}} = \left(-\frac{\partial \overline{A}_\phi}{\partial z},\ \overline{B}_\phi,\ \frac{1}{s}\frac{\partial}{\partial s}(s\overline{A}_\phi) \right) . \tag{7.8}$$

The dimensionless governing equations, resulting from (1.103) have the form

$$\frac{\partial \overline{B}_\phi}{\partial t} = -\mathrm{R}_\omega g \frac{\partial \overline{A}_\phi}{\partial z} + \frac{\partial^2 \overline{B}_\phi}{\partial z^2} + \varepsilon^2 \frac{\partial}{\partial s}\left[\frac{1}{s}\frac{\partial}{\partial s}(s\overline{B}_\phi)\right] , \tag{7.9a}$$

$$\frac{\partial \overline{A}_\phi}{\partial t} = \mathrm{R}_\alpha \alpha \overline{B}_\phi + \frac{\partial^2 \overline{A}_\phi}{\partial z^2} + \varepsilon^2 \frac{\partial}{\partial s}\left[\frac{1}{s}\frac{\partial}{\partial s}(s\overline{A}_\phi)\right] , \tag{7.9b}$$

where

$$\mathrm{R}_\omega = \frac{G_0 h_0^2}{\eta_\mathrm{T}} , \qquad \mathrm{R}_\alpha = \frac{\alpha_0 h_0}{\eta_\mathrm{T}} \tag{7.9c,d}$$

are the turbulent magnetic Reynolds numbers that characterise the intensity of induction effects due to differential rotation and the mean helicity of turbulence, respectively. We have neglected the vertical shear $\partial\Omega/\partial z$ which can easily be restored, and assumed for simplicity that $\eta_\mathrm{T} = \mathrm{const}$. A term containing α has been neglected in (7.9a) for the sake of simplicity (but can easily be restored), so the equations are written in the $\alpha\omega$–approximation.

The kinematic, axially symmetric asymptotic solution in a thin disc has the form

$$\begin{pmatrix} \overline{B}_\phi \\ \overline{A}_\phi \end{pmatrix} = \mathrm{e}^{\Gamma t} \left[Q(\varepsilon^{-1/3}s) \begin{pmatrix} \mathcal{B}(z;s) \\ \mathcal{A}(z;s) \end{pmatrix} + \ldots \right] ,$$

where Γ is the growth rate, $(\mathcal{B},\mathcal{A})$ represent the suitably normalised local solution (obtained for fixed s), and Q is the amplitude of the solution which can be identified with the field strength at a given radius.

THE LOCAL SOLUTION

The local solution (with s fixed) arises in the lowest order in ε. Its governing equations, obtained from (7.9a,b) by putting $\varepsilon = 0$, contain only derivatives with respect to z, with coefficients depending on s as a parameter (hence, the notation of the arguments of b and a with semicolon separating z and s):

$$\gamma(s)\mathcal{B} = -\mathrm{R}_\omega\, g(s)\,\partial_z\mathcal{A} + \partial_{zz}\mathcal{B}\,, \tag{7.10a}$$

$$\gamma(s)\mathcal{A} = \mathrm{R}_\alpha\, \alpha(s,z)\,\mathcal{B} + \partial_{zz}\mathcal{A}\,, \tag{7.10b}$$

were $\gamma(s)$ is the local growth rate. The boundary conditions often applied at the disc surface $z = \pm h(s)$ correspond to vacuum outside the disc. For axisymmetric fields and to the lowest order in ε they are (see below)

$$\mathcal{B} = 0 \quad \text{and} \quad \partial_z\mathcal{A} = 0 \quad \text{at } z = \pm h(s)\,. \tag{7.11a,b,c}$$

Since α is an odd function of z, kinematic modes have either even (quadrupole) or odd (dipole) parity, with the following symmetry conditions at the disc midplane (see, e.g., Ruzmaikin *et al.*, 1988):

$$\partial_z\mathcal{B} = 0 \quad \text{and} \quad \mathcal{A} = 0 \quad \text{at } z = 0 \quad \text{(quadrupole)}\,, \tag{7.12a}$$

or

$$\mathcal{B} = 0 \quad \text{and} \quad \partial_z\mathcal{A} = 0 \quad \text{at } z = 0 \quad \text{(dipole)}\,. \tag{7.12b}$$

In order to clarify the nature of the dynamo modes in a thin disc, here we consider an approximate solution of (7.10a,b) in the form of expansion in free-decay modes $\mathcal{B}_n(z)$ and $\mathcal{A}_n(z)$ obtained for $\mathrm{R}_\alpha = \mathrm{R}_\omega = 0$:

$$\gamma_n\mathcal{B}_n = \partial_{zz}\mathcal{B}_n\,, \qquad \gamma_n\mathcal{A}_n = \partial_{zz}\mathcal{A}_n\,,$$

where γ_n (< 0) is the decay rate of the nth mode. For the boundary conditions (7.11a–c) and (7.12a) that select quadrupolar modes, the resulting orthonormal set of basis functions is given by

$$\begin{pmatrix} \mathcal{B}_{2n} \\ \mathcal{A}_{2n} \end{pmatrix} = \begin{pmatrix} \sqrt{2}\cos\left[\pi(n+\frac{1}{2})z/h\right] \\ 0 \end{pmatrix},$$

$$\begin{pmatrix} \mathcal{B}_{2n+1} \\ \mathcal{A}_{2n+1} \end{pmatrix} = \begin{pmatrix} 0 \\ \sqrt{2}\sin\left[\pi(n+\frac{1}{2})z/h\right] \end{pmatrix},$$

$$\gamma_{2n} = \gamma_{2n+1} = -\pi^2(n+\tfrac{1}{2})^2\,, \qquad n = 0, 1, \dots\,.$$

The free-decay eigenvalues are all doubly degenerate, and two vector eigenfunctions, one with odd index and the other with even one, correspond to each eigenvalue, one with $\mathcal{B}_{2n+1} = 0$, and the other with $\mathcal{A}_{2n} = 0$. The eigenfunctions are normalised to have $\int_0^h(\mathcal{B}_n^2 + \mathcal{A}_n^2)\,\mathrm{d}z = 1$.

The solution of (7.10a,b) is represented as

$$\begin{pmatrix} \mathcal{B} \\ \mathcal{A} \end{pmatrix} \approx \mathrm{e}^{\gamma t}\sum_{n=0}^{\infty} c_n \begin{pmatrix} \mathcal{B}_n \\ \mathcal{A}_n \end{pmatrix},$$

where c_n are constants. We substitute this series into (7.10a,b), multiply by $(\mathcal{B}_k,\ \mathcal{A}_k)$ and integrate over z from 0 to h to obtain an algebraic system of homogeneous equations for c_k whose solvability condition yields an algebraic equation for γ. For our current purposes, it is sufficient to retain the smallest possible number of modes, which results in a system of two equations for c_0 and c_1 and a quadratic equation for γ whose positive solution is given by

$$\gamma \approx -\tfrac{1}{4}\pi^2 + \sqrt{W_{01}W_{10}}\,, \tag{7.13a}$$

where

$$W_{01} = \int_0^h \alpha\, b_0\, a_1\, \mathrm{d}z = 1 \quad \text{for } \alpha = \sin \pi z/h\,, \tag{7.13b}$$

$$W_{10} = -\mathrm{D}\int_0^h b_0\, a_1\, \mathrm{d}z = -\frac{\pi}{4}\mathrm{D}\,, \tag{7.13c}$$

and $\mathrm{D} = \mathrm{R}_\alpha \mathrm{R}_\omega$ is the dynamo number (see Section 1.5.3).

To assess the accuracy of (7.13a), we note that it yields $\gamma = 0$ for $\mathrm{D} = \mathrm{D}_{\mathrm{cr}} = -\pi^3/4 \approx -8$, very close to the accurate value obtained numerically (Ruzmaikin *et al.*, 1988). This solution indicates that the dominant mode is non-oscillatory ($\mathrm{Im}\,\gamma = 0$); this is confirmed by other analytical and numerical solutions of the dynamo equations in thin discs.

A similar solution can be obtained for dipolar modes. The free decay modes of dipolar symmetry have $\gamma_n = -n^2\pi^2$, $n = 1, 2, \ldots$, so that the lowest dipolar mode decays four times faster than the lowest quadrupolar mode. The reason for that is that the azimuthal field of dipolar parity has zero not only at $|z| = h$ but also at $z = 0$ and so a smaller scale than the quadrupolar solution. This immediately implies that quadrupolar modes, with $\overline{B}_\phi(z) = \overline{B}_\phi(-z)$, $\overline{B}_s(z) = \overline{B}_s(-z)$, $\overline{B}_z(z) = -\overline{B}_z(-z)$, should be dominant in galactic discs. The dominant symmetry of galactic magnetic fields is thus expected to be different from that in stars and planets, where dipolar fields are preferred. This prediction is confirmed by observations (see Section 7.5.2).

THE GLOBAL SOLUTION

The vacuum boundary conditions are often used in analytical and semi-analytical studies of disc dynamos because of their (relative) simplicity. Most importantly, they have a local form in the lowest order in ε – see (7.11a–c). However, this advantage is lost as soon as the next order in ε is considered, which is needed in order to obtain a governing equation for the field distribution along radius, Q. To this order, non-local magnetic connection between different radii has to be included, i.e. the fact that magnetic lines leave the disc at some radius, pass through the surrounding vacuum and return to the disc at another radius. In this section we discuss the radial dynamo equation, and for this purpose we have to consider vacuum boundary conditions to the first order in ε.

If the disc is surrounded by vacuum, there are no electric currents outside the disc, i.e. $\nabla \times \overline{\mathbf{B}} = \mathbf{0}$, so that the outer magnetic field is potential, $\overline{\mathbf{B}} = -\nabla\Phi$ (see Section 1.1.4). Then axial symmetry implies that the azimuthal field vanishes outside the disc. Since magnetic field must be continuous on the disc boundary, this yields the following boundary condition at the disc surface $z = \pm h(s)$:

$$\overline{B}_\phi|_{z=\pm h} = 0\,. \tag{7.14}$$

The vacuum boundary condition for the poloidal field (determined by $\overline{A}_\phi$) was derived in local Cartesian coordinates by Soward (1978). Priklonsky *et al.* (2000) rederived it in cylindrical geometry in the form

$$\frac{\partial \overline{A}_\phi}{\partial z} - \frac{\varepsilon}{s}\mathcal{L}\left(\overline{A}_\phi\right) = 0 \quad \text{at } z = \pm h(s)\,, \tag{7.15a}$$

where the integral operator $\mathcal{L}(\overline{A}_\phi)$ is defined as

$$\mathcal{L}\left(\overline{A}_\phi\right) = \int_0^\infty W(s,s')\frac{\partial}{\partial s'}\left(\frac{1}{s'}\frac{\partial}{\partial s'}s'\overline{A}_\phi\right)\,\mathrm{d}s' \tag{7.15b}$$

with the kernel

$$W(s,s') = ss'\int_0^\infty J_1(ks)J_1(ks')\,\mathrm{d}k\,, \tag{7.15c}$$

where $J_1(x)$ is the Bessel function. Willis *et al.* (2003) obtained another, equivalent form of the integral operator involving Green's function of the Neumann problem for the Laplace equation.

The integral part of the boundary condition (7.15a) can be transferred into a non-local term in the equation for Q which then becomes an integro-differential equation of the form (Priklonsky *et al.*, 2000)

$$[\Gamma - \gamma(s)]\,q(s) = \varepsilon p(s)\mathcal{L}\left\{q(s)\right\}\,, \tag{7.16a}$$

where

$$q(s) = Q(s)\mathcal{A}(h;s)\,, \qquad p(s) = \frac{\mathcal{A}(h,s)\mathcal{A}_*(h,s)}{\langle\mathbf{X}|\mathbf{X}_*\rangle}\,, \qquad \mathbf{X} = \begin{pmatrix}\mathcal{B}(z;s)\\ \mathcal{A}(z;s)\end{pmatrix}. \tag{7.16b,c,d}$$

Here $\mathbf{X}$ is the eigenvector of the lowest-order boundary value problem discussed in Section 7.4.1, the asterisk denotes the eigenvector of its adjoint problem, and

$$\langle\mathbf{X}|\mathbf{X}_*\rangle = \int_0^h \mathbf{X}\cdot\mathbf{X}_*\,\mathrm{d}z\,.$$

The solution of (7.16a) subject to the boundary conditions

$$q(0) = 0 \qquad \text{and} \qquad q \to 0 \quad \text{as} \quad s \to \infty$$

provides yet another eigenvalue problem, for which the eigenvalue is the global growth rate Γ and the eigenfunction is $q(s)$ which determines the radial profile of the global eigenfunction Q. As shown by Willis *et al.* (2003), the effect of the integral term in (7.16a) can be described as enhanced radial diffusion.

Equation (7.16a) is complicated enough as to provoke an irresistible desire to simplify it. Such a simplification, employed by Baryshnikova *et al.* (1987) (see also Ruzmaikin *et al.*, 1988) consists of neglecting the term containing ε in the boundary condition (7.15a). This makes the boundary condition local and leads to the following equation for $Q(s)$:

$$[\Gamma - \gamma(s)]Q = \varepsilon^2 \frac{\partial}{\partial s}\left(\frac{1}{s}\frac{\partial}{\partial s} sQ\right), \tag{7.17}$$

similar to (7.16a), but with the integral term replaced by the diffusion operator. Formally, (7.17) can be obtained from (7.16a) by replacing the integral kernel by the delta-function, $W(s, s') \to \delta(s - s')$. In other words, this simplification neglects any nonlocal coupling between different parts of the disc via the halo, but includes the local diffusive coupling within the disc. We note in this connection that the kernel $W(s, s')$ is indeed singular, although the singularity is only logarithmic in reality, $W(s, s') \sim \ln|s - s'|$.

The above simplification greatly facilitates the analysis of the global dynamo solutions and all applications of the thin-disc asymptotics to galaxies and accretion discs neglect the nonlocal effects. Equation (7.17) can be readily solved using a variety of analytical and numerical techniques (Ruzmaikin *et al.*, 1988), but some features of the solution are lost together with nonlocal effects. The most important failure is that the asymptotic scaling of the solution with ε is affected, with the radial scale becoming $\varepsilon^{-1/2}h_0$ instead of the correct value $\varepsilon^{-1/3}h_0$. However, the difference is hardly significant numerically for the realistic values $\varepsilon \approx 10^{-1}$–$10^{-2}$. We note that the thin-disc asymptotics are reasonably accurate for $\varepsilon \lesssim 10^{-1}$ (Baryshnikova *et al.*, 1987; Willis *et al.*, 2003).

Another consequence of the nonlocal effects is that solutions of (7.16a) possess algebraic tails far away from the dynamo active region, $q \sim s^{-4}$, whereas solutions of (7.17) have exponential tails typical of the diffusion equation. This affects the speed of propagation of magnetic fronts during the kinematic growth of the magnetic field: with the nonlocal effects, the front propagation is exponentially fast, whereas the local radial diffusion alone results in a linear propagation.

These topics are discussed in detail by Willis *et al.* (2003) who compare numerical solutions of (7.16a) and (7.17). Whether or not the nonlocal effects can be neglected depends on the goals of the analysis. There are several reasons why this simplification appears to be justified. The neglect of nonlocal effects does not seem to affect significantly any observable quantities, whereas the parameters of spiral galaxies and of their magnetic fields are known with a rather limited accuracy anyway. Moreover, the halos of spiral galaxies can be described as vacuum only in a very approximate sense, and the finite conductivity of the halo will weaken the nonlocal effects.

NON-AXISYMMETRIC, NONLINEAR AND NUMERICAL SOLUTIONS

The above asymptotic theory can readily be extended to non-axisymmetric solutions. This generalisation is discussed by Krasheninnikova *et al.* (1989) and Ruzmaikin *et al.* (1988). Starchenko & Shukurov (1989) developed WKBJ asymptotic solutions of the mean-field galactic dynamo equations valid for $|\mathrm{D}| \gg 1$. A similar asymptotic regime for one-dimensional dynamo equations (7.10a,b) is discussed in Section 9.IV of Zeldovich *et al.* (1983).

Another useful approximate approach, known as the "no–z" approximation, was suggested by Subramanian & Mestel (1993). In this approximation, derivatives across the disc in (7.9a,b) or their three-dimensional analogues are replaced by division by the disc semi-thickness, $\partial_z \to 1/h$, and the resulting equations in s and ϕ are solved, e.g. by the WKBJ method or numerically. This approach appears to be rather crude at first sight, but it is quite efficient because the structure of the magnetic field across a thin disc is rather simple, at least for the lowest mode. A refinement of the approximation to improve its accuracy is discussed by Phillips (2001). Mestel and Subramanian (1991) and Subramanian & Mestel (1993) apply these solutions to study the effects of spiral arms on galactic magnetic fields. This approximation was also extensively used in numerical simulations of galactic dynamos (Moss 1995; see Moss *et al.*, 2001 for an example).

Nonlinear asymptotics of (7.10a,b) for $|\mathrm{D}| \gg 1$ are discussed by Kvasz *et al.* (1992), where it is supposed that the nonlinearity affects significantly magnetic field distribution across the disc, and to the lowest approximation the steady state of the dynamo is established locally. This, however, may not be the case. The radial coupling is significant already at the kinematic stage where it results in the establishment of a global eigenfunction as described by (7.16a) or (7.17). Nonlinear effects are more likely to affect the global eigenfunction, and so have to affect the radial equation. Poezd *et al.* (1993) have derived a nonlinear version of (7.17) assuming the standard form of α–quenching with the α–coefficient modified by magnetic field as

$$\widetilde{\alpha} = \frac{\alpha}{1 + \overline{B}^2/B_0^2}, \qquad (7.18)$$

where B_0 is a suitably chosen saturation level most often identified with a state where magnetic and turbulent kinetic energy densities are of the same order of magnitude (see discussions in Sections 2.7.2 and 6.3). As a result, the magnetic field can grow when $\overline{B} \ll B_0$, but then the growth slows down as the quenched dynamo number obtained with $\widetilde{\alpha}$ approaches its critical value $\mathrm{D_{cr}}$, and the field growth saturates at $\overline{B} \approx B_0$. In terms of the thin-disc asymptotic model, this implies that $\gamma(s)$ in (7.16a) and (7.17) ought to be replaced by $\gamma(s)(1 - Q^2/B_0^2)$, so that the nonlinear version of (7.17) with the nonlinearity (7.18) has been derived in the form

$$\frac{\partial Q}{\partial t} = \gamma(s)\left(1 - \frac{Q^2}{B_0^2}\right) Q + \varepsilon^2 \frac{\partial}{\partial s}\left[\frac{1}{s}\frac{\partial}{\partial s} sQ(s)\right], \qquad (7.19)$$

provided the local solution has been normalised in such a way that Q is a field strength averaged across the disc at a given radius. The derivation of this equation by averaging the governing equations across the disc can be found in Poezd *et al.* (1993). This equation and its nonaxisymmetric version have been extensively applied to galactic dynamos (see Beck *et al.*, 1996, and references therein).

The detailed physical mechanism of the saturation of the dynamo action is still unclear. Cattaneo *et al.* (1996) suggest that the saturation is associated with the suppression of the Lagrangian chaos of the gas flow by the magnetic field. This mechanism, attractive in the context of convective systems (where the flow becomes random due to intrinsic reasons, e.g. instabilities), can hardly be effective in galaxies where the flow is random because of the randomness of its driving force (the supernova explosions).

Most numerical solutions of galactic dynamo equations that extend beyond the thin-disc approximation rely on the "embedded disc" approach (Stepinski & Levy, 1988; Elstner *et al.*, 1990). Instead of using complicated boundary conditions at the disc surface, this approach considers a disc embedded into a halo whose size is large enough as to make unimportant boundary conditions posed at the remote halo boundary. Since turbulent magnetic diffusivity in galactic halos is larger than in the disc (Sokoloff & Shukurov, 1990; Poezd *et al.*, 1993), meaningful embedded disc models are compatible with thin-disc asymptotic solutions obtained with vacuum boundary conditions and confirm the asymptotic results. The embedded disc approach was also used to study dynamo-active galactic halos (Brandenburg *et al.*, 1992, 1993, 1995; Elstner *et al.*, 1995). Further extensions of disc dynamo models include the effects of magnetic buoyancy (Moss *et al.*, 1999), accretion flows (Moss *et al.*, 2000) and external magnetic fields (Moss & Shukurov, 2001, 2004).

An implication of the nonlinear model for the thin-disc dynamo is that the local solution is unaffected by nonlinear effects whose main role is to modify the radial field structure. An important consequence of this is that it can be reasonably expected that the pitch angle of magnetic lines, $p_B = \arctan \overline{B}_s/\overline{B}_\phi$, is weakly affected by nonlinear effects, and so represents an important feature of the solution that can be directly compared with observations (Baryshnikova *et al.*, 1987). This expectation seems to be confirmed by observations (Section 7.5.1). Nevertheless, the modification of the magnetic pitch angle by nonlinear effects has never been studied in detail, which seems to be a regrettable omission.

DYNAMO CONTROL PARAMETERS IN SPIRAL GALAXIES

A remarkable feature of spiral galaxies is that they are (almost) transparent to electromagnetic waves over a broad range of frequencies, so the kinematics of the ISM is rather well understood, and therefore most parameters essential for dynamo action are well restricted by observations. This leaves less room for doubt and less freedom for speculation than in the case of other natural dynamos. Another advantage is that observations of polarised radio emission at a linear resolution of 1–3 kpc (typical of the modern observation of nearby galaxies) reveal exactly that field which is modelled by the mean-field dynamo theory (given volume and ensemble averages are identical).

The mean-field dynamo is controlled by two dimensionless parameters quantifying the differential rotation and the so-called α–effect, as defined in (7.9c,d). Using (7.1) and (7.3) and assuming a flat rotation curve, $\Omega = U_0/s$, we obtain the following estimates for the solar vicinity of the Milky Way:

$$\mathrm{R}_\omega \approx -3\,\frac{U_0}{u_0}\,\frac{h_0^2}{\ell s} \approx -20\,, \qquad \mathrm{R}_\alpha \approx 3\,\frac{U_0}{u_0}\,\frac{\ell}{s} \approx 1\,, \tag{7.20a,b}$$

where $U_0 = s_0\Omega_0$ is the typical rotational velocity. Since $|\mathrm{R}_\omega| \gg \mathrm{R}_\alpha$, differential rotation dominates in the production of the azimuthal magnetic field (i.e. the $\alpha\omega$–dynamo approximation is well applicable), and the dynamo action is essentially controlled by a single parameter, the dynamo number

$$\mathrm{D} = \mathrm{R}_\alpha \mathrm{R}_\omega \approx 10\,\frac{h_0^2}{u_0^2}\,s\Omega\,\frac{\partial\Omega}{\partial s} \approx -10\left(\frac{U_0 h_0}{u_0 s}\right)^2 \approx -20\,, \tag{7.21}$$

where the numerical estimate refers to the Solar vicinity. Thus, $|\mathrm{D}|$ does exceed the critical value for the lowest, non-oscillatory quadrupole dynamo mode, which then can be expected to dominate in the main parts of spiral galaxies. It is often useful to consider the local dynamo number $\mathrm{D}(s)$, a function of galactocentric radius s, obtained when the s–dependent, local values of the relevant parameters are used in equation (7.9c,d) or (7.21) instead of the characteristic ones.

The local regeneration (e-folding) rate of the regular magnetic field γ is related to the magnetic diffusion time along the smallest dimension of the gas layer and to the dynamo number (if $|\mathrm{R}_\omega| \gg \mathrm{R}_\alpha$). Using the perturbation solution of Section 7.4.1, the following expression (written in dimensional form) can be used as a rough estimate:

$$\gamma \sim \frac{\eta_{\rm T}}{h^2}\left(\sqrt{|{\rm D}|}-\sqrt{|{\rm D_{cr}}|}\right), \quad \text{for } |{\rm D}| \gtrsim {\rm D_{cr}}, \tag{7.22}$$

where $\mathrm{D_{cr}} \approx -8$ and numerical factor of order unity has been omitted. This yields the local e-folding time $\gamma^{-1} \approx 5\times10^8$ yr for the Solar neighbourhood. When the radial diffusion is included, i.e. equation (7.16a) or (7.17) is solved, the growth rate decreases, yielding a global e-folding time of $\Gamma^{-1} \approx 10^9$ yr near the Sun. Thus, the large-scale magnetic field near the Sun can be amplified by a factor of about 10^4 during the galactic lifetime, 10^{10} yr, and the Galactic seed field had to be rather strong, about 10^{-10} G. The fluctuation dynamo can produce such a statistical residual magnetic field at the scale of the leading eigenfunction either in the young galaxy (Section VII.13 in Ruzmaikin *et al.*, 1988; Widrow, 2002) or in the protogalaxy (Kulsrud *et al.*, 1997).

The above growth rate, estimated for the Solar neighbourhood of the Milky Way, is often erroneously adopted as a value typical of spiral galaxies in general. It is then important to note that the regeneration rate is significantly larger in the inner Galaxy (the local dynamo number rapidly grows as s becomes smaller, $\mathrm{D}(s) \propto G\Omega$ – see Figure 7.3) and in other galaxies. For example, Baryshnikova *et al.* (1987) estimate the global growth time of the leading axisymmetric mode in the galaxy M51 as 5×10^7 yr.

Gaseous discs of spiral galaxies are flared, ie., $h \propto s + \text{const}$ at $s \gtrsim 10$ kpc, whereas u_0 only slightly varies with s. For a flat rotation curve, $\Omega \propto s^{-1}$, equation (7.21) then shows that the local dynamo number does not vary much with galactocentric radius s and remains supercritical, $|\mathrm{D}(s)| \geq |\mathrm{D_{cr}}|$ out to a large radius. It is therefore not surprising that regular magnetic fields have been detected in all galaxies where observations have sufficient sensitivity and resolution (Wielebinski & Krause, 1993; Beck *et al.*, 1996; Beck, 2000, 2001).

A standard estimate of the steady-state strength of magnetic field produced by the mean-field dynamo follows from the balance of the Lorentz force due to the large-scale magnetic field and the Coriolis force that causes deviations from mirror symmetry (Ruzmaikin *et al.*, 1988; Shukurov, 1998):

$$\begin{aligned} \overline{B} &\approx \left[4\pi\rho u_0\Omega\ell\left(\left|\frac{\mathrm{D}}{\mathrm{D_{cr}}}\right|-1\right)\right]^{1/2} \\ &\approx 2\,\mu\mathrm{G}\left(\left|\frac{\mathrm{D}}{\mathrm{D_{cr}}}\right|-1\right)^{1/2}\left(\frac{n}{1\,\mathrm{cm^{-3}}}\right)^{1/2}\left(\frac{u_0}{10\,\mathrm{km\,s^{-1}}}\right)^{1/2}, \end{aligned} \tag{7.23}$$

where $\rho \approx 1.7\times10^{-21}\,\mathrm{kg\,m^{-3}}$ is the density of interstellar gas and n its number density, $n = \rho/m_{\rm H}$ with $m_{\rm H}$ the proton mass. This estimate yields values that are in good agreement with observations, but its applicability perhaps has to be reconsidered in view of the current controversy about the nonlinear behaviour of mean-field dynamos (see Section 7.4.3).

It is now clear what information is needed to construct a useful dynamo model for a specific galaxy: its rotation curve, the scale height of the gas layer, the turbulent scale and speed, and the gas density. All these parameters are observable, even though their observational estimates may be incomplete or controversial. One of successes of the mean-field dynamo theory is its application to spiral galaxies, where even simplest, quasi-kinematic models presented above are able to reproduce all salient features of the observed fields, both in terms of generic properties and for specific galaxies (Ruzmaikin *et al.*, 1988). We discuss this in Section 7.5.

Recent observational progress has allowed to explore the effects of galactic spiral patterns on magnetic fields (Beck, 2000). The corresponding dynamo models require the knowledge of the arm-interarm contrast in all the relevant variables (Shukurov & Sokoloff, 1998; Shukurov, 1998; Shukurov *et al.*, 2004).

7.4.2. THE FLUCTUATION DYNAMO AND SMALL-SCALE MAGNETIC FIELDS

Similarly to mean-field dynamos, the theory of the fluctuation dynamo is well understood in the kinematic regime, but nonlinear effects remain controversial. In this section we present results obtained with kinematic models of the fluctuation dynamo and those derived with simplified nonlinearity. The pioneering kinematic model of the fluctuation dynamo was developed by Kazantsev (1967), and many more recent developments are based on it. Detailed reviews of the theory and references can be found in Section 8.IV of Zeldovich *et al.* (1983), Chapter 9 of Zeldovich *et al.* (1990) and in Brandenburg & Subramanian (2005).

The growth time of the random magnetic field in a random velocity field of a scale ℓ is as short as the eddy turnover time, $\ell/u_0 \approx 10^7\,\mathrm{yr}$ in the warm phase for $\ell = 0.1\,\mathrm{kpc}$. The magnetic field produced by the dynamo action is a statistical ensemble of magnetic flux ropes and ribbons whose length is of the order of the flow correlation length, $\ell \approx 0.05$–$0.1\,\mathrm{kpc}$. Their thickness is of the order of the resistive scale, $\ell\,\mathrm{Rm}^{-1/2}$, in a single-scale velocity field, where Rm is the magnetic Reynolds number. A phenomenological model of dynamo in Kolmogorov turbulence yields the rope thickness of $\ell\,\mathrm{Rm}^{-3/4}$ (Subramanian, 1998). The dynamo action can occur provided $\mathrm{Rm} > \mathrm{Rm}_{\mathrm{cr}}$, where the critical magnetic Reynolds number is estimated as $\mathrm{Rm}_{\mathrm{cr}} = 30$–$100$ in simplified models of homogeneous, incompressible turbulence. Recent studies have revealed the possibility that small-scale magnetic fields can have peculiar fine structure because the magnetic dissipation scale in the interstellar gas is much smaller than that of turbulent motions, i.e. because the magnetic Prandtl number is much larger than unity (Schekochihin *et al.*, 2002).

Subramanian (1999) suggested that a steady state, reached via the back-action of the magnetic field on the flow, can be established by the reduction of the effective magnetic Reynolds number down to the value critical for the dynamo action, an idea similar to the concept of α–quenching in the mean-field theory. Then the thickness of the ropes in the steady state can be estimated as $\ell\,\mathrm{Rm}_{\mathrm{cr}}^{-1/2}$ or $\ell\,\mathrm{Rm}_{\mathrm{cr}}^{-3/4}$. Using a model nonlinearity in the induction equation with incompressible velocity field, Subramanian (1999) showed that the magnetic field strength within the ropes and ribbons b_0 saturates at the equipartition level with kinetic energy density, $b_0^2/8\pi \approx \frac{1}{2}\rho u_0^2$. The average magnetic energy density is estimated as $\overline{b^2}/8\pi \approx \frac{1}{2}\mathrm{Rm}_{\mathrm{cr}}^{-1}\rho u_0^2$, implying the volume filling factor of the ropes of order $f_V \sim \mathrm{Rm}_{\mathrm{cr}}^{-1} \approx 0.01$. In the case of magnetic sheets, we similarly obtain $f_V \sim \mathrm{Rm}_{\mathrm{cr}}^{-1/2} \approx 0.1$. Correspondingly, the mean magnetic energy generated by the small-scale dynamo in the steady state is about several percent of the turbulent kinetic energy density, in agreement with numerical simulations.

Shukurov & Berkhuijsen (2003) interpret thin, random filaments of zero polarised intensity observed in polarisation maps of the Milky Way (known as depolarisation canals) as a result of Faraday depolarisation in the turbulent interstellar gas. This interpretation has resulted in a tentative estimate of the Taylor microscale of the interstellar turbulence

$$\ell_{\mathrm{T}} = \ell\,\widetilde{\mathrm{Rm}}^{-1/2} \approx 0.6\,\mathrm{pc}\,,$$

where $\widetilde{\mathrm{Rm}}$ is the effective magnetic Reynolds number in the ISM. This yields the following estimate:

$$\widetilde{\mathrm{Rm}} \approx 10^4\,.$$

Of course, this is a very tentative estimate, and further analyses of observations and theoretical developments will be needed to refine it. The value of $\widetilde{\mathrm{Rm}}$ obtained is significantly larger than $\mathrm{Rm}_{\mathrm{cr}}$ obtained in idealised models. This might be due to the transonic nature of interstellar turbulence as the gas compressibility appears to hinder dynamo action. Kazantsev *et al.* (1985) have shown that the e-folding time of magnetic field in the acoustic-wave turbulence (i.e. a compressible flow) is as long as $\widetilde{\mathrm{M}}^4\ell/u_0$, where $\widetilde{\mathrm{M}}$ ($\gtrsim 1$) is the Mach number.

Using parameters typical of the warm phase of the ISM, this theory predicts that the small-scale dynamo would produce magnetic flux ropes and ribbons of the length (or the curvature radius) of about ℓ = 50–100 pc and thickness 5–10 pc for $\widetilde{\mathrm{Rm}} = 10^2$ and 0.5–10 pc for $\widetilde{\mathrm{Rm}} = 10^4$. The field strength within them, if at equipartition with the turbulent energy, has to be of order 2–5 μG in the warm phase and perhaps slightly less in the hot gas. Note that some heuristic models of the small-scale dynamo admit solutions with magnetic field strength within the ropes being significantly above the equipartition level, e.g. because the field configuration locally approaches a force-free one, $|(\nabla \times \mathbf{B}) \times \mathbf{B}| \ll |\mathbf{B}|^2/\ell$, where ℓ is the field scale (Belyanin *et al.*, 1993).

The small-scale dynamo is not the only mechanism producing random magnetic fields (e.g. Section 4.1 in Beck *et al.*, 1996, and references therein). Any mean-field dynamo action producing magnetic fields at scales exceeding the turbulent scale also generates small-scale magnetic fields. Similarly to the mean magnetic field, this component of the turbulent field presumably has a filling factor close to unity in the warm gas and its strength is expected to be close to equipartition with the turbulent energy at all scales.

The overall structure of the interstellar turbulent magnetic field in the warm gas can be envisaged as a quasi-uniform fluctuating background with several percent of the volume occupied by flux ropes and ribbons of a length 50–100 pc containing a well-ordered magnetic field. This basic distribution would be further complicated by compressibility, shock waves, MHD instabilities (such as Parker instability), the fine structure at subviscous scales, etc.

The site of the mean-field dynamo action is plausibly the warm phase rather than the other phases of the ISM. The warm gas has a large filling factor (so it can occupy a percolating global region), it is, on average, in a state of hydrostatic equilibrium, so it is an ideal site for both the small-scale and mean-field dynamo action. Molecular clouds and dense H I clouds have too small a filling factor to be of global importance. Fletcher & Shukurov (2001) argue that, globally, molecular clouds can be only weakly coupled to the magnetic field in the diffuse gas, but Beck (1991) suggests that a significant part of the large-scale magnetic flux can be anchored in molecular clouds. The timescale of the small-scale dynamo in the hot phase is $\ell/u_0 \approx 10^6\,\mathrm{yr}$ for $u_0 = 40\,\mathrm{km\,s^{-1}}$ and $\ell = 0.04\,\mathrm{kpc}$ (the width of the hot, "chimneys" extended vertically in the disc). This can be shorter than the advection time due to the vertical streaming, $h/U_z \approx 10^7\,\mathrm{yr}$ with $h = 1\,\mathrm{kpc}$ and $U_z = 100\,\mathrm{km\,s^{-1}}$. Therefore, the small-scale dynamo action should be possible in the hot gas. However, the growth time of the mean magnetic field must be significantly longer than ℓ/u_0, reaching a few hundred Myr. Thus, the hot gas can hardly contribute significantly to the mean-field dynamo action in the disc and can drive the dynamo only in the halo (Sokoloff & Shukurov, 1990). The main role of the fountain flow in the disc dynamo is to enhance magnetic connection between the disc and the halo (see Section 7.4.3).

7.4.3. MAGNETIC HELICITY BALANCE IN THE GALACTIC DISC

Conservation of magnetic helicity $\chi = \langle \mathbf{A} \cdot \mathbf{B} \rangle$ (where $\mathbf{B} = \boldsymbol{\nabla} \times \mathbf{A}$) in a perfectly conducting medium has been identified as an important constraint on mean-field dynamos that plausibly explains the catastrophic quenching of the α–effect discussed elsewhere in this volume (Blackman & Field, 2000; Kleeorin *et al.*, 2000, 2003; Brandenburg & Subramanian, 2005). In a closed system, magnetic helicity can only evolve on the (very long) molecular diffusion timescale; in galaxies, this timescale by far exceeds the Hubble time. The large-scale galactic magnetic fields have significant magnetic helicity of the order of $LB_sB_\phi \approx -\frac{1}{4}LB^2$, where $L \gtrsim 1\,\mathrm{kpc}$ is the field scale, $B_s/B_\phi = \tan p_B$ with $p_B \approx -15°$ the magnetic pitch angle. Since the initial (seed) magnetic field was weak, and so had negligible magnetic helicity, the large-scale magnetic helicity in a closed system must be balanced by the small-scale helicity of the opposite sign, $\approx \ell_\mathrm{h}\overline{b^2}$, where ℓ_h is an appropriate dominant scale of magnetic helicity. This immediately results in an upper limit on the steady-state mean magnetic field (Brandenburg & Subramanian, 2005, and references therein)

$$\frac{\overline{B}^2}{\overline{b^2}} \lesssim 4\,\frac{\ell_\mathrm{h}}{L} \approx 0.4\,, \tag{7.24}$$

where the numerical value is obtained for $\ell_\mathrm{h} = 0.1\,\mathrm{kpc}$ and $L = 1\,\mathrm{kpc}$. The result of Vainshtein & Cattaneo (1992), $\overline{B}^2/\overline{b^2} \sim \mathrm{Rm}^{-1/2}$ is recovered for $\ell_\mathrm{h} \sim L\mathrm{Rm}^{-1/2}$. The observed relative strength of the mean field in spiral galaxies is given by $\overline{B}^2/\overline{b^2} \approx 0.5$. The upper limit on the strength of the mean magnetic field (7.24) appears to be much lower than the observed field only if $\ell_\mathrm{h} \ll 0.1\,\mathrm{kpc}$. For $\ell_\mathrm{h} = 0.1\,\mathrm{kpc}$, the observed field strength is compatible with magnetic helicity conservation.

Blackman & Field (2000) and Kleeorin *et al.* (2000) suggested that the losses of the small-scale magnetic helicity through the boundaries of the dynamo region play the key role in the mean-field dynamo action. This is an appealing idea, especially because the mean-field dynamos rely on magnetic flux loss through the boundaries (Section 9.II in Zeldovich *et al.*, 1983; Section VII.5 in Ruzmaikin *et al.*, 1988). A similar situation occurs with the magnetic moment, which is a conserved quantity, and it only grows in a dynamo system of a finite size because the dynamo just redistributes it expelling magnetic moment out from the dynamo active region (Moffatt, 1978). However, these are the mean magnetic flux and moment that need to be transferred through the boundaries. Transport by turbulent magnetic diffusion is sufficient for these purposes. The new aspect of the magnetic helicity balance is that healthy mean-field dynamo action requires asymmetry between the transports of the magnetic helicities of the large- and small-scale magnetic fields.

A useful framework to assess the effects of magnetic helicity flow through the boundaries of the dynamo region was proposed by Brandenburg *et al.* (2002) who have presented the balance equation of magnetic helicity in the form

$$\frac{\mathrm{d}\chi_B}{\mathrm{d}t} + \frac{\mathrm{d}\chi_b}{\mathrm{d}t} = -2\eta\chi_J - 2\eta\chi_j - Q_B - Q_b\,, \tag{7.25}$$

where $\chi_B = \langle\overline{\mathbf{A}}\cdot\overline{\mathbf{B}}\rangle$ and $\chi_b = \langle\mathbf{a}\cdot\mathbf{b}\rangle$ are the magnetic helicities of the mean and random magnetic fields, respectively, η is the molecular magnetic diffusivity, $\chi_J = \overline{\mathbf{J}}\cdot\overline{\mathbf{B}}$ and $\chi_j = \langle\mathbf{j}\cdot\mathbf{b}\rangle$ are the current helicities (with $\overline{\mathbf{J}} = \boldsymbol{\nabla}\times\overline{\mathbf{B}}$ the current density). The first two terms on the right-hand side of equation (7.25) are responsible for the Ohmic losses whereas the last two terms represent the boundary losses. For illustrative purposes and following Brandenburg *et al.* (2002), we adopt the following assumptions:

(i) The magnetic fields are fully helical, so $M_B = \frac{1}{2}k_B|\chi_B|$ and $M_b = \frac{1}{2}k_b|\chi_b|$, where M_B and M_b are the average energy densities of the mean and random magnetic fields and k_B and k_b are their wavenumbers, respectively. Furthermore, $\chi_J = k_B^2\chi_B$ and $\chi_j = k_b^2\chi_b$.

(ii) The mean and random magnetic fields have widely separated scales, $k_B \ll k_b$.

(iii) Approximate equipartition is maintained between the mean and random magnetic fields, $M_B \approx M_b$.

Then
$$\left|\frac{\chi_B}{\chi_b}\right| = \frac{k_b}{k_B}\frac{M_B}{M_b} = \frac{k_b^2}{k_B^2}\frac{\chi_J}{\chi_j}\,,$$

and so $|\chi_B| \gg |\chi_b|$ and $|\chi_J| \ll |\chi_j|$. Assuming for definiteness that $\chi_B, \chi_J > 0$, we have $\chi_b, \chi_j < 0$, and (7.25) can be approximated by

$$\frac{\mathrm{d}M_B}{\mathrm{d}t} = 2\eta k_b k_B M_b + \tfrac{1}{2}k_B(Q_B + Q_b)\,. \tag{7.26}$$

It is important to note that the effective advection velocities for the large-scale and small-scale magnetic fields are *not* equal to each other. Both small-scale and large-scale magnetic fields are advected from the disc by the galactic fountain flow. With a typical vertical velocity of order $U_z = 100$–$200\,\mathrm{km\,s^{-1}}$ and the surface covering factor of the hot gas $f = 0.2$–0.3, the effective vertical advection speed is $fU_z = 30$–$70\,\mathrm{km\,s^{-1}}$. However, the large-scale magnetic field is subject to turbulent pumping (turbulent diamagnetism). Given that the turbulent magnetic diffusivity in the disc and the halo are $\eta_\mathrm{T}^{(\mathrm{d})} = 10^{26}\,\mathrm{cm^2\,s^{-1}}$ and $\eta_\mathrm{T}^{(\mathrm{h})} = 2\times10^{27}\,\mathrm{cm^2\,s^{-1}}$ (Poezd *et al.*, 1993), respectively, and that the transition layer between the disc and the halo has a thickness of $\Delta z = 1\,\mathrm{kpc}$, the resulting advection speed is $-U_\mathrm{d} = \frac{1}{2}|\boldsymbol{\nabla}\eta_\mathrm{T}| \approx 2$–$3\,\mathrm{km\,s^{-1}}$. Thus, the vertical advection velocities of the large-scale and small-scale magnetic fields are $fU_z + U_\mathrm{d}$ and fU_z, respectively.

Now we can estimate the magnetic helicity fluxes through the disc surface as

$$Q_B = -\left(U_B + \frac{1}{k_B\tau_\eta}\right)M_B\,, \qquad Q_b = U_b M_b\,, \tag{7.27a,b}$$

where $\tau_\eta = 1/(4\eta_T k_B^2)$ is the timescale of the (turbulent) diffusive transport of the mean magnetic field through the boundary, and U_B and U_b are effective advection velocities for the large-scale and small-scale magnetic helicities, respectively. The latter can be estimated from the following arguments. Consider advection of magnetic field through the disc surface $z = h$ by a flow with a speed U, $\partial_t\overline{B}^2 = -U\partial_z\overline{B}^2$.

Assuming for simplicity that U is independent of z, we obtain by integration over z: $2h\dot{M}_B = -2U\overline{B}^2(h)$, where $M_B = (2h)^{-1}\int_{-h}^{h}\overline{B}^2\,\mathrm{d}z$. With $M_B = -k_B\chi_B/2$, this shows that advection of magnetic field at a speed U produces the large-scale helicity loss at a rate $\dot{\chi}_B \equiv Q_B = (2U/k_B h)\overline{B}^2(h)$. Here $\overline{B}(h)$ is the large-scale field strength at the disc surface, which is given by $\overline{B}^2(h) \equiv \xi M_B$, where $\xi < 1$ because the large-scale magnetic field at the surface must be weaker than that deep in the disc. For example, we have $\xi \ll 1$ for vacuum boundary conditions, where $\overline{B}_\phi(h) = 0$ and so

$$U_B = \frac{2\xi}{k_B h}(fU_z + U_\mathrm{d})\,. \tag{7.28}$$

Unlike the large-scale magnetic field, the small-scale magnetic fields are not necessarily weaker at the disc surface, so similar arguments yield

$$U_b = \frac{2}{k_b h}fU_z \neq U_B\,.$$

Thus, there are several reasons for the magnetic helicity fluxes through the disc surface to be different at small and large scales: most importantly, the large-scale magnetic field at the surface can be much smaller than that deep in the disc ($\xi \ll 1$) and, in addition, turbulent diamagnetism introduces further difference ($U_\mathrm{d} \neq 0$).

Equation (7.26) has the following solution

$$\frac{M_B}{M_b} = \frac{4\eta k_b + U_b}{4\eta_T k_B + U_B}\left\{1 - \exp\left[-\frac{1}{2}\left(\frac{1}{\tau_B} + k_B U_B\right)t\right]\right\}, \tag{7.29}$$

which satisfies the initial condition $M_B(0) = 0$.

For $t \ll \tau_B$, this solution captures the exponential growth of the mean magnetic field at a timescale τ_B, $M_B \propto t/2\tau_B$.

For $U_B = U_b = 0$, we obtain $M_B/M_b \approx \eta k_b/\eta_T k_B \sim \mathrm{Rm}^{-1}$ for $t \to \infty$ – this corresponds to the catastrophic quenching of the α–effect associated with approximate magnetic helicity conservation in a medium with (weak) Ohmic losses alone.

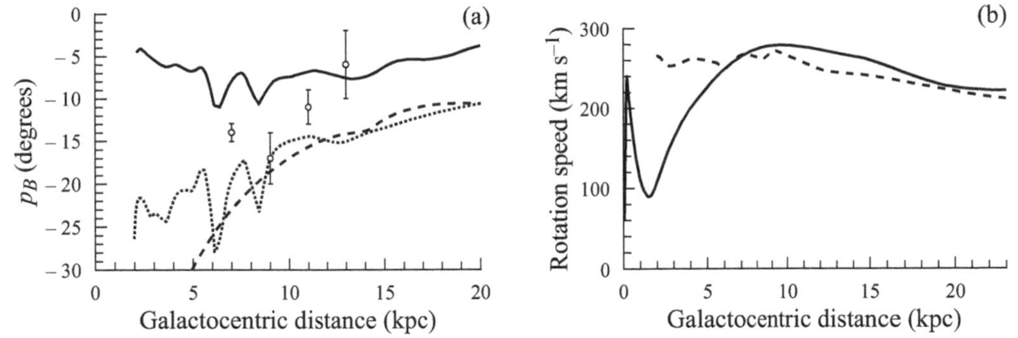

Figure 7.5 - (a) The pitch angle of magnetic field in the galaxy M31 as obtained from radio polarisation observations (circles with error bars) (Fletcher *et al.*, 2004), and from (7.33) using the rotation curve of Deharveng & Pellet (1975) and Haud (1981) (dashed) and Braun (1991) (dotted); $h(r)$ is twice the H I scale height of Braun (1991). Results from a nonlinear dynamo model for M31 (Beck *et al.*, 1998) are shown with solid line. (b) The rotation curve of M31 from Deharveng & Pellet (1975) and Haud (1981) (solid) and from Braun (1991) (dashed).

However, for $U_b \gg 4\eta k_b$ (a condition safely satisfied for any realistically small η) and $U_B \gg 4\eta_T k_B \approx 8\,\mathrm{km\,s^{-1}}$, we obtain

$$\left.\frac{M_B}{M_b}\right|_{t\to\infty} = \frac{U_b}{U_B} \sim \frac{k_B}{\xi k_b} \approx \frac{1}{10\xi}, \tag{7.30}$$

where we recall that $\xi < 1$ and neglect U_d. Thus, states with $M_B \approx M_b$ cannot be excluded, and this equipartition state is reached at the timescale of order $\tau_B \approx 4 \times 10^8\,\mathrm{yr}$.

These arguments suggest that the growth rate of the mean magnetic field is limited from above by the flux of the mean magnetic helicity through the boundary of the dynamo region, whereas the upper limit for its steady state strength is controlled by the rate at which the small-scale magnetic helicity is transferred through the boundaries, equation (7.30).

Another limit on the mean field strength arises from the balance of the Lorentz and Coriolis forces in the disc, equation (7.23). The steady-state strength of the mean magnetic field is the minimum of the two values. These arguments suggest that the restrictions on the mean-field dynamo action from magnetic helicity conservation can be removed as soon as one allows for the disc-halo connection and fountain flows in spiral galaxies. Of course, these heuristic arguments have to be confirmed by quantitative analysis.

7.5. OBSERVATIONAL EVIDENCE FOR THE ORIGIN OF GALACTIC MAGNETIC FIELDS

7.5.1. MAGNETIC PITCH ANGLE

Regular magnetic fields observed in spiral galaxies have field lines in the form of a spiral with a pitch angle in the range $p_B = -(10°\text{–}30°)$, with negative values indicating a trailing spiral (e.g. Beck *et al.*, 1996). As discussed in Section 7.4.1, the value of the pitch angle is a useful diagnostic of the mechanism maintaining the magnetic field.

Consider the simplest from of mean-field dynamo equations (7.10a,b) appropriate for a thin galactic disc, but now written in terms of dimensional variables for $\overline{B}_s$ and $\overline{B}_\phi$:

$$\frac{\partial \overline{B}_s}{\partial t} = -\frac{\partial}{\partial z}(\alpha \overline{B}_\phi) + \eta_{\rm T}\frac{\partial^2 \overline{B}_s}{\partial z^2}\,, \qquad \frac{\partial \overline{B}_\phi}{\partial t} = G\overline{B}_s + \eta_{\rm T}\frac{\partial^2 \overline{B}_\phi}{\partial z^2}\,. \tag{7.31a,b}$$

Any regular magnetic field maintained by the dynamo must have a non-zero pitch angle: for $\overline{B}_s \equiv 0$ (a purely azimuthal magnetic field), equation for $\overline{B}_\phi$ in (7.31a,b) reduces to a diffusion equation $\partial \overline{B}_\phi/\partial t = \eta_{\rm T}\partial^2 \overline{B}_\phi/\partial z^2$ which only has decaying solutions, $\overline{B}_\phi \propto \exp(-\eta_{\rm T} t/h^2)$. The same applies to a purely radial magnetic field.

Consider exponentially growing solutions, $\overline{B}_{s,\phi} = \mathcal{B}_{s,\phi}\exp(\gamma t)$, and replace ∂_z by $1/h$ and ∂_{zz} by $-1/h^2$ (as in the "no–z" approximation) to obtain from (7.31a,b) two algebraic equations,

$$\left(\gamma + \eta_{\rm T}/h^2\right)\mathcal{B}_s + \alpha\mathcal{B}_\phi/h = 0\,, \quad -G\mathcal{B}_s + \left(\gamma + \eta_{\rm T}/h^2\right)\mathcal{B}_\phi = 0\,, \tag{7.32a,b}$$

which have non-trivial solutions only if the determinant vanishes, which yields $(\gamma + \eta_{\rm T}/h^2)^2 \approx -\alpha G/h$, and (7.22) follows with $\rm D_{cr} = 1$. The resulting estimate of the magnetic pitch angle is given by

$$\tan p_B = \frac{\mathcal{B}_s}{\mathcal{B}_\phi} \approx -\sqrt{\frac{\alpha}{-Gh}} = -\sqrt{\frac{\rm R_\alpha}{|\rm R_\omega|}} \sim -\frac{\ell}{h}\left|\frac{\partial \ln\Omega}{\partial \ln s}\right|^{-1/2}. \tag{7.33}$$

For $\ell/h \approx 1/4$ and a flat rotation curve, $\partial \ln\Omega/\partial \ln s = -1$, we obtain $p_B \approx -15°$, and this is the middle of the range observed in spiral galaxies. More elaborate treatments discussed by Ruzmaikin *et al.* (1988b) confirm this estimate of p_B and yield a more accurate value of $\rm D_{cr}$. For example, the perturbation solution of Section 7.4.1 yields

$$p_B \approx -\frac{1}{2}\pi^{1/2}\sqrt{\frac{\rm R_\alpha}{|\rm R_\omega|}}\,. \tag{7.34}$$

If the steady state is established by reducing R_α to its critical value as to obtain $\mathrm{R}_\alpha \mathrm{R}_\omega = \mathrm{D}_{\mathrm{cr}}$, then the pitch angle in the nonlinear steady state becomes

$$\tan p_B \approx -\frac{1}{2}\pi^{1/2}\frac{\sqrt{|\mathrm{D}_{\mathrm{cr}}|}}{|\mathrm{R}_\omega|}\,. \tag{7.35}$$

The magnetic pitch angle in M31 determined from observations and dynamo theory is shown in Figure 7.5. Although the model curves show noticeable differences from the observed pitch angles, the general agreement is encouraging. The situation is typical: magnetic pitch angles of spiral galaxies are in a good agreement with predictions of dynamo theory (Beck *et al.*, 1996).

This picture does not explain why the pitch angles of galactic magnetic fields are invariably close (though not equal) to those of the spiral pattern in the parent galaxy. A plausible explanation is that magnetic pitch angles are further affected by streaming motions associated with the spiral pattern to make the match almost perfect (Moss, 1998). We note, however, that the pitch angle of the large-scale magnetic field near the Sun differs significantly from that of the local (Orion) arm; it is not clear whether this misalignment is of a local or global nature.

As shown by Moss *et al.* (2000), magnetic pitch angle can be affected by an axisymmetric radial inflow (as well as outflow):

$$\tan p_B \approx -\frac{1}{2}\pi^{1/2}\sqrt{\frac{\mathrm{R}_\alpha}{|\mathrm{R}_\omega|}}\left(1-\frac{1}{2}\mathcal{R}\sqrt{\frac{\pi}{-\mathrm{D}}}\right)\,, \quad \text{where } \mathcal{R} = \frac{h^2}{2\eta_{\mathrm{T}}}\left(\frac{u_s}{s}-\frac{\partial u_s}{\partial s}\right)\,,$$

which is useful to compare with equations (7.33) and (7.34). This effect is important if $u_s \gtrsim 2\eta_{\mathrm{T}}/h \approx 1\,\mathrm{km\,s^{-1}}$ (*cf.* Section 7.5.5).

Twisting of a horizontal primordial magnetic field by galactic differential rotation leads to a tightly wound magnetic structure with magnetic field direction alternating with radius at a progressively smaller scale $\Delta s \sim s_0/|G|t$ with $\tan p_B \sim -(|G|t)^{-1}$, where $s_0 \approx 10\,\mathrm{kpc}$ is the scale of variation in Ω (see Section 3.3 in Moffatt, 1978; Kulsrud, 1999 for a detailed discussion). The winding-up proceeds until a time $t_0 \approx 5\times10^9\,\mathrm{yr}$ such that $|G|t_0 \sim |C_\omega|^{1/2}$, where $C_\omega = Gs_0^2/\eta_{\mathrm{T}} = \mathrm{R}_\omega s_0^2/h^2 \approx 10^3$–$10^4$. At later times, the alternating magnetic field rapidly decays because of diffusion and reconnection. The resulting maximum magnetic field strength achieved at t_0 is given by

$$B_{\mathrm{max}} \sim B_0|C_\omega|^{1/2}\,, \tag{7.36}$$

where B_0 is the external magnetic field; the magnetic field reverses at a small radial scale $\Delta s \sim s_0|C_\omega|^{-1/2} \approx 0.1\,\mathrm{kpc}$. The magnetic pitch angle at t_0 is of the order of $|p_B| \sim |C_\omega|^{-1/2} \lesssim 1^\circ$, i.e. much smaller than the observed one. This picture cannot be reconciled with observations (*cf.* Kulsrud, 1999). It can be argued that streaming motions could make magnetic lines more open and parallel to the galactic spiral arms. However, then magnetic field will reverse on a small scale not only along radius, but also along azimuth. Such magnetic structures are quite different from what is observed. The moderate magnetic pitch angles observed in spiral galaxies are a direct indication that the regular magnetic field is not frozen into the interstellar gas and has to be maintained by the dynamo (Beck, 2000).

7.5.2. THE EVEN (QUADRUPOLE) SYMMETRY OF MAGNETIC FIELD IN THE MILKY WAY

One of the most convincing arguments in favour of the galactic dynamo theory comes from the symmetry of the observed regular magnetic field with respect to the Galactic equator in the Milky Way. The direction of the magnetic field is determined from Faraday rotation measures of the cosmic sources of polarised emission, pulsars and extragalactic radio sources. Since the Galactic magnetic field has a significant random component and extragalactic radio sources can have their own (intrinsic) Faraday rotation, any meaningful conclusions about the Galactic magnetic field must rely on statistically significant samples of Faraday rotation measures. Even though the quadrupole symmetry of the galactic magnetic fields has been widely accepted as a firmly established fact since mid-1970's, its objective observational verification has been obtained only recently. The main problem here is that it is difficult to separate local (small-scale) and global magnetic structures in the observed picture. However, wavelet analysis of the Faraday rotation measures of extragalactic radio sources has definitely confirmed that the horizontal components of the local regular magnetic field have even parity being similarly directed on both sides of the midplane (Frick *et al.*, 2001, see Figure 7.5.1).

The quadrupole symmetry is naturally explained by dynamo theory where even parity is strongly favoured against odd parity because the even field has twice larger scale in the vertical coordinate (see Section 7.4.1).

Primordial magnetic field twisted by differential rotation can have even vertical symmetry if it is parallel to the disc plane. However, then the field is rapidly destroyed by twisting and reconnection as described in Section 7.5.1. If, otherwise, the primordial field is parallel to the rotation axis and amplified by the vertical rotational shear $\partial_z\Omega$ (which, however, is insignificant within galactic discs, $|z| \leq h$), it can avoid catastrophic decay (Section 3.11 in Moffatt, 1978), but then it will have odd parity in z, which is ruled out by the observed parity of the Milky Way field.

The derivation of the regular magnetic field of the Milky Way from Faraday rotation measures of pulsars and extragalactic radio sources, RM, is complicated by the contribution of local magnetic perturbations, so it is difficult to decide which features

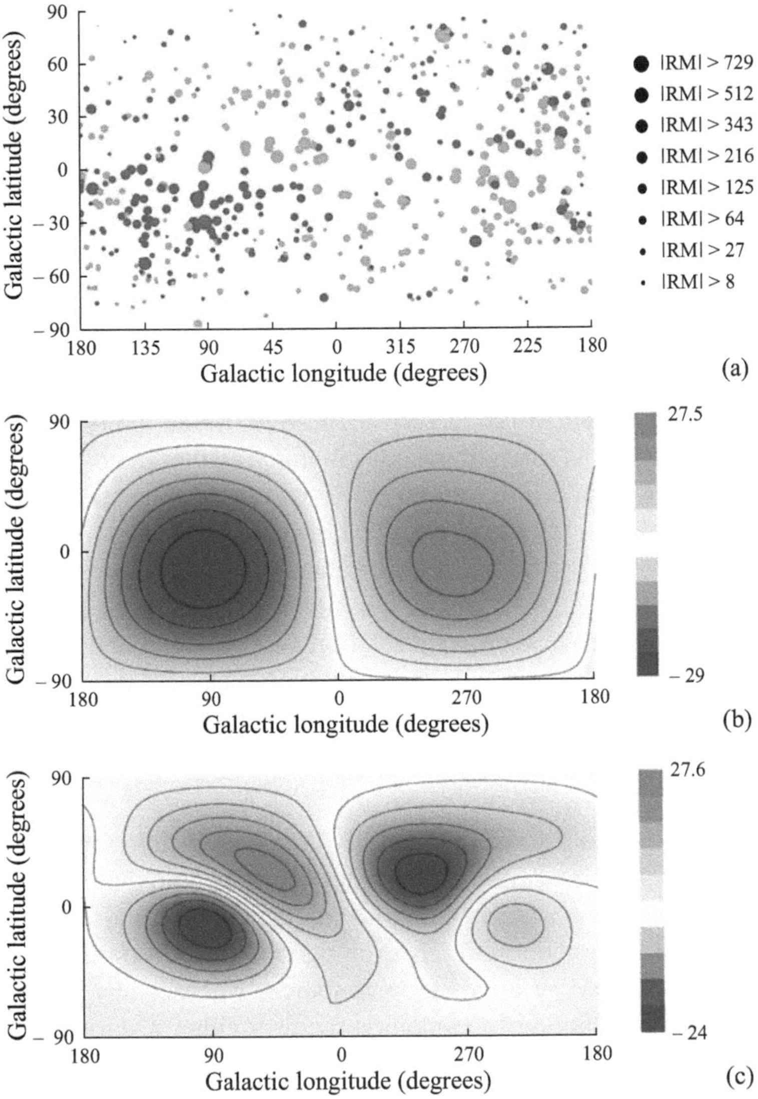

Figure 7.6 - (a) Faraday rotation measures of 551 extragalactic radio sources from the catalogue of Simard-Normandin & Kronberg (1980) shown in the (l, b)–plane, where (l, b) are the Galactic longitude and latitude in a reference frame centred at the Sun with the Galactic center in the direction $l = 0$ and Galactic midplane at $b = 0$. Positive (negative) RMs are shown with red (blue) circles whose radius indicates $|\mathrm{RM}|$ $(\mathrm{rad\,m^{-2}})$ as shown to the right of the panel. The lower two panels (b) & (c) show the wavelet transform of these data at scales 76° (b) and 35° (c) (Frick *et al.*, 2001). The transform at 76° has been obtained with the region of the Radio Loop I removed (this radio feature is a nearby supernova remnant). The wavelet transform at the scale 35° is dominated by local magneto–ionic features. (**See colour insert**.)

of the RM sky are due to the regular magnetic field and which are produced by localised magneto–ionic perturbations (e.g. supernova remnants). Therefore, the same observational data have lead different authors to different conclusions (see Frick *et al.*, 2001, for a review). Odd parity of the Galactic magnetic field has been suggested by Andreassian (1980, 1982) and, for the inner Galaxy, by Han *et al.* (1997). Quantitative methods of analysis (as opposed to the "naked-eye" fitting of more or less arbitrarily selected models) are especially appropriate in this case.

Unfortunately, it is difficult to determine the parity of magnetic field in external galaxies. In galaxies seen edge-on, the disc is depolarised, whereas Faraday rotation in the halo is weak. Beck *et al.* (1994) found weak evidence of even magnetic parity in the lower halo of NGC 253. The arrangement of polarisation planes in the halo of NGC 4631 (Beck, 2000) is very suggestive of odd parity, but this does not exclude even parity in the disc. In galaxies inclined to the line of sight, the amount of Faraday rotation produced by an odd (antisymmetric) magnetic field differs from zero because Faraday rotation and emission occur in the same volume; as a result, emission originating in the far half of the galactic layer will have small or zero net rotation (because $B_{\|}$ has a reversal in the middle of the layer), whereas emission from the near half will have significant rotation produced by the unidirectional magnetic field in that half. Therefore, Faraday rotation measures produced by even and odd magnetic structures of the same strength only differ by a factor of two (Krause *et al.*, 1989a; Sokoloff *et al.*, 1998) and it is difficult to distinguish between the two possibilities.

An interesting method to determine the parity of magnetic field in an external galaxy has been suggested by Han *et al.* (1998). These authors note that the contribution of the galaxy to the RM of a background radio source will be equal to the intrinsic RM of the galaxy if the magnetic field has even parity. For odd parity, the galaxy will not contribute to the RM of a background source, whereas any intrinsic RM will remain. The implementation of the method requires either a statistically significant sample of background sources or a single extended background source.

7.5.3. THE AZIMUTHAL STRUCTURE

Non-axisymmetric magnetic fields in a differentially rotating object are subject to twisting and enhanced dissipation as described in Section 7.5.1. The dynamo can compensate for the losses, but axisymmetric magnetic fields are still easier to maintain (Rädler, 1986). A few lowest non-axisymmetric modes with azimuthal wave numbers

$$m \lesssim \frac{s_0}{h}|\mathrm{R}_\omega|^{-1/4} \approx 2 \qquad (7.37)$$

can be maintained in thin galactic discs where $h \ll s_0$ (Section VII.8 in Ruzmaikin *et al.*, 1988). The WKBJ solution of the galactic $\alpha\omega$–dynamo equations by Starchenko & Shukurov (1989) shows that the bisymmetric mode ($m = 1$) can grow provided

$$\frac{u}{\ell\Omega}\left(\frac{h}{\ell}\right)^2 \left|\frac{\mathrm{d}\ln\Omega}{\mathrm{d}\ln s}\right| \lesssim 25\,,$$

which seems to be the case in some galaxies. These results indicate that it is natural to expect significant deviations from axial symmetry in magnetic fields of many spiral galaxies. However, the *dominance* of non-axisymmetric modes in most galaxies would be difficult to explain because the axisymmetric mode has the largest growth rate under typical conditions.

Early interpretations of Faraday rotation in spiral galaxies were in striking contrast with this picture, indicating strong dominance of bisymmetric magnetic structures ($m = 1$), $\mathbf{B} \propto \exp(\mathrm{i}\phi)$ with ϕ the azimuthal angle (Sofue *et al.*, 1986), and this was considered to be a severe difficulty of the dynamo theory and an evidence of the primordial origin of galactic magnetic fields. It was suggested by Ruzmaikin *et al.* (1986) (see also Sawa & Fujimoto, 1986; Baryshnikova *et al.*, 1987) that the bisymmetric magnetic structures can be interpreted as the $m = 1$ dynamo mode. However, despite effort, dynamo models could not explain the apparent widespread dominance of bisymmetric magnetic structures. Paradoxically, what seemed to be a difficulty of the dynamo theory has turned out to be its advantage as observations with better sensitivity and resolution and better interpretations have led to a dramatic revision of the observational picture. The present-day understanding is that modestly distorted axisymmetric magnetic structures occur in most galaxies, wherein the dominant axisymmetric mode is mixed with weaker higher azimuthal modes (Beck *et al.*, 1996; Beck, 2000). Among nearby galaxies, only M81 remains a candidate for a dominant bisymmetric magnetic structure, but the data are old and this result needs to be reconsidered (Krause *et al.*, 1989b); the interesting case of M51 is discussed below. Deviations from precise axial symmetry can result from the spiral pattern, asymmetry of the parent galaxy, etc. Dominant bisymmetric magnetic fields can be maintained by the dynamo action near the corotation radius due to a linear resonance with the spiral pattern (Mestel & Subramanian, 1991; Subramanian & Mestel, 1993; Moss, 1996) or nonlinear trapping of the field by the spiral pattern (Bykov *et al.*, 1997).

Twisting of a horizontal magnetic field by differential rotation generally produces a bisymmetric magnetic field, $m = 1$. Twisting of a horizontal primordial magnetic field can also produce an axisymmetric configuration near the galactic centre if the initial state is asymmetric (Sofue *et al.*, 1986; Nordlund & Rögnvaldsson, 2002), with a maximum of the primordial field displaced from the disc's rotation axis where the gas density is normally maximum. Thus, the maximum of the primordial field required by this scenario has to occur at a different position than the maximum in the gas density. This can only happen if the primordial field is not frozen into the gas – otherwise the field strength scales as a positive power of gas density. The fact that magnetic fields in most spiral galaxies are nearly axisymmetric within large radius (in fact, in the whole galaxy) would require that this strong asymmetry in the initial state occurs systematically for all the galaxies, which would be difficult to explain.

7.5.4. A COMPOSITE MAGNETIC STRUCTURE IN M51 AND MAGNETIC REVERSALS IN THE MILKY WAY

A striking example of a complicated magnetic structure that can hardly be explained by any mechanism other than the dynamo has been revealed in the galaxy M51 by Berkhuijsen *et al.* (1997). These authors used radio polarisation observations of the galaxy at wavelengths 2.8, 6.2, 18.0 and 20.5 cm (smoothed to a resolution of 3.5 kpc). The disc of this galaxy is not transparent to polarised radio emission at the two longer wavelengths. Therefore, it was possible to determine the magnetic field structure separately in two regions along the line the sight, which can be identified with the disc and halo of the galaxy. As shown in Figure 7.7, the regular magnetic fields in the disc is reversed in a region about 3 by 8 kpc in size extended along azimuth at galactocentric radii $s = 3$–6 kpc and azimuthal angles 300°–0 (shown with red arrows). A significant deviation from axial symmetry in the disc has been detected out to $s = 9$ kpc (in the azimuth range 160–260°), although it is too weak to result in a magnetic field reversal. The field reversal occurs around the corotation radius in M51, $s \approx 6$ kpc (i.e. the radius where the angular velocity of the spiral pattern is equal to that of the gas).

A nonlinear dynamo model for M51 was developed by Bykov *et al.* (1997) who used the rotation curve of M51, with the pitch angle of the spiral arms $-15°$ and corotation radius 6 kpc. Figure 7.7 shows one of their solutions where a region with reversed magnetic field persists in the disc near the corotation radius of the spiral pattern. Near the corotation, a non-axisymmetric (bisymmetric) magnetic field can be trapped by the spiral pattern and maintained over the galactic lifetime. The effect is favoured by a smaller pitch angle of the spiral arms, thinner gaseous disc, weaker rotational shear and stronger spiral pattern. This nonaxisymmetric structure is arguably similar to the structure observed in M51.

The regular magnetic field in the halo of M51 has a structure very different from that in the disc – the halo field is nearly axisymmetric and even directed oppositely to that in the disc in most of the galaxy. An external magnetic field should have a rather peculiar form to be twisted into such a configuration!

Distinct azimuthal magnetic structures in the disc and the halo can be readily ex-

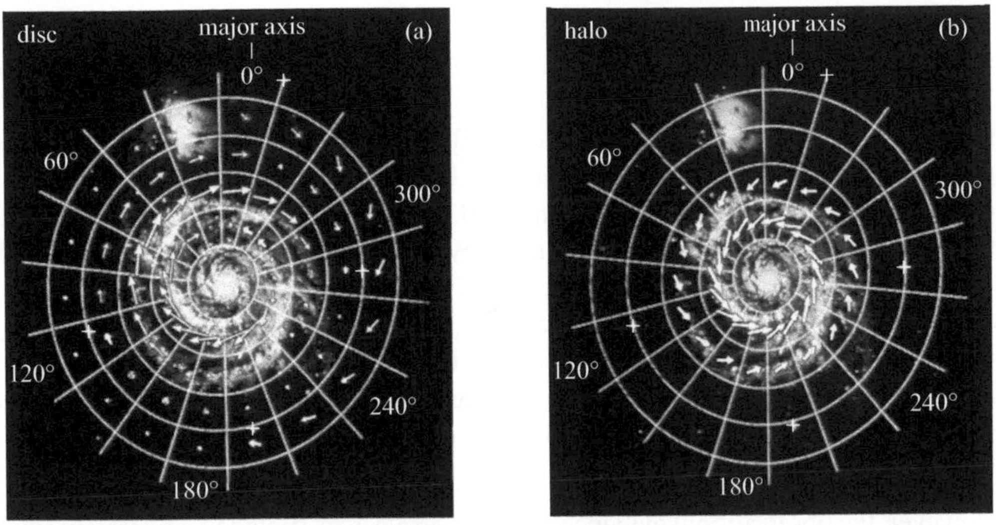

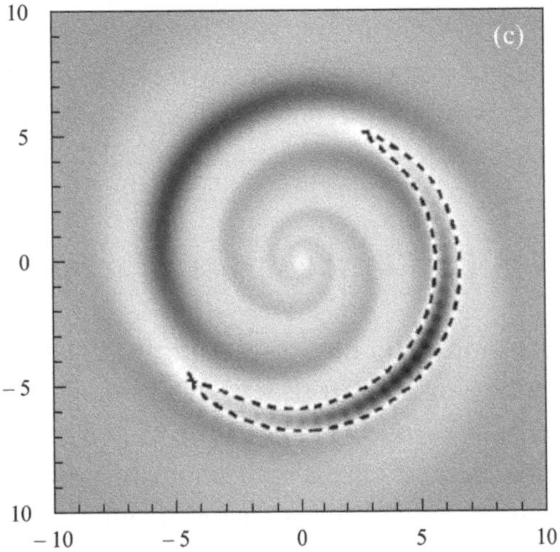

Figure 7.7 - The global magnetic structure of the galaxy M51, in the disc (a) and halo (b). Arrows show the direction and strength of the regular magnetic field on a polar grid shown superimposed on the optical image (Berkhuijsen *et al.*, 1997). The grid radii are 3, 6, 9, 12 and 15 kpc. (c) Magnetic field strength from the dynamo model for the disc of M51 (Bykov *et al.*, 1997) is shown with shades of grey (darker shade means stronger field). Magnetic field is reversed within the zero–level contour shown dashed; scale is given in kpc. The magnetic structure rotates rigidly together with the spiral pattern visible in the shades of grey. (**See colour insert.**)

plained by dynamo theory as non-axisymmetric magnetic fields can be maintained only in the thin disc but not in the quasi-spherical halo where $h \approx s_0$ and $|\mathrm{R}_\omega| \gg 1$ in (7.37). Moreover, dynamo action in the disc and the halo can proceed almost independently of each other producing distinctly directed magnetic fields (Sokoloff & Shukurov, 1990).

Another case of a regular magnetic field with unusual structure is the Milky Way where magnetic field reversals are observed along the galactocentric radius in the inner Galaxy between the Orion and Sagittarius arms at $s \approx 7.9\,\mathrm{kpc}$ and, possibly, in the outer Galaxy between the Orion and Perseus arms at $s \approx 10.5\,\mathrm{kpc}$ (Section 3.8.2 in Beck *et al.*, 1996, and Frick *et al.*, 2001; see, however, Brown & Taylor 2001). The reversals were first interpreted as an indication of a global bisymmetric magnetic structure (Sofue & Fujimoto, 1983), but it has been shown that dynamo-generated axisymmetric magnetic field can have reversals at the appropriate scale (Ruzmaikin *et al.*, 1985; Poezd *et al.*, 1993). Both interpretations presume that the reversals are of a global nature, i.e. they extend over the whole Galaxy to all azimuthal angles (or radii in the case of the bisymmetric structure). This leads to a question why reversals at this radial scale are not observed in any other galaxy (Beck, 2000). Poezd *et al.* (1993) argue that the lifetime of the reversals is sensitive to subtle features of the rotation curve and the geometry of the ionised gas layer (see also Belyanin *et al.*, 1994) and demonstrate that they are more probable to survive in the Milky Way than in, e.g. M31.

However, the observational evidence of the reversals is restricted to a relatively small neighbourhood of the Sun, of at most 3–5 kpc along azimuth. It is therefore quite possible that the reversals are local and arise from a magnetic structure similar to that in the disc of M51 as shown in Figure 7.7. The reversed field in the Solar neighbourhood has the same radial extent of 2–3 kpc as in M51 and also occurs near the corotation radius. This possibility has not yet been explored; its observational verification would require careful analysis of pulsar Faraday rotation measures.

7.5.5. THE RADIAL MAGNETIC STRUCTURE IN M31

An important clue to the origin of galactic magnetic fields is provided by the magnetic ring in M31 (Beck, 1982), which was predicted by dynamo theory (Ruzmaikin & Shukurov, 1981). Both the large-scale magnetic field and the gas density in this galaxy have a maximum in the same annulus $8 \lesssim s \lesssim 12\,\mathrm{kpc}$, with the apparent enhancement in the magnetic field strength by about 30% (Fletcher *et al.*, 2004). The kinematic dynamo model of Ruzmaikin & Shukurov (1981) was based on the double-peaked rotation curve of shown in Figure 7.5, where rotational shear is strongly reduced at $s = 2$–$6\,\mathrm{kpc}$. As a result, R_ω is small and even positive in this radial range, so $|\mathrm{D}| < |\mathrm{D_{cr}}|$ and the dynamo cannot maintain any regular magnetic field at $s = 2$–$6\,\mathrm{kpc}$.

An attractive aspect of this theory is that both magnetic and gas rings are attributed to the same feature of the rotation curve. Angular momentum transport by viscous stress leads to matter inflow at a rate

$$\dot{M} = 2\pi\Sigma\nu_\mathrm{T}(\partial \ln\Omega/\partial \ln s) \approx 0.1\,M_\odot\,\mathrm{yr}^{-1} \approx 6\times 10^{21}\,\mathrm{kg\,s^{-1}}\,,$$

where $\nu_\mathrm{T} \sim \eta_\mathrm{T}$ is the turbulent viscosity, resulting in the radial inflow at a speed $u_s = \dot{M}/2\pi s\Sigma$ with Σ the gas surface density. In the nearly-rigidly rotating parts,

u_s is reduced and matter piles up outside such a region producing gas ring. Gravitational torques from spiral arms can further enhance the inflow (see Moss *et al.*, 2000, for a discussion), so the total radial velocity is expected to be $u_s \approx 1\,\mathrm{km\,s^{-1}}$ at $s = 10\,\mathrm{kpc}$.

The double-peaked rotation curve of M31 is consistent with the existence of both magnetic and gas rings. The situation is different with the more recent rotation curve of Braun (1991) which does not have a double-peaked shape (Figure 7.5). The difference between the two rotation curves arises mainly from the fact that Braun allows for significant displacements of spiral arm segments from the galactic midplane: this results in a revision of the segments' galactocentric distances for regions away from the major axis. We note that the CO velocity field at the major axis (Loinard *et al.*, 1995) is compatible with a double-peaked rotation curve.

With Braun's rotation curve, the magnetic field can concentrate into a ring mainly because the gas is in the ring and $B \propto \rho^{1/2}$ as shown in (7.23). The dynamo model of Moss *et al.* (1998) based on the rotation curve of Braun (1991) has difficulties in reproducing a magnetic ring as well pronounced as implied by the observed amount of Faraday rotation. This has lead to an idea that magnetic field can be significant at $s = 2$–$6\,\mathrm{kpc}$ in M31. This has prompted Han *et al.* (1998) to search for magnetic fields at $s = 2$–$6\,\mathrm{kpc}$ that could have escaped detection because of reduced density of cosmic ray electrons at those radii. These authors have found that two out of three background polarised radio sources seen through that region of M31 have Faraday rotation measures compatible with the results of Moss *et al.* (1998). They further conclude that this indicates an even symmetry of the regular magnetic field. This is encouraging, but a statistically representative sample of background sources has to be used to reach definite conclusions because of their unknown intrinsic RM.

With a double-peaked rotation curve, a primordial magnetic field with a uniform radial component could have been twisted to produce a magnetic ring by virtue of (7.36). In this case the primordial and dynamo theories have similar problems and possibilities regarding the magnetic ring in M31.

Lou & Fan (2000) attribute the magnetic ring in M31 to an axisymmetric mode of MHD density waves. Because of the axial symmetry of the wave, the magnetic field in the ring must be purely azimuthal, $p_B = 0$, in contrast to the observed structure with a significant pitch angle (Figure 7.5). Furthermore, the ring can hardly represent a wave packet as envisaged in this theory because then its group velocity must be comparable to the Alfvén velocity of $30\,\mathrm{km\,s^{-1}}$ (Lou & Fan, 1998) and so the ring should be travelling at this speed along radius to traverse $30\,\mathrm{kpc}$ in $10^9\,\mathrm{yr}$, a distance much larger than the ring radius. The implication would be that the ring is a transient with a short lifetime of order $3 \times 10^8\,\mathrm{yr}$. And, of course, the theory cannot explain the origin of an azimuthal magnetic field required to launch the wave packet.

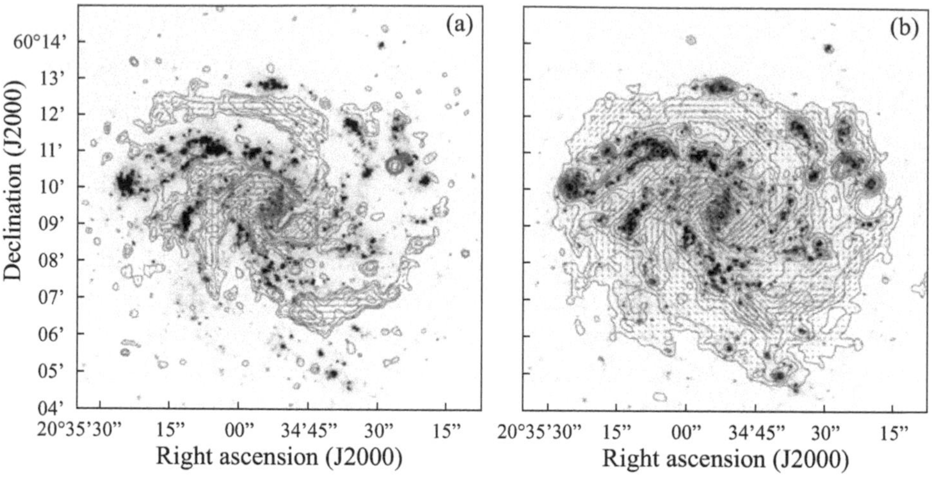

Figure 7.8 - (a) Magnetic arms in the galaxy NGC 6946: polarised intensity at the wavelength $\lambda = 6\,\mathrm{cm}$ (blue contours), a tracer of the large-scale magnetic field $\overline{B}$, superimposed on the galactic image in the Hα spectral line of ionised hydrogen (grey scale). Red dashes indicate the orientation of the B–vector of the polarised emission (parallel to the direction of intrinsic magnetic field if Faraday rotation is negligible), with length proportional to the fractional polarisation – see (7.5). The spiral arms visualised by Hα are the sites where gas density is maximum. The large-scale magnetic field is evidently stronger between the arms where gas density is lower. (b) As in the left panel, but now for the total synchrotron intensity, a tracer of the total magnetic field $B^2 = \overline{B}^2 + \overline{b^2}$. The total field is enhanced in the gaseous arms. Given that the large-scale field concentrates between the arms, this means that the random field is significantly stronger in the arms, a distribution very different from that of the large-scale field (images courtesy of R. Beck, MPIfR, Bonn). (**See colour insert**.)

7.5.6. STRENGTH OF THE REGULAR MAGNETIC FIELD

Interstellar regular magnetic fields are observed to be close to the energy equipartition with interstellar turbulence. This directly indicates that the regular magnetic field is coupled to the turbulent gas motions. [Note that $\ell\Omega$ does not differ much from the turbulent velocity u_0 in (7.23).] To appreciate the importance of this conclusion, consider primordial magnetic field twisted by differential rotation. Its maximum strength given by (7.36) as $B_{\max} \approx 10^2 B_0$ is controlled by the strength of the primordial field B_0, and so this theory, if applicable, would result in stringent constraints on extragalactic magnetic fields.

A striking evidence of the nontrivial behaviour of the large-scale galactic magnetic field are the so-called magnetic arms, discovered by Beck & Hoernes (1996) in the nearby galaxy NGC 6946. We show in Figure 7.8 a map of polarised radio emission from this galaxy (a tracer of the large-scale magnetic field strength) superimposed on the map in the Hα spectral line (a tracer of ionised gas). It is evident that magnetic field is stronger *between* the spiral arms of this galaxy, i.e. where the gas density (both total and ionised) is lower. This behaviour is just opposite to what is expected of a frozen-in magnetic field that scales with a power of gas density. The phenomenon of magnetic arms confirms in a spectacular manner that the large-scale magnetic field is not frozen into the interstellar gas, and therefore cannot be primordial. Shukurov (1998), Moss (1998) and Rohde *et al.* (1999) suggest an explanation of the magnetic arms in terms of the mean-field dynamo theory. In brief, they argue that dynamo number can be larger between the gaseous spiral arms, resulting in stronger dynamo action.

The theory of MHD density waves relates magnetic field excess $\Delta\overline{B}$ in spiral arms to the enhancement in stellar density, $\Delta\overline{B}/\langle\overline{B}\rangle = \Delta\Sigma/\langle\Sigma\rangle$ (Lou & Fan, 1998), where Σ is the stellar surface density, $\Delta\Sigma$ is its excess in the spiral arms, and angular brackets denote azimuthal averaging. Arm intensities in magnetic field and stellar surface density in NGC 6946 have been estimated by Frick *et al.* (2000) who applied wavelet transform techniques to radio polarisation maps and to the galaxy image in broadband red light, a tracer of stellar mass density. Their results indicate that the mean relative intensity of magnetic spiral arms remains rather constant with galactocentric radius at a level of 0.3–0.6. On the contrary, the relative strength of the stellar arms systematically grows with radius from very small values in the inner galaxy to 0.3–0.7 at $s = 5$–$6\,\mathrm{kpc}$, and then decreases to remain at a level of 0.1–0.3 out to $s = 12\,\mathrm{kpc}$. The distinct magnitudes and radial trends in the strengths of magnetic and stellar arms in NGC 6946 do not seem to support the idea that the magnetic arms are due to MHD density waves.

7.6. ELLIPTICAL GALAXIES

Elliptical galaxies do not rotate fast enough, so they are ellipsoidal systems without prominent disc components. The stellar population of elliptical galaxies is old and the interstellar gas is dilute (Mathews & Brighenti, 2003). Therefore, both relativistic and thermal electrons have low density, and any synchrotron emission and Faraday rotation can only be weak. Nevertheless, there are several lines of evidence, albeit mostly indirect, suggesting significant magnetic fields in ellipticals (Moss & Shukurov, 1996; Mathews & Brighenti, 1997). The magnetic field should be random, producing unpolarised synchrotron emission and fluctuating Faraday rotation. The root mean square (r.m.s.) Faraday rotation measure attributable to the ISM of the ellipticals is $\sigma_{\mathrm{RM}} = 5$–$100\,\mathrm{rad\,m^{-2}}$.

7.6.1. TURBULENT INTERSTELLAR GAS IN ELLIPTICAL GALAXIES

Interstellar gas in elliptical galaxies is observed via its X-ray emission. Type I supernovae (SNe) (and also stellar winds and random motions of stars) heat the gas to the observed temperatures $T \approx 10^7\,\mathrm{K}$. It is natural to expect that a fraction δ of the energy is converted into turbulent motions of the gas. The turbulent scale $\ell \approx 400\,\mathrm{pc}$ is given by the diameter of a SN remnant as it reaches pressure balance with the ambient medium whose typical density is $n \approx 10^{-3}\,\mathrm{cm}^{-3}$. The balance between energy injection and dissipation rates yields a turbulent velocity of $u_0 \approx 20\,\mathrm{km\,s}^{-1}$ for $\delta = 0.1$, assuming the energy dissipation time $\tau \approx \ell/u_0$ as for the Kolmogorov turbulence. This estimate of u_0 is compatible with the constraint $u_0 \lesssim 50\,\mathrm{km\,s}^{-1}$ resulting from the observed X-ray luminosity. Another driver of turbulence is the random motions of stars. These generate random vortical motions at a smaller scale and velocity, $\ell_* \approx 3\,\mathrm{pc}$ and $u_* \approx 3\,\mathrm{km\,s}^{-1}$, respectively (Moss & Shukurov, 1996).

The driving force produced by an expanding quasi-spherical SN remnant is potential. The above estimates assume that the motions driven by the SNe are vortical, so $\tau = \ell/u_0$ applies. In spiral galaxies, the potential (acoustic) motions are efficiently converted into vortical turbulence mainly due to the inhomogeneity of the ISM. The ISM in elliptical galaxies is hot and, presumably, rather homogeneous at kpc scales. Therefore, Moss & Shukurov (1996) suggested that SNe will drive sound-wave turbulence whose correlation time τ is $(u_0/c_\mathrm{s})^{-2}\ell/c_\mathrm{s} \approx 3 \times 10^7\,\mathrm{yr}$ rather than ℓ/u_0, where $c_\mathrm{s} \approx 300\,\mathrm{km\,s}^{-1}$ is the speed of sound. However, Mathews & Brighenti (1997) noted that sound waves quickly dissipate, and so cannot form a pervasive turbulent velocity field. The nature of turbulence in elliptical galaxies needs to be studied further.

7.6.2. THE FLUCTUATION DYNAMO IN ELLIPTICAL GALAXIES

As in most astrophysical objects, Rm in elliptical galaxies by far exceeds 100, so fluctuation dynamo action in ellipticals is quite plausible (see Section 7.4.2). The e-folding time of the random field in a vortical random flow is of the order of the eddy turnover time, $\tau = 2 \times 10^7\,\mathrm{yr}$. The magnetic field is concentrated into flux ropes and ribbons whose length and thickness are of order $\ell \approx 400\,\mathrm{pc}$ and $\ell\,\mathrm{Rm}_\mathrm{cr}^{-1/2} \approx 40\,\mathrm{pc}$. In the ropes, magnetic field is plausibly in equipartition with the turbulent kinetic energy, $b \approx 0.3\,\mu\mathrm{G}$.

Moss & Shukurov (1996) discuss a two-stage dynamo action by smaller scale vortical turbulence driven by random motions of the stars and, at larger scales, by the acoustic turbulence. However, the very existence of the acoustic turbulence in elliptical galaxies is questionable (Mathews & Brighenti, 1997).

Faraday rotation measure produced within a single turbulent cell with the above parameters is $\mathrm{RM}_0 \approx 0.81 b n_e \ell$, so the net Faraday rotation from an ensemble of turbulent cells, observed at a resolution D such that $D \gg \ell$, is given by $\sigma_{\mathrm{RM}} \sim \mathrm{RM}_0 \sqrt{N/N_D} \sim \mathrm{RM}_0 \sqrt{L\ell}/D$, where $N = L/\ell$ is the number of cells along the path length L and $N_D = (D/\ell)^2$ is the number of cells in the resolution element. Thus, $\mathrm{RM}_0 \approx 0.1\,\mathrm{rad\,m^{-2}}$ and $\sigma_{\mathrm{RM}} \approx 1\,\mathrm{rad\,m^{-2}}$. Faraday rotation can be stronger in the central regions where $\sigma_{\mathrm{RM}} \approx 5\,\mathrm{rad\,m^{-2}}$ at a distance $8\,\mathrm{kpc}$ from the galactic centre. These estimates agree fairly with the available observations.

Magnetic field generation in elliptical galaxies was discussed by Lesch & Bender (1990), but they considered a mean-field dynamo that needs overall rotation which is not present in elliptical galaxies. The fluctuation dynamo in elliptical (radio) galaxies was simulated by De Young (1980), but these simulations apparently had $\mathrm{Rm} < \mathrm{Rm_{cr}}$ as they resemble transient amplification of magnetic field by velocity shear rather than genuine dynamo action.

7.7. ACCRETION DISCS

Accretion discs represent another type of flat, rotating astrophysical objects, where magnetic fields are involved in or drive many important processes. The defining property of the accretion disc is a systematic radial flow of matter which often feeds activity at a massive central object (a protostar, compact star, of black hole). The mass of the disc is usually negligible in comparison with that of the central object. The gas in the disc is usually cool enough for its radial pressure gradient to be negligible, so that the gas rotation in nearly Keplerian, $\Omega \propto s^{-3/2}$. The disc thickness is controlled by hydrostatic equilibrium in the gravitational field of the central object; if the disc is cool, it is then very thin far enough from the centre. Accretion discs have to be turbulent to provide the accretion rate implied by the central activity observed. Altogether, accretion discs are turbulent, rotating, stratified objects and, thus, they can (or even must) be a site of dynamo activity at both large and small scales (if only the gas is ionised); the theory of disc dynamos presented above can be readily applied to these objects. In fact, accretion flows, at a speed of order $1\ \mathrm{km\ s^{-1}}$, are common in spiral galaxies as well, where they are driven mostly by the nonaxisymmetric gravitational torque of the stellar spiral arms (Moss *et al.*, 2000, and references therein).

Pudritz (1981a,b) developed an $\alpha\omega$–dynamo model for accretion discs, and Stepinski & Levy (1991) and Mangalam & Subramanian (1994a,b) discussed a thin-disc asymptotic solution similar to that discussed in Section 7.4.1 above (in the form suggested by Ruzmaikin *et al.*, 1985, 1988). MHD simulations of accretion discs performed by Brandenburg *et al.* (1995) and Stone *et al.* (1996) confirmed the efficiency of dynamo action and showed that the turbulence can be driven by the magneto–rotational instability mediated by the magnetic field. Numerical simulations of an outflow from an accretion disc that hosts a mean-field dynamo confirmed that the dynamo can produce a magnetic configuration suitable for launching vigorous outflows that occur near accretion discs (von Rekowski *et al.*, 2003; Moss & Shukurov, 2004; see below). Campbell (1999, 2000, 2001, 2003) and Campbell & Caunt (1999) discuss accretion disc models and disc winds based on magnetic fields produced by the mean-field dynamo. The role of dynamos in accretion discs is discussed by Brandenburg & Subramanian (2005) and Pudritz *et al.* (2006). The effects of accretion on *galactic* dynamos are discussed by Moss *et al.* (2000) and Moss & Shukurov (2001); the effects of vertical outflows on mean-field disc dynamos are explored by Bardou *et al.* (2001).

It might seem that the flow of matter towards the disc axis would lead to additional enhancement of magnetic field and therefore facilitate the dynamo action. However, this is not true: radial flow in a thin disc reduces the growth rate of the mean magnetic field (Moss *et al.*, 2000). Without any significant field at infinity (an appropriate condition for a dynamo system), the dynamo has to replenish magnetic field advected by the radial flow away from the region where it is generated most efficiently; this hampers the dynamo action. Magnetic field compression cannot counterbalance this effect. A noticeable effect of the compression is to produce a sharper peak in magnetic field strength at smaller radii. As shown by Moss *et al.* (2000), an additional radial velocity U_s reduces the local growth rate $\gamma(s)$ in the radial dynamo equation (7.17) which is modified as $\gamma(s) \rightarrow \gamma(s) - \frac{1}{4}(U_s h_0/\eta_{\mathrm{T}})^2$. Thus, both inflow and outflow reduce the growth rate and hence suppress the dynamo action. Moss *et al.* (2000) suggest that the suppression of the dynamo action by radial flows driven by magnetic stress can saturate the dynamo independently of other nonlinearities; their estimate of the corresponding steady-state mean magnetic field due to this effect alone is about $10\,\mu\,\mathrm{G}$ in the galactic context, which is only slightly larger than what is observed in spiral galaxies. This indicates that this nonlinear effect may contribute significantly to the saturation of the dynamo, especially in the inner parts of galaxies.

An interesting feature of accretion-disc dynamos is that turbulence in accretion discs is driven by the magneto–rotational instability which itself relies on the presence of magnetic field. Brandenburg (2000) (see also section 11.4 in Brandenburg & Subramanian, 2005) argue that this results in the sign of the α–effect which is opposite to that in planets, stars and galaxies, i.e. $\alpha < 0$ for $z > 0$. Therefore, the dominant mean-field dynamo mode in the accretion discs has dipolar parity, which is believed to be more favourable for launching disc wind.

An important role played by magnetic fields in accretion discs is to drive a wind and also to collimate it into a narrow jet; such jets are observed in a large number of radio galaxies and, at much smaller scales, in protostellar objects. A large-scale magnetic field anchored in the accretion disc is a keystone of most models of disc winds and jets (Königl & Pudritz, 2000; Pudritz *et al.*, 2006, and references therein). It is usually assumed that magnetic field required to launch and collimate the outflows is a large-scale external magnetic field captured and then deformed into an hour-glass shape by the accreting gas. Since dynamos normally produce magnetic fields whose field lines close within the immediate vicinity of the disc, it is often assumed that magnetic configurations produced by the dynamo cannot support an outflow extending to the very large distances observed (Pelletier & Pudritz, 1992; section V.B in Königl & Pudritz, 2000). However, numerical simulations of von Rekowski *et al.* (2003) show that a dynamo-generated magnetic field can support a vigorous outflow, provided the field can be opened up by, say, a thermally driven wind.

Otherwise, the dynamo-generated magnetic field can be opened up by external magnetic fields. Accretion discs around protostellar objects can reasonably be expected to occur in strongly magnetised interstellar environment, so that it is not unreasonable to assume that there exists a strong external magnetic field ordered at scales much larger than the disc size. However, this assumption is much more questionable in the case of extragalactic jets whose length can be as large as hundreds kiloparsecs: the existence of magnetic fields at such scales is not supported by any observational or theoretical evidence. Moreover, it is unclear whether an external magnetic field can be dragged into the disc given the large magnitude of the effective (turbulent) magnetic diffusivity compatible with the turbulent nature of the disc. Since the large-scale dynamo action in most types of accretion discs seems to be unavoidable, interaction of an external magnetic field with the disc dynamo has to be considered, and wind launching in the resulting composite magnetic configurations has to be explored.

It was suggested by Reyes-Ruiz & Stepinski (1997) and Reyes-Ruiz (2000) that an external magnetic field can open magnetic lines of a field produced by the mean-field disc dynamo to make the magnetic configuration suitable for the magneto–centrifugal wind launching. However, these authors used a simple superposition of a dynamo-generated magnetic field and a uniform external field; this only makes sense for kinematic dynamos and neglects any effect of the external magnetic field on the dynamo action. A nonlinear mean-field dynamo model in an accretion disc embedded into an external magnetic field (that can be dragged into the disc by the accretion flow) was explored by Moss & Shukurov (2004) who showed that a relatively weak external magnetic field can open up magnetic lines produced by the dynamo. Magnetic fields in the inner parts of the disc are opened first (i.e. for a weaker external field); even for a relatively weak external magnetic field (whose energy density exceeds about only 10% of the thermal energy density in the outer parts of the disc), the geometry of the poloidal magnetic field above the disc surface is almost independent of the details of the dynamo model and favourable for launching a centrifugally driven wind. Remarkably, the radial profile of the poloidal magnetic field strength on the disc surface is similar to the well-known self-similar wind solution of Blandford & Payne (1982) if the standard disc model of Shakura & Sunyaev (1973) is used. This radial profile is also consistent with that obtained for dipolar modes of kinematic mean-field dynamos from the asymptotic solutions of Soward (1978, 1992a, 2003).

7.8. CONCLUSIONS

The observational picture of galactic magnetic fields is compatible with the mean-field dynamo theory in its simplest, quasi-kinematic form. It is important to note that there is not much freedom in varying parameters of galactic dynamos as observations constrain them fairly tightly. Therefore, this agreement is not a result of a free-hand parameter adjustments. Moreover, galactic dynamo theory has demonstrated its predictive power. For example, it has been clear to dynamo theorists since the early 1970's that the partially ionised Galactic disc must have the scale height 0.4–0.5 kpc, i.e. significantly larger than that of the neutral hydrogen layer (see Section VI.2 in Ruzmaikin *et al.*, 1988, for a review), but the existence of this component of the interstellar medium was accepted by a broader astrophysical community only 15 years later (Lockman, 1984).

The agreement of the mean-field dynamo theory with observations discussed in Section 7.5 cannot be considered as a proof of its correctness – history of physics is familiar with concepts, such as the ether, that have proved to be irrelevant despite their perfect agreement with numerous experimental facts, before a single experiment refuted them. Nevertheless, the spectacular success of the dynamo theory when applied to galaxies warrants its careful treatment when compared with other theories. In particular, any rival theory has to be able to explain *at least* the same set of observational data as the dynamo theory. In this sense, there are no fair rivals to the dynamo theory.

The main current difficulty of the galactic dynamo theory is our lack of understanding of its nonlinear form. It is important to avoid an unjustified extension to real galaxies of results obtained for highly idealised systems. In particular, the disc of a spiral galaxy is *not* a closed system. The significance of the disc-halo connection for the mean-field galactic dynamos, touched upon in Section 7.4.3, should be carefully investigated.

Another outstanding problem in theory and observations of galactic magnetic fields is the effect of the multi-phase structure of the interstellar medium on the magnetic field. The effect of magnetic fields on the multi-phase structure is also expected to be very significant, and also poorly understood. The strength of the magnetic field in the hot phase is not known. The detailed nature of the balance between cosmic rays and magnetic field has to be clarified: it is unclear whether this balance is maintained pointwise (at each location) or only on average (e.g. at scales exceeding the diffusion scale of cosmic ray particles). The answer to this question is essential for the interpretation of the synchrotron emission from spiral galaxies. Progress in this direction will eventually make theory of galactic magnetic fields an integral part of galactic astrophysics.

CHAPTER 8

SURVEY OF EXPERIMENTAL RESULTS

Philippe Cardin & Daniel Brito

8.1. INTRODUCTION

Magnetohydrodynamic experiments with liquid metals and natural dynamos share a common property, a very low magnetic Prandtl number ($\mathrm{Pm} = \nu/\eta$, see Chapter2). Typical values of the magnetic Prandtl number for liquid metals in the core of terrestrial planets are indeed very small ($< 10^{-4}$), despite the extreme pressure and temperature conditions (Chapters 4 & 5 and Poirier, 1988). In gaseous planets, the hydrogen gas presents a metallic phase at large pressure, with again a low value of Pm (Chapter 5 and Guillot, 1999). Further, both stellar convection zones (Chapter 6) and diluted gas in magnetospheres are in the low Pm regime. As liquid metals share this low Pm property, we believe that most of natural dynamos may be modelled and studied in a laboratory. A number of groups in the world have therefore focused their liquid experiments on fundamental aspects of the magnetohydrodynamics of natural objects.

To date, two experiments using liquid sodium – in Riga (in Latvia) and in Karlsruhe (in Germany) – have succeeded in observing dynamo action. Sodium was chosen primarily because it is the highest conductor of electricity among liquid metals at laboratory conditions. With liquid sodium, it is possible to reach magnetic Reynolds numbers ($\mathrm{Rm} = UL/\eta$) of the order of 10 in an experiment of metric size, where

U is the typical velocity of the fluid, L is the typical scale of the experiment and η the magnetic diffusivity (see Table VI for numerical values), the velocity and size being chosen to optimize Rm. The velocity of a fluid inside a container of a given size is, however, directly related to the mechanical power available to perform the experiment. In the turbulent regime, the power P may be expressed as the cubic power of the velocity

$$P \propto \rho L^2 U^3 , \tag{8.1}$$

where ρ is the density of the fluid, as measured for example in the Riga experiment (see Section 8.3.6 and Figure 8.15). In the presence of global rotation or other external forcing, a different scaling of the power may be obtained, but leads to the same conclusion (Cardin *et al.*, 2002): a huge power of a few hundreds of kilowatts has to be injected in a metric size experiment of sodium in order to reach values of Rm of the order of 100. This is why dynamo experiments generally require heavy infrastructures.

As other liquid metals, liquid sodium has very low magnetic, Pm, and thermal, Pr, Prandtl numbers (see Table VI). As such the magnetic diffusion is much larger than thermal diffusion; and thermal diffusion is in turn much larger than viscous diffusion. It is worth noting that the magnetic diffusion time, $\tau_\eta \;\; \propto L^2/\eta$, of an experimental dynamo of metric size with liquid sodium is of the order of a few seconds, a very long time compared to the dynamical timescale (the turnover time for a fluid particle below $10^{-1}\,\mathrm{s}$ in a $1\,\mathrm{m}$ size dynamo experiment). This means that if a growing magnetic field is observed for more than a few seconds during an experiment, this would be sufficient to demonstrate dynamo action. Experiments performed performed during thousand of magnetic diffusion times are, in principle, feasible.

The intrinsic molecular properties of liquid metals makes numerical simulations of magnetohydrodynamic dynamos very difficult. From the low Pr and Pm numbers associated with liquid metals, one expects the temporal and spatial scales of the magnetic field, the velocity field, and the temperature field to be very different. As $\mathrm{Rm} = \mathrm{RePm}$, dynamo experiments with large $\mathrm{Rm} \sim \mathcal{O}(10)$ will necessarily correspond to a very large hydrodynamic Reynolds number ($\mathrm{Re} \approx \; 10^7 - 10^8$), meaning that experimental dynamos will have undoubtedly strong turbulent flows. In presence of a strong magnetic field, or rotation, the statistical and geometrical properties of these turbulent magnetohydrodynamic flows may be different from those of pure hydrodynamical turbulence; for example in the Earth's core (Chapter 4) the Reynolds number is presumably very high ($\mathrm{Re} \gg 1$), but the nonlinear term in the Navier-Stokes equation is of second order compared to Coriolis and Lorentz forces [the Rossby number (see page 11) $\mathrm{Ro} \ll 1$, and the interaction parameter (see page 71) $\mathrm{N} \gg 1$] and therefore the nonlinear chaotic behaviour of the system might be induced by the nonlinearities in the induction equation or in the energy equation.

As the flows are expected to be turbulent, it might be interesting to consider turbulent diffusivities to describe small scales in numerical modelling (Glatzmaier & Roberts, 1995; Matsushima *et al.*, 1999; Phillips & Ivers, 2003; Buffett, 2003). However, little is known about these processes in magnetohydrodynamics cases and experiments are certainly needed to check the validity of this idea and to propose scaling laws for turbulent diffusivities (see discussion in Section 8.3.9). More generally, experiments are often very useful, as they enable the verification of theoretical considerations or the confirmation of numerical calculations. Such experiments can also be used to test the validity of assumptions necessary to understand the physics of dynamos. A great advantage of experiments as compared to numerical simulations is the fact they are performed with real liquid metals, with known physical properties. Properties that are often not achievable in numerical simulations. Such experiments shed light also on new unexpected effects and new unexpected regimes. However, the main drawback of the experimental approach is the limitation of the measurements. Measurements in the bulk of the flow, in which the dynamo process operates, are particularly problematic. Experiments require an important theoretical and numerical work to complete the understanding of measurements. Note that experiments with liquid metals can also lead to improvements in technology and instrumentation, which may indirectly benefit to the industry for example of metallurgy.

We will consider that a dynamo exists when a non-zero magnetic field is solution to the induction equation [see equation (1.14), Chapter 1]. Clearly, $\mathbf{B} = \mathbf{0}$ is also solution. In the laboratory, it is very difficult to have a strictly zero external magnetic field around the experiment. The magnetic field of the Earth itself produces a background magnetic field of few tenths of Gauss ($1 - 5 \times 10^{-4}$ T, see Chapter 4). From a theoretical point of view, it is very difficult to differentiate between a self-sustained magnetic field of an experimental dynamo from the one produced by a simple amplification of the ambient field by the velocity flow. The same problem exists regarding the observation of planetary magnetic fields (Chapter 5); for instance, the magnetic field of Io is believed to be produced by magnetoconvection in the core of Io in the presence of the jovian external magnetic field (Sarson *et al.*, 1997). In practice, during an experiment, one assumes dynamo action based on the observation that there is a large self-induced magnetic field compared to the ambient field that persists for a long time, compared to the magnetic diffusion time. Moreover, there is generally a very clear transition between the induction and the self-induction of a magnetic field as the experimental parameters are varied.

To date there have been many experiments that have been conducted with the objective of describing astrophysical and geophysical objects. However, we will restrict this review to those experiments directly devoted to the understanding of dynamo action. Other experiments have been carried out to understand the basic dynamics on which a dynamo can start. These cover aspects such as thermal convection in a

Table VI - Liquid sodium properties at 393 K (120 °C).

Density	ρ	932×10^3	$\mathrm{kg\,m^{-3}}$
Dynamic viscosity	μ	6.2×10^{-4}	$\mathrm{kg\,m^{-1}\,s^{-1}}$
Kinematic viscosity	$\nu \equiv \mu/\rho$	6.77×10^{-7}	$\mathrm{m^2\,s^{-1}}$
Fusion temperature	T_f	97.8	°C
Thermal expansion coefficient (at 298 K)	α	7.1×10^{-5}	$\mathrm{K^{-1}}$
Thermal conductivity	k	85.9	$\mathrm{W\,K^{-1}\,m^{-1}}$
Specific heat	c_P	1373	$\mathrm{J\,kg^{-1}\,K^{-1}}$
Thermal diffusivity	$\kappa \equiv k/(\rho\, c_P)$	6.71×10^{-5}	$\mathrm{m^2\,s^{-1}}$
Electrical conductivity	σ	9.35×10^6	$\Omega^{-1}\,\mathrm{m^{-1}}$
Magnetic diffusivity	$\eta \equiv (\mu_0\,\sigma)^{-1}$	8.53×10^{-2}	$\mathrm{m^2\,s^{-1}}$
Prandtl number	$\mathrm{Pr} \equiv \nu/\kappa$	10^{-2}	
Magnetic Prandtl number	$\mathrm{Pm} \equiv \nu/\eta$	7.9×10^{-6}	

rotating sphere, precession, boundary effects or instabilities, magnetohydrodynamics turbulence. They are reported in a review by Nataf (2003).

This chapter will be divided in two parts. The first part describes a set of experiments using the same definition for all dimensionless numbers in order to allow an easier comparison between these. It ends with a discussion on sodium technology and measurements. The experimental results are discussed in terms of dynamo mechanisms in the second part. Future challenges of experimental dynamo modelling are exposed in the conclusion.

8.2. DESCRIPTION OF THE EXPERIMENTS

A survey of magnetohydrodynamics experiments devoted to study high magnetic Reynolds numbers flows is presented here in a chronological order. Although numerous experiments greatly improved our knowledge on dynamo mechanisms, it is only in 1999, that great advances were achieved. It was then that the dynamo effect was first measured in a flow of liquid sodium, quasi-simultaneously in Riga (Gailitis *et al.*, 2000) and Karlsruhe (Müller & Stieglitz, 2000). In addition to these two successful dynamo experiments, a number of high magnetic Reynolds experiments of a second generation have recently been performed in order to investigate magnetic field amplification in less constrained flows than in Riga or Karlsruhe. Some new projects of sodium experiments are presented at the end of this survey. In the following sections, every experiment is described and presented in a schematic diagram, completed by a table giving its main characteristics and relevant dimensionless numbers.

Specific conditions associated with the use of sodium in dynamo experiments are listed at the end of this section. Progress and limitations of measurement techniques in dynamo experiments are eventually discussed.

8.2.1. A RAPIDLY ROTATING DISC IN A CYLINDER OF SODIUM

B. Lehnert can be considered as a pioneer of magnetohydodynamics experiments with liquid metals such as mercury (Lehnert, 1951; Lehnert & Little, 1956) and sodium (Lehnert, 1958). His most relevant experiment for dynamo mechanisms is the latter one, performed in liquid sodium. He constructed a cylindrical vessel filled with sodium, a rotating copper disc driving the flow inside. Lehnert successfully verified the so-called ω–effect by measuring the conversion of an imposed poloidal magnetic field $\mathbf{B}_0$ (generated by a coil below the vessel) into a toroidal one by an axisymmetric flow of liquid sodium.

▷ $\mathrm{N} = \dfrac{\sigma B_0^2\, R}{\rho\,(\omega_{\mathrm{disc}} R_{\mathrm{disc}})} \approx 0 - 1.$

▷ $\mathrm{Rm} = \mu_0 \sigma (\omega_{\mathrm{disc}} R_{\mathrm{disc}}) R \leq 10.$

▷ Prediction: induced field of the same order as the imposed magnetic field with the same apparatus if $\rightarrow \omega_{\mathrm{disc}} \approx 3000\,\mathrm{rpm}$.

▷ Power of the rotating motor, $P_{\mathrm{motor}} \leq 3\,\mathrm{kW}$.
▷ $\omega_{\mathrm{disc}} \leq 500\,\mathrm{rpm}$ (rounds per minute).
▷ $B_0 \leq 0.03\,\mathrm{T}$ at the height of the disc.

<u>Measurements</u>:
▷ Power input for a constant rotation rate of the motor.
▷ Induced magnetic field measured by a probe coil in the bulk of the fluid.

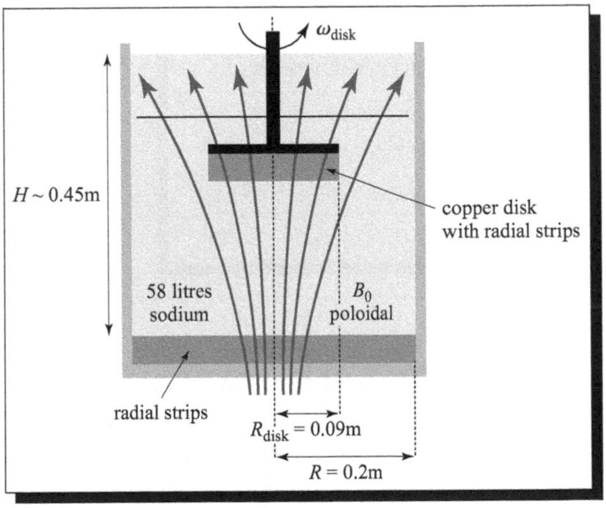

8.2.2. A DYNAMO WITH TWO SOLID ROTATING CYLINDERS

Lowes & Wilkinson (1963, 1968) were the first to achieve a solid dynamo experiment in the laboratory following an idea of Herzenberg (1958). Two ferromagnetic cylinders of iron alloy with their axes at right angle were rotated independently in a housing of the same material (Lowes & Wilkinson, 1963). Electric contacts between the housing and the rotating cylinders were done with liquid mercury. With adequate directions of rotation and sufficiently high angular velocities for the cylinders, the ambient magnetic field was amplified and eventually, at the critical Rm, the magnetic field of the system (cylinder 1 and cylinder 2) became self-sustained provided the cylinders were kept rotating at the same velocities. The dynamo mechanism was the following: the induced toroidal magnetic field (via the ω–effect) of cylinder 1 provides the external poloidal magnetic field of cylinder 2, the induced toroidal magnetic field (via the ω–effect again) of cylinder 2 providing in turn the external poloidal magnetic field of cylinder 1. Note that, with a slightly modified experimental set-up, but still with ferromagnetic materials, Lowes & Wilkinson (1968) could witness reversals of the self-sustained magnetic field in their system.

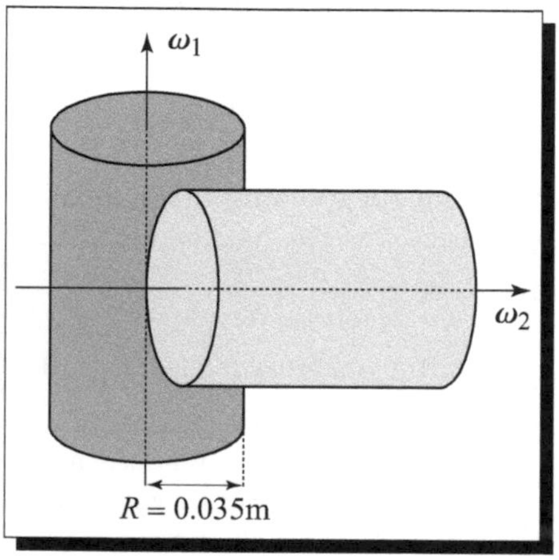

Measurements:

▷ Induced magnetic field.

▷ Differences in electric potential between cylinders.

▷ $B_{\text{induced}} \leq 0.1\,\text{T}$.

▷ $\text{Rm} = \mu^*_{\text{iron}}\sigma\omega R^2 \leq 200$.

▷ $P_{\text{motor}} \approx 2 \times 100\,\text{W}$.

▷ $\omega_1 \approx \omega_2 \leq 2000 - 3000\,\text{rpm}$.

▷ B_0 = ambiant magnetic field.

▷ $\mu_{\text{iron alloy}} \approx 150\mu_0$.

▷ Depth between the two axes of the cylinder = $0.08\,\text{m}$ in the 1963 experiment.

8.2.3. THE α–BOX EXPERIMENT

▷ $\text{Re} = \dfrac{UL}{\nu} \approx 5 \times 10^5$.

▷ $\text{N} = \dfrac{\sigma B_0^2 L}{\rho U} \approx 0 - 30$.

▷ $\text{Rm} = \mu_0\,\sigma\,U\,L \leq 2$.

A joint Postdam-Riga experiment was constructed to measure the so called α–effect in a small container: liquid sodium was run through a system of orthogonally wounded channels of stainless steels (Steenbeck *et al.*, 1968). The set-up was designed to drive the sodium through an helicoidal flow under an imposed magnetic field. Differences in electric potential between the bottom and the top of the container were measured with a pair of electrodes in the direction of the applied magnetic field.

Figure 8.1 - The experiment of Lowes & Wilkinson at the University of Newcastle, England, in 1963 (photographs courtesy of F. Lowes).

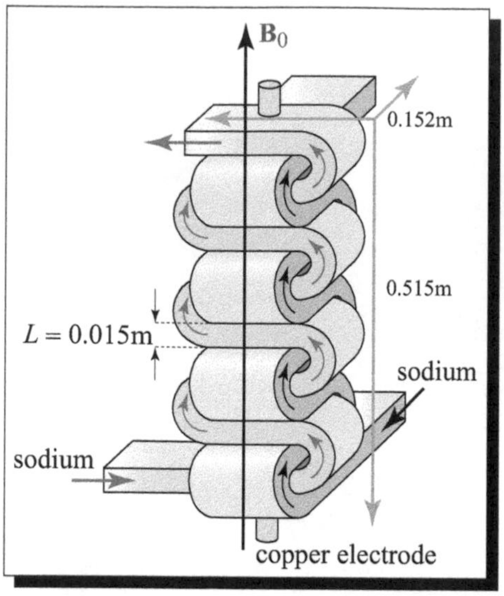

▷ Number of horizontal channels along the total height = 28 .

▷ Sodium velocity, $U_{\rm max} = 11\,{\rm m\,s^{-1}}$.

▷ $B_0 \leq 0.3\,{\rm T}$.

Measurements:

▷ Differences in electric potential between the top and bottom with electrodes.

8.2.4. A PRECESSING EXPERIMENT IN LIQUID SODIUM

Gans (1970) initiated magnetohydrodynamic experiments in the presence of a global rotation of the system (and therefore in presence of the Coriolis force). He built a precessing experiment in liquid sodium following the work of Malkus (1968) in water. A cylinder filled with liquid sodium rotated in an axisymmetric imposed magnetic field along its rotation axis. The whole set-up was spun-up simultaneously on a rotating table, at right angle to the rotation axis of the cylinder. The experiment was built with the theoretical idea that the precession of the Earth's core may be one of the main source of energy of the dynamo (Malkus, 1994; Kerswell, 1996; Noir *et al.*, 2003). Unfortunately, due to technical difficulties, Gans (1970) could not cover the full expected parameter regime with this experiment.

▷ $\omega_{\rm cyl} \leq 3600\,{\rm rpm}$.

▷ $\Omega_{\rm tab} \leq 50\,{\rm rpm}$.

▷ $B_0 = 0.023,\ 0.046\,{\rm T}$ by a d.c. coil.

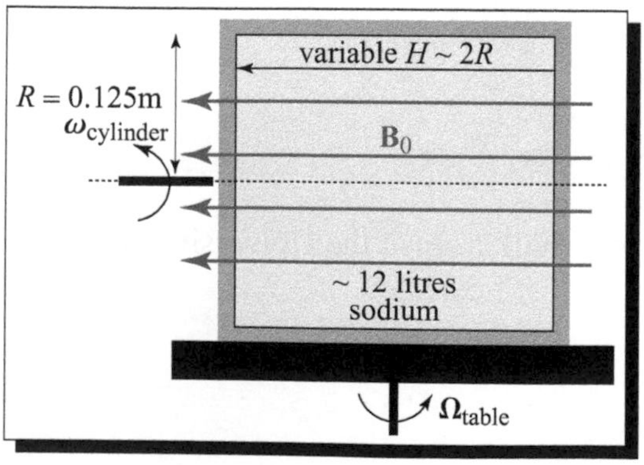

Measurements:

▷ Torque imposed by the rotation of the cylinder induced magnetic field.

▷ $\mathrm{E} = \dfrac{\nu}{\Omega_{\rm tab} R^2} \approx 10^{-7}$.

▷ $\mathrm{N}\ \dfrac{\sigma B_0^2 R}{\rho(\omega_{\rm cyl} R)} \approx 10^{-2}$.

▷ $\mathrm{Rm} = \mu_0\, \sigma\, \omega_{\rm cyl}\, R^2 \leq 70$.

8.2.5. THE FIRST PONOMARENKO TYPE EXPERIMENT

Following a theoretical prediction of dynamo action in an endless helical stream of screw type (Ponomarenko, 1973; see also Chapter 1, Section 1.6.1) for a relatively low critical magnetic Reynolds, an experimental set up was assembled in Riga and run in Leningrad in 1986 (Gailitis *et al.*, 1987). More than 150 litres of sodium was powered by electromagnetic pumps and circulated through a cylinder with an helicoidal diverter at the top. The external field was imposed by a 3–phase generator as theoretically required by the Ponomarenko dynamo. Although the experiment was run successfully, it had to be stopped probably close to the dynamo onset (see Section 8.3.5 and Figure 8.11) due to mechanical vibrations in the central thin wall of stainless steel in the center of the device.

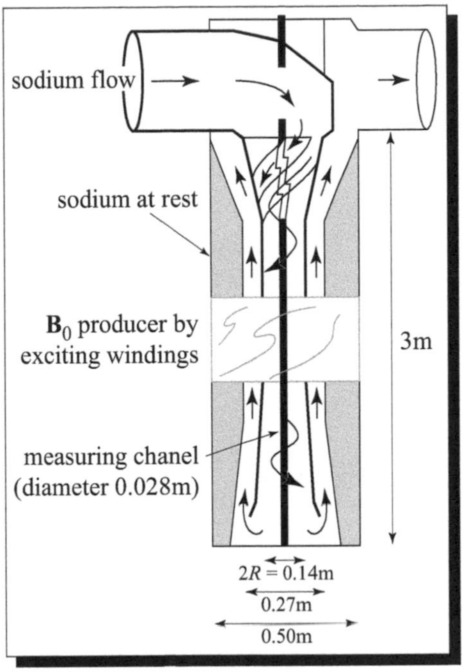

▷ $T_{\text{experiment}} \approx 200\,^\circ\text{C}$.

▷ Flow Sodium rates
$Q_{\text{sodium}} = 280 - 660\,\text{m}^3\,\text{h}^{-1}$.

Measurements:

▷ Sodium flow rate (electromagnetic flow meter).

▷ Induced magnetic field inside the channel.

▷ $U_{\max} = \dfrac{Q_{\max}}{\pi R^2} \approx 12\,\text{m}\,\text{s}^{-1}$.

▷ $\text{Rm} = \mu_0\,\sigma\,U\,R\ \leq 8$.

8.2.6. THE VORTICES OF GALLIUM

In the mid-ninety's two experiments were run using liquid gallium. Gallium has a low point of fusion (30 °C), is a fairly high conductor of electricity (3 times larger than mercury, but 3 times smaller than sodium), and is easier and safer to handle in a laboratory compared to sodium or mercury.
Motivated by the description of a geophysicaly relevant regime, in which the Lorentz and Coriolis forces are comparable, Brito *et al.* (1995, 1996) run an isolated vortex generated by a rotating disc at the bottom, the vortex being also rotated on a table (to include the effect of the Coriolis force on the flow). A transverse magnetic field was imposed perpendicular to the axis of rotation. A set of measurements was performed: gallium velocity inside the vortex, differences in electric potential at the vortex boundary, induced magnetic field outside the vortex and gallium temperature. These measurements accompanied by a numerical model of the electric current circulation in the bulk of the vortex allowed to quantitatively describe the dynamics of that geostrophic vortex under the presence of a transverse applied field. Brito *et al.* (1996) derive a quantitative scaling law of the Joule dissipation as a function of the forcing and the imposed magnetic field.

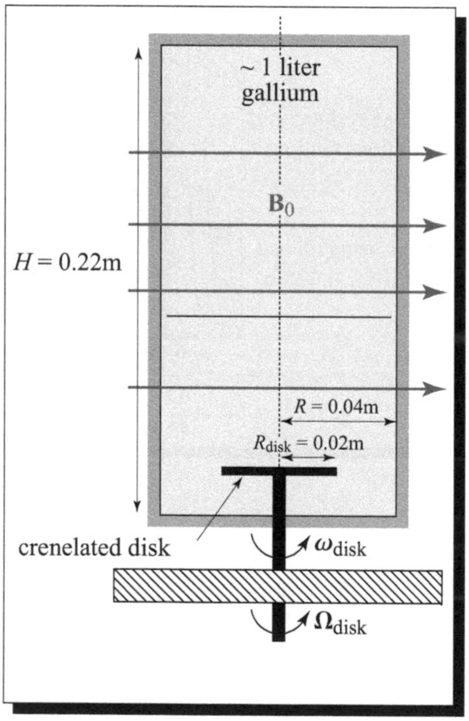

▷ $P_{\text{motor}} = 1.3\,\text{kW}$.
▷ $\omega_{\text{disc}} \leq 3000\,\text{rpm}$.
▷ $\Omega_{\text{table}} \leq 90\,\text{rpm}$.
▷ $B_0 \leq 0.075\,\text{T}$.

Measurements:
▷ Gallium velocity field (Venturi tubes at the top).
▷ Induced magnetic field (Hall probe).
▷ Differences in electric potential (copper electrodes).
▷ Gallium temperature (thermistor).
▷ Torque applied by the rotating motor.

▷ $\text{Ro} = \dfrac{(\omega_{\text{disc}} R_{\text{disc}})}{\Omega_{\text{table}}\, R} \approx 0.7 - 15.$

▷ $\text{N} = \dfrac{\sigma B_0^2\, R}{\rho\, (\omega_{\text{disc}} R_{\text{disc}})} \approx 0 - 1.$

▷ $\text{E} = \dfrac{\nu}{\Omega_{\text{table}} R^2} \approx 10^{-4} - 10^{-6}.$

▷ $\Lambda = \dfrac{\sigma B_0^2}{\rho \Omega_{\text{table}}} \approx 10^{-3} - 1.5.$

▷ $\text{Rm} = \mu_0 \sigma (\omega_{\text{disc}} R_{\text{disc}}) R \leq 0.1.$

Odier *et al.* (1998, 2000) generated the so-called Von Kármán flow in gallium in a cylinder with two corotating discs in presence of an imposed magnetic field parallel or orthogonal to the vertical axis of the cylinder. Measurements of the magnetic field inside the cylinder allowed to describe precisely the advection and expulsion of the imposed magnetic field.

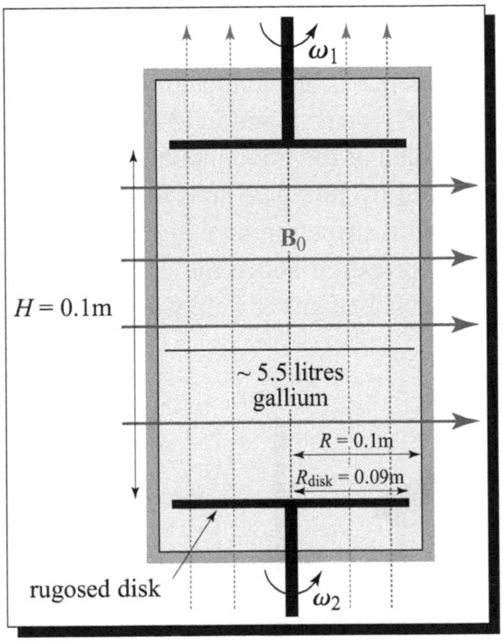

▷ $P_{\text{motor}} = 2 \times 11\,\text{kW}$.

▷ $\omega_{\text{disc}} \leq 3000\,\text{rpm}$.

▷ $B_0 \leq 0.002\,\text{T}$.

Measurements:

▷ Induced magnetic field at various depth in the equator plane (Hall probes).

▷ Dynamic pressure at the cylinder boundary (piezoelectric transducer).

▷ $\mathrm{N} = \dfrac{\sigma B_0^2\, R}{\rho\,(\omega_{\text{disc}} R_{\text{disc}})} \approx 0 - 10^{-4}$.

▷ $\mathrm{Re} = \dfrac{(\omega_{\text{disc}} R_{\text{disc}}) R}{\nu} \leq 10^8$.

▷ $\mathrm{Rm} = \mu_0 \sigma \omega_{\text{disc}} R_{\text{disc}} R \leq 3$.

8.2.7. THE RIGA DYNAMO

After the promising results of the Leningrad experiment (see Section 8.2.5), the latvian team built a new experiment: the shape and sizes of the central channel were changed, an important effort was done to optimize the velocity profiles both with experiments in water and numerical modelling (Stefani *et al.*, 1999; see Section 8.3.5). The shape of the propeller was also optimised and eventually, the main change was the replacement of the electromagnetic pumps by two powerful motors driving the propeller at the top of the device (Gailitis *et al.*, 2002b).

The first experimental evidence of dynamo action was obtained in Riga at the end of 1999 (Gailitis *et al.*, 2000): an imposed field as close as possible to the expected one – from theoretical studied of the Ponomarenko dynamo – was amplified during an experiment, as measured by flux gate-sensors along the vertical of the device. The magnetic fields spatial distribution and frequency were studied as a function of the rotation rate of the propeller above the critical magnetic Reynolds number. Saturation of the self-sustained magnetic field was observed (Gailitis *et al.*, 2001).

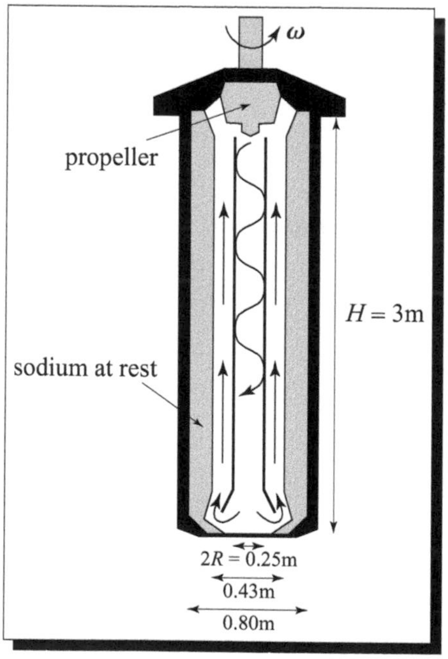

▷ $P_{\text{motor}} \leq 120\,\text{kW}$.

▷ $\omega \leq 2200\,\text{rpm}$.

▷ B_0 is an helicoidal field along the vertical axis of the device.

Measurements:

▷ Induced magnetic field with flux gates and Hall sensor at different heights along the vertical.

▷ Motor power delivered as a function of rotation rate.

▷ Monitoring of the sodium temperature.

▷ $\text{Rm} = \mu_0 \sigma \omega R^2 \leq 42$.

Figure 8.2 - The Riga dynamo experiment (photograph courtesy F. Stefani).

8.2.8. THE KARLSRUHE DYNAMO

Self-excitation was also observed in the Karlsruhe device, which was based on a theoretical two-scale periodic kinematic dynamo of G.O. Roberts (1972), see Section 1.5. It was designed jointly by Busse (Bayreuth) and Müller (Karlsruhe) (Busse *et al.*, 1996). A set of 52 spin-generators were assembled in a large container, a pair of spin-generators being distinctively shown in the figure (Müller & Stieglitz, 2000). Each generator contains a central tube in which the sodium is flowing unidirectionally with a flow rate V_C and an outer part in which the sodium flows with an helicoidal forced motion and a flow rate V_H. The sodium is going up and down in his neighbouring generator. The gap between the 52 helicoidal cylinders is filled with liquid sodium at rest. Three electromagnetic pumps forced the sodium to flow in and out of the container, one pump running the sodium through the central tubes, and the two other ones through the helicoidal outer part.

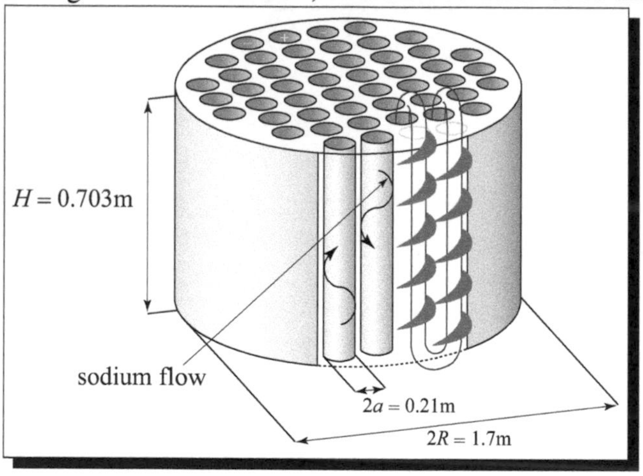

Sodium flow rates were monitored. Beyond the critical rate for both flows (in the central and the outer parts), magnetic measurements showed that the ambient magnetic field was rapidly amplified and saturated after a transient time (Stieglitz & Müller, 2001; Müller *et al.*, 2004).

▷ Flow sodium rates $Q_{\text{sodium}} = 70 - 120\,\text{m}^3\,\text{h}^{-1}$.

▷ $P_{\text{threepumps}} \leq 500\,\text{kW}$.

▷ B_0 ambient magnetic field.

Measurements:

▷ Induced magnetic field (three components) at various locations inside and outside the container (Hall probes).

▷ Induced magnetic field with compass needles outside the container.

▷ Flow rates of sodium.

▷ Sodium temperature.

▷ $U_{\text{max}} = Q_{\text{sodiummax}}/(\pi a^2) \approx 1\,\text{m}\,\text{s}^{-1}$

▷ $\text{Rm} = \mu_0 \sigma U_{\text{max}} R^2 \leq 10$

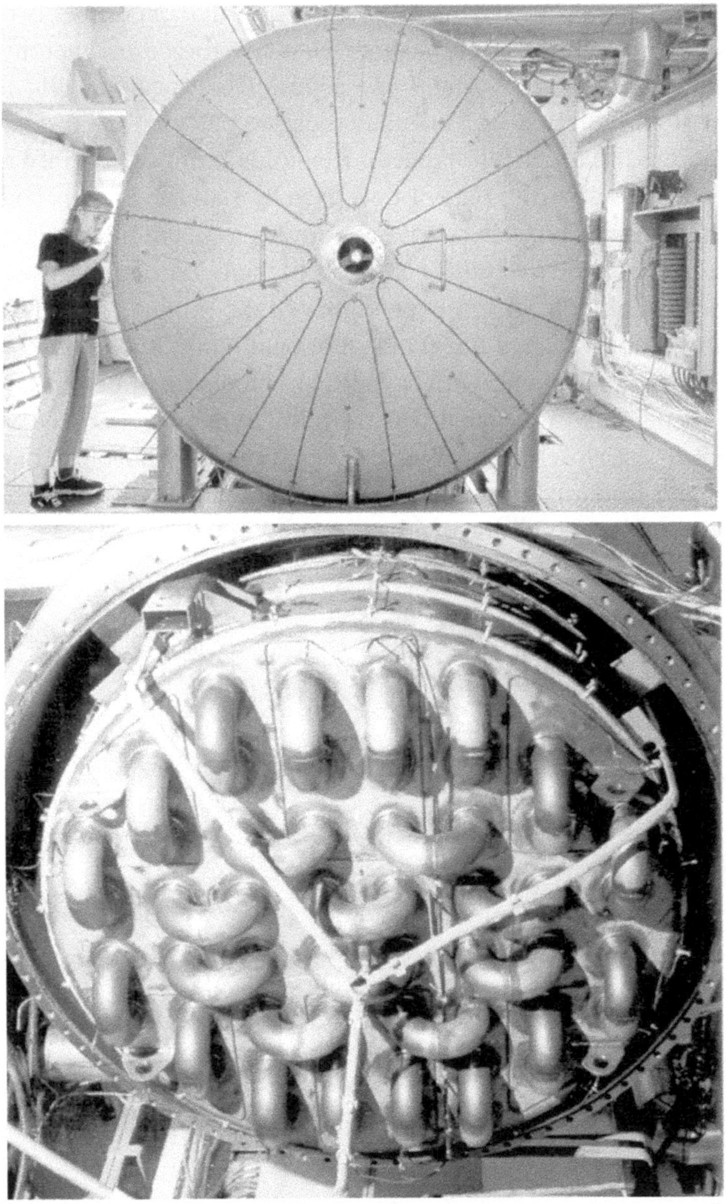

Figure 8.3 - The Karlsruhe dynamo experiment (photographs courtesy U. Müller, R. Stieglitz).

8.2.9. THE COLLEGE PARK EXPERIMENTS

The Lathrop group have been running sodium experiments for a few years in College Park, Maryland. They have performed convective experiments (Peffley *et al.*, 2000a; Shew *et al.*, 2002), as well as mechanically forced magnetohydrodynamics flow (Peffley *et al.*, 2000b; Sisan *et al.*, 2003). This mechanically forced spherical experiment is motivated by kinematic dynamo calculations of Dudley & James (1989) which predict a critical magnetic Reynolds number possibly reachable in a laboratory with the following type of flow used by Lathrop and his collaborators. In this device, two mixing propellers drive the flow in a sphere filled with sodium (co or counter rotating propellers). Baffles attached to the outer boundary are added to the rotating sphere in order to increase the vigour of the mixing. An imposed magnetic field is either parallel (dashed lines) or orthogonal (solid lines) to the rotating shaft. This team has been using a pulse decay measurements of an externally applied field to quantify how far an experiment was from the dynamo onset (see Section 8.3.5). In an attempt to get closer to the dynamo transition, they have tried number of various set-ups and have, for example, changed the shape of the propeller, as well as changed the boundary conditions with equatorial copper discs at the equator of the sphere (Shew *et al.*, 2001).

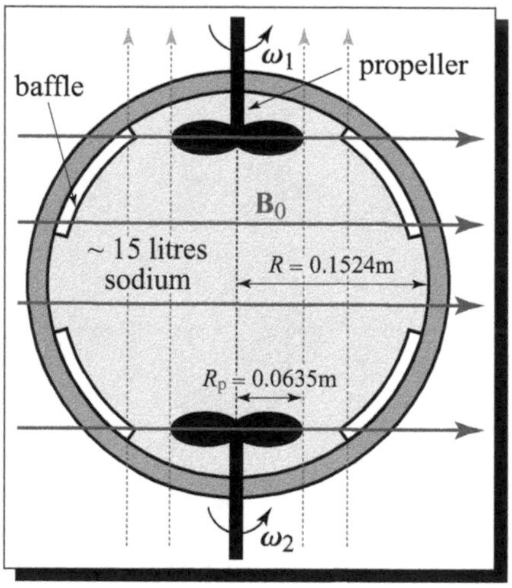

▷ $P_{\text{motor}} \leq 15\,\text{kW}$.
▷ $\omega_1 \approx \omega_2 \leq 3000\,\text{rpm}$.
▷ $B_0 \leq 0.2\,\text{T}$.

Measurements:

▷ Induced magnetic field after imposed pulses, measured by Hall probes.
▷ Mechanical power as a function of rotation rate.
▷ Monitoring of the sodium temperature.

▷ $N = \dfrac{\sigma B_0^2 R}{\rho \omega R_p} \approx 0 - 17.$
▷ $\text{Rm} = \mu_0 \sigma \omega R_p R \leq 30.$

8.2.10. VON KÁRMÁN SODIUM EXPERIMENTS

The Von Kármán Sodium or VKS experiments performed in Cadarache (France) follows from the Von Kármán Gallium (see Section 8.2.6) experiments, at higher magnetic Reynolds number (Bourgoin *et al.*, 2002; Marié *et al.*, 2002). Again, the VKS flow is of the same type as that of Dudley & James (1989) with possibly a relatively low critical magnetic Reynolds number. The VKS team has placed a considerable amount of energy in tuning this experiments, both experimentally (with water and with gallium with similar set-ups) and numerically [by using kinematic dynamo calculations based on velocity flows measured in water (Bourgoin *et al.*, 2002; Marié *et al.*, 2002; Marié *et al.*, 2003), see Section 8.3.5]. In particular, they examined the optimised ratio of poloidal versus toroidal velocity for the dynamo action. So far, they have observed an amplification of the imposed magnetic field but not reached a self-sustained dynamo (Bourgoin *et al.*, 2002; Pétrélis *et al.*, 2003).[13]

▷ $P_{\text{motor}} = 2 \times 75\,\text{kW}$.

▷ $\omega_{\text{disc}} \leq 1500\,\text{rpm}$.

▷ $B_0 \leq 0.002\,\text{T}$.

Measurements:

▷ Induced magnetic field inside the flow using a 3D Hall probe.

▷ Dynamic pressure at the wall.

▷ LDV velocity measurements in water experiments.

▷ $\text{Rm} = \mu_0 \sigma \omega_{\text{disc}} R_{\text{disc}} R \leq 50$

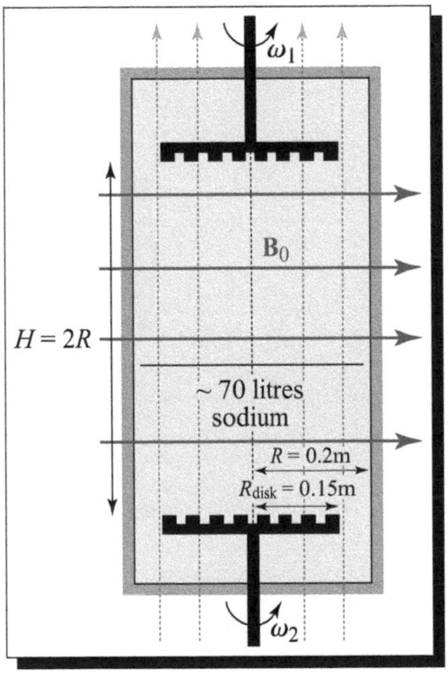

13 Editorial comment: see recent developments in the concluding chapter of the book.

8.2.11. DERVICHE TOURNEUR SODIUM PROJECT

The Geodynamo team (Grenoble, France) has constructed an experiment devoted to study the magnetostrophic regime in a sphere (an experiment called DTS or "Derviche Tourneur Sodium"). This regime implies that the Lorentz and Coriolis forces are dominant on the flow. An inner and an outer sphere are allowed to rotate independently, and the magnetised inner sphere carries a permanent dipolar magnetic field (Cardin *et al.*, 2002). The particularity of this project lies in the crucial roles of Coriolis or rotational forces, that are assumed to be very important in the generation of most planetary magnetic fields (Chapters 4 & 5).

Experiments in water with a similar geometry of the experimental set-up are performed and compared to direct numerical simulation (Schaeffer & Cardin, 2005); available results indicate that such a spherical-Couette flow might be favourable for dynamo action. Dynamo action is not expected in the present experimental set-up, as the DTS experiment is fairly small in size.

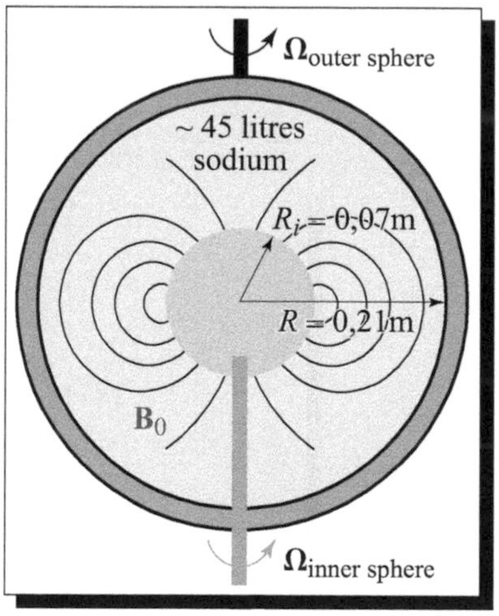

▷ $P_{\text{motor}} = 2 \times 11\,\text{kW}$.
▷ $\Omega_{\text{in}} = \Omega_{\text{out}} \leq \pm 3000\,\text{rpm}$.
▷ $B_0 = \leq 0.022\,\text{T}$ at mid-depth of the shell.
Measurements:
▷ Ultrasonic Doppler velocimetry for the sodium flow.
▷ Induced magnetic field outside the shell.
▷ Differences in electric potential at the outer sphere boundary.

▷ $\text{E} = \dfrac{\nu}{\Omega_{\text{out}} R^2} \approx 10^{-8}$.

▷ $\Lambda = \dfrac{\sigma B_0^2}{\rho \Omega_{\text{out}}} \leq 0.2$.

▷ $\text{Rm} = \mu\sigma[\Omega_i - \Omega_o\,(R - R_i/2)]R \leq 20$.

8.2.12. THE MADISON PROJECT

Forest and his collaborators in Madison, Wisconsin, have been preparing an experimental dynamo that is also based on a Dudley & James type flow in a sphere (Forest *et al.*, 2002). The experiment is close to the geometry of the experiment of Lathrop's group but larger in size. This group has placed a significant effort in the hydrodynamic experimental modelling of the flow, in particular they have worked to select the most appropriate propeller to drive a flow as close as possible to the Dudley & James one. Velocity flows measured in water experiments were used in a kinematic dynamo model. This indicated that the present size of the experiment and the power of the motors should allow to obtain an homogeneous dynamo in sodium.

Measurements:
▷ Laser Doppler velocimetery in the analogous experiment in water.
▷ Poloidal induced magnetic field with an array of 64 Hall probes at the surface of the sphere.
▷ Toroidal induced magnetic field with external toroidal coils?
▷ $\mathrm{N} = \dfrac{\sigma B_0^2 R}{\rho U} \approx 0 - 1.$
▷ $\mathrm{Rm} = \mu_0 \sigma U_{\max} R \leq 120.$

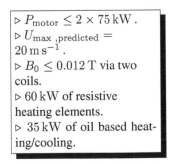

▷ $P_{\mathrm{motor}} \leq 2 \times 75\,\mathrm{kW}$.
▷ $U_{\max\ ,\mathrm{predicted}} = 20\,\mathrm{m\,s^{-1}}$.
▷ $B_0 \leq 0.012\,\mathrm{T}$ via two coils.
▷ 60 kW of resistive heating elements.
▷ 35 kW of oil based heating/cooling.

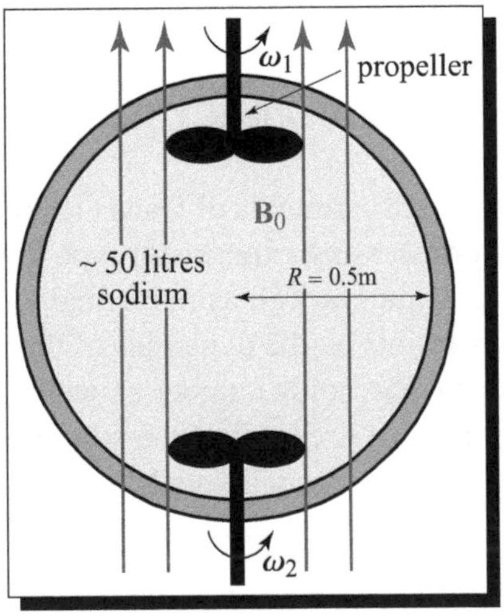

8.2.13. THE PERM PROJECT

A new kind of experimental dynamo project is under study in Perm, Russia. This project also relies on the Ponomarenko dynamo, more precisely on dynamo action caused by a strongly time-dependent helical flow. The idea is to used a toroidal channel filled with liquid sodium ($\approx$ 100 litres) with an helicoidal diverter inside (Frick *et al.*, 2002). The torus would be accelerated to a very high velocity of rotation of order 3000 rounds per minute (rpm) and then stopped abruptly. The dynamo effect should then be observed during the spin-down time of the torus. Experiments in water and kinematic calculations are promising for the dynamo experiment (Frick *et al.*, 2002; Dobler *et al.*, 2003): the dynamo in such a torus requires a short time of braking of less than 0.2 second. A thin and very high conductive shell is required for the torus (copper) and an appropriate seed magnetic field could be assembled with an arrangement of permanent magnets around the torus.

8.2.14. THE SOCORRO PROJECT

A dynamo experiment is under development in Socorro, New-Mexico, around the group led by S. Colgate. The objective being to create an $\alpha\omega$–type dynamo experiment. The experiment uses a Taylor Couette flow between two cylinders rotating at different angular velocities to model the ω–effect (Colgate *et al.*, 2002). The α–effect is produced by the rising of two jets of liquid sodium at the base of the experiment. Experiments in water and kinematic dynamo calculations are currently performed and indicate that self-excitation of a magnetic field might be reachable in such an experimental device.

8.2.15. A NEW PRECESSING PROJECT IN SODIUM

Following the experiments of Gans (1970), the Léorat group is at present studying a cylindrical precessing experiment type flow, in Meudon, France. A preliminary experiment in water, as well as numerical kinematic calculations (Léorat *et al.*, 2001), placed constraints on the dynamics of the precessing flow in a cylinder at a high hydrodynamical Reynolds number as well as on the power dissipated by such a flow. The water project is completed, a sodium experiment with a large precessing cylinder, of metric size, is anticipated.

8.2.16. TECHNOLOGY AND MEASUREMENTS IN DYNAMO EXPERIMENTS

LIQUID SODIUM AND ITS PROPERTIES

As seen throughout the survey of experiments, liquid sodium is now broadly used in high magnetic Reynolds experiments and appears to be the preferred fluid for dynamo modelling in the laboratory. Its main physical properties are shown in Table VI. As mentioned in the introduction, its electrical conductivity is very large [see Nataf (2003) for comparison of physical properties of gallium and mercury], but its low density and melting point also make it very attractive to use in the laboratory. The large production of sodium (23000 tons yr^{-1} in France for example) makes it quite inexpensive (10 euros/kg for sodium) compared to other metals (1000 euros/kg for gallium, for example). The main difficulty with sodium is its strong reactivity with water, air and plenty of other materials such as alcohol, concrete, etc. For example, sodium reduces water with production of hydrogen, which may spontaneously explodes in air. At high temperature ($\geq 250\,°\mathrm{C}$), droplets of sodium may even burn in air with small flames generating solid oxydes at the surface of liquid sodium or aerosol in the surrounding atmosphere. This explains why dynamo experiments are usually conducted in installations dedicated to nuclear technology, with a solid experience in sodium handling (Forschungszentrum Karlsruhe GmbH, Karlsruhe experiment; Institute of Physics, Salaspils, Riga experiment; Comissariat à l'énergie atomique, Cadarache, VKS experiment) or in particular buildings (free of water) specially devoted to such experiments (such as in Grenoble).

EXTRACTION OF POWER AND SEALING

As discussed in the introduction, a large amount of power must be introduced in dynamo experimental set-ups. This power is ultimately converted into heat through viscous or magnetic dissipation. If this heat is not extracted, the sodium temperature quicly rises, decreasing the value of Rm (as the electrical resistance of sodium increases with temperature). Typical experiments are conducted during a limited period of time, of the order of a minute, like in Riga or Cadarache. The experiment is then kept at rest during a few minutes or hours until the sodium cools down, generally to around $120\,°\mathrm{C}$. In Karlsruhe, the circulation of sodium through powerful heat exchangers allowed to perform the experiment during a few hours without stopping; this kind of sodium circulation through exchangers is also under development in a new experimental set-up of VKS. In Grenoble, in a smaller device in which $20\,\mathrm{kW}$ only are injected, a strong flow of cool/hot air circulation around the rotating sphere is planned to monitor the temperature of the experiment. Another possibility is to use an oil circulation in order to extract heat around the container. This is, for example, used in the Madison experiment.

Leakage of sodium in a dynamo experiment involving a vigorous flow may be very damaging. However, dynamic sealings in sodium are not entirely satisfactory, they are still under development; instead for example, in the VKS experiment small leakage of sodium is permitted, in College Park the joint around the rotating shaft is replaced after every experimental run. As another example, the first dynamo run in Riga (November 1999) also had to be stopped because of a sodium leakage at the top of the container. Experiments in which no specific sealings are needed such as the Karlsruhe dynamo or the precession experiments (Gans, 1970; Léorat *et al.*,

2001) are in that respect very appealing. An electromagnetic coupling has been successfully tested in Grenoble, in order to rotate the inner sphere. This solution, which also avoids sealing in sodium, might offer a promising alternative.

MEASUREMENTS

Quantitative measurements in classical fluid dynamic experiments are usually difficult. They become very challenging in magnetohydodynamics experiments using sodium, in particular because electromagnetic waves cannot be used. The temperature of sodium – between 120 °C and 200 °C – places severe constraints and prevents the use of classical measurement technics. In the following, we present the sort of measurements performed nowadays in dynamo experiments. Most of them need a very efficient electronic system to process measured signals, protection from electromagnetic noise, digitalisation for computer analysis. Even if the quality of the measuring probes is important, data processing remains crucial.

Induced magnetic field: The key measurement in a magnetohydrodynamic experiment remains the magnetic field! It is systematically measured in all experiments. These measurements are usually done outside the flow. Local measurements in the flow may perturb the flow, and in addition magnetometer probes operate at low temperature and generally need a controlled temperature to work properly. The probes are usually of two types: the Hall effect probes ($|\mathbf{B}|$ greater than a few microteslas) measure stationary and time varying magnetic field (the bandwidth is generally controlled by the electronics). The principle of the second type of probes is based on the measurement of an induced electric current produced by a time varying magnetic field in a coil (sensitivity and precision are directly connected to the coil and the electronics). Both measurements are unidirectional. These probes are generally small (less than a few millimetres). Given that a probe provides a local measurement of one component of the magnetic field, it is very difficult to build a good spatial description of the magnetic field; an array of probes is necessary in order to have a spatial description (e.g. Forest *et al.*, 2002 for example). Large coils (of the size of the experiment) are sometimes used to impose a magnetic field on the flow. The same coils may be used to measure the oscillating or decaying induced field.

Dynamic Pressure measurements: Dynamic pressures can be measured by piezoelectric probes in contact with the fluid. Their typical sizes are a few millimetres in diameter. They can be very sensitive up to 1 Pa. This technique measures time variations of the pressure (from a few Hz to a few tenths of kHz); they are used as indirect measurement of time variations of the velocity field. Pressure temporal spectra are then used to characterise turbulence in the flow.

Electric potentials: Electric potentials may be measured with copper electrodes in contact with the liquid sodium. The sensitivity and bandwidth of these measurements are given by the ones of the measuring voltmeter. These potentials are difficult to interpret because they are related to electric currents which may have two sources, electric or electromotive fields (Steenbeck *et al.*, 1968; see Section 8.2.3). The temporal evolution of the measured currents with the electrodes at the edge of the container may also be directly related to the dynamic of the flow (Brito *et al.*, 1995).

Fluid velocity measurements: The velocity field, while being a key measurement in magnetohydrodynamic experiments, is only very rarely measured. This is mainly because sodium is opaque. Experimentalists usually use an indirect volumic measurement of the velocity field via the control of the torque (or power) delivered by rotating motors (Gailitis *et al.*, 2001, see Figure 8.15). Control of sodium flow rates through pumps also permits an averaged measurement of the velocity field, as in Karlsruhe for example (Stieglitz & Müller, 2001; e.g. Figure 8.16).

The intrusive hot film probes technique gives very good result in term of local variations of the velocity field. They are based on the measurement of the electrical resistance of a conducting wire, which varies with its averaged temperature that is controlled by the flow around the wire. As far as we know, this technique has not been used in a dynamo experiment, whereas it is largely used in MHD turbulence experiments (e.g. Alémany *et al.*, 1979).

A promising non-intrusive technique to measure velocity fields in fluid dynamics experiment is the Doppler Ultrasound Velocimetry: it is based on the ultrasonic back scattering of oxydes (or other particles) in suspension in liquid sodium (for example). This technique which is successful in water, and gallium, should work as well in sodium (Brito *et al.*, 2001; Eckert & Gerbeth, 2002). Laser Doppler velocimetry is also broadly used in water experiments (Forest *et al.*, 2002; Marié *et al.*, 2003): water models of sodium experiments enable velocity field measurements below the onset of dynamo (see Section 8.3.5).

Temperature: Temperature measurements (usually performed at the boundaries) are easily achieved. They are generally based on measurements of the electrical resistance of a material which varies with temperature. They may indicate the dissipation rate (or Joule dissipation) in the MHD flow (Brito *et al.*, 1996). Temperature probes can also be used to track motions of thermal dynamic structures acting as passive tracers in front of temperature probes.

8.3. WHAT HAVE WE LEARNT FROM THE EXPERIMENTAL APPROACH?

In the second part of this chapter, we discuss the results of the various experiments described in the first part. Each subsection is devoted to a particular aspect or related aspects of the dynamo mechanisms. We will discuss to what extent experiments validate, or not, dynamo theory.

8.3.1. THE ω–EFFECT

A vortex of liquid metal permeated by an external magnetic field induces an azimuthal (toroidal) magnetic field parallel to the flow. This geometrical conversion of magnetic field lines is known as the ω–effect (see Chapter 1). Lehnert (1958, see Section 8.2.1) measured an induced azimuthal field up to 25% of the value of the imposed axial poloidal magnetic field. He produced a meridional map of the average induced field (Figure 8.4) with some singularities at the side boundary (reversed field), which maybe due to singularities in the fluid flow. The same effect was measured by Brito *et al.* (1995) in a geostrophic vortex of liquid gallium (see Section 8.2.6) and quantitatively understood with loops of electric currents and electric potentials within the flow: the transverse imposed magnetic field produced electric Foucault currents parallel to the axis of the vortex, which in turn induced a magnetic field diffusing outside the tank, where it was measured (Figure 8.5). The induced electric currents were produced by shear layers as shown both by the measurements and by a numerical model (Brito *et al.*, 1995). Note that a solid body rotation would only produce an ω–effect in its edge, within the hydrodynamic shear boundary layer. This effect should really be referred to as the "ω–gradient–effect" to emphasise the importance of the differential rotation. These effects are clearly linear in Rm (Figure 8.5). More recently, the VKS experiments (8.2.10) also verified this mechanism in gallium and sodium for larger Rm (Odier *et al.*, 1998; Bourgoin *et al.*, 2002; Marié *et al.*, 2002; Figure 8.8).

The Lowes & Wilkinson (1963) solid dynamo experiments (Section 8.2.2) also relies upon the ω–effect. Each solid cylinder transforms an axial component of the magnetic field into an azimuthal one; the position of both cylinders is chosen so that the azimuthal component of the magnetic field associated with a cylinder is axial to the other one. Electric currents are produced at the edge of the rotating cylinders in a thin layer of mercury which connects the main solid piece to the solid cylinder and return in the solid parts creating an induced azimuthal field.

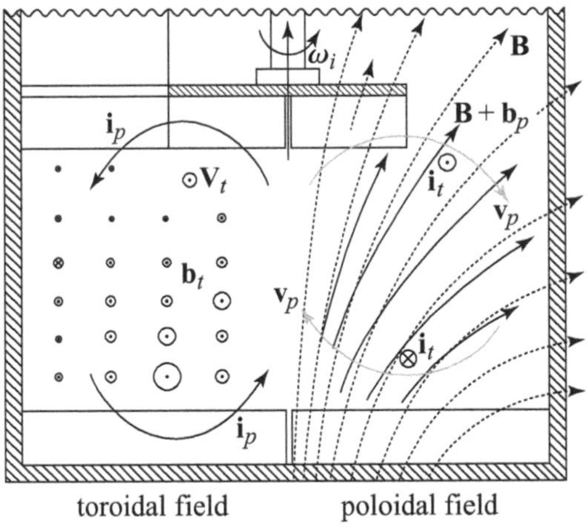

Figure 8.4 - Meridional map of the induced magnetic field demonstrating the ω–effect in the Lehnert's experiment (see Section 8.2.1). An imposed poloidal magnetic field $\mathbf{B}$ (dashed lines) is twisted, by the toroidal velocity flow $\mathbf{v}_t$, into a toroidal magnetic field $\mathbf{b}_t$ (shown on the left, the size of the symbols being proportional to $|\mathbf{b}_t|/|\mathbf{B}|$). $\mathbf{i}$ stands for electric currents. The resulting poloidal magnetic field lines $\mathbf{B} + \mathbf{b}_p$ (solid lines) seem to be expulsed (from Lehnert, 1958).

8.3.2. MAGNETIC FIELD EXPULSION

When the magnetic Reynolds number is high, the magnetic field can be expelled from very active dynamical regions by the so-called "magnetic field expulsion" (Gubbins & Roberts, 1987). This process may be understood as a skin effect: in the reference frame of the moving fluid (for example, a rotating frame at frequency ω associated to a vortex of radius R), we consider a magnetic field which oscillates in time. The magnetic field penetrates the metal in a skin of size $\sqrt{\eta/\omega} = R/\sqrt{\mathrm{Rm}}$. Electric currents are consequently produced in the skin layer, which in turn produce an induced magnetic field, opposite in direction to the applied magnetic field in the heart of the vortex. The resulting magnetic field is then expulsed from the heart of the vortex. Lehnert (1958) (see Section 8.2.1) observed this effect in the poloidal part of the magnetic field as seen in Figure 8.4, in presence of the motion of the liquid sodium for $\mathrm{Rm} \gtrsim 5$). The magnetic field lines were deflected outside the sodium tank. Another evidence of this phenomena has been observed in the VKS experiment (see Section 8.2.10): at Rm above 30, a departure from linearity associated to the ω–effect (Figure 8.6) is observed. The induced magnetic field increases less rapidly than predicted. In these cases, it is nevertheless difficult to differentiate the precise effect of the magnetic field expulsion from a dynamic change of the flow at large values of Rm. This second explanation, however, appears unlikely as the

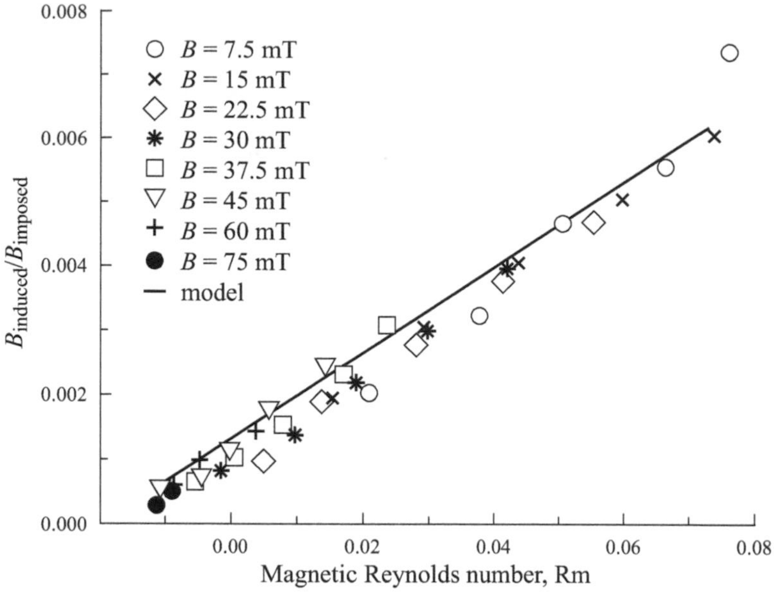

Figure 8.5 - Induced magnetic field B_{induced} by a vortex of gallium in an imposed magnetic field B_{imposed} (see Section 8.2.6). The ω–effect is a linear function of Rm (from Brito *et al.*, 1995).

interaction parameter N is rather small in both experiments. Note that the effect of a rotating magnetic field on a liquid metal flow also has been studied for its application in metallurgy (mixing techniques), these studies being generally focused on large interaction parameters (see Witkowski *et al.*, 1998, for example).

8.3.3. THE α–EFFECT

When there is production of an electric current parallel to an imposed magnetic field that process is called, in very general terms, the α–effect. Historically, this effect was introduced to model the effect of small scales on large scales in the two–scale concept introduced in Section 1.5, but this effect is often used in a generalised sense. It is then usually referred to as the "Parker effect" for general flows (when two scales are not easy to define). The expression "macroscopic α–effect is also sometimes used.

As soon as the α–effect was theoretically derived (Steenbeck *et al.*, 1966), the same team built the α–box (see Section 8.2.3) in order to prove its existence in the laboratory. Figure 8.7 from Steenbeck *et al.* (1968) shows that measurements of differences in electric potentials between the top and the bottom of the box are linear

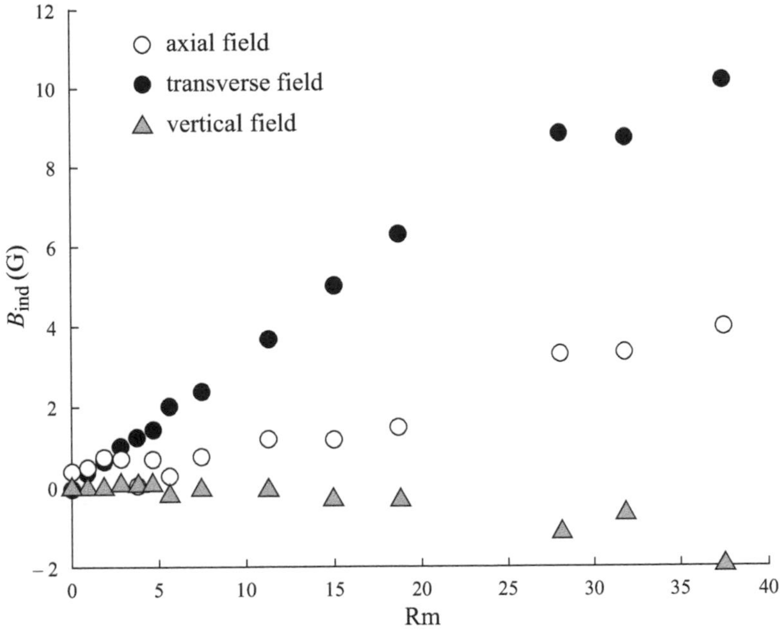

Figure 8.6 - Induced magnetic field b_{ind} measured in the VKS experiment (see Section 8.2.10) versus the magnetic Reynolds number Rm. An axial magnetic field (in the direction of the rotation axis of the discs) of $5.4 \times 10^{-4}\,\mathrm{T}$ is imposed. The linearity of the transverse (and azimuthal) component demonstrates the ω–effect and the departure from linearity may be associated to the expulsion of the magnetic field (from Marié *et al.*, 2002).

with the squared velocity and with the magnetic field as expected. However, it is not straightforward to interpret these electric potential measurements is terms of electric currents. If a wire had been connected between the two electrodes, assuming these electric potentials were due to an average induced electric current aligned with the applied magnetic field in the volume of the α–box, a back of the envelope calculation shows that a current of a few thousand amperes would have circulated between these (inducing a measurable magnetic field). Unfortunately, this type of measurements could not have been performed at that time and it is therefore possible that more complicated geometries of the currents inside the box (especially in the presence of stainless steel boundaries) were responsible for the measured electric potentials. Nevertheless, the clear dependence in $|\mathbf{u}|^2$ is a clear indication of a second order effect in Rm, and thus an α–effect. Open questions however remain after that experiment: what would have been the measurements had there been one cell instead of 28 (question regarding scale separation), what was the role of helicity in the α–effect (as the flow in the pipes does not have much hydrodynamic helicity)?

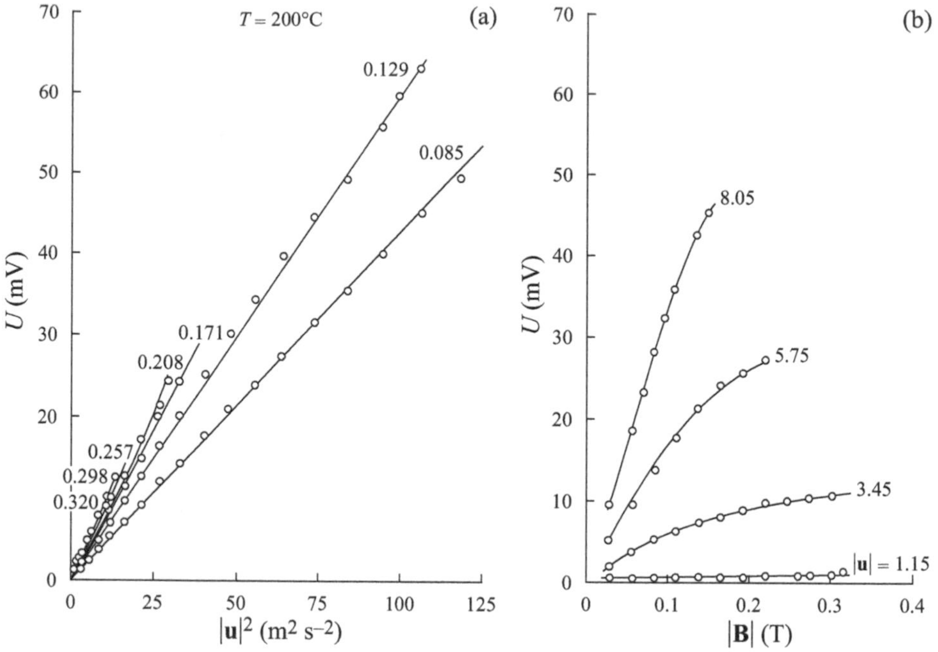

Figure 8.7 - (a) Differences in electric potential measured in the α box (see Section 8.2.3) as function of the squared velocity ($\mathrm{m}^2\,\mathrm{s}^{-2}$) for different value of the imposed magnetic field (in T). (b) Differences in electric potential measured in the α box as function of the imposed magnetic field for different value of the velocity (in $\mathrm{m\,s}^{-1}$). The two linear dependances in $|\mathbf{B}_0|$ and $|\mathbf{u}|^2$ are the experimental evidences for the α–effect (from Steenbeck *et al.*, 1968).

A macroscopic α–effect has been seen in the VKS experiment (see Section 8.2.10). Pétrélis *et al.* (2003) measured an induced magnetic field perpendicular to the applied magnetic field which is quadratic in Rm, for small Rm, as shown in Figure 8.8. Considering symmetry arguments, they have also shown that this magnetic field was associated to an electric current parallel to the applied magnetic field and that its sign was determined by the helicity sign. Although there was no clear scale separation in their experiment, their observation may be understood as a macroscopic α–effect or Parker effect.

The good agreement (see Section 8.3.6) between the experimental measurements in the Karlsruhe dynamo and the theoretical prediction of Rädler *et al.* (1998) using an α–effect in a mean-field approach (Chapter 1) is an indirect evidence for the presence of an α–effect in the Karlsruhe experiment (see Figure 8.16). There were unfortunately no direct measurements in the Karlsruhe apparatus, which would have allowed a detailed description of the α–effect.

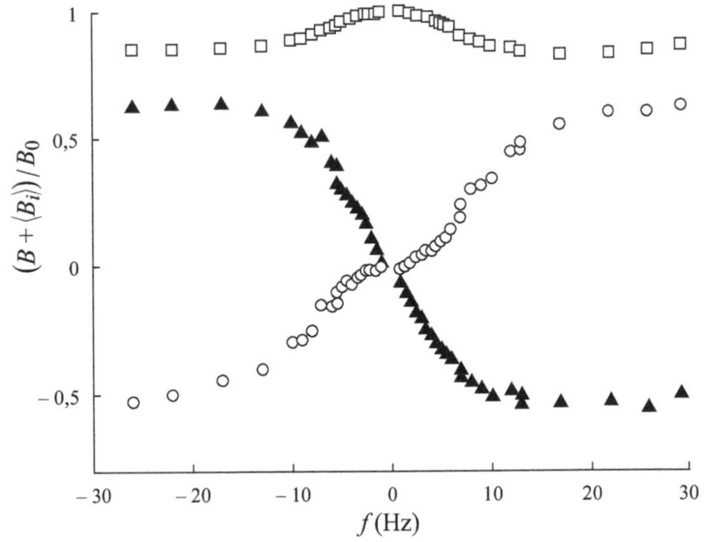

Figure 8.8 - Components of the total mean magnetic field as a function of the rotation frequency of a disc in the VKS experiment (see Section 8.2.10). The magnetic field B_0 is imposed along the y–axis and the disc is rotating along the z–axis. ○ symbols are used for $\langle B_x \rangle / B_0$; □ for $(\langle B_y \rangle + B_0)/B_0$; ▲ for $\langle B_z \rangle / B_0$. For low rotation rates, B_z is linear with the velocity (ω–effect) while B_x exhibits a quadratic behaviour (second order induction effect or α–effect). Departures from this law are clearly seen for frequencies larger than 5 Hz. This saturation may be interpreted as a quenching effect (see Sections 2.7 and 8.3.4) (from Pétrélis *et al.*, 2003).

8.3.4. QUENCHING EFFECTS

In various experiments, both linear and quadratic induction effects tend to saturate for large magnetic Reynolds number. We refer to this as the quenching effects (see Section 2.7). This effect is generally associated with magnetic field expulsion from the moving part of the fluid (see Section 8.3.2) or with a change in the dynamic of the liquid metal due to the Lorentz forces when the interaction parameter N is large. A clear evidence for a quenching effect can be seen in Figure 8.9 from Steenbeck *et al.* (1968). Electric potentials decrease approximatively as an hyperbola function of $|\mathbf{B}_0|^2$ (where $\mathbf{B}_0$ is the applied magnetic field). One may conclude that the α–effect is reduced as the magnetic field increases.

A quenching effect can also be observed regarding the ω–effect. Figure 8.10 shows the hyperbolic magnetic brake of a vortex by an applied transverse magnetic field. This behaviour really reflects an effect of the magnetic field on the dynamics, as the magnetic Reynolds number, Rm, is too low to yield field expulsion.

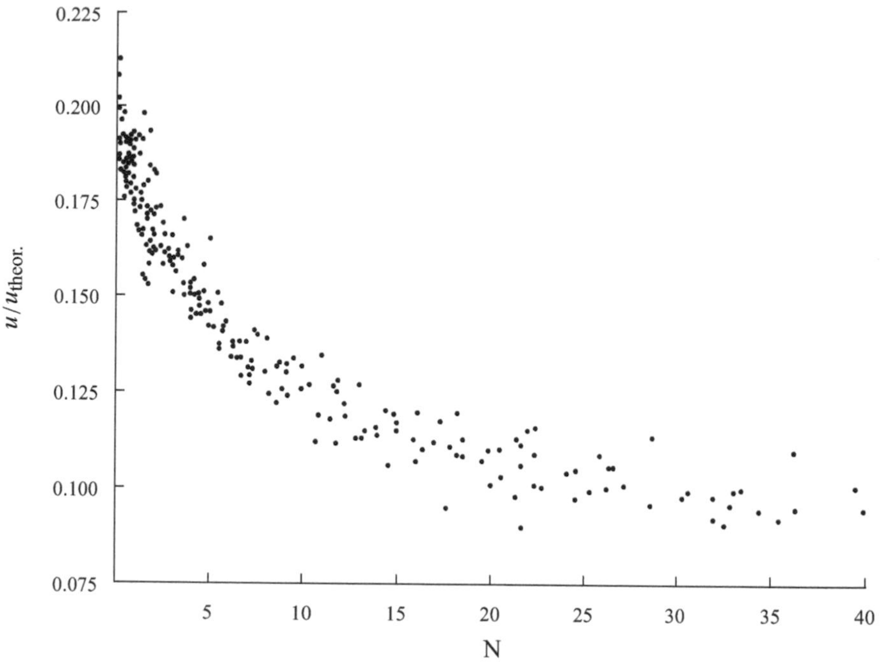

Figure 8.9 - Differences in electric potential measured in the α–box experiments (see Section 8.2.3) as function of the interaction parameter N. The α–effect is quenched (hyperbolic decrease) when the magnetic field increases (from Steenbeck *et al.*, 1968).

The departure from linear and quadratic variations of the induced magnetic field components in the VKS experiment, for high rotation rate of the discs, may also be interpreted in terms of quenching effects (see Figure 8.8). Careful experimental analysis of these effects – reducing the efficiency of the α and ω–effects as the magnetic field grows – are needed in order to better understand saturation mechanisms.

8.3.5. THE EXPERIMENTAL APPROACH TO A KINEMATIC DYNAMO

The kinematic approach implies that a given flow (unaltered by the Lorentz force) is considered. Its ability to induce a self-sustained magnetic field is measured. If this approach is successful, the growing magnetic field corresponds to the eigenvector which eigenvalue becomes positive at the dynamo onset. However, in the subcritical dynamo regime, it is possible to measure the negative eigenvalue of a given magnetic field and study its variation as the forcing is increased to get closer to its critical

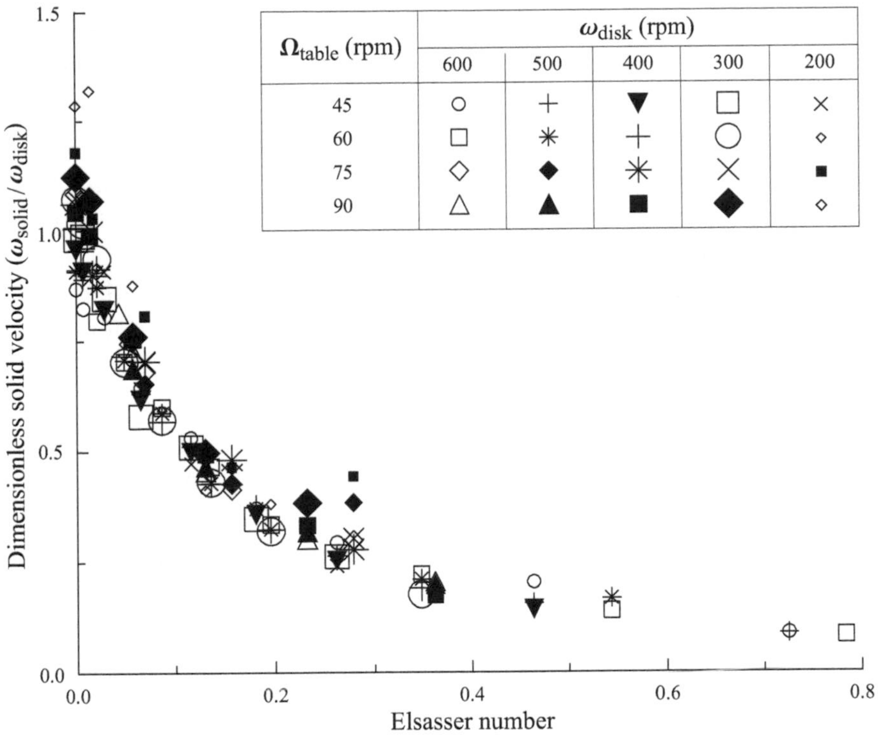

Figure 8.10 - Variation of the angular velocity ω_{solid} of a vortex of liquid gallium over the imposed velocity of the disc ω_{disc} as function of the Elsasser number Λ (see Section 8.2.6). The decrease in amplitude of the angular velocity decreases the amplitude of the induced magnetic field (by ω–effect, see Section 8.3.1). This may be seen as a ω quenching effect (after Brito *et al.*, 1995).

value for dynamo action. This technique may provide a good approach to estimate the value of the dynamo onset for a given configuration.

This approach was first used by the Gailitis *et al.* (1987) group in the Leningrad experiment (see Section 8.2.5). An oscillating magnetic field, close to the Ponomarenko dynamo eigenvector, was applied and the magnetic response was measured inside the container. Figure 8.11 shows that the imposed magnetic field was significantly amplified by the flow. This amplification was linear in Rm, up to the maximum value of the tested flow rate, before the experiment had to be stopped. One may argue that their experiment was ended just before dynamo action could be achieved.

Alémany *et al.* (2000) also used this technique to investigate dynamo action in the secondary pump of the Superphenix fast breeder reactor. Linearly extrapolating the magnetic field decay rate, they found that the pump velocity (500 rpm) was only four times below critical. One may however moderate their conclusion, as there is no theoretical ground to use a linear extrapolation of the decay rate far from the dynamo onset.

Lathrop and his collaborators have studied the time relaxation of an applied magnetic field (Peffley *et al.*, 2000a,b) on the mechanically forced experiment described in Section 8.2.9. For a given flow, they imposed a dipolar magnetic field (either axial $m = 0$ or equatorial $m = 1$) of small amplitude (a few mT) for 1 to 10 s. They turned off the imposed magnetic field and measured the exponential decay. With no motion, the exponential decay corresponds, as expected, to the Joule decay time for a sphere. Increasing Rm (Figure 8.12), the decay time increases for $m = 0$ or decreases for $m = 1$. This result disagrees with the kinematic numerical result of Dudley & James (1989) which for this type of flow predict the growth an equatorial dipole. This experimental result shows that the non-axisymmetric part of the flow (due to the propeller, baffles or turbulent fluctuations) plays a significant role in the generation of an axisymmetric magnetic field (because of Cowling's theorem, see Chapter 1). The broadwidth of the variance of the decay time rates of the magnetic field, for flows at large values of Rm, is also a good indicator of the turbulence in the magnetic field generation process (Peffley *et al.*, 2000a). Note that tests have also been performed with time dependent applied magnetic fields in order to measure the imaginary part of the eigenvalue.

The experimental kinematic approach which consists in approaching the critical eigenvalue of a given field is strongly limited by the type of geometry of the applied magnetic field which can be envisaged. Although there is almost no time constraint on a kinematic-type dynamo experiment, the geometry of a magnetic eigenvector derived theoretically can only be reproduced in the laboratory if it is quite simple.

In order to get some kinematic predictions on the onset of dynamo action, water experiments producing the same velocity flow as in sodium experiments are broadly performed. In water, velocity measurements are much easier than in sodium, and the mean flow may be described with a good resolution. Once the experimental velocity field has been measured, it is used as an input in a numerical simulation to solve the kinematic dynamo problem. This approach has been used by many groups; Riga/Dresden group (Stefani *et al.*, 1999), Karlsruhe (Stieglitz & Müller, 2001), VKS group (Marié *et al.*, 2003), Wisconsin (Forest *et al.*, 2002), Perm (Dobler *et al.*, 2003), Grenoble (Schaeffer & Cardin, 2005), Léorat *et al.* (2001).

Numerical kinematic calculations are then used to determine how efficient an averaged flow can be to amplify an initial magnetic field and to produce a dynamo. Experimentally, in water, it is quite convenient to change the geometry of the device or the shape of the propellers for example, and check numerically with the new measured velocity fields if the critical value of Rm has been decreased. Numerically, it also convenient, once the velocity field is known, to change the boundary conditions,

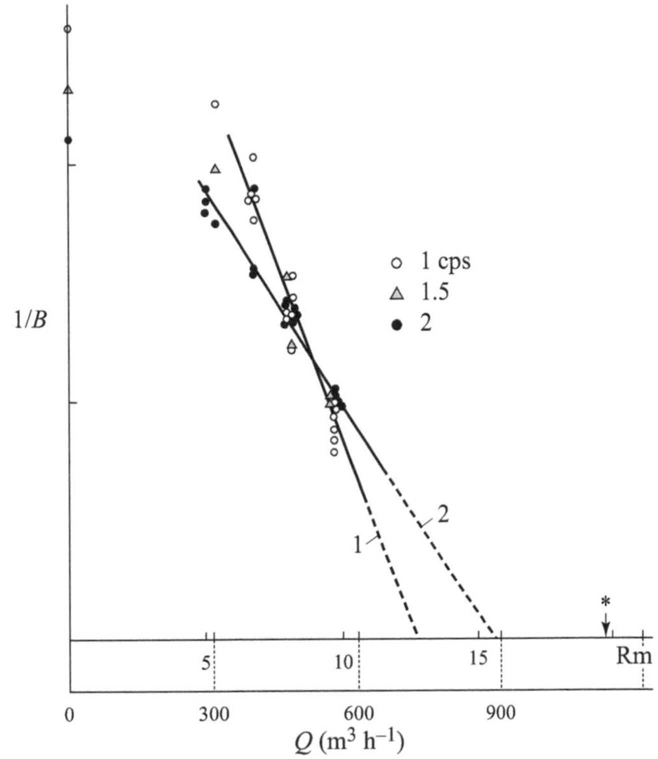

Figure 8.11 - Measurement of the inverse ratio of the induced magnetic field $1/B$ signal versus flow rates Q (in $\mathrm{m^3\,h^{-1}}$) for three frequencies of the applied magnetic field $\mathbf{B}_0$ in the first attempt to perform a Ponomarenko experiment in Leningrad (see Section 8.2.5). The linear extrapolation of the experimental results may indicate the critical value of the flux rate for dynamo action. Note that the extrapolation would lead to a critical Rm lower than the theoretical prediction, shown here with a star around $\mathrm{Rm} = 19$ (from Gailitis *et al.*, 1987).

for example considering insulating or electrically conducting boundaries. Such optimisation has been successfully used by the Riga group. Unfortunately, many other groups have shown that a tiny difference in the averaged velocity field may change drastically the sign of the eigenvalues (Forest *et al.*, 2003; Marié *et al.*, 2003). Does it mean that dynamo action is not robust and really depends on very small changes in the velocity field? This is a good an interestin open question, and it is important to note that in this approach, only averaged velocities are considered. It may be not sufficient, as fluctuations play an important role in the dynamics. Another point is that only the measured velocity field reflects only the large scale flow. The small scales of the velocity field, which may for example be important to produce an α–effect, are not measured and not considered numerically.

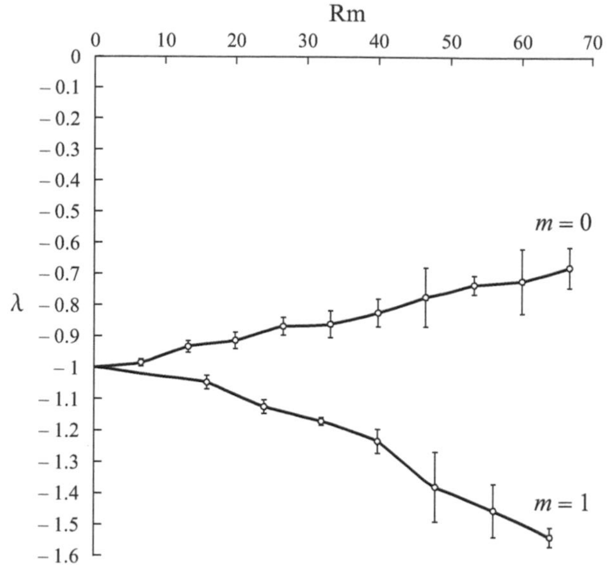

Figure 8.12 - Growth rates (inverse of exponential decay times) λ of an applied magnetic field – after its suppression – as a function of magnetic Reynolds number (Rm) in the College Park experiment (see Section 8.2.9). $m = 0$ refers to an imposed magnetic field aligned with the axis of rotation of the propellers, and $m = 1$ to a perpendicular field. The growth rates are normalised by the ohmic decay rate. In the numerical model of kinematic dynamo of Dudley & James (1989), the $m = 1$ curve increases instead and crosses the critical axis for a value of $\mathrm{Rm} \approx 55$ (from Peffley *et al.*, 2000b)

8.3.6. THE ONSET OF DYNAMO ACTION

Two liquid metal experiments have exhibited a self-induced magnetic field (Gailitis *et al.*, 2000, Müller & Stieglitz, 2000). Both experiments have been built (see Sections 8.2.7,8.2.8) in order to reproduce well known kinematic dynamo flows.

THE RIGA DYNAMO

Figure 8.13 shows the measured magnetic field as a function of time for different speeds of the propeller in the Riga dynamo (Gailitis *et al.*, 2000). As expected, the growing magnetic field is a propagating wave along the axis of the experiment (Ponomarenko, 1973). The decay rate and the frequency of the growing magnetic field mode were measured and compared to the predicted ones (Figure 8.14). Predictions have been done relying on a numerical kinematic approach using the averaged

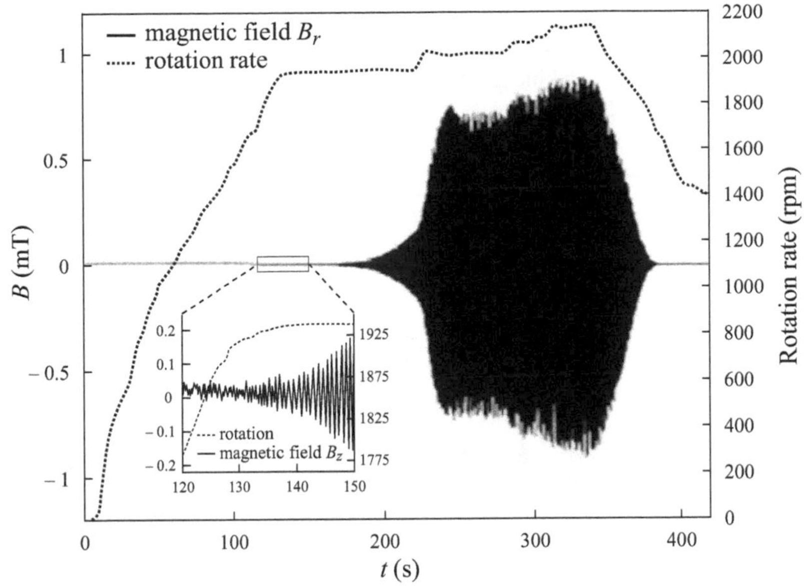

Figure 8.13 - Time evolution of the induced magnetic field (solid line) in the Riga dynamo (see Section 8.2.7). The rotation rate of the propeller is reported (dashed line) and shows the critical rate (around 1925 rpm as shown in the close up), from Gailitis *et al.*, 2002b).

velocity field (Stefani *et al.*, 1999; Gailitis *et al.*, 2002b). The onset is correctly described by the numerical approach (to within 10% of precision), despite the fact the flow turbulence (of a few percents, Gailitis, private communication) is omitted in the central pipe of the experiment. The frequency of the dynamo solution does not seem to be influenced by the magnetic field saturation (Figure 8.14). Does it imply that the back reaction is very small in the Riga dynamo? The answer to this question is still under investigation. However, measurements of the magnetic field along the axis of the experiment show that dynamo action is mainly produced at the top of the experiment close to the propeller (Gailitis *et al.*, 2001). The onset of dynamo action could also be seen in the evolution of the power dissipated in the experiment as shown in Figure 8.15. Below the onset, the power needed to maintain the rotation rate of the propeller varies a the cubic power of the rotating rate, while there a clear deviation from this law is observed above the onset (Gailitis *et al.*, 2001).

THE KARLSRUHE DYNAMO

A self-induced magnetic field was observed in the Karlsruhe experiment (Stieglitz & Müller, 2001) for imposed flow rates Q_{sodium} (see Section 8.2.8) comparable to

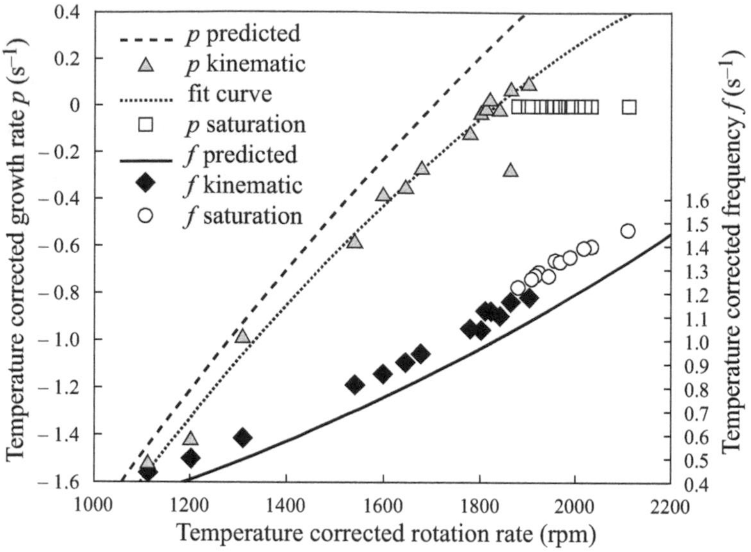

Figure 8.14 - Measurements and predictions of critical growth rates and associated frequency of the magnetic mode for different rotation rates in the Riga dynamo. Experimental data agree within 10% with numerical predictions. Above the onset, the frequency seems equal to the one predicted by the linear theory (from Gailitis *et al.*, 2002b).

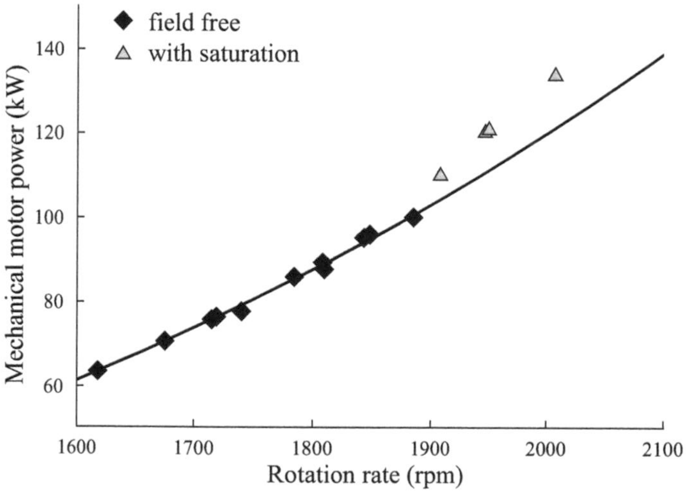

Figure 8.15 - Motor power delivered versus propeller rotation rate in the Riga dynamo. Below the onset, dissipation increases as the cubic power of the rotation rate (solid line). Extra power is needed to maintain the rotation rate above the dynamo onset (from Gailitis *et al.*, 2001).

the predicted onsets (Figure 8.16), the exact experimental onset being lower than the one predicted by Rädler *et al.* (1998) and by Tilgner (1997) by only 10%. However, numerical predictions used very simplified models of the actual experimental fluid flow (for example only straight parts of the tubes were included in the modelling). Measurements of the induced magnetic field, as well as the pressure drop in the piping system, seems to show a smooth, as opposed to a sharp, Hopf bifurcation (Müller *et al.*, 2004). Before the onset, the measured induced magnetic field may be understood as the amplification (by a factor 10) of the Earth's magnetic field (see Section 8.3.5). Typical growth rates of $10\,s^{-1}$ could be deduced from Stieglitz *et al.* (2001) during the transient after the onset; these are ten times greater than the ones predicted by Tilgner in 1997. Nevertheless, the spatial distribution of the experimental saturated magnetic field (Stieglitz & Müller, 2002) is in agreement with the one predicted by the numerical studies of Tilgner (1997) ad Rädler *et al.*(2002). The growing magnetic field in the Karlsruhe experiment however varies with the applied external imposed magnetic field and may change its sign depending on initial conditions (Müller *et al.*, 2004). This behaviour was also reproduced numerically by Tilgner & Busse (2002).

As in the case of the Riga dynamo, the Karlsruhe dynamo numerical modelling agrees remarkably well with the prediction of the kinematic or mean field approach. The mean field approach can be seen to be very successful in the Karlsruhe dynamo experiment, but that could be expected given that the experiment has been explicitely built to produce a two-scale dynamo, suitable to the mean field approach: the velocity field is small scale (size of the helicoidal tube) whereas the magnetic field is dominated by the large scale (size of the dynamo device). The Karlsruhe experiment may be seen as an experimental evidence in favour of the validity of the mean field theory in MHD.

8.3.7. THE EFFECT OF TURBULENCE

Laminar description of the velocity flow has enabled a good prediction of the onset in both successful dynamo experiments. Nevertheless, these flows have to be turbulent considering the large values of the Reynolds number Re. It is not easy to understand why turbulence does not affect more drastically the onset of dynamo action.

For smaller experiments, the kinematic numerical studies devoted to predict the onset of dynamo action with the averaged large scale flow have predicted dynamo action in the parameter regime in which experiments were performed, but dynamo action was never observed (Peffley *et al.*, 2000; Bourgoin *et al.*, 2002). One may argue that the averaged flow, which is numerically used for kinematic computations, is unlikely to be realised during the real experiment. However the same approach

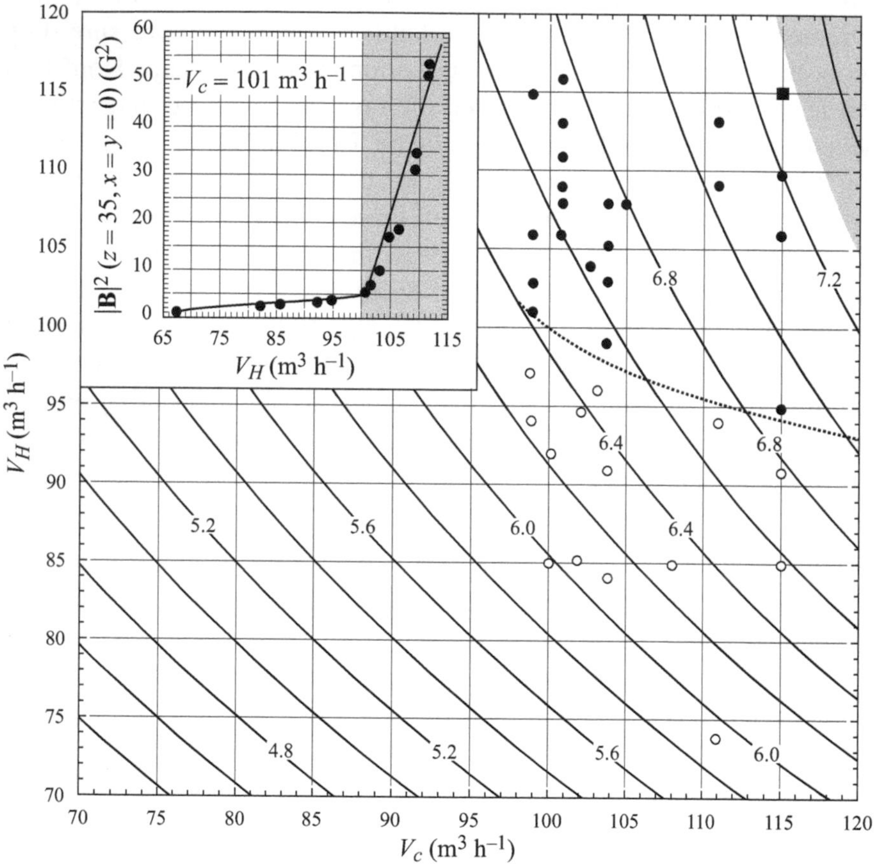

Figure 8.16 - Phase diagram of the Karlsruhe dynamo (see Section 8.2.8). V_C and V_H (in $\mathrm{m^3\,h^{-1}}$) are the flow rates in the central pipes and in the helicoidal pipes respectively. Open circles (○) correspond to non dynamo states, while full circles (•) are dynamo states. A dashed line separates those. In the upper corner, the evolution of magnetic energy ($|\mathbf{B}|^2$) is presented against the helicoidal flux V_H for a constant V_C. The linear fit above and below the onset very precisely determines the critical value for the onset. Mean field theory predictions of the dynamo state are shown in grey (upper right corner) and a typical onset determined by the numerical kinematic approach appears as a filled square (from Stieglitz & Müller, 2001).

was successfully used in the case of the Riga and Karlsruhe dynamo experiments. One may also argue that the characteristic turbulent time is much smaller than the exponential growth rate of the dynamo. It is known, from kinematic dynamo studies, that only small changes in the velocity field may strongly change the growth rate of the magnetic field. If a turbulent experiment could therefore exhibit a favourable dynamo flow for a certain time, it may not last long enough to start a dynamo.

Sisan *et al.* (2003) in the College Park's experiment (see Section 8.2.9) have clearly identified the effect of an imposed magnetic field on the dynamic regime of their experiment. For a given magnetic Reynolds number ($\mathrm{Rm} \approx 7.5$), they varied the intensity of the imposed magnetic field. As the interaction parameter N increases, the measured induced magnetic field exhibits different time and amplitude variations, which may reveal different magneto-turbulent regimes. Five distinct regimes of the induced magnetic field are identified in Figure 8.17.

On the edge of the experimental dynamos context, MHD turbulence experiments have been built to study fundamental properties of the flow (applied to metallurgy), see Moreau (1998). In general terms, the presence of a strong magnetic field tends to form quasi-two-dimensional flows aligned with the magnetic field (Moreau, 1990) that can exhibit 2D turbulent properties (Alémany *et al.*, 1979). A recent experimental study of such MHD turbulence has been carried out by Messadek & Moreau (2002) on unstable shear flows at low Rm. MHD turbulence enlarges the thickness of the shear zone by two orders of magnitude. This enhances the momentum transport and mixing across the layer. As in many other experiments, the flow turbulence is characterised by measurements of magnetic and kinetic spectra.

8.3.8. SPECTRA

Turbulence theories generally predict the behaviour of scalar fields in a fluid flow in terms of spectral decomposition. Although these spectra are generally in the spatial domain, it is not convenient to measure the spatial distribution of a field during an experiment. Instead, time variations of theses fields are measured, and the Taylor (or ergodicity) hypothesis connecting time and spectral variations for homogeneous turbulence (Frisch, 1995; Lesieur, 1997) is assumed.

Kinetic energy spectra are generally deduced from pressure measurements while magnetic energy spectra come from the measurement of a component of the magnetic field. In presence of an external magnetic field and with a low Rm flow, the dependence between the two spectra proceeds from the induction equation. If one supposes the kinetic energy to vary as $\mathcal{K} \propto k^{\alpha}$ (k is the wavenumber and $\alpha < 1$), using the induction equation, one can show that the magnetic energy varies as $\mathcal{M} \propto k^{\alpha-2}$ (Moffatt, 1978; Moreau, 1990).

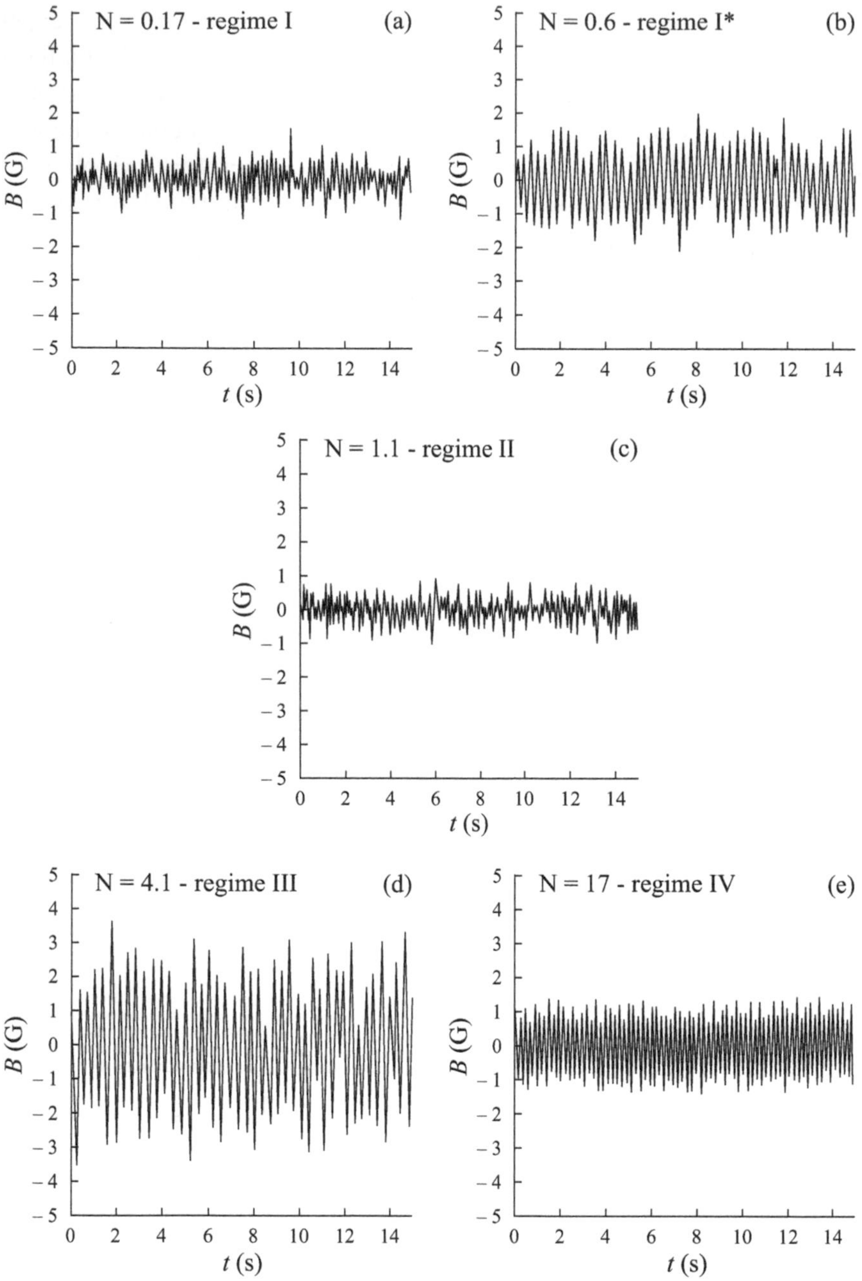

Figure 8.17 - Time evolution of the induced magnetic field B for different values of the interaction parameter N in the mechanically forced experiment in College Park (see Section 8.2.9). The rotation rate of the propellers is fixed ($\mathrm{Rm} = 7.5$). Different regimes may be identified by considering the frequency and amplitude of the measured magnetic field (from Sisan *et al.*, 2003).

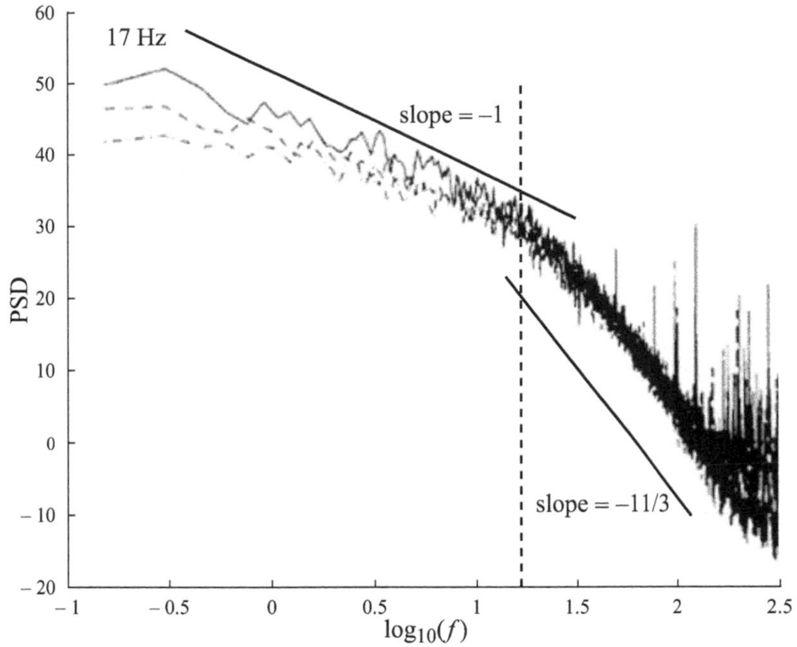

Figure 8.18 - Power spectral density function of the induced magnetic field in the VKS experiment (see Section 8.2.10). Above the frequency (dashed line) associated to the rotating rate of the propeller, the spectra is in agreement with the Kolmogorov prediction (low N). Below this frequency, the spectra is hyperbolic (from Bourgoin *et al.*, 2002).

Depending on the intensity of the imposed magnetic field, which is measured by the interaction parameter N, one may get two types of spectra dependence. For low N, the turbulence is of the Kolmogorov type with $\mathcal{K} \propto k^{-5/3}$ and $\mathcal{M} \propto k^{-11/3}$. This has been seen in many experiments. The VKS team (Odier *et al.*, 1998; Bourgoin *et al.*, 2002) has documented this regime, in which the magnetic field behaves as a passive vector. In Figure 8.18, magnetic measurements show a clear $-11/3$ power law above the frequency of the driving disc, while the authors proposed an hyperbolic range of frequencies for the induced magnetic field fluctuations below this frequency. Although such a k^{-1} behaviour is also reported in the Karlsruhe experiment (Müller *et al.*, 2004) and in the Maryland experiment (D. Lathrop, private communication), the physical mechanism which leads to such a power law is not yet fully understood.

For strong magnetic field (high N), Alémany *et al.* (1979) found $\mathcal{K} \propto k^{-3}$ and $\mathcal{M} \propto k^{-5}$ in an experiment in which turbulence was produced by the motion of a grid and the velocity was measured using quartz-coated hot film probes. As shown by the spectrum dependence, the Joule effect strongly influences the energy dissipation rate, and leads to an anisotropic flow during the decay of turbulence. The -3 exponent of the kinetic energy may be deduced from the balance between the angular transfer time and the Joule dissipation time. A second experiment has been performed to study this regime, but under stationary forcing. Messadek & Moreau (2002) found the $-5/3$ exponent for the spectral kinetic energy at low N and -3 exponent at high N.

Under a dynamo state, power spectral density of the magnetic field have been measured in the Karlruhe experiment (Stieglitz *et al.*, 2002, Müller *et al.*, 2004). Figure 8.19 shows typical spectra of the induced magnetic field inside the container (see Section 8.2.8). Above the critical flow rate ($V_C \approx 120\,\mathrm{m}^3\,\mathrm{h}^{-1}$), a self-induced magnetic field is generated and a peak appears in the magnetic spectrum around 1Hz. One would like to interpret the frequency of this peak as the injecting magnetic energy scale using the ergodicity hypothesis. However, the frequency associated to the helicoidal flow may be evaluated to 5Hz. This value is too large to explain the power peak. Moreover, the frequency of the power peak changes with the supercriticality of the dynamo and not with the volumetric rate of the helicoidal tube. Note that Müller *et al.* (2004) proposed an interpretation in term of Alfvén waves travelling along the cylinders. For larger frequencies (above the peak), the Joule damping of the magnetic field leads to a large negative exponent in the power spectrum (from -3 to -5, and sometimes even smaller). The -5 exponent is in agreement with the results of Alémany *et al.* (1979). For smaller frequencies (below the peak), they found a f^{-1} type behaviour like in Bourgoin *et al.* (2002) study. In the context of their dynamo, they link this observation to the prediction of Pouquet *et al.* (1976) based on theoretical arguments of inverse cascade of magnetic helicity.

Experimental spectra at large Rm are always difficult to interpret. It is difficult to get a clear power law for a decade in frequency and to infer an exponent without any theoretical background is rather conjectural. As already mentioned, conversion from temporal to spatial field are based on the ergodicity hypothesis, which remains an hypothesis in MHD flows. Furthermore, theories are generally done for $\mathrm{Pm} = 1$ (Biskamp, 1993) and the compatibility of theoretical predictions with with liquid metal experiments is not straightforward.

8.3.9. THE β–EFFECT AND TURBULENT VISCOSITY

The β–effect is a turbulent effect associated to the $\boldsymbol{\nabla} \times (\mathbf{u} \times \mathbf{B})$ in the induction equation and can be modelled as a magnetic dissipative effect. In some regimes, the β–effect could modify the magnetic diffusivity of conducting fluids. Reighard & Brown (2001) have measured the apparent magnetic diffusivity of sodium as a function of the magnetic Reynolds number Rm. They found a reduction from the

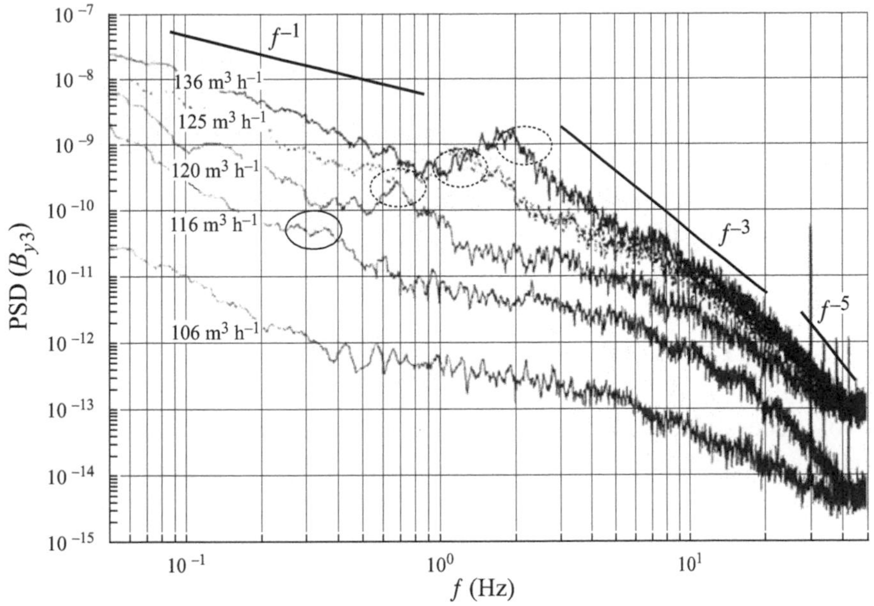

Figure 8.19 - Power spectral density of a component of the magnetic field (perpendicular to the axis of the tubes at the center of the container) in the Karlsruhe dynamo experiment (see Section 8.2.8). The helicoidal flux rates are set to $100\,m^3\,h^{-1}$ and the central flux rate is increased from 106 to $136\,m^3\,h^{-1}$, the onset of dynamo action occurring around $120\,m^3\,h^{-1}$. A central peak appears above the onset, separating the spectra in two parts: an hyperbolic range for low frequency and a steeper exponent (-3 to -5) for higher frequency (from Müller *et al.*, 2004).

molecular value of the electrical conductivity of 4% at Rm of order 10. No such effect has ever been measured in other sodium experiments. Nevertheless, the mean field theory developed by Rädler *et al.* (2002) evaluated a β–effect between 1 to 10% of the molecular value in the Karlsruhe dynamo flow. The good agreement between the mean field approach and the experimental results may be interpreted an indirect observation of the β–effect. Moreover, Tilgner & Busse (2002) with their kinematic approach, need to assume an increased value of the magnetic diffusivity in order to explain correctly the precise position of the onset of the Karlsruhe dynamo. In that case, they associate the enhanced magnetic diffusivity to the averaged diffusivity of sodium and stainless steel, rather than to the effect of turbulence.

Similarly, the nonlinear term in velocity in the momentum equation may be modelled as a dissipative viscous effect. The turbulent viscosity (and more sophisticated models) is largely used in geophysical and astrophysical numerical fluid dynamics (meteorological or oceanographic models for example). Direct experimental measurements of the turbulent viscosity are quite difficult, in particular as this would require very precise maps of the velocity field within the bulk of the flow. In a Couette experiment, Lathrop *et al.* (1992) interpreted the measured torque delivered by the rotating motor in terms of turbulent viscosity. They proposed a law for the turbulent viscosity as a function of the hydrodynamic Reynolds number, Re. More recently, Brito *et al.* (2004) have shown experimental evidences for turbulent viscosity in the context of a rotating flow.

8.3.10. SATURATION OF THE DYNAMO

The two successful experimental dynamos exhibited a saturated state of the self-sustained magnetic field after a period of exponential growth (this exponential growth is shown in Figure 8.13 in the case of the Riga dynamo). In both experiments, the injected mechanical power required to drive the flow was measured as a function of the averaged velocity of the fluid flow. Figure 8.15 shows an increase of around $10\,\mathrm{kW}$ after the onset of the dynamo regime in the Riga experiment. If one assumes that this increase of power is directly dissipated by Joule effect, one can estimate a typical dissipation length scale L_d:

$$P_j \propto \frac{B^2 L^3}{\mu_0^2 \sigma L_d^2} \approx 10^4\,\mathrm{W} \qquad \text{which yields} \qquad L_d \approx 10^{-3}\,\mathrm{m} \tag{8.2}$$

with $B = 1\,\mathrm{mT}$, $L = 1\,\mathrm{m}$. This Joule dissipation scale is much larger than the viscous dissipation scale ($L_d \approx 10^{-6}\,\mathrm{m}$, if $\mathrm{Re} \approx 1$ with $U \approx 1\,\mathrm{m\,s^{-1}}$) and may be the main dissipative process in the dynamo state. At the dissipation scale, Rm is small and the results obtained for low Rm turbulence should apply, such as the spectrum dependence in k^{-3} for the kinetic energy (see Section 8.3.8). Figure 8.19 shows indeed a steep tail of the spectra (for large frequencies), which may be the signature of the low Rm turbulence.

The balance between the nonlinear velocity term and the Lorentz force may allow to predict the saturated magnetic field intensity. Pétrélis & Fauve (2001) and Tilgner & Busse (2002) had to introduce a turbulent viscosity (at least 10^4 times the molecular one) to explain the observed value of the saturated magnetic field in Karlsruhe and Riga. Their approach excludes any laminar viscous balance which would lead to an intensity of the saturated magnetic field much too low compared to the experimental measurements.

The saturation mechanism may also be associated with a change in the "large scale" fluid flow dynamics after the onset of the dynamo. In Riga, observations of the saturated magnetic field show indeed a dependence along the height of the experiment (Gailitis *et al.*, 2001) and sodium originally at rest (at the edge of the container, see

Section 8.2.7) is driven into motion after the onset (Gailitis, private communication). The same idea was proposed by Tilgner & Busse (2001) in the context of the Karlsruhe experiment with the presence of vortices of sodium in between the tubes of the experiment (see Section 8.2.8).

8.4. CONCLUSIONS

Two experimental dynamos (Gailitis *et al.*, 2000; Müller & Stieglitz, 2000) have been observed in the laboratory. They experimentally demonstrated the existence of a self-sustained dynamo regime in fluid flow, in good agreement with the theoretical predictions. These two experiments have reported dynamo action for the parameters predicted by analytical and numerical methods, even if the experiments could not exactly reproduce the idealised models (boundary conditions, presence of stainless steel, small scale flow, turbulence). These two successes are really associated to the choice of very robust flows to produce dynamo action. In the dynamo state, these two experiments exhibit results (power, spectra, etc.) that are not fully understood yet and many questions remain regarding the presence of a large magnetic field. Homogeneous experiments, appear to be needed to address these questions.

Surprisingly, in the case of homogeneous experiments, the dynamo onset seems to vary significantly with small variations in the velocity field as mentioned above. We can conclude that the robustness of these flows to produce a self-sustained magnetic field is weaker. Lathrop and his collaborators investigated different configurations to try and achieve dynamo action in their experimental set-up. Shew *et al.* (2001) report on tests with change of propellers, addition of copper rings or plates at the equator, change of baffles in the sodium tank. A clear variation in the exponential decay times of the imposed magnetic fields associated with these changes is observed but it is difficult to infer general properties on dynamo mechanism from these tests. Shew *et al.* (2002) built an updated version of the Lowes & Wilkinson dynamo, in which the external solid housing is replaced by liquid sodium. No dynamo has been observed in this configuration. The VKS team also changed the electrical conductivity of the boundaries by adding a copper housing to their vessel, without observing dynamo action (Bourgoin *et al.*, 2002), although a numerical kinematic study predicted a reduction the critical Rm, by a factor of two, when the boundaries were changed from insulating to perfectly conducting (Marié *et al.*, 2002). Martin *et al.* (2000) and Frick *et al.* (2002) have tried to increase the magnetic permeability of the liquid metal by using ferromagnetic iron beads or small particles. Frick *et al.* (2002) proposed a linear law for low concentration of small particles (0.01 to $0.1\,\mathrm{mm}$ of diameter) which may increase the magnetic Reynolds number by a factor of two.

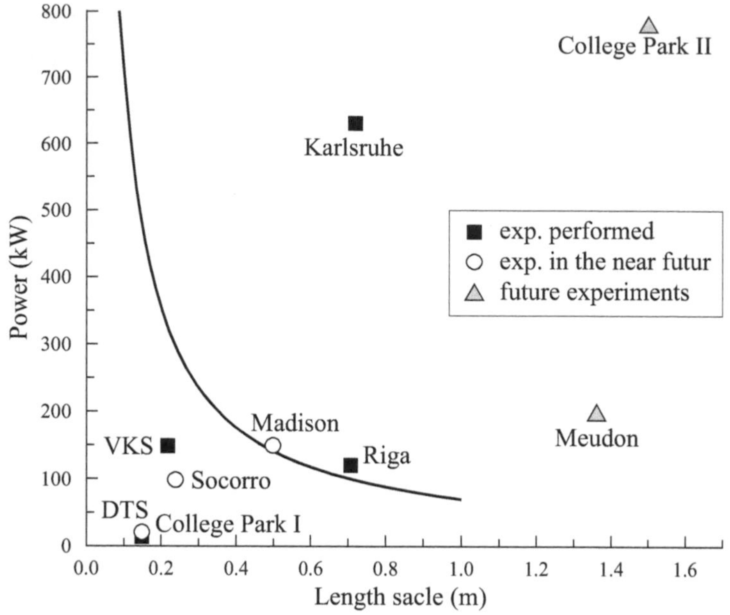

Figure 8.20 - Summary of the present experiments and projects presented in term of characteristic length scale (x–axis) versus the mechanical power (in kW) injected to run the sodium flow. Triangles: possible future experiments; Circles: experiments to be run in the near future; Squares: experiments already performed. The curve is a temptative line that may separate the successful experiments from the non-dynamo ones.

Building an experimental dynamo is a very long and hard enterprise. This is why most of the homogeneous dynamo experiments, or projects, presented throughout this survey expect to observe dynamo action in the near future. Out of the groups that are currently working in this area, we may predict that two of them will soon observe dynamo action: The VKS team is planning to run a second version of their Von Kàrmàn experiment in a larger container, with new optimised propellers and a cooling system unit. The College park group is building a very large rotating spherical experiment with 15 tons of sodium, which should have all the ingredients to self-sustain a magnetic field, if one dares to compare it with natural planetary dynamos.

Size and power are the two main factors to determine the actual cost of an experiment. Small experiments are easier to build and to modify. Up to now, only large experiments have been able to self-induce a magnetic field in liquid sodium. Clearly, in order to reach a given magnetic Reynolds number, one has to find a trade-off between the typical scale of the flow and the typical velocity (and consequently power).

Too much power injected in a small experiment may yield cooling problems. On the contrary, large scale experiments need a significant power input to reach high fluid velocities. Nataf (2003) produced an interesting representation of the power versus size of dynamo experiments. We show an updated version of this graph in Figure 8.20 including newer projects. Note however that this graph does not take into account the type of flow generated into the vessel, which may be a crucial point when the vigour of the flow is close the onset of dynamo action.

In the coming years, the main challenge in this field is to understand MHD turbulence. Studies of spectra will be valuable and enable a classification of different regimes. As noted in Section 8.3.8, experimental data are generally measured as a function of time and we rely on the ergodicity hypothesis to interpret these spectra in term of spatial behaviour of the MHD turbulence. Theories with very low Prandtl numbers will be needed to help the interpretation of liquid sodium experiments.

Finally, we note that global rotation could be a key ingredient for dynamo action. The presence of rotation provides a prefered direction in the flow and the isotropic turbulence can be replaced with quasi-geostrophic turbulence. This is also the case with precessional flows, in which Gans (1970) has observed a large amplification of the magnetic field, still unexplained. Quasi-geostrophic dynamos have recently been computed based on shear flows taking into account the properties of a rapidly rotating flow (Schaeffer & Cardin, 2006). These preliminary results are encouraging for experimental dynamo modelling of rotating planets (low Pm, low E), because they exhibit robust dynamos that can be understood under the $\alpha\omega$–dynamo formalism. We leave the reader with an interesting open question: does the rotation increase the robustness and the ability of the flow to produce the dynamo action?

CHAPTER 9

PROSPECTS

Emmanuel Dormy & Andrew Soward

Dynamos on an impressive range of scales have been addressed in this book ranging from the experimental scale of a few meters to galactic discs, i.e. some 10^{20} m. Even if one restricts attention to natural dynamos, the Earth's core or the Sun would constitute objects on the nano, or even a pico, scale when measured in Galactic units! As we have seen in this book, it is quite remarkable that a common formalism (and similar physical processes) governs the dynamics of dynamos on such a wide disparity of scales. Not only do the same governing equations (of course, under different parameter regimes) apply to these objects, but the same magnetic field generating instability appears to develop; albeit in the "slow dynamo" regime for the small planetary systems, and in "fast dynamo" regime for larger objects. This ability of the dynamo instability to develop in such a large variety of natural bodies is probably its most striking property.

Thanks to this large variety of applications, progress on understanding dynamo action is being achieved in the various fields of geophysics, planetary physics, solar physics and astrophysics. After many years of largely independent development, these fields of research have been increasingly interactive over the last few years. This has led to fruitful developments which are a sign of maturity for the subject.

Researchers trying to address the origin of magnetic fields in natural objects usually combine the three approaches of observations, theory and numerics as presented in this book. Each is progressing independently at a rapid pace.

Our knowledge of the present geomagnetic field and its secular variation increases as satellite measurements become more numerous and start to cover longer periods of time. Recently Olsen *et al.* (2006) were able to provide a representation of the static (core and crustal) field up to spherical harmonic degree $\ell = 40$, and of the first time derivative up to $\ell = 15$ relying on measurements from CHAMP, Ørsted, and SAC-C magnetic satellite data. On longer timescales, paleomagnetism yields an ever more refined description of the past magnetic field of the Earth and its evolution. Recently, the intensity variation of the geomagnetic field has been carefully reconstructed over the past two million years (Valet *et al.*, 2005). This has led to a description of the field intensity variations spanning the five reversals which have occurred during this period. They appear to be characterised by a slow decrease of the field intensity (over 60-80 kyr) followed by a rapid recovery of the field strength immediately after the reversals. On much longer time scales, research on geomagnetic reversal chronology has recently pointed to the existence of three superchrons (Pavlov & Gallet, 2005); one during the Cretaceous (some 100 Myr ago, see Chapter 4), another during the Late Paleozoic (some 300 Myr ago) and a third during the Lower and Middle Ordovician (some 475 Myr ago). The existence of these repeated superchrons as well as the variation of the reversal rate in between (from 0 to 10 per Myr) provide important constraints on geodynamo modelling. Observation of our star, the Sun, is also improving with data from the Hinode satellite, which will soon be supplemented by the STEREO and SDO satellites due to be flying together sometime in the next three years. This should lead to a more detailed description of the magnetic field in the corona and its connection with the solar dynamo.

The numerical simulation of natural dynamos has seen rapid developments during the last ten years and much of this progress has been reported in this book. A numerical model, originally developed for the Sun (Gilman & Glatzmaier, 1981; Glatzmaier, 1985a,b), has been extremely successful in the geodynamo context (see Chapter 4) and recently has been extended to address some issues in planetary physics, such as the banded structure of Jupiter (Heimpel *et al.*, 2005) and the low intensity of Mercury's magnetic field (Christensen, 2006). Despite these successes, future progress depends on a combination of numerical, analytical and observational approaches.

Beyond the natural applications described, another community from experimental physics is contributing to the development of our understanding of dynamo action. Following the success of the first two experimental fluid dynamos in Riga and Karlsruhe, many groups throughout the world are developing and testing new experimental setups; these are reviewed in Chapter 8. Indeed, progress in this area is so rapid that a new successful experimental result was obtained during the fruition of this book by the VKS group (see Section 8.2.10). They have reported a self-excited dynamo in an unconstrained turbulent flow of liquid Sodium (Monchaux *et al.*, 2007). The experiment is of course somewhat dissimilar (both in geometry and in parameter regime) from natural dynamos; yet we stress that in many respects it is closer to the natural dynamo applications than numerical models, particularly in respect to the value of the magnetic Prandtl number. In this respect, this finding is exciting as it opens the way to an experimental understanding of dynamo action. In addition, the connection with natural dynamos is reinforced by the existence of parameter regimes for which the field reverses either regularly of chaotically (see Berhanu *et al.*, 2007). The possibility of dynamo experiments performed with an unconstrained flow, exhibiting rich temporal behaviour, will hopefully lead to a productive period of research on natural dynamos.

We should not conclude this chapter without commenting on our environment. While the objective of this book has been to cover the mathematical background of natural dynamos, the existence of magnetic fields in astrophysical bodies affects our every-day living conditions. Indeed, it is well known that the Earth's magnetic field acts as a shield that deflects the solar wind. The weakening of this shield, associated with the recent decay of the Earth's dipole strength, could therefore be a worry. The decay however appears to be compatible with a local fluctuation rather than a long term trend which could be an environmental concern (Dormy, 2006). Not surprisingly, the Earth's climate shows links with solar activity, which might even be the origin of quasi-periodic glacial events (Braun *et al.*, 2005). Furthermore, the Earth's magnetic field may be connected to the climate, because variations in the geometry of the geomagnetic field might result in enhanced cosmic-ray induced nucleation of clouds and so drive the climate (Courtillot *et al.*, 2007). On the solar side, there are suggestions (Dikpati *et al.*, 2006a, 2006b; Hathaway & Wilson, 2006) that the current Solar cycle (due to peak in 2010 or 2011) will be characterised by a strong increase in the solar activity. This could have serious consequences on the Earth and the climate (Clark, 2006). Tobias *et al.* (2006) however noted that episodes of high activity have, in the past, often been followed by Grand Minima (periods of severely reduced magnetic activity).

While the links between the Earth's magnetic field, the Solar magnetic field and the climate/environmental concerns are very likely, we are still far from relating them to dynamo action. We hope the reader will be convinced, as we are, that the beauty of this subject is such that it is enough to justify its study, independently of any environmental concerns.

APPENDIX A

VECTORS AND COORDINATES

CARTESIAN COORDINATES

$$\boldsymbol{\nabla}\Phi = \frac{\partial \Phi}{\partial x}\mathbf{e}_x + \frac{\partial \Phi}{\partial y}\mathbf{e}_y + \frac{\partial \Phi}{\partial z}\mathbf{e}_z\,, \tag{A.1}$$

$$\boldsymbol{\nabla}\cdot\mathbf{V} = \frac{\partial V_x}{\partial x} + \frac{\partial V_y}{\partial y} + \frac{\partial V_z}{\partial z}\,, \tag{A.2}$$

$$\boldsymbol{\nabla}\times\mathbf{V} = \left[\frac{\partial V_z}{\partial y} - \frac{\partial V_y}{\partial z}\right]\mathbf{e}_x + \left[\frac{\partial V_x}{\partial z} - \frac{\partial V_z}{\partial x}\right]\mathbf{e}_y + \left[\frac{\partial V_y}{\partial x} - \frac{\partial V_x}{\partial y}\right]\mathbf{e}_z\,, \tag{A.3}$$

$$\Delta\Phi = \frac{\partial^2 \Phi}{\partial x^2} + \frac{\partial^2 \Phi}{\partial y^2} + \frac{\partial^2 \Phi}{\partial z^2}\,, \tag{A.4}$$

$$\Delta\mathbf{V} = [\Delta V_x]\,\mathbf{e}_x + [\Delta V_y]\,\mathbf{e}_y + [\Delta V_z]\,\mathbf{e}_z\,, \tag{A.5}$$

$$\boldsymbol{\nabla}\mathbf{V} = \begin{bmatrix} \dfrac{\partial V_x}{\partial x} & \dfrac{\partial V_x}{\partial y} & \dfrac{\partial V_x}{\partial z} \\ \dfrac{\partial V_y}{\partial x} & \dfrac{\partial V_y}{\partial y} & \dfrac{\partial V_y}{\partial z} \\ \dfrac{\partial V_z}{\partial x} & \dfrac{\partial V_z}{\partial y} & \dfrac{\partial V_z}{\partial z} \end{bmatrix}. \tag{A.6}$$

CYLINDRICAL POLAR COORDINATES

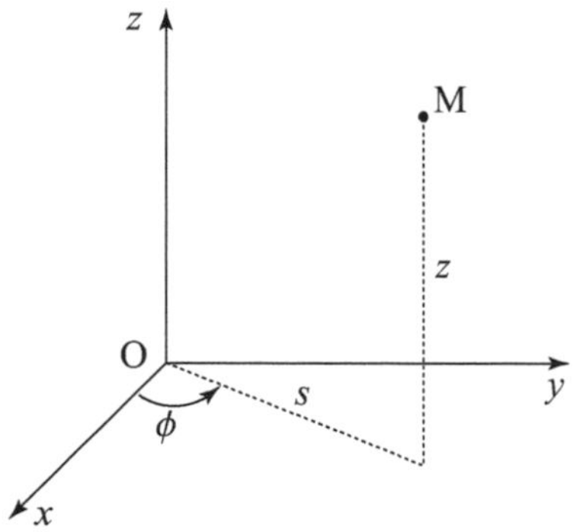

$$\nabla \Phi = \frac{\partial \Phi}{\partial s}\mathbf{e}_s + \frac{1}{s}\frac{\partial \Phi}{\partial \phi}\mathbf{e}_\phi + \frac{\partial \Phi}{\partial z}\mathbf{e}_z , \tag{A.7}$$

$$\nabla \cdot \mathbf{V} = \frac{1}{s}\frac{\partial}{\partial s}(sV_s) + \frac{1}{s}\frac{\partial V_\phi}{\partial \phi} + \frac{\partial V_z}{\partial z} , \tag{A.8}$$

$$\begin{aligned} \nabla \times \mathbf{V} &= \left[\frac{1}{s}\frac{\partial V_z}{\partial \phi} - \frac{\partial V_\phi}{\partial z}\right]\mathbf{e}_s \\ &+ \left[\frac{\partial V_s}{\partial z} - \frac{\partial V_z}{\partial s}\right]\mathbf{e}_\phi \\ &+ \left[\frac{1}{s}\frac{\partial}{\partial s}(s\,V_\phi) - \frac{1}{s}\frac{\partial V_s}{\partial \phi}\right]\mathbf{e}_z , \end{aligned} \tag{A.9}$$

$$\Delta \Phi = \frac{\partial^2 \Phi}{\partial s^2} + \frac{1}{s}\frac{\partial \Phi}{\partial s} + \frac{1}{s^2}\frac{\partial^2 \Phi}{\partial \phi^2} + \frac{\partial^2 \Phi}{\partial z^2} , \tag{A.10}$$

$$\Delta \mathbf{V} = \nabla(\nabla \cdot \mathbf{V}) - \nabla \times (\nabla \times \mathbf{V}) , \tag{A.11}$$

$$\nabla \mathbf{V} = \begin{bmatrix} \dfrac{\partial V_s}{\partial s} & \dfrac{1}{s}\dfrac{\partial V_s}{\partial \phi} - \dfrac{V_\phi}{s} & \dfrac{\partial V_s}{\partial z} \\ \dfrac{\partial V_\phi}{\partial s} & \dfrac{1}{s}\dfrac{\partial V_\phi}{\partial \phi} + \dfrac{V_s}{s} & \dfrac{\partial V_\phi}{\partial z} \\ \dfrac{\partial V_z}{\partial s} & \dfrac{1}{s}\dfrac{\partial V_z}{\partial \phi} & \dfrac{\partial V_z}{\partial z} \end{bmatrix} . \tag{A.12}$$

SPHERICAL POLAR COORDINATES

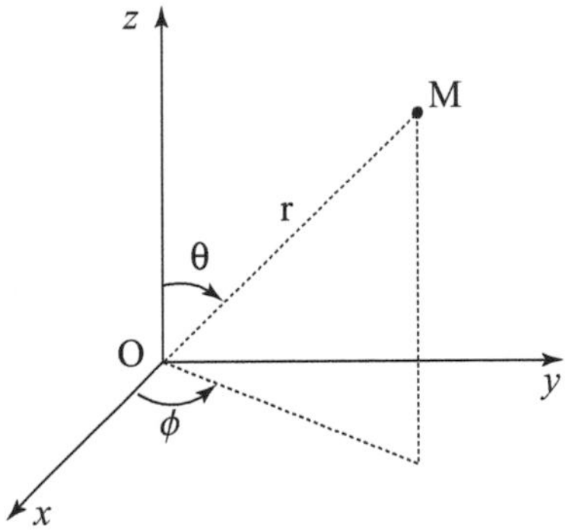

$$\nabla\,\Phi = \frac{\partial \Phi}{\partial r}\mathbf{e}_r + \frac{1}{r}\,\frac{\partial \Phi}{\partial \theta}\mathbf{e}_\theta + \frac{1}{r\,\sin\theta}\,\frac{\partial \Phi}{\partial \phi}\mathbf{e}_\phi\,, \tag{A.13}$$

$$\nabla\cdot\mathbf{V} = \frac{1}{r^2}\frac{\partial}{\partial r}\left(r^2\,V_r\right) + \frac{1}{r\,\sin\theta}\frac{\partial}{\partial\theta}\left(\sin\theta\,V_\theta\right) + \frac{1}{r\,\sin\theta}\,\frac{\partial V_\phi}{\partial\phi}\,, \tag{A.14}$$

$$\begin{aligned}\nabla\times\mathbf{V} \;&=\; \frac{1}{r\,\sin\theta}\left[\frac{\partial}{\partial\theta}\left(\sin\theta\,V_\phi\right) - \frac{\partial V_\theta}{\partial\phi}\right]\mathbf{e}_r\\ &+\; \frac{1}{r}\left[\frac{1}{\sin\theta}\,\frac{\partial V_r}{\partial\phi} - \frac{\partial}{\partial r}\left(r\,V_\phi\right)\right]\mathbf{e}_\theta\\ &+\; \frac{1}{r}\left[\frac{\partial}{\partial r}\left(r\,V_\theta\right) - \frac{\partial V_r}{\partial\theta}\right]\mathbf{e}_\phi\,,\end{aligned} \tag{A.15}$$

$$\Delta\Phi = \frac{\partial^2\Phi}{\partial r^2} + \frac{2}{r}\frac{\partial\Phi}{\partial r} + \frac{1}{r^2\sin\theta}\,\frac{\partial}{\partial\theta}\left(\sin\theta\,\frac{\partial\Phi}{\partial\theta}\right) + \frac{1}{r^2\sin^2\theta}\frac{\partial^2\Phi}{\partial\phi^2}\,, \tag{A.16}$$

$$L_2\,\Phi = -\frac{1}{\sin\theta}\frac{\partial}{\partial\theta}\left(\sin\theta\frac{\partial\Phi}{\partial\theta}\right) - \frac{1}{\sin^2\theta}\frac{\partial^2\Phi}{\partial\phi^2}\,, \tag{A.17}$$

$$\Delta\mathbf{V} = \nabla\left(\nabla\cdot\mathbf{V}\right) - \nabla\times\left(\nabla\times\mathbf{V}\right)\,, \tag{A.18}$$

$$\nabla\mathbf{V} = \left[\begin{array}{ccc} \dfrac{\partial V_r}{\partial r} & \dfrac{1}{r}\dfrac{\partial V_r}{\partial\theta} - \dfrac{V_\theta}{r} & \dfrac{1}{r\sin\theta}\,\dfrac{\partial V_r}{\partial\phi} - \dfrac{V_\phi}{r} \\ \dfrac{\partial V_\theta}{\partial r} & \dfrac{1}{r}\dfrac{\partial V_\theta}{\partial\theta} + \dfrac{V_r}{r} & \dfrac{1}{r\,\sin\theta}\left(\dfrac{\partial V_\theta}{\partial\phi} - V_\phi\cos\theta\right) \\ \dfrac{\partial V_\phi}{\partial r} & \dfrac{1}{r}\dfrac{\partial V_\phi}{\partial\theta} & \dfrac{1}{r\,\sin\theta}\left(\dfrac{\partial V_\phi}{\partial\phi} + V_r\,\sin\theta + V_\theta\,\cos\theta\right) \end{array}\right]. \tag{A.19}$$

VECTOR IDENTITIES

$$\nabla \cdot (\nabla \times \mathbf{V}) = 0\,, \tag{A.20}$$

$$\nabla \times (\nabla \times \mathbf{V}) = \nabla(\nabla \cdot \mathbf{V}) - \Delta \mathbf{V}\,, \tag{A.21}$$

$$\nabla \cdot (\nabla \Phi) = \Delta\, \Phi\,, \tag{A.22}$$

$$\nabla \times (\nabla \Phi) = \mathbf{0}\,, \tag{A.23}$$

$$\nabla \cdot (\Phi \mathbf{V}) = \Phi\, \nabla \cdot \mathbf{V} + \mathbf{V} \cdot \nabla \Phi\,, \tag{A.24}$$

$$\nabla \times (\Phi \mathbf{V}) = \Phi \nabla \times \mathbf{V} + (\nabla \Phi) \times \mathbf{V}\,, \tag{A.25}$$

$$\nabla \cdot (\mathbf{V} \times \mathbf{W}) = \mathbf{W} \cdot (\nabla \times \mathbf{V}) - \mathbf{V} \cdot (\nabla \times \mathbf{W})\,, \tag{A.26}$$

$$\begin{aligned} \nabla \times (\mathbf{V} \times \mathbf{W}) = {} & \mathbf{V}\,(\nabla \cdot \mathbf{W}) - \mathbf{W}\,(\nabla \cdot \mathbf{V}) \\ & + (\mathbf{W} \cdot \nabla)\mathbf{V} - (\mathbf{V} \cdot \nabla)\mathbf{W}\,, \end{aligned} \tag{A.27}$$

$$\begin{aligned} \nabla\,(\mathbf{V} \cdot \mathbf{W}) = {} & \mathbf{V} \times (\nabla \times \mathbf{W}) + \mathbf{W} \times (\nabla \times \mathbf{V}) \\ & + (\mathbf{V} \cdot \nabla)\,\mathbf{W} + (\mathbf{W} \cdot \nabla)\,\mathbf{V}\,. \end{aligned} \tag{A.28}$$

GREEN'S FORMULAE

$$\int_{\mathcal{D}} ((\nabla \Phi) \cdot \mathbf{V} + \Phi(\nabla \cdot \mathbf{V}))\, \mathrm{dV} = \int_{\partial \mathcal{D}} \Phi(\mathbf{V} \cdot \mathbf{n})\, \mathrm{dS}\,, \tag{A.29}$$

$$\int_{\mathcal{D}} (\nabla \Phi \cdot \nabla \Psi + \Phi \Delta \Psi)\, \mathrm{dV} = \int_{\partial \mathcal{D}} \Phi \nabla \Psi \cdot \mathrm{d}\mathbf{S}\,, \tag{A.30}$$

$$\int_{\mathcal{D}} (\Phi \Delta \Psi - \Psi \Delta \Phi)\, \mathrm{d}V = \int_{\partial \mathcal{D}} (\Phi \nabla \Psi - \Psi \nabla \Phi) \cdot \mathrm{d}\mathbf{S}\,, \tag{A.31}$$

$$\int_{\mathcal{D}} (\mathbf{V} \cdot (\nabla \times \mathbf{W}) - (\nabla \times \mathbf{V}) \cdot \mathbf{W})\, \mathrm{dV} = \int_{\partial \mathcal{D}} (\mathbf{V} \times \mathbf{n}) \cdot \mathbf{W}\, \mathrm{dS}\,. \tag{A.32}$$

APPENDIX B

POLOIDAL–TOROIDAL DECOMPOSITION

The poloidal–toroidal decomposition is often used in this book in the spherical geometry (other decompostions can be constructed relying on other basis vectors). It relies on the fact any solenoidal vector field can be expressed as

$$\mathbf{V} = \underbrace{\boldsymbol{\nabla} \times \boldsymbol{\nabla} \times (\mathbf{r}\,\beta)}_{\text{poloidal component}} + \underbrace{\boldsymbol{\nabla} \times (\mathbf{r}\,\gamma)}_{\text{toroidal component}}, \tag{B.1}$$

where $\mathbf{r}$ is the position vector $\mathbf{r} = r\mathbf{e}_r$. The scalar functions β and γ are rendered unique by requiring

$$\langle \beta \rangle_r = \langle \gamma \rangle_r = 0\,, \tag{B.2}$$

where $\langle \cdot \rangle_r$ denotes the average on the spherical surface of radius r.

To demonstrate (B.1), we start from the representation

$$\mathbf{V} = \alpha\,\mathbf{r} + \boldsymbol{\nabla} \times \beta\,\mathbf{r} + \boldsymbol{\nabla} \times \boldsymbol{\nabla} \times \gamma'\,\mathbf{r} \tag{B.3}$$

of any vector field in $\mathbb{R}^3$, where α, β, γ' are functions of r, θ, ϕ, and demonstrate that $\alpha\mathbf{r}$ may be expressed in the form

$$\alpha\mathbf{r} = \boldsymbol{\nabla} \times \boldsymbol{\nabla} \times \gamma''\,\mathbf{r} \tag{B.4}$$

for some function $\gamma''(r, \theta, \phi)$.

Since $\mathbf{V}$ is solenoidal, it follows from (B.3) that $\boldsymbol{\nabla} \cdot (\alpha\mathbf{r}) = 0$ which in turn implies

$$r\frac{\partial \alpha}{\partial r} + 3\alpha = 0 \qquad \text{and thus} \qquad \alpha = \frac{\widehat{\alpha}}{r^3}\,, \tag{B.5a,b}$$

where $\widehat{\alpha}$ is a function of θ and ϕ only, We express $\widehat{\alpha}(\theta,\phi)$ in terms of its mean part $\overline{\alpha} = \langle\widehat{\alpha}\rangle_r$, a constant, and its remaining fluctuating part as

$$\widehat{\alpha}(\theta,\phi) = \overline{\alpha} + L_2\Big(\widehat{\delta}(\theta,\phi)\Big)\,, \tag{B.6}$$

where L_2 is the operator defined in (A.17). Then on substitution into (B.3) we obtain

$$\mathbf{V}\cdot\mathbf{r} = \frac{1}{r^2}\,\Big(\overline{\alpha} + L_2(\widehat{\delta})\Big) + \underbrace{(\boldsymbol{\nabla}\times\beta\,\mathbf{r})\cdot\mathbf{r}}_{=0} + \underbrace{(\boldsymbol{\nabla}\times\boldsymbol{\nabla}\times\gamma'\,\mathbf{r})\cdot\mathbf{r}}_{=L_2(\gamma')}\,. \tag{B.7}$$

Now, since $\boldsymbol{\nabla}\cdot\mathbf{V} = 0$ everywhere, it follows that the spherical surface average of $\mathbf{V}\cdot\mathbf{r}$, namely

$$\langle\mathbf{V}\cdot\mathbf{r}\rangle_r = \frac{1}{r^2}\,\overline{\alpha} + \frac{1}{r^2}\,\underbrace{\langle L_2(\widehat{\delta})\rangle_r}_{=0} + \underbrace{\langle L_2(\gamma')\rangle_r}_{=0}\,, \tag{B.8}$$

vanishes, with the consequence

$$\overline{\alpha} = 0\,. \tag{B.9}$$

Thus, as anticipated in (B.4), the first term in (B.3) can be rewritten in the form

$$\alpha\,\mathbf{r} = \frac{1}{r^3}\,L_2\Big(\widehat{\delta}(\theta,\phi)\Big)\,\mathbf{r} = \boldsymbol{\nabla}\times\boldsymbol{\nabla}\times\gamma''\,\mathbf{r}\,, \tag{B.10}$$

where $\gamma'' = \widehat{\delta}/r$. It follows that any solenoidal vector field can be expressed in the form (B.1) as

$$\mathbf{V} = \boldsymbol{\nabla}\times\beta\,\mathbf{r} + \boldsymbol{\nabla}\times\boldsymbol{\nabla}\times\gamma\,\mathbf{r}\,, \tag{B.11}$$

where $\gamma = \gamma' + \gamma''$.

The functions β and γ are obtained by inverting

$$L_2(\beta) = (\boldsymbol{\nabla}\times\mathbf{V})\cdot\mathbf{r} \qquad \text{and} \qquad L_2(\gamma) = \mathbf{V}\cdot\mathbf{r} \tag{B.12a,b}$$

subject to $\langle\beta\rangle_r = \langle\gamma\rangle_r = 0$ [see (B.2)], which renders them unique. This inversion is usually done using spherical harmonics $Y_\ell^m \propto P_\ell^m \mathrm{e}^{\mathrm{i}m\phi}$, which are the eigenfunctions of L_2 with

$$L_2 Y_\ell^m = \ell(\ell+1)\,Y_\ell^m\,. \tag{B.13}$$

Relative to spherical polar coordinates, the poloidal–toroidal decomposition (B.1), equivalently (B.11), has the following explicit form:

$$V_r = \frac{1}{r}\,L_2(\gamma)\,, \tag{B.14a}$$

$$V_\theta = \frac{1}{\sin\theta}\frac{\partial\beta}{\partial\phi} + \frac{\partial}{\partial\theta}\left[\frac{1}{r}\frac{\partial}{\partial r}(r\gamma)\right]\,, \tag{B.14b}$$

$$V_\phi = -\frac{\partial\beta}{\partial\theta} + \frac{1}{\sin\theta}\frac{\partial}{\partial\phi}\left[\frac{1}{r}\frac{\partial}{\partial r}(r\gamma)\right]\,. \tag{B.14c}$$

APPENDIX C

TAYLOR'S CONSTRAINT

TAYLOR'S CONSTRAINT IN A DEFORMED SPHERICAL CAVITY

Taylor's constraint is generally discussed within the context of fluid contained within a sphere or spherical shell as in Sections 4.3.1 and 4.3.2. However, Taylor's idea continues to apply when the boundaries are non-spherical, as for example in the case of boundaries with an undulating surface, (bumps, topography) as explored by Bell & Soward (1996). Furthermore, the degenerate case of plane parallel boundaries often employed by dynamo modellers (see, for example, Jones & Roberts, 2000) is worthy of separate attention and has been discussed by Rotvig & Jones (2002).

In this Appendix we investigate the steady velocity field $\mathbf{u}(\mathbf{x})$ and pressure distribution $p(\mathbf{x})$, which in dimensionless units solves the inhomogeneous problem

$$2\,\mathbf{e}_z \times \mathbf{u} = -\boldsymbol{\nabla} p + \mathbf{F}\,, \qquad \boldsymbol{\nabla}\cdot\mathbf{u} = 0 \qquad \text{(C.1a,b)}$$

subject to $\mathbf{u}\cdot\mathbf{n} = 0$ on the boundary Σ. Here the force $\mathbf{F}(\mathbf{x})$ is assumed to summarise all the remaining terms in the equation of motion. In Taylor's original application $\mathbf{F}$ was envisaged to be the Lorentz force but, from a more general point of view, $\mathbf{F}$ could include terms like the buoyancy force and in unsteady applications even the inertia term.

We introduce the coordinate $\mathbf{x}_\perp \equiv (x,\, y)\,(= \mathbf{x} - z\,\mathbf{e}_z)$ in the plane $z = 0$ and assume that every line $\mathbf{x}_\perp$ = constant, parallel to the rotation z–axis, which intersects the boundary, does so at two points $z = h_+(\mathbf{x}_\perp)$ (upper surface Σ_+) and $z = -h_-(\mathbf{x}_\perp)$ (lower surface Σ_-) separated by the distance $h = h_+ + h_- \geq 0$. The entire container surface Σ is the composition $\Sigma \equiv \Sigma_+ \cup \Sigma_-$, e.g. for a sphere Σ_+ and Σ_- would be the boundaries of the Northern and Southern hemispheres respectively. We introduce the outward normals

$$\mathbf{n}_\pm(\mathbf{x}_\perp) = \pm\mathbf{e}_z - \boldsymbol{\nabla} h_\pm \qquad \text{for} \qquad (\mathbf{x}_\perp, \pm h_\pm) \in \Sigma_\pm \tag{C.2a}$$

and the vector

$$\mathbf{t}(\mathbf{x}_\perp) = \mathbf{n}_+ \times \mathbf{n}_- \qquad \text{tangent to both } \Sigma_\pm\text{:} \qquad \mathbf{t}\cdot\mathbf{n}_\pm = 0\,. \tag{C.2b,c}$$

They have the useful properties:

$$\mathbf{e}_z \times \mathbf{t} = -(\mathbf{n}_+ + \mathbf{n}_-) = \boldsymbol{\nabla} h\,, \qquad \mathbf{e}_z\cdot\mathbf{n}_\pm = \pm 1\,, \tag{C.2d,e}$$

$$\pm\mathbf{n}_\mp \times \boldsymbol{\nabla} h = \mathbf{t}\,, \qquad \mathbf{t}\cdot\boldsymbol{\nabla} h = 0\,, \tag{C.2f,g}$$

$$\boldsymbol{\nabla}\times\mathbf{n}_\pm = \mathbf{0}\,, \qquad \boldsymbol{\nabla}\cdot\mathbf{t} = 0\,, \tag{C.2h,i}$$

which we use repeatedly together with the vector identities of Appendix A in the analysis below. We call a curve with $\mathbf{t}$ tangent to it at every point a geostrophic contour $\mathcal{C}$. Since they have the property $\mathbf{t}\cdot\boldsymbol{\nabla} h = 0$ [see (C.2g)], they lie on surfaces $h(\mathbf{x}_\perp) = $ constant (i.e. cylinders with generators aligned to the z–axis) and may be identified collectively as $\mathcal{C}(h)$. That part of the surface lying within the fluid $-h_- \le z \le h_+$ we refer to as the geostrophic cylinder $\mathcal{S}(h)$ with upper and lower bounding geostrophic contours $\mathcal{C}_\pm(h)$ lying on the surfaces $\Sigma_\pm$.

Crucial to Taylor's idea is the value of the volume flux $\mathcal{Q} \equiv \int_{\mathcal{S}(h)} \mathbf{u}\cdot \mathrm{d}\mathbf{S}$ out across the geostrophic cylinder $\mathcal{S}(h)$, where $\mathrm{d}\mathbf{S} = \mathrm{d}\mathbf{l}\times\mathbf{e}_z\,\mathrm{d}z$ is the surface element on $\mathcal{S}(h)$, $\mathrm{d}\mathbf{l} = \widehat{\mathbf{t}}\,\mathrm{d}l$ is the line element on every $\mathcal{C}(h)$ and $\widehat{\mathbf{t}} = \mathbf{t}/|\mathbf{t}|$ is the unit tangent vector. Since there is no inflow across the caps $\Sigma_\pm$ within $\mathcal{C}_\pm(h)$, mass continuity (C.1b) requires that $\mathcal{Q} = 0$. We now apply the depth integral $\langle\cdots\rangle = \int_{-h_-}^{h_+}\cdots\,\mathrm{d}z$ to (C.1a) and integrate about a geostrophic contour $\mathcal{C}(h)$. Since $\oint_{\mathcal{C}(h)} \mathrm{d}\mathbf{l}\cdot\boldsymbol{\nabla} p = 0$ (p is single valued) and since $\oint_{\mathcal{C}(h)} 2\,(\mathbf{e}_z\times\langle\mathbf{u}\rangle)\cdot\mathrm{d}\mathbf{l} = 2\mathcal{Q} = 0$, we obtain the following generalised form

$$\oint_{\mathcal{C}(h)} \langle\mathbf{F}\rangle\cdot\mathrm{d}\mathbf{l} = 0 \tag{C.3}$$

of Taylor's constraint.

THE SUFFICIENCY OF TAYLOR'S CONSTRAINT (C.3)

Though Taylor's constraint (C.3) is a necessary condition for a solution of the boundary value problem (C.1), it remains to establish that it is sufficient, which we accomplish here by constructing the explicit solution. The analysis below is based on

Soward (1970; old unpublished notes, which were partially reported by Rotvig & Jones, 2007).

We start by noting that the inhomogeneous equations (C.1) have the particular solution $\mathbf{u} = -\frac{1}{2}\boldsymbol{\nabla} \times \boldsymbol{\mathcal{F}}$, $p = \mathbf{e}_z \cdot \boldsymbol{\mathcal{F}}$, where

$$\boldsymbol{\mathcal{F}}(\mathbf{x}_\perp, z) \equiv \tfrac{1}{2}\left[\int_{-h_-}^{z} \mathbf{F}(\mathbf{x}_\perp, z')\,\mathrm{d}z' - \int_{z}^{h_+} \mathbf{F}(\mathbf{x}_\perp, z')\,\mathrm{d}z'\right] \tag{C.4}$$

has the property $\partial_z \boldsymbol{\mathcal{F}} = \mathbf{F}$. It is also easy to show with the help of the various identities listed in (C.2) that in the absence of the body force the geostrophic flow, which solves $2\,\mathbf{e}_z \times \mathbf{u}_G = -\boldsymbol{\nabla} p_G$, $\boldsymbol{\nabla} \cdot \mathbf{u}_G = 0$ and meets the boundary condition $\mathbf{u}_G \cdot \mathbf{n} = 0$ on Σ, is $\mathbf{u}_G = -(\mathrm{d}\Psi_G/\mathrm{d}h)\,\mathbf{t}$, $p_G = 2\Psi_G$, where $\Psi_G(h)$ is an arbitrary function of h. The method employed by Greenspan (1968, pp. 43, 44) to obtain that geostrophic solution of the homogeneous problem suggests a solution construction of the full inhomogeneous problem in the form

$$\mathbf{u} - \mathbf{u}_G = -\tfrac{1}{2}\boldsymbol{\nabla} \times \boldsymbol{\mathcal{F}} + \tfrac{1}{2}(\mathbf{v}_+ + \mathbf{v}_-)\,, \qquad \mathbf{v}_\pm = \mp\mathbf{n}_\mp \times \boldsymbol{\nabla}\Psi_\pm\,, \tag{C.5a,b}$$

$$p - p_G = \mathbf{e}_z \cdot \boldsymbol{\mathcal{F}} + \Psi_+ + \Psi_-\,, \tag{C.5c}$$

where, as yet, $\Psi_\pm$ are unknown functions of $\mathbf{x}_\perp$ alone. From (C.5b) and use of (C.2) it follows that $\mathbf{e}_z \times \mathbf{v}_\pm = -\boldsymbol{\nabla}\Psi_\pm$ and $\boldsymbol{\nabla} \cdot \mathbf{v}_\pm = 0$. Accordingly the entire form (C.5) satisfies the inhomogeneous equations (C.1). The application of the boundary condition is more tricky but is needed to fix the unknown functions $\Psi_\pm(\mathbf{x}_\perp)$. To that end, however, we note that the forms of $\mathbf{v}_\pm$ are motivated by the fact that $\mathbf{v}_\pm \cdot \mathbf{n}_\mp = 0$ on the boundaries $\Sigma_\mp$, though in general $\mathbf{v}_\pm \cdot \mathbf{n}_\pm \neq 0$ on $\Sigma_\pm$.

To make further progress we utilise the identity

$$\boldsymbol{\nabla} \times \boldsymbol{\mathcal{F}} = \boldsymbol{\nabla}_\perp \times \boldsymbol{\mathcal{F}} + \mathbf{e}_z \times \mathbf{F}\,, \tag{C.6a}$$

where $\boldsymbol{\nabla}_\perp$ is the gradient normal to the rotation axis, i.e.

$$\boldsymbol{\nabla} \equiv \boldsymbol{\nabla}_\perp + \mathbf{e}_z\,\partial_z\,. \tag{C.6b}$$

Direct differentiation of (C.4) gives

$$\boldsymbol{\nabla}_\perp \times \boldsymbol{\mathcal{F}} = \tfrac{1}{2}\left[\int_{-h_-}^{z} \boldsymbol{\nabla}_\perp \times \mathbf{F}\,\mathrm{d}z - \int_{z}^{h_+} \boldsymbol{\nabla}_\perp \times \mathbf{F}\,\mathrm{d}z + \boldsymbol{\nabla}h_- \times \mathbf{F}_- - \boldsymbol{\nabla}h_+ \times \mathbf{F}_+\right], \tag{C.6c}$$

where $\mathbf{F}_\pm(\mathbf{x}_\perp) \equiv \mathbf{F}(\mathbf{x}_\perp, \pm h_\pm)$. We supplement the results (C.6a,c) with

$$\langle \boldsymbol{\nabla}_\perp \times \mathbf{F} \rangle = \boldsymbol{\nabla} \times \langle \mathbf{F} \rangle - \boldsymbol{\nabla}h_- \times \mathbf{F}_- - \boldsymbol{\nabla}h_+ \times \mathbf{F}_+\,, \tag{C.7a}$$

which together show that the values of $\boldsymbol{\nabla}\times\boldsymbol{\mathcal{F}}$ on the boundaries $\Sigma_\pm$ are

$$(\boldsymbol{\nabla}\times\boldsymbol{\mathcal{F}})_\pm = \pm\tfrac{1}{2}\boldsymbol{\nabla}\times\langle\mathbf{F}\rangle \pm \mathbf{n}_\pm\times\mathbf{F}_\pm\,. \tag{C.7b}$$

Armed with this result, substitution of the expression (C.5a) for $\mathbf{u}$ into the boundary condition $\mathbf{u}\cdot\mathbf{n}=0$ on Σ yields

$$0 = 2\,\mathbf{u}_\pm\cdot\mathbf{n}_\pm = \boldsymbol{\nabla}\cdot\left[\pm\tfrac{1}{2}\mathbf{n}_\pm\times\langle\mathbf{F}\rangle - \Psi_\pm\mathbf{t}\right], \tag{C.8}$$

where, of course, $\mathbf{u}_\pm(\mathbf{x}_\perp)\equiv\mathbf{u}(\mathbf{x}_\perp,\,\pm h_\pm)$. The solenoidal condition (C.8) is satisfied by

$$\tfrac{1}{2}\mathbf{n}_\pm\times(\langle\mathbf{F}\rangle+\boldsymbol{\nabla}\Phi_\pm) = \pm\Psi_\pm\mathbf{t} = \mathbf{n}_\pm\times\mathbf{n}_\mp\Psi_\pm\,, \tag{C.9a}$$

where as yet the functions $\Phi_\pm(\mathbf{x}_\perp)$ are undetermined.

The scalar product of (C.9a) with $\mathbf{n}_\mp$ gives

$$\mathbf{t}\cdot\boldsymbol{\nabla}\Phi_\pm = -\mathbf{t}\cdot\langle\mathbf{F}\rangle\,. \tag{C.9b}$$

Integration of (C.9b) along geostrophic contours $\mathcal{C}(h)$ determines $\Phi_\pm$ up to a constant of integration or, more precisely, a function of h. Significantly the functions $\Phi_\pm(\mathbf{x}_\perp)$ obtained this way are single valued due to the fact that $\langle\mathbf{F}\rangle$ satisfies Taylor's constraint (C.3). Accordingly we write our solution in the form

$$\Phi_\pm(\mathbf{x}_\perp) = \Phi(\mathbf{x}_\perp) + \Phi_{G\pm}(h)\,, \tag{C.10}$$

where the single valued function $\Phi(\mathbf{x}_\perp)$ is any solution of (C.9b), while $\Phi_{G\pm}$ are arbitrary functions of h.

Finally the values of the functions $\Psi_\pm(\mathbf{x}_\perp)$ in (C.5) are obtained on taking the scalar product of (C.9a) with $\mathbf{t}$, which gives

$$\Psi_\pm = \pm\tfrac{1}{2}\mathbf{n}_\pm\cdot\mathbf{A} + \Psi_{G\pm}\,, \qquad \Psi_{G\pm} = \tfrac{1}{2}\frac{\mathrm{d}\Phi_{G\pm}}{\mathrm{d}h}\,, \tag{C.11a,b}$$

where

$$\mathbf{A} = -|\mathbf{t}|^{-2}\,\mathbf{t}\times(\langle\mathbf{F}\rangle+\boldsymbol{\nabla}\Phi)\,. \tag{C.11c}$$

Evidently the velocity contributions $\mathbf{v}_{G\pm}=\mp\mathbf{n}_\mp\times\boldsymbol{\nabla}\Psi_{G\pm}=-\,(\mathrm{d}\Psi_{G\pm}/\mathrm{d}h)\,\mathbf{t}$ to (C.5b) are geostrophic and can be absorbed within $\mathbf{u}_G$ by the simple expedient of replacing the sum $\Psi_G+\frac{1}{2}(\Psi_{G-}+\Psi_{G+})$ by Ψ_G. Interestingly $\mathbf{A}(\mathbf{x}_\perp)$ is the solution of

$$\mathbf{t}\times\mathbf{A} = \langle\mathbf{F}\rangle + \boldsymbol{\nabla}\Phi\,, \qquad \mathbf{t}\cdot\mathbf{A} = 0\,. \tag{C.12a,b}$$

Thus given $\langle\mathbf{F}\rangle$ satisfying Taylor's constraint (C.3), Φ could be obtained from (C.9b) and (C.10), then $\mathbf{A}$ from (C.11c) and $\Psi_\pm$ from (C.11a). Hence everything that (C.5) requires is determined, (apart from the geostrophic flow), whenever $\mathbf{t}\neq\mathbf{0}$. The condition $\mathbf{t}\neq\mathbf{0}$ is likely to hold almost everywhere except at isolated points or curves, e.g. for a sphere at the poles and equator. In many mathematical models the boundaries are planar and parallel. Then $\mathbf{t}=\mathbf{0}$ everywhere and geostrophic degeneracy ensues as considered below.

PLANE PARALLEL BOUNDARIES

We consider briefly the case of plane parallel boundaries $\Sigma_\pm$, separated everywhere by the constant z–axial distance h, with outward normals which satisfy

$$\pm\mathbf{n}_\pm = \mathbf{n} = \text{constant} \qquad \text{and imply} \qquad \mathbf{t} = \mathbf{0}. \qquad \text{(C.13a,b)}$$

Note that though in many applications $\mathbf{n}$ is parallel to $\mathbf{e}_z$ ($\mathbf{e}_z \times \mathbf{n} = \mathbf{0}$), this restriction is not necessary and following Rotvig & Jones (2002) we make no such assumption here. For the case (C.13), much of our previous analysis continues to hold but with a few modifications.

We begin by noticing that, in the case of non-parellel boundaries ($\mathbf{t} \neq \mathbf{0}$) just considered, the vectors $\mathbf{v}_\pm$ introduced in (C.5) did not satisfy all the boundary conditions, specifically $\mathbf{v}_\pm \cdot \mathbf{n}_\pm \neq 0$ on $\Sigma_\pm$, which is why we refrained from referring to those contributions as geostrophic flows. That is no longer the case for our degenerate case of plane parallel boundaries. So without loss of generality we let $\Psi_\pm(\mathbf{x}_\perp) = \Psi_G(\mathbf{x}_\perp)$ and call $\mathbf{v}_\pm(\mathbf{x}_\perp) = \mathbf{u}_G(\mathbf{x}_\perp)$ the geostrophic velocity. Since there are no geostrophic contours ($\mathbf{t} = \mathbf{0}$), the function Ψ_G is an arbitrary function of $\mathbf{x}_\perp$. Accordingly, in place of (C.5), we write

$$\mathbf{u} - \mathbf{u}_G = -\tfrac{1}{2}\boldsymbol{\nabla} \times \boldsymbol{\mathcal{F}} + w\,\mathbf{e}_z\,, \qquad \mathbf{u}_G = \mathbf{n} \times \boldsymbol{\nabla}\Psi_G\,, \qquad \text{(C.14a,b)}$$

$$p - p_G = \mathbf{e}_z \cdot \boldsymbol{\mathcal{F}}\,, \qquad p_G = 2\Psi_G\,, \qquad \text{(C.14c,d)}$$

in which we have also allowed for the possibility that there is some additional axial mean velocity $w(\mathbf{x}_\perp)\mathbf{e}_z$ to accommodate the loss of one of the two independent arbitrary vector functions $\Psi_\pm(\mathbf{x}_\perp)\,\mathbf{n}_\pm$ under the geostrophic degeneracy of the plane layer.

Substitution of (C.14a,b) into the boundary condition $\mathbf{u} \cdot \mathbf{n} = 0$ on Σ yields

$$0 = 2\,\mathbf{u}_\pm \cdot \mathbf{n}_\pm = -\,\mathbf{n} \cdot \big(\tfrac{1}{2}\boldsymbol{\nabla} \times \langle\mathbf{F}\rangle\big) \pm 2\,w \qquad \text{(C.15)}$$

in place of (C.8); remember that $\mathbf{t} = \mathbf{0}$. It follows that

$$\mathbf{n} \cdot (\boldsymbol{\nabla} \times \langle\mathbf{F}\rangle) = 0 \qquad \text{and} \qquad w = 0 \qquad \text{(C.16a,b)}$$

everywhere. The former result $\mathbf{n} \cdot (\boldsymbol{\nabla} \times \langle\mathbf{F}\rangle) = 0$ also follows directly from Taylor's constraint $\oint_{\mathcal{C}} \langle\mathbf{F}\rangle \cdot \mathrm{d}\mathbf{l} = 0$, which in our degenerate plane geometry holds for arbitrary closed curves $\mathcal{C}$ on any plane with normal $\mathbf{n}$. Furthermore (C.16a) is satisfied [as in (C.9a)] by

$$\mathbf{n} \times (\langle\mathbf{F}\rangle + \boldsymbol{\nabla}\,\Phi) = 0\,, \qquad \text{(C.17)}$$

where Φ is an arbitrary function of $\mathbf{x}_\perp$ alone.

Additional insight is obtained by evaluating the depth integral of (C.1), which gives

$$2\,\mathbf{e}_z \times \langle \mathbf{u} \rangle \,=\, -\,\boldsymbol{\nabla} \langle p \rangle \,-\, (p_+ \,-\, p_-)\,\mathbf{n} + \langle \mathbf{F} \rangle\,, \qquad\qquad \boldsymbol{\nabla} \cdot \langle \mathbf{u} \rangle \,=\, 0\,, \quad \text{(C.18a,b)}$$

where, as usual, $p_\pm(\mathbf{x}_\perp) \;\equiv\; p(\mathbf{x}_\perp,\, \pm h_\pm)$. On taking the scalar product of (C.18a) with $\mathbf{e}_z$, we note that the z–component of $\langle \mathbf{F} \rangle$ determines, the pressure jump

$$p_+ \,-\, p_- \,=\, \mathbf{e}_z \cdot \langle \mathbf{F} \rangle \qquad \text{(C.19a)}$$

between the boundaries. Then consideration of $\mathbf{e}_z \times [\mathbf{n} \times (\langle \mathbf{F} \rangle + \boldsymbol{\nabla}\Phi)]$ in conjunction with (C.17) and (C.19a) leads to the identity

$$(p_+ \,-\, p_-)\,\mathbf{n} \,=\, (\mathbf{e}_z \cdot \langle \mathbf{F} \rangle)\,\mathbf{n} \,=\, \langle \mathbf{F} \rangle \,+\, \boldsymbol{\nabla}\,\Phi \qquad \text{(C.19b)}$$

analogous to (C.12). The existence of some $\Phi(\mathbf{x}_\perp)$ that enables (C.19b) to be satisfied is the essence of Taylor's constraint (C.16a) in our planar geometry.

With the help of (C.19b), the depth integral equation (C.18a) reduces to

$$\mathbf{e}_z \,\times\, \langle \mathbf{u} \rangle \,=\, -\,\boldsymbol{\nabla}\,(h\Psi_\perp)\,, \qquad \text{(C.20a)}$$

where

$$2h\Psi_\perp \,=\, \langle p \rangle \,+\, \Phi \qquad \text{and} \qquad \langle p \rangle \,=\, \langle \mathbf{e}_z \cdot \boldsymbol{\mathcal{F}} \rangle \,+\, 2h\Psi_G \quad \text{(C.20b,c)}$$

results from the integration of (C.14c,d). Since $\Psi_\perp(\mathbf{x}_\perp)$ is arbitrary, albeit linked to the arbitrary function $\Psi(\mathbf{x}_\perp)$ via (C.20b,c), the mean velocity $h^{-1}\langle \mathbf{u}_\perp \rangle \,=\, \mathbf{e}_z \,\times\, \boldsymbol{\nabla}\,\Psi_\perp$, where $\langle \mathbf{u}_\perp \rangle \equiv \langle \mathbf{u} \rangle - (\mathbf{e}_z \cdot \langle \mathbf{u} \rangle)\,\mathbf{e}_z$ is the projection of $\langle \mathbf{u} \rangle$ onto the (x, y)–plane, is independent of the force $\mathbf{F}$. In contrast, the depth integrated velocity component normal to the boundary is influenced directly by $\mathbf{F}$ via (C.14a), which gives the generally non-zero value

$$\mathbf{n} \cdot \langle \mathbf{u} \rangle \,=\, -\,\tfrac{1}{2}\,\mathbf{n} \cdot \langle \boldsymbol{\nabla} \,\times\, \boldsymbol{\mathcal{F}} \rangle\,. \qquad \text{(C.21a)}$$

In turn, that fixes the mean axial velocity component:

$$h^{-1}\mathbf{e}_z \cdot \langle \mathbf{u} \rangle \,=\, h^{-1}\langle \mathbf{u}_\perp \rangle \cdot \boldsymbol{\nabla} h_+ \,-\, \tfrac{1}{2}\,h^{-1}\mathbf{n} \cdot \langle \boldsymbol{\nabla} \,\times\, \boldsymbol{\mathcal{F}} \rangle\,. \qquad \text{(C.21b)}$$

The essence of our results are summarised succinctly by (C.14), in which $w \,=\, 0$, together with Taylor's constraint (C.16a). However the representation (C.19b) of $\langle \mathbf{F} \rangle$, in terms of the potential Φ, sheds further light on the nature of the mean velocity $h^{-1}\langle \mathbf{u} \rangle$ as revealed by the component $h^{-1}\langle \mathbf{u}_\perp \rangle \,=\, \mathbf{e}_z \,\times\, \boldsymbol{\nabla}\,\Psi_\perp$ in the (x, y)–plane [mainly through the link (C.20b,c) between the stream functions $\Psi_\perp$ and Ψ_G] and the component (C.21b) in the z–direction.

Appendix D

Units

Scientific units used in this book are the standardized SI units (Système International). Since part of the literature on the subject (mostly in astrophysics) uses the Gaussian or CGS electromagnetic units (emu) system, we describe here the main differences between these units.

Physical Quantity	**SI unit**	**Gaussian CGS emu unit**	
Length	1 m	10^2	cm
Mass	1 kg	10^3	g
Magnetic indiction **B**	1 Tesla (T)=1 kg A s^{-2}	10^4	Gauss (G)
Magnetic Field **H**	1 A m^{-1}	$4\pi \times 10^{-3}$	Oersted (Oe)
Magnetic flux	1 Weber (Wb)	10^8	Maxwell (Mx)
Electric field	1 V m^{-1}	10^6	volt cm^{-1}
Charge	1 Coulomb (C)	10^{-1}	coulomb
Charge density	1 C m^{-3}	10^{-7}	coul cm^{-3}
Electric current	1 Ampere (A)	10^{-1}	ampere
Electric current density	1 A m^{-2}	10^{-5}	amp cm^{-2}
Resistance	1 Ohm (Ω)	10^9	cm s^{-1}
Resistivity	1 Ω m	10^{11}	cm^2 s^{-1}
Conductance	1 siemens (S)	10^{-9}	cm^{-1} s
Conductivity	1 S m^{-1}	10^{11}	emu
Magnetic permeability	1 H m^{-1}	$10^7/4\pi$	G Oe^{-1}
Force	1 Newton (N)	10^5	dynes
Energy	1 Joule	10^7	erg

Other unit conversions:

1 astronomical unit (AU)	1.495979×10^{11}	m
1 parsec (pc)	3.085678×10^{16}	m
1 light year	9.460530×10^{15}	m
1 solar radius ($R_\odot$)	6.9599×10^{8}	m
1 solar mass ($M_\odot$)	1.9891×10^{30}	kg
1 eV	$1.602\,177 \times 10^{-19}$	Joule
0 °C	273.15	K

APPENDIX E

ABBREVIATIONS

Notation	Meaning	Context
ABC–flow	Arnold–Beltrami–Childress	Fast dynamos (e.g. p.46)
BRA	Boussinesq–Reynolds Ansatz	Turbulence (e.g. p.236)
CMB	Core-Mantle Boundary	Geophysics
CZ	Convective Zone	Solar physics
DNS	Direct Numerical Simulation	Turbulence
DTS	Derviche Tourneur Sodium	Experiments (e.g. p.378)
EDQN	Eddy-Damped Quasi-Normal Markovian	Turbulence (e.g. p.90)
EMF	ElectroMotive Force	General (e.g. p.34)
GP–flow	Galloway–Proctor	Fast dynamos (e.g. p.31)
GS	Grid Scale	Turbulence (e.g. p.230)
GSV	Geomagnetic Secular Variation	Geophysics
ICB	Inner-Core Boundary	Geophysics
LES	Large Eddy Simulation	Turbulence
MAC	Magnetic–Archimedean–Coriolis	Geophysics
MFE	Mean Field Electrodynamics	Dynamo
MHD	MagnetoHydroDynamic	General (e.g. p.6)
MRI	Magneto–Rotational Instability	MHD (e.g. p.310)
ODE	Ordinary Differential Equation	General
PDE	Partial Differential Equation	General
ICB	Inner Core Boundary	Geophysics
ISM	InterStellar medium	Astrophysics
SGS	Sub-Grid Scales	Turbulence (e.g. p.230)
SN	Supernova star	Astrophysics
STF	Stretch–Twist–Fold	Fast dynamos (e.g. p.47)
UPOs	Unstable Periodic Orbits	Dynamical systems (e.g. p.117)
VKS	Von Kármán Sodium	Experiments (e.g. p.377)
WKBJ	Wentzel–Kramers–Brillouin–Jeffreys	Asymptotics (e.g. p.157)

REFERENCES

Abramowitz M. & Stegun I.A., *Handbook of Mathematical Functions*, Dover, New York (1965).

Acheson D.J., *Elementary Fluid Dynamics*, Oxford University Press (1990).

Acheson D.J. & Hide R., Hydromagnetics of rotating fluids, *Rep. Prog. Phys.*, **36**, 159–221 (1973).

Aldridge K., Dynamics of the core at short periods: theory, experiments and observations, in *Earth's core and lower mantle* (C.A. Jones, A.M. Soward & K. Zhang, Eds.), Taylor & Francis, London, 180–210 (2003).

Alémany A., Moreau R., Sulem L. & Frisch U., Influence of an external magnetic field on homogenous turbulence, *J. de Méca.*, **18**, 277–313 (1979).

Alémany A., Marty P., Plunian F. & Soto J. Experimental investigation of dynamo effect in the secondary pumps of the fast breeder reactor Superphenix, *J. Fluid Mech.*, **403**, 263–276 (2000).

Alfè D., Gillan M.J. & Price G.D., Composition and temperature of the Earth's core constrained by combining *ab initio* calculations and seismic data, Earth Planet. Sci. Lett., **195**, 91–98 (2002a).

Alfè D., Gillan M.J., Vočadlo L., Brodholt J. & Price G.D., The *ab initio* simulation of the Earth's core, Phil. Trans. R. Soc. Lond., A **360**, 1227–1244 (2002b).

Anderson O.L., The power balance at the core-mantle boundary, Phys. Earth Planet. Inter., **131**, 1–17 (2002).

Anderson O.L. & Isaak D.G., Another look at the core density deficit of Earth's outer core, Phys. Earth Planet. Inter., **131**, 19–27 (2002).

Andreassian R.R., On the structure of the Galactic magnetic field, *Astrofizika,* **16**, 707–713 (1980).

Andreassian R.R., A study of the magnetic field of the Galaxy, *Astrofizika,* **18**, 255–262 (1982).

Anufiev A.P., Jones C.A. & Soward A.M., The Boussinesq and anelastic liquid approximations for convection in the Earth's core, Phys. Earth Planet. Inter., **152**, 163–190 (2005).

Ardes M., Busse F.H. & Wicht J., Thermal Convection in Rotating Spherical Shells, *Phys. Earth Planet. Inter.*, **99**, 55-67 (1997).

Armstrong J.W., Rickett B.J. & Spangler S.R., Electron density power spectrum in the local interstellar medium, *Astrophys. J.,* **443**, 209–221 (1995).

Ashwin P., Covas E. & Tavakol R., Transverse instability for non-normal parameters, *Nonlinearity*, **12**, 562–577 (1999).

Ashwin P., Rucklidge A.M. & Sturman R., Two-state intermittency near a symmetric interaction of saddle-node and Hopf bifurcations: a case study from dynamo theory, *Physica*, D **194**, 30–48 (2004).

Aubert J., Brito D., Nataf H.-C., Cardin P. & Masson J.-P., A systematic experimental study of rapidly rotating spherical convection in water and liquid gallium, Phys. Earth Planet. Inter., **128**, 51–74 (2001).

Aubert J. & Wicht J., Axial vs. equatorial dipolar dynamo models with implications for planetary magnetic fields, *Earth Planet. Sci. Lett.*, **221**, 409–419 (2004).

Babcock H.W., The topology of the Sun's magnetic field and the 22-year cycle, *Astrophys. J.*, **133**, 572–587 (1961).

Backus G., A class of self-sutaining dissipative spherical dynamos. *Ann. Phys.*, **4**, 372–447 (1958).

Backus G., Parker R. & Constable C., *Foundations of Geomagnetism*, Cambridge University Press (1996).

Balbus S.A. & Hawley J.F., Instability, turbulence and enhanced transport in accretion disks, *Rev. Mod. Phys.*, **70**, 1–53 (1998).

Baliunas S.L. & Jastrow R., Evidence for long-term brightness changes of solar-type stars, *Nature*, **348**, 520–523 (1990).

Baliunas S.L. & Vaughan A.H., Stellar activity cycles, *Annu. Rev. Astron. Astrophys.*, **23**, 379–412 (1985).

Baliunas S.L. & 26 others, Chromospheric variations in main-sequence stars, *Astrophys. J.*, **438**, 269–287 (1995).

Bardou A., von Rekowski B., Dobler W., Brandenburg A. & Shukurov A., The effects of vertical outflows on disk dynamos, *Astron. Astrophys.*, **370**, 635–648 (2001).

Barenghi C.F., Nonlinear planetary dynamos in a rotating spherical shell. II. The post-Taylor equilibration for α^2-dynamos, Geophys. Astrophys. Fluid. Dyn., **67**, 27–36 (1992).

Barnes J.R. & Collier Cameron A., Starspot patterns on the M dwarfs HK Aqr and RE 1816 +541, *Mon. Not. R. Astron. Soc.*, **326**, 950–958 (2001).

Baryshnikova Yu., Ruzmaikin A., Sokoloff D. & Shukurov A., Generation of large-scale magnetic fields in spiral galaxies, *Astron. Astrophys.*, **177**, 27–41 (1987).

Bassom A.P. & Gilbert A.D., Nonlinear equilibration of a dynamo in a smooth helical flow, *J. Fluid Mech.*, **343**, 375–406 (1997).

Batchelor G.K., *An Introduction to Fluid Dynamics*, Cambridge University Press (1967).

Bayly B.J., Maps and dynamos, in *Lectures on Solar and planetary dynamos* (ed. M.R.E. Proctor, A.D. Gilbert), 305–329. Cambridge University Press (1994).

Bayly B.J. & Childress S., Construction of fast dynamos using unsteady flows and maps in three dimensions, *Geophys. Astrophys. Fluid Dyn.*, **44**, 211–240 (1988).

Bayly B.J. & Childress S., Unsteady dynamo effects at large magnetic Reynolds numbers, *Geophys. Astrophys. Fluid Dyn.*, **49**, 23–43 (1989).

Beck R., The magnetic field in M31, *Astron. Astrophys.,* **106**, 121–132 (1982).

Beck R., Magnetic fields and interstellar gas clouds in the spiral galaxy NGC 6946, *Astron. Astrophys.,* **251**, 15–26 (1991). Erratum: *Astron. Astrophys.,* **258**, 605 (1992).

Beck R., Magnetic fields in normal galaxies, *Phil. Trans. R. Soc. Lond.*, A **358**, 777–796 (2000).

Beck R., Galactic and extragalactic magnetic fields, *Sp. Sci. Rev.,* **99**, 243–260 (2001).

Beck R., Berkhuijsen E.M., Moss D., Shukurov A. & Sokoloff D., The nature of the magnetic belt in M31, *Astron. Astrophys.,* **335**, 500–509 (1998).

Beck R., Brandenburg A., Moss D., Shukurov A. & Sokoloff D., Galactic magnetism: recent developments and perspectives, *Ann. Rev. Astron. Astrophys.,* **34**, 155–206 (1996).

Beck R., Carilli C.L., Holdaway M.A. & Klein U., Multifrequency observations of the radio continuum emission from NGC 253. I. Magnetic fields and rotation measures in the bar and halo, *Astron. Astrophys.,* **292**, 409–424 (1994).

Beck R. & Hoernes P., Magnetic spiral arms in the galaxy NGC 6946, *Nature,* **379**, 47–49 (1996).

Beck R., Shukurov A., Sokoloff D. & Wielebinski R., Systematic bias in interstellar magnetic field estimates, *Astron. Astrophys.,* **411**, 99–107 (2003) [astro-ph/0307330].

Beck R., Brandenburg A., Moss D., Shukurov A. & Sokoloff D., Galactic Magnetism: Recent Developments and Perspectives, Annu. Rev. Astron. Astrophys.,, **34**, 155–206 (1996).

Beer J., Tobias S.M. & Weiss N.O., An active Sun throughout the Maunder Minimum, *Solar Phys.*, **181**, 237–249 (1998).

Bell P.I. & Soward A.M., The influence of surface topography on rotating convection, *J. Fluid Mech.*, **313**, 147–180 (1996).

Belvedere G., Pidatella, R.M. & Proctor, M.R.E., Nonlinear dynamics of a stellar dynamo in a spherical shell, *Geophys. Astrophys. Fluid Dyn.*, **51**, 263–286 (1990).

Belyanin M., Sokoloff D., Shukurov A., Simple models of nonlinear fluctuation dynamo, *Geophys. Astrophys. Fluid. Dyn.,* **68**, 237–261 (1993).

Belyanin M.P., Sokoloff D.D. & Shukurov A.M., Asymptotic steady-state solutions to the nonlinear hydromagnetic dynamo equations, *Russ. J. Math. Phys.,* **2**, 149–174 (1994).

Berhanu M., Monchaux R., Fauve S., Mordant N., Petrelis F., Chiffaudel A., Daviaud F., Dubrulle B., Marie L., Ravelet F., Bourgoin M., Odier Ph., Pinton J.-F. & Volk R., Magnetic field reversals in an experimental turbulent dynamo, *Europhysics Letters*, **77**, 59001 (2007).

Berkhuijsen E.M., Horellou C., Krause M., Neininger N., Poezd A.D., Shukurov A. & Sokoloff D.D., Magnetic fields in the disk and halo of M51, *Astron. Astrophys.,* **318**, 700–720 (1997).

Biskamp D., *Non linear magnetohydrodynamics*, Cambridge University Press, U.K. (1993).

Biskamp D., *Magnetohydrodynamic Turbulence*, Cambridge University Press, U.K. (2003).

Bisshopp F.E. & Niiler P.P., Onset of convection in a rapidly rotating fluid sphere, *J. Fluid Mech.*, **23**, 459–469 (1965).

Bjerknes V., Solar hydrodynamics, *Astrophys. J.*, **64**, 93–121 (1926).

Blackman E.G. & Field, G.B., Constraints on the magnitude of α in dynamo theory, Astrophys. J., **534**, 984–988 (2000).

Blackman E.G. & Field, G.B., Coronal activity from dynamos in astrophysical rotators, *Mon. Not. R. Astron. Soc.,* **318**, 724–732 (2000).

Blandford R.D. & Payne D.G., Hydromagnetic flows from accretion discs and the production of radio jets, *Mon. Not. R. Astron. Soc.,* **199**, 883–903 (1982).

Bloxham J., The effect of thermal core-mantle interactions on the palaeomagnetic secular variation, Phil. Trans. R. Soc. Lond., A **358**, 1171–1179 (2000).

Bloxham J. & Jackson A., Fluid flow near the surface of the Earth's outer core, *Rev. Geophys.*, **29**, 97–120 (1991).

Bloxham J. & Jackson A., Time-dependent mapping of the magnetic field at the core-mantle boundary, J. Geophys. Res.,**97**, 19537–19563 (1992).

Bloxham J., Zatman S. & Dumberry M., The origin of geomagnetic jerks, Nature, **420**, 65–68 (2002).

Boldyrev S. & Cattaneo F., Magnetic-Field Generation in Kolmogorov Turbulence, *Phys. Rev. Lett.*, **92**, 144501 (2004).

Bolt B.A., *Inside the Earth*, Freeman, San Fransisco (1982).

Boulares A. & Cox D.P., Galactic hydrostatic equilibrium with magnetic tension and cosmic-ray diffusion, *Astrophys. J.,* **365**, 544–558 (1990).

Bourgoin M., Marié L., Pétrélis F., Gasquet C., Guigon A., Luciani J.-B., Moulin M., Namer F., Burguete J., Chiffaudel A., Daviaud F., Fauve S., Odier P. & Pinton J.-F., Magnetohydrodynamics measurements in the von Kármán sodium experiment, *Phys. Fluids*, **14**, 3046–3058 (2002).

Braginsky S.I., Structure of the *F* layer and reasons for convection in the Earth's core, *Sov. Phys. Dokl.*, **149**, 8–10 (1963).

Braginsky S.I., Self-excitation of a magnetic field during the motion of a highly conducting fluid, *Sov. Phys. JETP*, **20**, 1462–1471 (1964a).

Braginsky S.I., Magnetohydrodynamics of the Earth's core, *Geomag. & Aeron.*, **4**, 698–712 (1964b).

Braginsky S.I., Magnetic waves in the Earth's core, *Geomag. & Aeron.*, **7**, 851–859 (1967).

Braginsky S.I., Torsional magnetohydrodynamic vibrations in the Earth's core and variations in day length, Geomag. Aeron., **10**, 1–8 (1970).

Braginsky S.I., Nearly axially symmetric model of the hydromagnetic dynamo of the Earth. I, Geomag. Aeron., **15**, 122–128 (1975).

Braginsky S.I., The nonlinear dynamo and model-Z, in *Lectures on Solar and Planetary Dynamos* (Proctor M.R.E. & Gilbert A.D. Eds.) Cambridge University Press, 267–304 (1994).

Braginsky S.I. & Meytlis, V.P., Local turbulence in the Earth's core, *Geophys. Astrophys. Fluid Dyn.*, **55**, 71–87 (1990).

Braginsky S.I. & Roberts P.H., A model-Z geodynamo, Geophys. Astrophys. Fluid. Dyn., **38**, 327–349 (1987).

Braginsky S.I. & Roberts P.H., Equations governing convection in Earth's core and the geodynamo, Geophys. Astrophys. Fluid. Dyn., **79**, 1–97 (1995).

Braginsky S.I. & Roberts P.H., On the theory of convection in the Earth's core, in *Advances in nonlinear dynamos*, Ferriz-Mas & Nuñez Eds., Taylor & Francis (2003).

Brandenburg A., Dynamo-generated turbulence and outflows from accretion discs, *Phil. Trans. Roy. Soc.*, **A**358, 759–776 (2000).

Brandenburg A., The inverse cascade and nonlinear α-effect in simulations of isotropic helical hydromagnetic turbulence, Astrophys. J., **550**, 824–840 (2001).

Brandenburg A., Dobler, W. & Subramanian, K., Magnetic helicity in stellar dynamos: new numerical experiments, *Astron. Nachr.,* **323**, 99–122 (2002).

Brandenburg A., Donner K.-J., Moss D., Shukurov, A., Sokoloff D.D. & Tuominen I., Dynamos in discs and halos of galaxies, *Astron. Astrophys.,* **259**, 453–461 (1992).

Brandenburg A., Donner K.-J., Moss D., Shukurov, A., Sokoloff D.D. & Tuominen I., Vertical magnetic fields above the discs of spiral galaxies, *Astron. Astrophys.,* **271**, 26–50 (1993).

Brandenburg A., Jennings R.L., Nordlund Å., Rieutord, M., Stein, R.F. & Tuominen, I., Magnetic structures in a dynamo simulation, J. Fluid Mech., **306**, 325–352 (1996).

Brandenburg A., Krause F., Meinel R., Moss D. & Tuominen I., The stability of nonlinear dynamos and the limited role of kinematic growth rates, *Astron. Astrophys.*, **213**, 411–422 (1989).

Brandenburg A., Moss D., Rüdiger G. & Tuominen I., Hydromagnetic $\alpha\Omega$–type dynamos with feedback from large scale motions, Geophys. Astrophys. Fluid. Dyn., **61**, 179–198 (1991).

Brandenburg A., Moss D. & Shukurov A., Galactic fountains as magnetic pumps, *Mon. Not. R. Astron. Soc.,* **276**, 651–662 (1995).

Brandenburg A., Nordlund Å., Stein R.F. & Torkelsson U., Dynamo-generated turbulence and large-scale magnetic fields in a Keplerian shear flow, *Astrophys. J.*, **446**, 741–754 (1995).

Brandenburg A. & Schmitt D., Simulations of an alpha-effect due to magnetic buoyancy, *Astron. Astrophys.*, **338**, L55–L58 (1998).

Brandenburg A. & Subramanian K., Astrophysical magnetic fields and nonlinear dynamo theory, Phys. Rep., **417**, 1–209 (2005).

Braun D.C. & Fan Y., Helioseismic measurements of the subsurface meridional flow, *Astrophys. J.*, **508**, L105–L108 (1998).

Braun H., Christl M., Rahmstorf S., Ganopolski A., Mangini A., Kubatzki C., Roth K. & Kromer B., Possible solar origin of the 1,470–year glacial climate cycle demonstrated in a coupled model, *Nature*, **438**, 208–211 (2005).

Braun R., The distribution and kinematics of neutral gas in M31, *Astrophys. J.,* **372**, 54–66 (1991).

Brito D., Aurnou J. & Cardin P., Turbulent viscosity measurements relevant to planetary core mantle dynamics, *Phys. Earth Planet. Inter.*, **141/1**, 3–8 (2003).

Brito D., Cardin P., Nataf H.-C. & Marolleau G., Experimental study of a geostrophic vortex of gallium in a transverse magnetic field, *Phys. Earth Planet. Inter.*, **91**, 77–98 (1995).

Brito D., Cardin P., Nataf H.-C. & P. Olson, Experiments on Joule heating and dynamo efficiency, *Geophys. J. Int.*, **127**, 339–347 (1996).

Brito D., Nataf H.-C., Cardin P., Aubert J. & Masson J.-P., Ultrasonic Doppler velocimetry in liquid gallium, *Exp. Fluids*, **31**, 653–663 (2001).

Brooke J., Moss D. & Phillips A., Deep minima in stellar dynamos, *Astron. Astrophys.*, **395**, 1013–1022 (2002).

Brooke J., Pelt J., Tavakol R. & Tworkowski A., Grand minima and equatorial symmetry breaking in axisymmetric dynamo models, *Astron. Astrophys.*, **332**, 339–352 (1998).

Brown J.C. & Taylor A.R., The structure of the magnetic field in the outer galaxy from rotation measure observations through the disk, *Astrophys. J., Lett.*, **563**, L31–L34 (2001).

Brummell N.H. & Hart J.E., High Rayleigh number β-convection, *Geophys. Astrophys. Fluid Dyn.*, **68**, 133–150 (1993).

Brun A.S. & Toomre J., Turbulent convection under the influence of rotation: sustaining a strong differential rotation, *Astrophys. J.*, **570**, 865–885 (2002).

Buffett B.A., A comparison of subgrid scale models for large eddy simulations of convection in the Earth's core, *Geophys. J. Int.*, **153**, 753–765 (2003).

Bullard E.C., The magnetic field in sunspots, *Vistas in Astronomy*, **1**, 685–691 (1955).

Bullard E.C., The stability of a homopolar dynamo, *Proc. Camb. Phil. Soc.,* **51**, 744-760 (1955).

Bullard E.C. & Gellman H., Homogeneous dynamos and terrestrial magnetism, *Phil. Trans. R. Soc. Lond.,* A **247**, 213–278 (1954).

Burke B.F. & Franklin K.L., Observations of a variable radio source associated with the planet Jupiter, *J. Geophys. Res.*, **60**, 213–217 (1955).

Bushby P.J., Strong asymmetry in stellar dynamos, *Mon. Not. R. Astron. Soc.*, **338**, 655–664 (2003a).

Bushby P.J., Modelling dynamos in rapidly rotating late-type stars, *Mon. Not. R. Astron. Soc.*, **342**, L15–L19 (2003b).

Bushby P.J., *Nonlinear dynamos in stars*, Ph.D. thesis, University of Cambridge (2004).

Bushby P.J., Zonal flows in a mean-field solar dynamo model, *Astron. Nachr.*, **326**, 3/4, 218–222 (2005).

Busse F.H., Thermal instabilities in rapidly rotating systems, *J. Fluid Mech.*, **44**, 441–460 (1970a).

Busse F.H., Differential rotation in stellar convection zones, *Astrophys. J.*, **159**, 629–639 (1970b).

Busse F.H., Differential rotation in stellar convection zones II, *Astron. & Astrophys.*, **28**, 27–37 (1973).

Busse F.H., A necessary condition for the geodynamo, *J.Geophys. Res.* **80**, 278–280 (1975a).

Busse F.H., A model of the geodynamo, *Geophys. J. R. Astr. Soc.*, **42**, 437–459 (1975b).

Busse F.H., Generation of planetary magnetism by convection, *Phys. Earth Planet. Inter.*, **12**, 350–358 (1976).

Busse F.H., On the prblem of stellar rotation, *Astrophys. J.*, **259**, 759–766 (1982).

Busse F.H., Is low Rayleigh number convection possible in the Earth's core? *Geophys. Res. Lett.*, **29**, GLO14597 (2002).

Busse F.H., Convective flows in rapidly rotating spheres and their dynamo action, Phys. Fluids, **14**, 1301–1314 (2002).

Busse F.H., On thermal convection in slowly rotating systems, *CHAOS*, **14**, 803–808 (2004).

Busse F.H. & Carrigan C.R., Convection induced by centrifugal buoyancy, *J. Fluid Mech.*, **62**, 579–592 (1974).

Busse F.H. & Carrigan C.R., Laboratory simulation of thermal convection in rotating planets and stars, *Science*, **191**, 81–83 (1976).

Busse F.H., Grote E. & Simitev R., Convection in rotating spherical shells and its dynamo action, in *Earth's core and lower mantle* (C.A. Jones, A.M. Soward & K. Zhang Eds.), Taylor & Francis, London, 130–152 (2003).

Busse F.H. & Hood L.L., Differential rotation driven by convection in a rotating annulus, *Geophys. Astrophys. Fluid Dyn.*, **21**, 59–74 (1982).

Busse F.H., Müller U., Stieglitz R. & Tilgner A., A two-scale homogeneous dynamo: an extended analytical model and an experimental demonstration under development, *Magnetohydrodynamics*, **32**, 235–248 (1996).

Busse F.H. & Simitev R., Inertial convection in rotating fluid spheres, *J. Fluid Mech.*, **498**, 23–30 (2004).

Bykov A., Popov V., Shukurov A., Sokoloff D., Anomalous persistence of bisymmetric magnetic structures in spiral galaxies, *Mon. Not. R. Astron. Soc.*, **292**, 1–10 (1997).

Cally P.S., A sufficient condition for instability in a sheared incompressible magnetofluid, *Solar Phys.*, **194**, 189–196 (2000).

Cally P.S., Nonlinear evolution of 2D tachocline instabilities, *Solar Phys.*, **199**, 231–249 (2001).

Cally P.S., Three-dimensional magneto-shear instabilities in the solar tachocline, *Mon. Not. R. Astron. Soc.*, **339**, 957–972 (2003).

Campbell C.G., Launching of accretion disc winds along dynamo-generated magnetic fields, *Mon. Not. R. Astron. Soc.*, **310**, 1175–1184 (1999).

Campbell C.G., An accretion disc model with a magnetic wind and turbulent viscosity, *Mon. Not. R. Astron. Soc.*, **317**, 501–527 (2000).

Campbell C.G., The stability of turbulent accretion discs with magnetically influenced winds, *Mon. Not. R. Astron. Soc.*, **323**, 211–222 (2001).

Campbell C.G., A semi-analytic solution for the radial and vertical structure of accretion discs with a magnetic wind, *Mon. Not. R. Astron. Soc.*, **345**, 123–143 (2003).

Campbell C.G. & Caunt S.E., An analytic model for magneto-viscous accretion discs, *Mon. Not. R. Astron. Soc.,* **306**, 122–136 (1999).

Cardin P., Brito D., Jault D., Nataf H.-C. & Masson J.-P., Towards a rapidly rotating liquid sodium dynamo experiment, *Magnetohydrodynamics*, **38**, 177–189 (2002).

Cardin P. & Olson P., Chaotic thermal convection in a rapidly rotating spherical shell: consequences for flow in the outer core, *Phys. Earth Planet. Inter.*, **82**, 235–259 (1994).

Carrigan C.R. & Busse F.H., An experimental and theoretical investigation of the onset of convection in rotating spherical shells, *J. Fluid Mech.*, **126**, 287–305 (1983).

Cattaneo F., On the effect of a weak magnetic field on turbulent transport, Astrophys. J., **434**, 200–205 (1994).

Cattaneo F. & Hughes D.W., Nonlinear saturation of the turbulent α-effect, *Phys. Rev. E*, **54**, R4532–R4535 (1996).

Cattaneo F., Hughes D.W. & Kim, E.-J., Suppression of chaos in a simplified nonlinear dynamo model, *Phys. Rev. Lett.,* **76**, 2057–2060 (1996).

Cattaneo F., Hughes D.W. & Thelen J.C., The nonlinear properties of a large-scale dynamo driven by helical forcing, *J. Fluid Mech.*, **456**, 219–237 (2002).

Cattaneo F. & Vainshtein S.I., Suppression of turbulent transport by a weak magnetic field, Astrophys. J., **376**, L21–L24 (1991).

Cavazzoni, C., Chiarotti, G.L., Scandolo, S., Tosatti, E., Bernasconi, M. & Parinello, M., Superioinc and metallic states of water and ammonia at giant planet conditions, *Science*, **283**, 44–46 (1999).

Chandrasekhar S., Hydromagnetic turbulence. II. An elementary theory, *Proc. R. Soc. Lond.*, A **233**, 330–350 (1955).

Chandrasekhar S., *Hydrodynamic and Hydromagnetic Stability*, Oxford: Clarendon Press (1961).

Charbonneau P. & Dikpati M., Stochastic fluctuations in a Babcock-Leighton model of the solar cycle, *Astrophys. J.*, **543**, 1027–1043 (2000).

Charbonneau P. & MacGregor K.B., On the generation of equipartition-strength magnetic fields by turbulent hydromagnetic dynamos, *Astrophys. J.*, **473**, L59–L62 (1996).

Chen C.X. & Zhang K., Nonlinear convection in a rotating Annulus with a finite gap, *Geophys. Astrophys. Fluid Dyn.*, **96**, 519–518 (2002).

Chen Q. & Glatzmaier G.A., Large eddy simulations of two–dimensional turbulent convection, Geophys. Astrophys. Fluid. Dyn., **99**, 355–375 (2005).

Childress S., *Théorie magnetohydrodynamique de l'effet dynamo*. Technical report, Dept. Mech. Fac. Sci., Paris (1969).

Childress S., Alpha–effect in flux ropes and sheets, *Phys. Earth Planet. Inter.*, **20**, 172–180 (1979).

Childress S., Fast dynamo theory, in *Topological aspects of the dynamics of fluids and plasmas* (ed. H.K. Moffatt, G.M. Zaslavsky, P. Comte, M. Tabor), 111–147. Kluwer Academic Publishers (1992).

Childress S. & Gilbert A.D., *Stretch, twist, fold: the fast dynamo,* Lecture Notes in Physics: Monographs. Springer (1995).

Childress S. & Soward A.M., Convection-driven hydromagnetic dynamo, *Phys. Rev. Lett.*, **29**, 837–839 (1972).

Choudhuri A.R., The solar dynamo as a model of the solar cycle, in *Dynamic Sun*, B.N. Dwivedi Ed., 103–127, Cambridge University Press, U.K. (2003).

Christensen U.R., Zonal flow driven by strongly supercritical convection in rotating spherical shells, *J. Fluid Mech.*, **470**, 115–133 (2002).

Christensen U.R., A deep dynamo generating Mercury's magnetic field, *Nature*, **444**, 1056–1058 (2006).

Christensen U.R., Aubert J., Cardin P., Dormy E., Gibbons S., Glatzmaier G.A., Grote E., Honkura Y., Jones C., Kono M., Matsushima M., Sakuraba A., Takahashi F., Tilgner A., Wicht J. & Zhang K., A numerical dynamo benchmark, Phys. Earth Planet. Inter., **128**, 25–34 (2001).

Christensen U.R. & Olson P., Secular variation in numerical geodynamo models with lateral variations of boundary heat flow, Phys. Earth Planet. Inter., **138**, 39–54 (2003).

Christensen U.R., Olson P. & Glatzmaier G.A., Numerical modelling of the geodynamo: a systematic parameter survey, Geophys. J. Int., **138**, 393–409 (1999).

Christensen U.R. & Tilgner A., Power requirement of the geodynamo from ohmic losses in numerical and laboratory dynamos, *Nature*, **429**, 169–171 (2004).

Chulliat A., Loper D.E. & Shimizu H., Buoyancy-driven perturbations in a rapidly rotating, electrically conducting fluid. Part III. Effect of the Lorentz force, *Geophys. Astrophys. Fluid Dyn.*, **98**, 507–535 (2004).

Clark S., The dark side of the Sun, *Nature*, **441**, 7092, 402–404 (2006).

Clemens D.P., Massachusetts–Stony Brook Galactic plane CO survey: the Galactic disk rotation curve, *Astrophys. J.*, **295**, 422–436 (1985).

Clever R.M. & Busse F.H., Nonlinear oscillatory convection, *J. Fluid Mech.*, **176**, 403–417 (1987).

Cline K.S., Brummell N.H. & Cattaneo F., On the formation of magnetic structures by the combined action of velocity shear and magnetic buoyancy, *Astrophys. J.*, **588**, 630–644 (2003).

Colgate S.A., Pariev V.I., Beckley H.F., Ferrel R., Romero V.D. & Weatherall J.C., The new mexico $\alpha\omega$ dynamo experiment: modelling astrophysical dynamos. *Magnetohydrodynamics*, **38**, 129–142 (2002).

Connerney J.E.P., Magnetic fields of the outer planets, *J. Geophys. Res. – Planet*, **98**, 18659–18679 (1993).

Cordero S. & Busse F.H., Experiments on convection in rotating hemispherical shell: Transition to quasi-periodic state, *Geophys. Res. Letts.*, **19**, 733–736 (1992).

Courtillot V. & Le Mouël J.-L., Time variations of the Earth's magnetic field: from daily to secular, *Ann. Rev. Earth Planet. Sci.*, **16**, 389–476 (1988).

Courtillot V., Gallet Y., Le Mouel J.L. Are there connections between the Earth's magnetic field and climate? *Earth Planet. Sci. Lett.*, **253**, 328–339 (2007).

Courvoisier A., Hughes D.W. & Tobias S.M., α Effect in a Family of Chaotic Flows, *Phys. Rev. Lett.*, **96**, 034503 (2006).

Covas E., Moss D. & Tavakol R., The influence of density stratification and multiple nonlinearities on solar torsional oscillations, *Astron. Astrophys.*, **416**, 775–782 (2004).

Covas E. & Tavakol R., Multiple forms of intermittency in partial differential equation dynamo models, *Phys. Rev. E*, **60**, 5435–5438 (1999).

Covas E., Tavakol R., Ashwin P., Tworkowski A. & Brooke J., In-out intermittency in partial differential equation and ordinary differential equation models, *Chaos*, **11**, 404–409 (2001a).

Covas E., Tavakol R., Moss D. & Tworkowski A., Torsional oscillations in the solar convection zone, *Astron. Astrophys.*, **360**, L21–L24 (2000).

Covas E., Tavakol R. & Moss D., Dynamical variations of the differential rotation in the solar convection zone, *Astron. Astrophys.*, **371**, 718–730 (2001b).

Covas E., Tavakol R., Tworkowski A. & Brandenburg A., Axisymmetric mean field dynamos with dynamic and algebraic α-quenchings, Astron. Astrophys., **329**, 350–360 (1998).

Covas E., Tworkowski A., Brandenburg A. & Tavakol R., Dynamos with different formulations of a dynamic α–effect, *Astron. Astrophys.*, **317**, 610–617 (1997).

Covas E., Tworkowski A., Tavakol R. & Brandenburg A., Robustness of truncated $\alpha\Omega$-dynamos with a dynamic α, *Sol. Phys.*, **172**, 3–9 (1997).

Cowling T.G., The magnetic field of sunspots, *Mon. Not. R. Astr. Soc.* **94**, 39–48 (1934).

Cowling T.G., Sunspots and the solar cycle, *Nature*, **255**, 189–190 (1975).

Cox D.P., The diffuse interstellar medium. In *The Interstellar Medium in Galaxies,* H.A. Thronson & J.M. Shull Eds., Kluwer, Dordrecht, 181–200 (1990).

Cox J.P., *Principles of Stellar Structure*, 2 volumes, Gordon & Breach, New York (1968).

Davidson P.A., *An Introduction to Magnetohydrodynamics*, Cambridge University Press, U.K. (2001).

De Young D., Turbulent generation of magnetic fields in extragalactic radio sources, *Astrophys. J.,* **241**, 81–97 (1980).

Dehant V., Van Hoolst T., de Viron O., Greff-Lefftz M., Legros H. & Defraigne P., Can a solid inner core of Mars be detected from observations of polar motion and nutation of Mars? *J. Geophys Res.-Planet*, **108**, 5127 (2003).

Deharveng J.M. & Pellet A., Étude cinématique et dynamique de M31 à partir de l'observation des régions d'émission, *Astron. Astrophys.,* **38**, 15–28 (1975).

De Wijs G.A., Kresse G., Vočadlo L., Dobson D., Alfè D., Gillan M.J. & Price G.D., The viscosity of liquid iron at the physical conditions of the Earth's core, *Nature*, **392**, 805–807 (1998).

Diamond P.H., Hughes D.W. & Kim E.-J., Self-consistent mean field electrodynamics in two and three dimensions, in *Fluid Dynamics and Dynamos in Astrophysics and Geophysics*, ed. A.M. Soward, C.A. Jones, D.W. Hughes & N.O. Weiss, Taylor & Francis, London (2004).

Dikpati M. & Charbonneau P., A Babcock-Leighton flux transport dynamo with solar-like differential rotation, *Astrophys. J.*, **518**, 508–520 (1999).

Dikpati M. & Gilman P., Simulating and predicting solar cycles using a flux-transport dynamo, *Astrophys. J.*, **649**, 498–514 (2006a).

Dikpati M., de Toma G. & Gilman P., Predicting the strength of solar cycle 24 using a flux-transport dynamo-based tool, *Geophys. Res. Lett.*, **33**, L05102 (2006b).

Dikpati M., de Toma G., Gilman P.A., Arge C.N. & White O.R., Diagnostics of polar field reversal in solar cycle 23 using a flux transport dynamo model, *Astrophys. J.*, **601**, 1136–1151 (2004).

Dobler W., Frick P. & Stepanov R., Screw dynamo in a time-dependent pipe flow, *Phys. Rev.*, E **67**-5, 056309 (2003).

Donald J.T. & Roberts P.H., The effect of anisotropic heat transport in Earth's core on the geodynamo, *Geophys. Astrophys. Fluid Dyn.*, **98**, 367–384 (2004).

Donati J.-F., Collier Cameron A., Hussain G.A.J. & Semel M., Magnetic topology and prominence patterns on AB Doradus, *Mon. Not. R. Astron. Soc.*, **302**, 437–456 (1999).

Dopita M.A. & Sutherland R.S., *Astrophysics of the Diffuse Universe,* Springer, Berlin (2003).

Dorch S.B.F. & Nordlund Å., On the transport of magnetic fields by solar-like stratified convection, *Astron. Astrophys.*, **365**, 562–570 (2001).

Dormy E., The origin of the Earths magnetic field: fundamental or environmental research? *Europhys. news*, **37**/2, 2006.

Dormy E., Cardin P. & Jault D., MHD flow in a slightly differentially rotating spherical shell, with conducting inner core, in a dipolar magnetic field, *Earth Planet. Sci. Lett.*, **160**, 15–30 (1998).

Dormy E., Jault D. & Soward A.M., A super-rotating shear layer in magnetohydrodynamic spherical Couette flow, *J. Fluid Mech.*, **452**, 263–291 (2002).

Dormy E., Soward A.M., Jones C.A., Jault D. & Cardin P., The onset of thermal convection in rotating spherical shells, *J. Fluid Mech.*, **501**, 43–70 (2004).

Dormy E., Valet J.-P. & Courtillot V., Numerical models of the geodynamo and observational constraints, *Geochem. Geophys. Geosyst.*, **1**, 62 (2000).

Drobyshevski E.M. & Yuferev V.S., Topological pumping of magnetic flux by three-dimensional convection, *J. Fluid Mech.*, **65**, 33–44 (1974).

Du Y., Tél T. & Ott E., Characterization of sign singular measures, *Physica*, D **76**, 168–180 (1994).

Dudley M.L. & R.W. James, Time-dependent kinematic dynamos with stationary flows, *Proc. R. Soc. Lond.*, A **425**, 407–429 (1989).

Eckert S., Gerbeth G. Velocity measurements in liquid sodium by means of ultrasound Doppler velocimetry. *Exp. Fluids*, **32**, 542–546 (2002).

Eddy J.A., The Maunder minimum, *Science*, **192**, 1189–1202 (1976).

Elsasser W.M., Induction Effects in Terrestrial Magnetism, *Phys. Rev.*, **69**-3-4, 106–116 (1946).

Elstner D., Meinel R. & Rüdiger G., Galactic dynamo models without sharp boundaries, *Geophys. Astrophys. Fluid. Dyn.,* **50**, 85–94 (1990).

Elstner D., Golla G., Rüdiger G. & Wielebinski R., Galactic halo magnetic fields due to a 'spiky' wind, *Astron. Astrophys.*, **297**, 77–82 (1995).

Encyclopaedia of Planetary Sciences, J.H. Shirley and R.W. Fairbridge Eds., Kluwer (1997).

Falgarone E. & T. Passot (Eds.), *Turbulence and Magnetic Fields in Astrophysics*, Springer (2003).

Fauterelle Y. & Childress S., Convective dynamos with intermediate and strong fields,*Geophys. Astrophys. Fluid Dyn.*, **22**, 235–279 (1982).

Fauve S. & F. Pétrélis, The dynamo effect, Peyresq Lectures on Nonlinea Phenomena, **II**,1–66, Sepulchre J.-A. Ed., World Scientific, Singapore (2003).

Fearn D.R., Nonlinear planetary dynamos, in *Lectures on Solar and Planetary Dynamos* (Proctor M.R.E. & Gilbert A.D. Eds.), Cambridge University Press, 219–244 (1994).

Fearn D.R., Hydromagnetic flow in planetary cores, *Rep. Prog. Phys.*, **61**, 175–235 (1998).

Fearn D.R. & Morrison G., The role of inertia in hydrodynamic models of the geodynamo, Phys. Earth Planet. Inter., **128**, 75–92 (2001).

Fearn D.R. & Proctor M.R.E., The stabilising role of differential rotation on hydromagnetic waves, J. Fluid Mech., **128**, 21–36 (1983).

Fearn D.R. & Proctor M.R.E., Dynamically consistent magnetic fields produced by differential rotation, J. Fluid Mech., **178**, 521–534 (1987).

Fearn D.R. & Proctor M.R.E., Magnetostrophic balance in non-axisymmetric, non-standard dynamo models, Geophys. Astrophys. Fluid. Dyn., **67**, 117–128 (1992).

Fearn D.R. & Rahman M.M., Instability of nonlinear α^2-dynamos, Phys. Earth Planet. Inter., **142**, 101–112 (2004a).

Fearn D.R. & Rahman M.M., The role of inertia in models of the geodynamo, Geophys. J. Int., **158**, 515–528 (2004b).

Fearn D.R., Roberts P.H. & Soward A.M., Convection, stability and the dynamo. In *Energy Stability and Convection* (Galdi G.P.& Straughan B. Eds.) Pitman Research Notes in Mathematics **168**, 60–324, Longman, Essex (1986).

Fenstermacher P.R., Swinney H.L. & Gollub J.P., Dynamical instabilities and the transition to chaotic Taylor vortex flow *J. Fluid Mech*, **94**, 103–128 (1979).

Fereday D.R., Haynes P.H., Wonhas A. & Vassilicos J.C., Scalar variance decay in chaotic advection and Batchelor-regime turbulence, *Phys. Rev.*, E **65**, article 035301 (2002).

Ferriz-Mas A., Schmitt D. & Schüssler M., A dynamo effect due to the instability of magnetic flux tubes, *Astron. Astrophys.*, **289**, 949–956 (1994).

Feudel U., Jansen W. & Kurths J., Tori and chaos in a nonlinear dynamo model for solar activity, *Int. J. Bifurc. Chaos*, **3**, 131–138 (1993).

Field G.B., Blackman E.G. & Chou H.S., Nonlinear α-effect in dynamo theory, Astrophys. J., **513**, 638–651 (1999).

Finn J.M. & Ott E., Chaotic flows and fast magnetic dynamos, *Phys. Fluids*, **31**, 2992–3011 (1988).

Finn J.M. & Ott E., The fast kinematic magnetic dynamo and the dissipationless limit, *Phys. Fluids*, B **2**, 916–926 (1990).

Fletcher A. & Shukurov A., Hydrostatic equilibrium in a magnetized, warped Galactic disc, *Mon. Not. R. Astron. Soc.,* **325**, 312–320 (2001) [astro-ph/0101387].

Fletcher A., Berkhuijsen E.M., Beck R. & Shukurov A., The magnetic field of M31 from multi-wavelength radio polarization observations, *Astron. Astrophys.*, *Astron. Astrophys.*, **414**, 53–67 (2004) [astro-ph/0310258].

Forest C.B., R.A. Bayliss, R.D. Kendrick, M.D. Nornberg, R. O'Connell, E.J. Spence Hydrodynamic and numerical modeling of a spherical homogeneous dynamo experiment, *Magnetohydrodynamics*, **38**, 107–120 (2002).

Fotheringham P., Fearn D.R. & Hollerbach R., Magnetic stability and nonlinear evolution of a selection of mean field dynamos, Phys. Earth Planet. Inter., **134**, 213–237 (2002).

Fraternali F., Oosterloo T., Boomsma R., Swaters R. & Sancisi R., High velocity gas in the halos of spiral galaxies. In *Recycling Intergalactic and Interstellar Matter,* IAU Symp. 217, P.-A. Duc, J. Braine & E. Brinks Eds., ASP (2003) [astro-ph/0310799].

Frick P., Beck R., Shukurov A., Sokoloff D., Ehle M. & Kamphuis J., Radio and optical spiral patterns in the galaxy NGC 6946, *Mon. Not. R. Astron. Soc.,* **318**, 925–937 (2000).

Frick P., Stepanov R., Shukurov A. & Sokoloff D., Structures in the rotation measure sky, *Mon. Not. R. Astron. Soc.,* **325**, 649–664 (2001) [astro-ph/0012459].

Frick P., Noskov V., Denisov S., Khripchenko S., Sokoloff D., Stepanov R. & Sukhanovsky A., Non-stationnary screw flow in a toroidal channel: way to laboratory dynamo experiment, *Magnetohydrodynamics*, **38**, 143–162 (2002).

Frisch, U., *Turbulence. The Legacy of A.N. Kolmogorov*, Cambridge University Press, U.K. (1995).

Gailitis A., Self-excitation of a magnetic field by a pair of annular vortices. *Magnetic. Gidrod.*, **6**, 19–22, [*Magnetohydrodynamics* **6**, 14–17] (1970).

Gailitis A., Karasev B., Kirilov I., Lielausis O., Luzhanskii S. & Ogorodnikov A., Experiment with a liquid metal model of an MHD dynamo, *Magnetohydrodynamics*, **23**, 349–353 (1987).

Gailitis A., Lielausis O., Dementév S., Platacis E., Cifersons A., Gerbeth G., Gundrum Th., Stefani F., Christen M., Hanel H. & Will G., Detection of a flow induced magnetic field eigenmode in the Riga dynamo facility, *Phys. Rev. Lett.*, **84**, 4365–4368 (2000).

Gailitis A., Lielausis O., Platacis E., Dementév S., Cifersons A., Gerbeth G., Gundrum Th., Stefani F., Christen M. & Will G., Magnetic field saturation in the Riga dynamo experiment, *Phys. Rev. Lett.*, **86**, 3024–3027 (2001).

Gailitis A., Lielausis O., Platacis E., Dement'ev E., Cifersons A., Gerbeth G., Gundrum T., Stefani F., Christen M. & Will G., Dynamo experiments at the Riga sodium facility, *Magnetohydrodynamics*, **38**, 5–14 (2002a).

Gailitis A., Lielausis O., Platacis E., Gerbeth G. & Stefani F., Laboratory experiments on hydromagnetics dynamos, *Rev. Mod. Phys.*, **74**, 973–989 (2002b).

Galloway D.J. & Frisch U., Dynamo action in a family of flows with chaotic streamlines, *Geophys. Astrophys. Fluid Dyn.*, **36**, 53–83 (1986).

Galloway D.J. & Proctor M.R.E., Numerical calculations of fast dynamos for smooth velocity fields with realistic diffusion, *Nature*, **356**, 691–693 (1992).

Gans R.F., On hydromagnetic precession in a cylinder, *J. Fluid Mech.*, **45**, 111–130 (1970).

Geiger G. & Busse F.H., On the onset of thermal convection in slowly rotating fluid shells, *Geophys. Astrophys. Fluid Dyn.*, **18**, 147–156 (1981).

Ghil M. & Childress S., *Topics in geophysical fluid dynamics: atmospheric dynamics, dynamo theory and climate dynamics,* New York, Springer (1987).

Gilbert A.D., Fast dynamo action in the Ponomarenko dynamo, *Geophys. Astrophys. Fluid Dyn.*, **44**, 214–258 (1988).

Gilbert A.D., Magnetic helicity in fast dynamos, *Geophys. Astrophys. Fluid Dyn.*, **96**, 135–151 (2002a).

Gilbert A.D., Advected fields in maps: I. Magnetic flux growth in the stretch–fold–shear map, *Physica*, D **166**, 167–196 (2002b).

Gilbert A.D., Dynamo theory, in *Handbook of mathematical fluid dynamics* (S. Friedlander, D. Serre Eds.), vol. 2, 355–441. Elsevier Science (2003).

Gilbert A.D., Advected fields in maps: II. Dynamo action in the stretch–fold–shear map, Geophys. Astrophys. Fluid. Dyn., **99**, 3, 241–269 (2005).

Gilbert A.D., Advected fields in maps: III. Passive scalar decay in baker's maps, *Dynamical Systems*, **21**, 25–71 (2006) [http://www.tandf.co.uk/journals/titles/03091929.asp].

Gilbert A.D. & Ponty Y., Dynamos on stream surfaces of a highly conducting fluid, *Geophys. Astrophys. Fluid. Dyn.*, **93**, 55–95 (2000).

Gilbert A.D. & Sulem P.-L., On inverse cascades in alpha effect dynamos, *Geophys. Astrophys. Fluid Dyn.*, **51**, 243–261 (1990).

Gilman P.A., Dynamically consistent nonlinear dynamos driven by convection in a rotating shell. II. Dynamos with cycles and strong feedbacks, *Astrophys. J. Suppl.*, **53**, 243–268 (1983).

Gilman P.A. & Benton E.R., Influence of an axial magnetic field on the steady linear Ekman boundary layer., *Phys. Fluids*, **11**, 2397–2401 (1968).

Gilman P.A. & Dikpati M., Analysis of instability of latitudinal differential rotation and toroidal field in the solar tachocline using a magnetohydrodynamic shallow-water model. I. Instability for broad toroidal field profiles, *Astrophys. J.*, **576**, 1031–1047 (2002).

Gilman P.A. & Fox P.A., Joint instability of latitudinal differential rotation and toroidal magnetic fields below the solar convection zone. II. Instability for toroidal fields that have a node between the equator and pole, *Astrophys. J.*, **510**, 1018–1044 (1999); Erratum, **534**, 1020 (2000).

Gilman P.A. & Glatzmaier G.A., Compressible convection in a rotating spherical shell. I. Anelastic equations, *Astrophysical J. Suppl. Ser.*, **45**, 335–349 (1981).

Gilman P.A. & Miller J., Dynamically consistent nonlinear dynamos driven by convection in a rotating spherical shell, Astrophys. J., Suppl.Ser., **46**, 211–238 (1981).

Glatzmaier G.A., Numerical simulations of stellar convective dynamos II. Field propagation in the convection zone, Astrophys. J., **291**, 300–307 (1985a).

Glatzmaier G.A., Numerical simulations of stellar convective dynamos III. At the base of the convection zone, Geophys. Astrophys. Fluid. Dyn., **31**, 137–150 (1985b).

Glatzmaier G.A., Geodynamo simulations: how realistic are they? *Ann. Rev. Earth Planet. Sci.*, **30**, 237–257 (2002).

Glatzmaier G.A., Coe R.S., Hongre L. & Roberts P.H., The role of the earth's mantle in controlling the frequency of geomagnetic reversals, *Nature*, **401**, 885–890 (1999).

Glatzmaier G.A. & Roberts P.H., A three-dimensional self-consistent computer simulation of a geomagnetic field reversal, *Nature*, **377**, 203–209 (1995a).

Glatzmaier G.A. & Roberts P.H., A three-dimensional convective dynamo solution with rotating and finitely conducting inner core and mantle, Phys. Earth Planet. Inter., **91**, 63–75 (1995b).

Glatzmaier G.A. & Roberts P.H., On the magnetic sounding of planetary inter., Phys. Earth Planet. Inter., **98**, 207–220 (1996a).

Glatzmaier G.A. & Roberts P.H., An anelastic evolutionary geodynamo simulation driven by compositional and thermal convection, *Physica*, D **97**, 81–94 (1996b).

Glatzmaier G.A. & Roberts P.H., Rotation and magnetism of Earth's inner core, *Science*, **274**, 1887–1891 (1996c).

Glatzmaier G.A. & Roberts P.H., Simulating the geodynamo, *Contemp. Phys.*, **38**, 269–288 (1997).

Glatzmaier G.A. & Roberts P.H., Dynamo theory then and now, *Int. J. Eng. Sci.*, **36**, 1325-1338 (1998).

Gog J.R., Oprea I., Proctor M.R.E. & Rucklidge A.M., Destabilization by noise of tranverse perturbations to heteroclinic cycles: a simple model and an example from dynamo theory, *Proc. R. Soc. Lond.*, A **455**, 4205–4222 (1999).

Goldbrum P., I.M. Moroz & R. Hide, A self-exciting single Faraday disk dynamo with battery bias, *Int. J. Bif. Chaos*, **10**, 1875–85 (2000).

Gough D.O., The anelastic approximation for thermal convection, *J. Atmos. Sci*, **26**, 448–456 (1969).

Gough D.O. & McIntyre M.E., Inevitability of a magnetic field in the Sun's radiative interior, *Nature*, **394** 755–757 (1998).

Greenspan H.P., *The Theory of Rotating Fluids.* Cambridge University Press, U.K. (1968).

Grote E. & Busse F.H., Dynamics of convection and dynamos in rotating spherical fluid shells, *Fluid Dyn. Res.*, **28**, 349–368 (2001).

Gruzinov A.V. & Diamond P.H., Self-consistent theory of mean field electrodynamics, *Phys. Rev. Lett.*, **72**, 1651–1654 (1994).

Gruzinov A.V. & Diamond P.H., Nonlinear mean field electrodynamics of turbulent dynamos, *Phys. Plasmas*, **3**, 1853–1857 (1996).

Gubbins D., Energetics of the Earth's core, *J. Geophys.*, **43**, 453–464 (1977).

Gubbins D., The distinction between geomagnetic excursions and reversals, Geophys. J. Int., **197**, F1-F3 (1999).

Gubbins D., The Rayleigh number for convection in the Earth's core, Phys. Earth Planet. Inter., **128**, 312 (2001).

Gubbins D., Alfè D., Masters G., Price D. & Gillan M.J., Can the Earth's dynamo run on heat alone?, Geophys. J. Int., **155**, 609–622 (2003).

Gubbins D., Alfè D., Masters G., Price D. & Gillan M.J., Gross thermodynamics of 2-component core convection, Geophys. J. Int., **157**, 1407–1414, (2004).

Gubbins D., Kent D. & Laj C. (Eds), Geomagnetic polarity reversals and long term secular variation, *Phil. Trans R. Soc. Lond.*, A **358**, 889–1223 (2000).

Gubbins D. & Roberts P.H., *Magnetohydrodynamics of the Earth's Core,* In: Geomagnetism (Jacobs J.A. Ed.) 1–183, Academic Press, London (1987).

Guckenheimer J. & Holmes P., *Nonlinear Oscillations, Dynamical Systems and Bifurcations of Vector Fields*, 2nd printing, Springer, New York (1986).

Guillot T., Interiors of Giant Planets Inside and Outside the Solar System, *Science*, **286**, 72–77 (1999).

Han J.L., Manchester R.N., Berkhuijsen E.M. & Beck R., Antisymmetric rotation measures in our Galaxy: evidence for an A0 dynamo, *Astron. Astrophys.,* **322**, 98–102 (1997).

Han J.L., Beck R. & Berkhuijsen E.M., New clues to the magnetic field structure of M31, *Astron. Astrophys.,* **335**, 1117–1123 (1998).

Hansen C.J. & Kawaler S.D., *Stellar Interiors. Physical Principles, Structure, and Evolution*, Springer, Heidelberg (1994).

Hart J.E., Glatzmaier G.A. & Toomre J., Space-laboratory and numerical simulations of thermal convection in a rotating hemispherical shell with radial gravity, *J. Fluid Mech.*, **173**, 519–544 (1986).

Hathaway D.H., Doppler measurements of the Sun's meridional flow, *Astrophys. J.*, **460**, 1027–1033 (1996).

Hathaway D.H. & Wilson R.M., Geomagnetic activity indicates large amplitude for sunspot cycle 24, *Geophys. Res. Lett.*, **33**, L18101 (2006).

Hathaway D.H., Wilson R.M. & Reichmann E.J., A synthesis of solar cycle prediction techniques, *J. Geophys. Res.*, **104**, 22375–22388 (1999).

Haud U., Gas kinematics in M31, *Astrophys.Space Sci.,* **76**, 477–490 (1981).

Hawley J.F., Gammie C.F. & Balbus S.A., Local three-dimensional simulations of an accretion disk hydromagnetic dynamo, *Astrophys. J.*, **464**, 690–703 (1996).

Heimpel M., Aurnou J. & Wicht J., Simulation of equatorial and high-latitude jets on Jupiter in a deep convection model, *Nature*, **438**, 193–196 (2005).

Heisenberg W., Zur statistischen Theorie der Turbulenz, *Zeit. Phys.*, **124**, 628–657 (1948).

Heller R., Merrill R.T. & McFadden P.L., The variation of the intensity of the Earth's magnetic field with time, Phys. Earth Planet. Inter., **131**, 237–249 (2002).

Henry T.J., Soderblom D.R., Donahue R.A. & Baliunas S.L., A survey of Ca II H and K chromospheric emission in southern solar-type stars, *Astron. J.*, **111**, 439–465 (1996).

Herrmann J. & Busse F.H., Convection in a Rotating Cylindrial Annulus. Part 4. Modulations and Transition to Chaos at Low Prandtl Numbers, *J. Fluid Mech.*, **350**, 209–229 (1997).

Herzenberg A., Geomagnetic dynamos, *Phil. Trans R. Soc. Lond.*, A **250**, 543–583 (1958).

Hide R., Structural instability of the Rikitake disk dynamo, *Geophs. Res. Lett.*, **22**, 1057–1059 (1995).

Hide R., The nonlinear differential equations governing a hierarchy of self-exciting coupled Faraday-disk homopolar dynamos, *Phys. Earth Planet. Inter.*, **103**, 281–291 (1997a).

Hide R., Nonlinear quenching of current fluctuations in a self-exciting homopolar dynamo, *Nonlinear Processes in Geophysics*, **4**, 201–205 (1997b).

Hide R., Generic nonlinear processes in self-exciting dynamos and the long-term behaviour of the main geomagnetic field, *Phil. Trans R. Soc.*, A **358**, 943–955 (2000).

Hide R., Lewis S.R. & Read P.L., Sloping convection: a paradigm for large-scale waves and eddies in planetary atmospheres, *Chaos*, **4**, 135–162 (1994).

Hide R. & Moroz I.M., Effects due to induced azimuthal eddy currents in the Faraday disk self-exciting homopolar dynamo with a nonlinear series motor: I Two special cases, *Physica*, D **134**, 287–301 (1999).

Hide R. & Palmer T.N., Generalisations of Cowling's Theorem, *Geophys. Astrophys. Fluid Dyn.*, **19**, 301–309 (1982).

Hide R., Skeldon A.C. & Acheson D. J., A study of two novel self-exciting single-disk homopolar dynamos: theory, *Proc. Roy. Soc. Lond.*, A **452,** 1369–1395 (1996).

Hirsching W. & Busse F.H., Stationary and chaotic dynamos in rotating spherical shells, *Phys. Earth Planet. Inter.*, **90**, 43–254 (1995).

Hollerbach R., Magnetohydrodynamic Ekman and Stewartson layers in a rotating spherical shell, *Proc. R. Soc. Lond.*, A **444**, 333–346 (1994a).

Hollerbach R., Imposing a magnetic field across a nonaxisymmetric shear layer in a rotating spherical shell, *Phys. Fluids*, **6**(7), 2540–2544 (1994b).

Hollerbach R., Magnetohydrodynamic shear layers in a rapidly rotating plane layer, *Geophys. Astrophys. Fluid Dyn.*, **82**, 237–253 (1996).

Hollerbach R., A spectral solution of the magneto-convection equations in spherical geometry, *Int. J. Num. Meth. Fluids*, **32**, 773–797 (2000).

Hollerbach R. & Ierley G., A modal α^2-dynamo in the limit of asymptotically small viscosity, Geophys. Astrophys. Fluid. Dyn., **56**, 133–158 (1991).

Hollerbach R. & Jones C.A., A geodynamo model incorporating a finitely conducting inner core, Phys. Earth Planet. Inter., **75**, 317–327 (1993).

Hollerbach R. & Jones C.A., On the magnetically stabilising role of the Earth's inner core, Phys. Earth Planet. Inter., **87**, 171–181 (1995).

Holm D.D., Averaged Lagrangians and the mean effects of fluctuations in ideal fluid dynamics, *Physica*, D **170**, 253–286 (2002).

Holme R. & Bloxham J., The magnetic fields of Uranus and Neptune: methods and models, *J. Geophys. Res. – Planet*, **101**, 2177–2200 (1996).

Howe R., Christensen-Dalsgaard J., Hill F., Komm R.W.. Larsen R.M., Schou J., Thompson M.J. & Toomre J., Deeply penetrating banded zonal flows in the solar convection zone, *Astrophys. J.*, **533**, L163–L166 (2000a).

Howe R., Christensen-Dalsgaard J., Hill F., Komm R.W., Larsen R.M., Schou J., Thompson M.J. & Toomre J., Dynamic variations at the base of the solar convection zone, *Science*, **287**, 2456–2460 (2000b).

Hoyng P., Turbulent transport of magnetic fields. III. Stochastic excitation of global modes, *Astrophys. J.*, **332**, 857–871 (1988).

Hoyt D.V. & Schatten K.H., Group sunspot numbers: a new solar activity reconstruction, *Solar Phys.*, **181**, 491–512 (1998).

Hubbard W.B., Burrows A. & Lunine, J.L., Theory of giant planets, *Ann. Rev. Astron. Astrophys.*, **40**, 103–136 (2002).

Hughes D.W., Magnetic Buoyancy, in *Advances in Solar System MHD* (E.R. Priest & A.W. Hood Eds.), Cambridge University Press, 77–104 (1991).

Hughes D.W. & Proctor M.R.E., Magnetic fields in the solar convection zone: magnetoconvection and magnetic buoyancy, *Annu. Rev. Fluid Mech.*, **20**, 187–223 (1988).

Hughes D.W. & Tobias S.M., On the instability of magnetohydrodynamic shear flows, *Proc. R. Soc. Lond.*, A **457**, 1365–1384 (2001).

Hulot G., Eymin C., Langlais B., Mandea M. & Olsen N., Small-scale structure of the geodynamo inferred from Oersted and Magsat satellite data, Nature, **416**, 620–623 (2002).

Im H.G., Lund T.S. & Ferziger, J.H., Large eddy simulation of turbulent front propagation with dynamic subgrid models, *Phys. Fluids*, **9** 3826–3833 (1997).

Iskakov A., Descombes S., Dormy E., An integro-differential formulation for magnetic induction in bounded domains: boundary element-finite volume method, *J. Comput. Phys.*, **197**, 540–554 (2004).

Iskakov A., Dormy E., On magnetic boundary conditions for non-spectral dynamo simulations, Geophys. Astrophys. Fluid. Dyn., **99**-6, 481–492 (2005).

Ivers D.J. & James R.W., Axisymmetric anti-dynamo theorems in compressible non-uniform conducting fluids, *Phil. Trans. R. Soc. Lond.*, A **312**, 317–324 (1984).

Jackson A., Time-dependency of tangentially geostrophic core surface motions, *Phys. Earth Planet. Inter.*, **103**, 293–311 (1997).

Jackson A., Intense equatorial flux spots on the surface of the Earth's core, *Nature*, **424**, 760–763 (2003).

Jackson A., Jonkers A.R. & Walker M.R., Four centuries of geomagnetic secular variation from historical records, Phil. Trans. R. Soc. Lond., A **358**, 957–990 (2000).

Jacobs J.A., *Reversals of the Earth's magnetic field*, Cambridge University Press (1994).

Jault D., Model Z by computation and Taylor's condition, Geophys. Astrophys. Fluid. Dyn., **79**, 99–124 (1995).

Jault D., Electromagnetic and topographic coupling, and LOD variations, in *Earth's core and lower mantle* (C.A. Jones, A.M. Soward & K. Zhang Eds.), Taylor & Francis, London, 56–76 (2003).

Jault D., Gire C. & Le Mouël J.L., Westward drift, core motions and exchanges of angular-momentum between core and mantle, *Nature*, **333**, 353–356 (1988).

Jault D. & Le Mouël J.-L., Exchange of angular momentum between the core and the mantle, *J. Geomag. Geoelectr.*, **43**, 111–129 (1991).

Jennings R.L. & Weiss N.O., Symmetry breaking in stellar dynamos, *Mon. Not. R. Astron. Soc.*, **252**, 249–260 (1991).

Jepps S.A., Numerical models of hydromagnetic dynamos, J. Fluid Mech., **67**, 625–645 (1975).

Jones C.A., Convection-driven geodynamo models, Phil. Trans. R. Soc. Lond., A **358**, 873–897 (2000).

Jones C.A., Dynamos in planets, in *Stellar Astrophysical Fluid Dynamics*, M.J. Thompson & J.C. Christensen-Dalsgaard Eds, Cambridge University Press, 159–176 (2003).

Jones C.A., Longbottom A.W. & Hollerbach, R., A self-consistent convection driven geodynamo model, using a mean field approximation, Phys. Earth Planet. Inter., **92**, 119–141 (1995).

Jones C.A. & Roberts P.H., Convection-driven dynamos in a rotating plane layer, *J. Fluid Mech.*, **404**, 311–343 (2000).

Jones C.A. & Roberts P.H., Turbulence models and plane layer dynamos, in *Fluid Dynamics and Dynamos in Astrophysics and Geophysics*, A.M. Soward, C.A. Jones Eds., D.W. Hughes and N.O. Weiss, CRC press, Boca Raton (2005), pp. 295–330.

Jones C.A., Soward A.M. & Mussa A.I., The onset of thermal convection in a rapidly rotating sphere, *J. Fluid Mech.*, **405**, 157–179 (2000).

Jones C.A., Weiss N.O. & Cattaneo F., Nonlinear dynamos: a complex generalization of the Lorenz equations, *Physica*, D **14**, 161–176 (1985).

Julien K., Legg S., McWilliams J. & Werne J., Hard turbulence in rotating Rayleigh-Benard convection, *Phys. Rev.*, E **53**, R5557–R5560 (1996).

Kageyama A., Ochi M.M. & Sato T., Flip-flop transitions of the magnetic intensity and polarity reversals in the magnetohydrodynamic dynamo, *Phys. Rev. Letts*, **82**, 5409–5412 (1999).

Kahn F.D. & Brett L., Magnetic reconnection in the disc and halo, *Mon. Not. R. Astron. Soc.*, **263**, 37–48 (1993).

Kazantsev A.P., Enhancement of a magnetic field by a conducting fluid, *Zh. Teor. Eksper. Fiz.*, **53**, 1806–1813 (1967); English translation: *Sov. Phys. JETP*, **26**, 1031–1034 (1968).

Kazantsev A.P., Ruzmaikin A.A. & Sokoloff D.D., Magnetic field transport in an acoustic turbulence, *Sov. Phys. JETP*, **61**, 285–292 (1985).

Kerswell R.R., Upper bounds on the energy dissipation in turbulent precession, J. Fluid Mech., 321, 335–370 (1996).

Kim E.J., Hughes D.W. & Soward A.M., An investigation into high conductivity dynamo action driven by rotating convection, *Geophys. Astrophys. Fluid Dyn.*, **91**, 303–332 (1999).

Kirk V., Breaking of symmetry in the Saddle-node–Hopf bifurcation, *Phys. Lett. A*, **154**, 243–248 (1991).

Kirk V., Merging of resonance tongues, *Physica*, D **66**, 267–281 (1993).

Kitchatinov L.L., Pipin V.V., Makarov V.I. & Tlatov A.G., Solar torsional oscillations and the grand activity cycle, *Solar Phys.*, **189**, 227–239 (1999).

Kitchatinov L.L., Rüdiger G. & Küker M., Λ–quenching as the nonlinearity in stellar turbulence dynamos, *Astron. Astrophys.*, **292**, 125–132 (1994).

Kivelson M.G., Khurana K.K., Russell C.T., Joy S.P., Volwerk M., Walker R.J., Zimmer C. & Linker J.A., Magnetized or unmagnetized: Ambiguity persists, following Galileo's encounters with Io in 1999 and 2000, *J. Geophys. Res-Space Sci.*, **106**, 26121–26135 (2001).

Klapper I. & Young L.-S., Bounds on the fast dynamo growth rate involving topological entropy, *Comm. Math. Phys.*, **173**, 623–646 (1995).

Kleeorin N., Moss D., Rogachevskii I. & Sokoloff D., Helicity balance and steady-state strength of the dynamo generated galactic magnetic field, *Astron. Astrophys.*, **361**, L5–L8 (2000).

Kleeorin N., Moss D., Rogachevskii I. & Sokoloff D., Nonlinear magnetic diffusion and magnetic helicity transport in galactic dynamos, *Astron. Astrophys.*, **400**, 9–18 (2003).

Kleeorin N. & Ruzmaikin, A.A., Dynamics of the average turbulent helicity in a magnetic field, *Magnetohydrodynamics*, **18**, 116–122 (1982).

Kleeorin N., Rogachevskii A., Ruzmaikin A., Soward A.M. & Starchenko S., Axisymmetric flow between differentially rotating spheres in a magnetic field with dipole symmetry, *J. Fluid Mech.*, **344**, 213–244 (1997).

Knobloch E. & Landsberg A.S., A new model of the solar cycle, *Mon. Not. R. Astron. Soc.*, **278**, 294–302 (1996).

Knobloch E., Rosner R. & Weiss N.O., Magnetic fields in late-type stars, *Mon. Not. R. Astron. Soc.*, **197**, 45P–49P (1981).

Knobloch E., Tobias S.M. & Weiss N.O., Modulation and symmetry changes in stellar dynamos, *Mon. Not. R. Astron. Soc.*, **297**, 1123–1138 (1998).

Königl A. & Pudritz R. E., Disk winds and the accretion-outflow connection. In *Protostars and Planets IV*, eds V. Mannings, A. P. Boss & S. S. Russell, Univ. Arizona Press, Tucson, p. 759 (2000).

Kono M. & Roberts P.H., Recent geodynamo simulations and observations of the geomagnetic field, *Rev. Geophys.*, **40**, 1013 (2002).

Kono M. & Tanaka H., Intensity of the geomagnetic field in geological time: A statistical study, in *The Earth's Central Part: Its Structure and Dynamics* (T. Yukutake, ed.), 75–94, Terrapub, Tokyo (1995).

Korpi M.J., Brandenburg A., Shukurov A., Tuominen I. & Nordlund Å., A supernova-regulated interstellar medium: simulations of the turbulent multiphase medium, *Astrophys. J., Lett.*, **514**, L99–L102 (1999a).

Korpi M.J., Brandenburg A., Shukurov A. & Tuominen I., Evolution of a superbubble in a turbulent, multi-phased and magnetized ISM, *Astron. Astrophys.*, **350**, 230–239 (1999b).

Kosovichev A.G. & Schou J., Detection of zonal shear flows beneath the Sun's surface from f-mode frequency splitting, *Astrophys. J.*, **482**, 207–210 (1997).

Kraichnan R.H., Consistency of the Alpha-Effect Turbulent Dynamo, *Phys. Rev. Lett.*, **42**, 1677–1680 (1979).

Krasheninnikova Yu.S., Ruzmaikin A.A., Sokoloff D.D. & Shukurov A., Configuration of large-scale magnetic fields in spiral galaxies, *Astron. Astrophys.*, **213**, 19–28 (1989).

Krause F. (Ed.) *The cosmic dynamo*, Dordrecht: Kluwer (1993).

Krause M., Beck R. & Hummel E., The magnetic field structures in two nearby spiral galaxies. I. The axisymmetric spiral magnetic field in IC342, *Astron. Astrophys.*, **217**, 1–17 (1989a).

Krause M., Beck R. & Hummel E., The magnetic field structures in two nearby spiral galaxies. II. The bisymmetric spiral magnetic field in M81, *Astron. Astrophys.*, **217**, 17–30 (1989b).

Krause F. & Rädler K.-H., *Mean field magnetohydrodynamics and dynamo theory*, Pergamon Press (New-York, 1980).

Kuang W. & Bloxham J., An Earth-like numerical dynamo model, Nature, **389**, 371–374 (1997).

Kuang W. & Bloxham J., Numerical modelling of magnetohydrodynamic convection in a rapidly rotating spherical shell: weak and strong field dynamo action, *J. Comp. Phys.* **153**, 51–81 (1999).

Küker M., Arlt R. & Rüdiger G., The Maunder minimum as due to Λ-quenching, *Astron. Astrophys.*, **343**, 977–982 (1999).

Kulsrud R.M., A critical review of galactic dynamos, *Ann. Rev. Astron. Astrophys.*, **37**, 37–64 (1999).

Kulsrud R.M., Cen R., Ostriker J.P. & Ryu D., The protogalactic origin for cosmic magnetic fields, *Astrophys. J.*, **480**, 481–491 (1997).

Kutzner C. & Christensen U.R., From stable dipolar towards reversing numerical dynamos, Phys. Earth Planet. Inter., **131**, 29–45 (2002).

Kuzanyan K.M., Bassom A.P. Soward A.M. & Sokoloff D.D., Non-axisymmetric $\alpha^2\Omega$–dynamo waves in thin stellar shells, *Geophys. Astrophys. Fluid Dyn.*, **99**, 309–336 (2005).

Kvasz L., Sokoloff D.D. & Shukurov A., A steady state of the disk dynamo, *Geophys. Astrophys. Fluid. Dyn.*, **65**, 231–244 (1992).

Labrosse S., Poirier J.-P. & Le Mouël J.-L., The age of the inner core, Earth Planet. Sci. Lett.,**190**, 111–123 (2001).

Lantz S.R. & Fan Y., Anelastic magnetohydrodynamic equations for modeling solar and stellar convection zones, *Astrophysical J. Suppl. Ser.*, **121**, 247–264 (1999).

Larmor J., How could a body such as the sun become a magnet? *Reports of the British Association for the Advancement of Science* (Rept. Brit. Assoc.), 159 (1919).

Lathrop D., Fineberg J., Swinney H., Transition to shear-driven turbulence in Couette-Taylor flow, *Phys. Rev.*, A **46**, 6390–6405 (1992).

Latour J., Spiegel E., Toomre J. & Zahn J.P., Stellar convection theory. I. The anelastic modal equations, *Astrophysical J.*, **207**, 233–243 (1976).

Lay T., Hernlund J., Garnero E.J. & Thorne M.S., A post-perovskite lens and D$''$ heat flux beneath the central Pacific, *Science*, **314**, 1272–1276 (2006).

Le Mouël J-L., Allègre C.J. & Narteau C., Multiple scale dynamo, *Proc. Nat. Acad. Sci. USA*, **94**, 5510–5514 (1997).

Léorat J., Pouquet A. & Frisch U., Fully developed MHD turbulence near critical magnetic Reynolds number, *J. Fluid Mech.*, **104**, 419–443 (1981).

Léorat J., Allemand P., Guermond J.-L. & Plunian F., Dynamo action, between numerical experiments and liquid sodium devices. in P. Chossat *et al.*Eds, Dynamo and dynamics, a mathematical challenge, Kluwer Academic Publisher, 25–34 (2001).

Lehnert B., Experiments on non-laminar flow of mercury in presence of a magnetic field, *Tellus*, **4**, 63–67 (1951).

Lehnert B., Magnetohydrodynamic waves under the action of the Coriolis force, *Astrophys. J.* **119**, 647–654 (1954).

Lehnert B., An experiment on axisymmetric flow of liquid sodium in a magnetic field, *Ark. Fys.*, **13**, 109–116 (1958).

Lehnert B. & Little N.C., Experiments on the effect of inhomogeneity and obliquity of a magnetic field in inhibiting convection, *Tellus*, **9**, 97–103 (1957).

Leighton R.B., A magneto-kinematic model of the solar cycle, *Astrophys. J.*, **156**, 1–20 (1969).

Lesch H. & Bender R., Magnetic fields in elliptical galaxies, *Astron. Astrophys.*, **233**, 417–421 (1990).

Lesieur M., Turbulence in fluids, *Kluwer* (1997).

Levy E.H., Planetary dynamos, *Earth, Moon Planet.*, **67**, 143–160 (1995).

Lockman F.J., The H I halo in the inner galaxy, *Astrophys. J.*, **283**, 90–97 (1984).

Lodders K. & Fegley B., *The Planetary Scientist's Companion,* Oxford University Press (1998).

Loinard L., Allen R.J., Lequeux J., An unbiased survey for CO emission in the inner disk of the Andromeda galaxy, *Astron. Astrophys.*, **301**, 68–74 (1995).

Loper D.E., General solution for the linearised Ekman-Hartmann layer on a spherical boundary, *Phys. Fluids*, **13**, 2995–2998 (1970).

Loper D.E., Some thermal consequences of a gravitationally powered dynamo. *J. Geophys. Res.*, **83**, 5961–5970 (1978).

Loper D.E., Chulliat A. & Shimizu H., Buoyancy-driven perturbations in a rapidly rotating, electrically conducting fluid. Part I. Flow and magnetic field, *Geophys. Astrophys. Fluid Dyn.*, **97**, 429–469 (2003).

Loper D.E. & Roberts P.H., Compositional convection and the gravitationally powered dynamo, in *Stellar and Planetary Magnetism* (A.M. Soward, ed.), Gordon & Breach, New York, 297–327 (1983).

Lorenzani S. & Tilgner A., Inertial instabilities of fluid flow in precessing spheroidal shells, *J. Fluid Mech.*, **492**, 363–379 (2003).

Lortz D., Exact solutions of the hydromagnetic dynamo problem, *Plasma Phys.*, **10**, 967–972 (1968).

Lou Y.-Q., Fan A., Coupled galactic density-wave modes in a composite system of thin stellar and gaseous discs, *Mon. Not. R. Astron. Soc.*, **297**, 84–100 (1998).

Lou Y.-Q., Fan A., Large-scale magnetohydrodynamic density-wave structures in the Andromeda nebula, *Mon. Not. R. Astron. Soc.*, **315**, 646–654 (2000).

Love J., Palaeomagnetic secular variation as a function of intensity, Phil. Trans. R. Soc. Lond., A **358**, 1191–1223 (2000a).

Love J., Statistical assessment of preferred transitional VGP longitudes based on palaeomagnetic lava data, Geophys. J. Int., **140**, 211-221 (2000b).

Lowes F.J. & Wilkinson I., Geomagnetic dynamo : a laboratory model, *Nature*, **198**, 1158–1160 (1963).

Lowes F.J. & Wilkinson I., Geomagnetic dynamo : an improved laboratory model, *Nature*, **219**, 717–718 (1968).

Lozinskaya T.A., *Supernovae and Stellar Winds in the Interstellar Medium,* A.I.P., New York (1992).

Maksymczuk J. & Gilbert A.D., Remarks on the equilibration of high conductivity dynamos, *Geophys. Astrophys. Fluid Dyn.*, **90**, 127–137 (1998).

Malkus W.V.R., *Boussinesq equations and convection energetics*, Woods Hole Oceanographic Institution, Geophysical Fluid Dynamics Proceedings (1964).

Malkus W.V.R., Precession of the Earth as cause of Geomagnetism, *Nature*, 259–264 (1968).

Malkus W.V.R., Reversing Bullard's dynamo, *Eos,* **53**, 617 (1972).

Malkus W.V.R., Energy sources for planetary dynamos, in Lectures on Solar and Planetary Dynamos (Proctor M.R.E. & Gilbert A.D. Eds.) Cambridge University Press, 161–179 (1994).

Malkus W.V.R. & Proctor M.R.E., The macrodynamics of α-effect dynamos in rotating fluids, J. Fluid Mech., **67**, 417–443 (1975).

Mangalam A. V. & Subramanian K., Dynamo generation of magnetic fields in accretion disks, *Astrophys. J.,* **434**, 509–517 (1994a).

Mangalam A. V. & Subramanian K., Effect of advected fields on accretion disk dynamos, *Astrophys. J., Suppl.Ser.,* **90**, 963–967 (1994b).

Marié L., Burguete J., Daviaud F. & Léorat J., Numerical study of homogeneous dynamo based on experimental von Kàrmàn type flows, *EPJ*, B **33**, 469–485 (2003).

Marié L., Pétrélis F., M. Bourgoin, J. Burguete, A. Chiffaudel, F. Daviaud, S. Fauve, P. Odier & J.-F. Pinton, Open questions about homogeneous fluid dynamos: The VKS experiment, *Magnetohydrodynamics*, **38**, 156–169 (2002).

Markiel J.A. & Thomas J.H., Solar interface dynamo models with a realistic rotation profile, *Astrophys. J.*, **523**, 827–837 (1999).

Mason J., Hughes D.W. & Tobias S.M., The competition in the solar dynamo between surface and deep-seated α–effects, *Astrophys. J.*, **580**, L89–L92 (2002).

Mathews W.G. & Brighenti F., Self-generated magnetic fields in galactic cooling flows, *Astrophys. J.,* **488**, 595–605 (1997).

Mathews W.G. & Brighenti F., Hot gas in and around elliptical galaxies, *Ann. Rev. Astron. Astrophys.,* **41**, 191–239 (2003).

Matsushima M., Expression of turbulent heat flux in the Earth's core in terms of a second moment closure model, *Phys. Earth Planet. Inter.*, **128**, 137–148 (2001).

Matsushima M., Scale similarity of MHD turbulence in the Earth's core, *Earth Planets Space*, **56**, 599–605 (2004).

Matsushima M., Nakajima T. & Roberts P.H., The anisotropy of local turbulence in the Earth's core, *Earth Planets Space*, **51**, 277–286 (1999).

Matthews P.C., Hughes D.W. & Proctor M.R.E., Magnetic buoyancy, vorticity, and three-dimensional flux-tube formation, *Astrophys. J.*, **448**, 938–941 (1995).

McFadden P.L. & Merrill R.T., Inhibition and geomagnetic field reversals, J. Geophys. Res., **98**, 6189–6199 (1993).

McFadden P.L. & Merrill R.T., Evolution of the geomagnetic reversal rate since 160Ma: Is the process continuous?, J. Geophys. Res., **105**, 28455–28460 (2000).

McKee C.F. & Ostriker J.P., A theory of the interstellar medium: three components regulated by supernova explosions in inhomogeneous substrate, *Astrophys. J.,* **217**, 148–169 (1977).

Melchior P.J., *T*he physics of the Earth's core, Pergamon, Oxford (1986).

Meneguzzi M., Frisch U. & Pouquet A., Helical and nonhelical turbulent dynamos, *Phys. Rev. Lett.*, **47**, 1060–1064 (1981).

Meneveau C. & Katz, J., Scale-invariance and turbulence models for large-eddy simulation, *Annu. Rev. Fluid Mech.*, **32**, 1–32 (2000).

Merrill R.T., McElhinny M.W. & McFadden P.L., *The Magnetic Field of the Earth*, Academic Press (1996).

Merrill R.T. & McFadden P.L., Geomagnetic polarity transitions, *Rev. Geophys.*, **37**, 201–226 (1999).

Merrill R.T. & McFadden P.L., The geomagnetic axial dipole field assumption, Phys. Earth Planet. Inter., **139**, 171–185 (2003).

Messadek K. & Moreau R., An experimental investigation of MHD quasi-two-dimensional turbulent shear flows, *J. Fluid Mech.*, **456**, 137–159 (2002).

Mestel L., *Stellar Magnetism*, Clarendon Press, Oxford (1999).

Mestel L. & Subramanian K., Galactic dynamos and density wave theory, *Mon. Not. R. Astron. Soc.,* **248**, 677–687 (1991).

Miesch M., *Turbulence and Convection in Stellar and Interstellar Environments*, PhD thesis, University of Colorado (1993).

Miner E.D., *Uranus*, Wiley and Sons (1998).

Moffatt H.K., The mean electromotive force generated by turbulence in the limit of perfect conductivity, J. Fluid Mech., **65**, 1–10 (1974).

Moffatt H.K., *Magnetic field generation in electrically conducting fluids,* Cambridge University Press (1978).

Moffatt H.K, A self-consistent treatment of simple dynamo systems, *Geophys. Astrophys. Fluid Dyn.*, **14**, 147–166 (1979).

Moffatt H.K., Transport effects associated with turbulence, with particular attention to the influence of helicity, *Rep. Prog. Phys.*, **46**, 621–664 (1983).

Moffatt H.K. & Proctor M.R.E., Topological constraints associated with fast dynamo action, *J. Fluid Mech.*, **154**, 493–507 (1985).

Monchaux R., Berhanu M., Bourgoin M., Moulin M., Odier Ph., Pinton J.-F., Volk R., Fauve S., Mordant N., Petrelis F., Chiffaudel A., Daviaud F., Dubrulle B., Gasquet C., Marie L. & Ravelet F., Generation of magnetic field by dynamo action in a turbulent flow of liquid sodium, *Phys. Rev. Lett.*, **98**, 044502 (2007).

Moore D.W., *Homogeneous fluids in rotation. Section A: Viscous effects,* in: Rotating Fluids in Geophysics (P.H. Roberts & A.M. Soward Eds.) 29–66, Academic Press, London (1978).

Moore D.W. & Saffman P.G., The structure of free vertical shear layers in a rotating fluid and the motion produced by a slowly rising body, *Phil. Trans.. R. Soc. Lond.*, A **264**, 597–634 (1969).

Moreau R., Magnetohydrodynamics, *Kluwer* (1990).

Moreau R., MHD Turbulence at the laboratory scale: Established ideas and new challenges, *Applied Scientific Research*, **58**, 131–147 (1998).

Moroz I.M., Synchronised dynamics in three coupled Faraday disk homopolar dynamos. In Fluid Dynamics and the Environment: Dynamical Approaches, Springer-Verlag 225–238 (2001a).

Moroz I.M., Behaviour of a self-exciting Faraday-disk homopolar dynamo with battery in the presence of an external magnetic field, *Int. J. Bif. Chaos*, **11**, 6, 1695–1705 (2001b).

Moroz I.M., On the behaviour of a self-exciting Faraday disk homopolar dynamo with a variable nonlinear series motor, *Int. J. Bif. Chaos*, **12**, 10, 2123–35 (2002).

Moroz I.M., The Malkus-Robbins dynamo with a linear series motor, *Int. J. Bif. Chaos*, **13**, 1, 147–161 (2003).

Moroz I.M., The Malkus-Robbins dynamo with a nonlinear series motor, *Int. J. Bif. Chaos*, **14**, 8, 2885–92 (2004a).

Moroz I.M., The extended Malkus-Robbins dynamo as a perturbed Lorenz equation, Special Issue of the International Journal of Nonlinear Dynamics on Reduced Order Models: Methods and Applications (2004b).

Moroz I.M. & Hide R., Effects due to azimuthal eddy currents in the Faraday disk self-exciting homopolar dynamo with series motor: II The general case, *Int. J. Bif. Chaos*, **10**, 2701–16 (2000).

Moroz I.M., Hide R. & Smith L.A., Synchronised chaos in coupled double disk homopolar dynamos, *Int. J. Bifur. Chaos*, **8**, 2125–33 (1998).

Moroz I.M., Hide R. & Soward A.M., On self-exciting coupled Faraday disk homopolar dynamos driving series motors, *Physica*, D **117**, 128–144 (1998).

Morrison G. & Fearn D.R., The infuence of Rayleigh number, azimuthal wavenumber and inner core radius on $2\frac{1}{2}$D hydromagnetic dynamos, Phys. Earth Planet. Inter., **112**, 237–258 (2000).

Moss D., On the generation of bisymmetric magnetic-field structures in spiral galaxies by tidal interactions, *Mon. Not. R. Astron. Soc.*, **275**, 191–194 (1995).

Moss D., Parametric resonance and bisymmetric dynamo solutions in spiral galaxies, *Astron. Astrophys.,* **308**, 381–386 (1996).

Moss D., The relation between magnetic and gas arms in spiral galaxies, *Mon. Not. R. Astron. Soc.,* **297**, 860–866 (1998).

Moss D. & Shukurov A., Turbulence and magnetic fields in elliptical galaxies, *Mon. Not. R. Astron. Soc.,* **279**, 229–239 (1996).

Moss D. & Shukurov A., Galactic dynamos with captured magnetic flux and an accretion flow, *Astron. Astrophys.,* **372**, 1048–1063 (2001) [astro-ph/0012436].

Moss D. & Shukurov A., Accretion disc dynamos opened up by external magnetic fields, *Astron. Astrophys.,* **413**, 403–414 (2004).

Moss D., Shukurov A. & Sokoloff D., Galactic dynamos driven by magnetic buoyancy, *Astron. Astrophys.,* **343**, 120–131 (1999).

Moss D., Shukurov A. & Sokoloff D., Accretion and galactic dynamos, *Astron. Astrophys.,* **358**, 1142–1150 (2000).

Moss D., Shukurov A., Sokoloff D., Beck R. & Fletcher A., Magnetic fields in barred galaxies. II. Dynamo models, *Astron. Astrophys.,* **380**, 55–71 (2001) [astro-ph/0107214].

Müller U. & Bühler L., *Magnetofluiddynamics in Channels and Containers,* Springer (2001).

Müller U. & Stieglitz R., Can the Earth's magnetic field field be simulated in the laboratory? *Naturwissenschaften*, **87**, 381–390 (2000).

Müller U., Stieglitz R. & Horanyi S., A two-scale hydomagnetic dynamo experiment, J. Fluid Mech., **498**, 31-71 (2004).

Nataf H.-C., Dynamo and convection experiments, *Earth's core and lower mantle* (C.A. Jones, A.M. Soward & K. Zhang Eds.), Gordon & Breach (2003).

Nellis W.J., Metallization of fluid hydrogen at 140 GPa (1.4Mbar): implications for Jupiter, *Planet. Space Sci.*, **48**, 671–677 (2000).

Nellis W.J., Holmes N.C., Mitchel A.C. *et al.*, Equation of state and electrical conductivity of 'synthetic Uranus' a mixture of water ammonia and isopropanol, at shock pressure up to 200 GPa (2 Mbar), *J. Chem. Phys.*, **107**, 9096–9100, (1997).

Niklas S., *Eigenschaften von Spiralgalaxien in hochfrequenten Radiokontinuum,* PhD Thesis, Max-Planck-Institut für Radioastronomie, Bonn (1995).

Nimmo F., Why does Venus lack a magnetic field? *Geology*, **30**, 987–990 (2002).

Noir J., Cardin P., Jault D. & Masson J.P., Experimental evidence of nonlinear resonance effects between retrograde precession and the tilt-over mode within a spheroid, *Geophys. J. Int.*, **154**, 407–416 (2003).

Nordlund Å. & Rögnvaldsson Ö., Magnetic fields in young galaxies, *Highlights Astron.,* **12**, 706–708 (2002).

Novikov V.G., Ruzmaikin A.A. & Sokolov D.D., Kinematic dynamo in a reflection-invariant random field, *Sov. Phys. JETP*, **58**, 527–532 (1983).

Noyes R.W., Hartmann L.W., Baliunas S.L., Duncan D.K. & Vaughan A.H., Rotation, convection and magnetic activity in lower main-sequence stars, *Astrophys. J.*, **279**, 763–777 (1984).

Nunez A., Pétrélis F. & Fauve S., Saturation of a Ponomarenko type fluid dynamo, in *Dynamo and dynamics, a mathematical challenge*, P. Chossat *et al.*Eds., 67–74, Kluwer Academic Publishers (2001).

Odier P., Pinton J.-F. & Fauve S., Advection of a magnetic field by a turbulent swirling flow, *Phys. Rev.*, E **58**, 7397–7401 (1998).

Odier P., Pinton J.-F. & Fauve S., Magnetic induction by coherent vortex Motion *Eur. Phys. J.*, B **16**, 373–378 (2000).

Ogura Y. & Phillips N., Scale Analysis of Deep and Shallow Convection in the Atmosphere, *J. Atmos. Sci.*, **19**, 173–179 (1962).

Olsen N., Lühr H., Sabaka T., Mandea M., Rother M., Toffner L. & Choi S., CHAOS–a model of the Earth's magnetic field derived from CHAMP, Ørsted, and SAC-C magnetic satellite data, *Geophys. J. Inter.*, **166**-1, 67–75 (2006).

Olson P., Thermal interaction of the core and mantle, in *Earth's core and lower mantle* (C.A. Jones, A.M. Soward & K. Zhang Eds.), Taylor & Francis, London, 1–38 (2003).

Olson P. & Christensen U.R., The time-averaged magnetic field in numerical dynamos with non-uniform boundary heat flow, Geophys. J. Int., **151**, 809–823 (2002).

Olson P. Christensen U.R. & Glatzmaier G.A., Numerical modelling of the geodynamo: mechanism of field generation, *J. Geophys. Res.*, **104**, 10383–10404 (1999).

Or A.C. & Busse F.H., Convection in a rotating cylindrical annulus. Part 2. Transitions to asymmetric and vacillating flow, *J. Fluid Mech.*, **174**, 313–326 (1987).

Orszag S.A., Analytical theories of turbulence, J. Fluid Mech., **41**, 363–386 (1970).

Orr W.McF., The stability or instability of the steady motions of a fluid, *Proc. R. Irish Acad.*, A **27**, 69–138 (1907).

Ossendrijver M.A.J.H., On the cycle periods of stellar dynamos, *Astron. Astrophys.*, **323**, 151–157 (1997).

Ossendrijver M.A.J.H., Grand-minima in a buoyancy-driven solar dynamo, *Astron. Astrophys.*, **359**, 364–372 (2000a).

Ossendrijver M.A.J.H., The dynamo effect of magnetic flux tubes, *Astron. Astrophys.*, **359**, 1205–1210 (2000b).

Ossendrijver M.A.J.H., The solar dynamo, *Astron. Astrophys. Rev.*, **11**, 287–367 (2003).

Ossendrijver M.A.J.H. & Hoyng P., Stochastic and nonlinear fluctuations in a mean field dynamo, *Astron. Astrophys.*, **313**, 959–970 (1996).

Otani N.F., A fast kinematic dynamo in two-dimensional time-dependent flows, *J. Fluid Mech.*, **253**, 327–340 (1993).

Ott E. & Sommerer J., Blowout bifurcations: the occurrence of riddled basins and on-off intermittency, *Phys. Lett.*, A **188**, 39–47 (1994).

Pais A. & Hulot G., Length of day decade variations, torsional oscillations and inner core superrotation: evidence from recovered core surface zonal flows, Phys. Earth Planet. Inter., **118**, 291–316 (2000).

Parker E.N., Hydromagnetic dynamo models, *Astrophys. J.*, **122**, 293–314 (1955).

Parker E.N., The generation of magnetic fields in astrophysical bodies. II. The galactic field, *Astrophys. J.,* **163**, 252–278 (1971).

Parker E.N., *Cosmical Magnetic Fields, their Origin and their Activity*, Clarendon Press, Oxford (1979).

Parker E.N., A solar dynamo surface-wave at the interface between convection and non-uniform rotation, *Astrophys. J.*, **408**, 707–719 (1993).

Pavlov V. & Gallet Y., A third superchron during Early Paleozoic, *Episodes*, **28**-2, 1–7 (2005).

Pedlosky J., *Geophysical Fluid Dynamics,* Springer (1979).

Peffley N., Cawthrone A. & Lathrop D., Toward a self-generating magnetic dynamo: the role of turbulence, *Phys. Rev.*, E **61**, 5287–5294 (2000a).

Peffley N., Goumilevski A., Cawthrone A. & Lathrop D., Characterization of experimental dynamos, *Geophys. J. Int.*, **142**, 52–58 (2000b).

Pelletier G. & Pudritz R.E., Hydromagnetic disk winds in young stellar objects and active galactic nuclei, *Astrophys. J.,* **394**, 117–138 (1992).

Pétrélis F., Bourgoin M., Marié L., Burguete J., Chiffaudel A., Daviaud F., Fauve S., Odier P. & Pinton J.-F., Nonlinear magnetic induction by helical motion in a liquid sodium turbulent flow, *Phys. Rev. Lett.*, **90**, 174501 (2003).

Pétrélis F. & Fauve S., Saturation of the magnetic field above the dynamo threshold, *Eur. Phys. J.*, B **22**, 273–276 (2001).

Phillips A., A comparison of the asymptotic and no-z approximations for galactic dynamos, *Geophys. Astrophys. Fluid. Dyn.,* **94**, 135–150 (2001).

Phillips A., Brooke J. & Moss D., The importance of physical structure in solar dynamo models, *Astron. Astrophys.*, **392**, 713–727 (2002).

Phillips C.G. & Ivers D.J., Strong field anisotropic diffusion models for the Earth's core, *Phys. Earth Planet. Inter.*, **140**, 13–28 (2003).

Phillips O.M., Energy transfer in rotating fluids by reflexion of inertial waves, *Phys. Fluids* **6**, 513–520 (1963).

Pino D., Mercader I. & Net M., Thermal and inertial modes of convection in a rapidly rotating annulus, *Phys. Rev.*, E **61**, 1507–1517 (2000).

Pino D., Net M., Sanchez J. & Mercader I., Thermal Rossby waves in a rotating annulus, *Phys. Rev.*, E **63**, 056312 (2001).

Pipin V.V., The Gleissberg cycle by a nonlinear $\alpha\Lambda$ dynamo, *Astron. Astrophys.*, **346**, 295–302 (1999).

Platt N., Spiegel E. & Tresser C., On-off intermittency: a mechanism for bursting, *Phys. Rev. Lett.*, **70**, 279–282 (1993a).

Platt N., Spiegel E. & Tresser C., The intermittent solar cycle, *Geophys. Astrophys. Fluid Dyn.*, **73**, 146–151 (1993b).

Plunian F., Marty P. & Alemany A., Chaotic behaviour of the Rikitake dynamo with symmetrical mechanical friction and azimuthal currents, *Proc. Roy Soc. Lond.*, A **454,** 1835–1842 (1998).

Plunian F., Marty P. & Alemany A., Kinematic dynamo action in a network of screw motions; application to the core of a fast breeder reactor, *J. Fluid Mech.*, **382**, 137–154 (1999).

Poezd A., Shukurov A. & Sokoloff D., Global magnetic patterns in the Milky Way and the Andromeda nebula, *Mon. Not. R. Astron. Soc.,* **264**, 285–297 (1993).

Poirier J.-P., Transport properties of liquid metals and viscosity in the Earth's core, *Geophys. J.*, **92**, 99–105 (1988).

Poirier J.-P., *Introduction to the Physics of the Earth's Interior*, Cambridge University Press, Cambridge, UK (2000).

Ponomarenko Yu.B., On the theory of hydromagnetic dynamo, *J. Appl. Mech. Tech. Phys.* **14**, 775–778 [*Zh. Prikl. Mekh. & Tekh. Fiz. (USSR)*, **6**, 47–51] (1973).

Ponty Y., Gilbert A.D. & Soward A.M., Kinematic dynamo action in large magnetic Reynolds number flows driven by shear and convection, *J. Fluid Mech.*, **435**, 261–287 (2001).

Ponty Y., Mininni P.D., Montgomery D.C., Pinton J.-F., Politano H. & Pouquet A., Numerical Study of Dynamo Action at Low Magnetic Prandtl Numbers, *Phys. Rev. Lett.*, **94**, 164502 (2005).

Pouquet A., Frisch U. & Léorat J., Strong MHD helical turbulence and the nonlinear dynamo effect, J. Fluid Mech., **77**, 321–354 (1976).

Priklonsky V., Shukurov A., Sokoloff D. & Soward A.M., Non-local effects in the mean-field disc dynamo. I. An asymptotic expansion, *Geophys. Astrophys. Fluid. Dyn.,* **93**, 97–114 (2000) [astro-ph/0309666].

Proctor M.R.E., On Backus' necessary condition for dynamo action in a conducting sphere, *Geophys. Astrophys. Fluid Dyn.*, **9**, 89–93 (1977a).

Proctor M.R.E., The role of mean cirdulation in parity selection by planetary magnetic fields, *Geophys. Astrophys. Fluid Dyn.*, **8**, 311–324 (1977b).

Proctor M.R.E., Numerical solutions of the nonlinear α-effect dynamo equations, J. Fluid Mech., **80**, 769–784 (1977c).

Proctor M.R.E., Necessary conditions for the magnetohydrodynamic dynamo, *Geophys. Astrophys. Fluid Dyn.*, **14**, 127–146 (1979).

Proctor M.R.E., Convection and magnetoconvection in a rapidly rotating sphere, in *Lectures on Solar and Planetary Dynamos* (M.R.E. Proctor & A.D. Gilbert Eds.) Cambridge University Press, 1–58 (1994).

Proctor M.R.E., Dynamo processes: the interaction of turbulence and magnetic fields, in *Stellar Astrophysical Fluid Dynamics* (M.J. Thompson & J. Christensen-Dalsgaard Eds.), Cambridge University Press (2003).

Proctor M.R.E., An extension of the Toroidal Theorem, *Geophys. Astrophys. Fluid Dyn.*, **98**, 235–240 (2004).

Proctor M.R.E. & Gilbert A.D. (Eds.), *Lectures on solar and planetary dynamos*, Cambridge University Press (1994).

Proudman I., The almost rigid rotation of a viscous fluid between concentric spheres, *J. Fluid Mech.*, **1**, 505–516 (1956).

Pudritz R.E., Dynamo action in turbulent accretion discs around black holes. II. The mean magnetic field. *Mon. Not. R. Astron. Soc.,* **195**, 897–914 (1981b)

Pudritz R.E., Ouyed R., Fendt C. & Brandenburg A., Disk winds, jets, and outflows: theoretical and computational foundations. In *Protostars and Planets V,* Reipurth B., Jewitt D. & Keil K. (Eds.), Univ. Arizona Press, Tucson (2006) [astro-ph/0603592].

Purucker M., Ravat D., Frey H., Voorhies C., Sabaka T. & Acuna M., An altitude-normalized magnetic map of Mars and its interpretation, *Geophys. Res. Lett.*, **27**, 2449–2452 (2000).

Rädler K.-H., On the effect of differential rotation on axisymmetric and non-axisymmetric magnetic fields of cosmical bodies. In *Plasma Astrophysics*, ESA Publ. SP-251, 569–574 (1986).

Rädler K.-H., Apstein E., Rheinhardt M. & Schüler M., The Karlsruhe dynamo experiment, a mean field approach, *Stud. Geophys. Geod.*, **42**, 224–231 (1998).

Rädler K.-H., Rheinhardt M., Apstein E. & Fuchs H., On the mean field theory of the Karlsruhe dynamo experiment I. Kinematic theory, *Magnetohydrodynamics*, **38**, 41–71 (2002a).

Rädler K.-H., Rheinhardt M., Apstein E. & Fuchs H., On the mean field theory of the Karlsruhe dynamo experiment II. Back-reaction of the magnetic field on the fluid flow, *Magnetohydrodynamics*, **38**, 73–94 (2002b).

Rädler K.-H. & Rheinhardt M., Can a disc dynamo work in the laboratory ? *Magnetohydrodynamics*, **38**-1-2, 211-217 (2002).

Reighard A. & Brown M., Turbulent conductivity measurements in a spherical liquid sodium flow, *Phys. Rev. Lett.*, **86**, 2794–2797 (2001).

von Rekowski B., Brandenburg A., Dobler W. & Shukurov A., Structured outflow from a dynamo active accretion disc, *Astron. Astrophys.,* **398**, 825–844 (2003).

Reyes-Ruiz M., An unconventional accretion disc dynamo, *Mon. Not. R. Astron. Soc.,* **319**, 1039–1046 (2000).

Reyes-Ruiz M. & Stepinski T.F., Accretion disc dynamos in the presence of a weak external magnetic field, *Mon. Not. R. Astron. Soc.,* **285**, 501–510 (1997).

Reyl C., Antonsen T.M. & Ott E., Vorticity generation by instabilities of chaotic fluid flows, *Physica*, D **111**, 202–226 (1998).

Ribes J.C. & Nesme-Ribes E., The solar sunspot cycle in the Maunder minimum AD 1645 – AD 1715, *Astron. Astrophys.*, **276**, 549–563 (1993).

Rice J.B., Doppler imaging of stellar surfaces – techniques and issues, *Astron. Nachr.*, **323**, 220–235 (2002).

Rieutord M. & Valdettaro L., Inertial waves in a rotating spherical shell, *J. Fluid Mech.*, **341**, 77–99 (1997).

Rikitake T., Oscillations of a system of disk dynamos, *Proc. Camb. Phil. Soc.*, **54**, 89–105 (1958).

Robbins K.A., A new approach to sub-critical instability and turbulent transitions in a simple dynamo, *Proc Camb. Phil. Soc.*, **82**, 309–325 (1977).

Roberts G.O., Spatially periodic dynamos, *Phil. Trans. R. Soc. Lond.*, A **266**, 535–558 (1970).

Roberts, G.O., Dynamo action of fluid motions with two-dimensional periodicity. *Phil. Trans. R. Soc. Lond.*, A **271**, 411–454 (1972).

Roberts P.H., On the thermal instability of a highly rotating fluid sphere, *Astrophys. J.* **141**, 240–250 (1965).

Roberts, P.H., *An Introduction to Magnetohydrodynamics,* Longmans, London (1967a).

Roberts, P.H., Singularities of Hartmann layers, *Proc. R. Soc. Lond.*, A **300**, 94–107 (1967b).

Roberts P.H., On the thermal instability of a rotating-fluid sphere containing heat sources, *Phil. Trans. R. Soc. Lond.*, A **263**, 93–117 (1968).

Roberts P.H., Kinematic Dynamo Models, *Phil. Trans. R. Soc. Lond.*, A **272**, 663–703 (1972).

Roberts P.H., *Geophys. Astrophys. Fluid Dyn.*, **44**, 3–31 (1988).

Roberts P.H., From Taylor state to model-Z ?, Geophys. Astrophys. Fluid. Dyn., **49**, 143–160 (1989).

Roberts P.H., Fundamentals of dynamo theory, in *Lectures on Solar and Planetary Dynamos* (M.R.E. Proctor & A.D. Gilbert Eds.) Cambridge University Press, 1–58 (1994).

Roberts P.H. & Glatzmaier G.A., Geodynamo theory and simulations, *Rev. Mod. Phys.*, **72**, 1081–1123 (2000).

Roberts P.H., Jones C.A. & Calderwood A.R., Energy fluxes and ohmic dissipation in the Earth's core, in *Earth's core and lower mantle* (C.A. Jones, A.M. Soward & K. Zhang Eds.), Taylor & Francis, London, 100–129 (2003).

Roberts P.H. & Soward A.M., Magnetohydrodynamics of the Earth's core, *Ann. Rev. Fluid Mech.*, **4**, 117–154 (1972).

Roberts P.H. & Soward A.M., *Rotating Fluids in Geophysics*, Academic Press, London (1978).

Roberts P.H. & Soward A.M., Dynamo Theory, Annu. Rev. Fluid Mech., **24**, 459–512 (1992).

Roberts P.H., Yu Z.J. & Russell C.T., On the 60–Year Signal from the Core, *Geophys. Astrophys. Fluid Dyn.*, **101** , in press (2007).

Rogachevskii I. & Kleeorin N., Intermittency and anomalous scaling for magnetic fluctuations, *Phys. Rev.*, E **56**, 417–425 (1997).

Rohde R., Beck R. & Elstner D., Magnetic arms in NGC 6946 generated by a turbulent dynamo, *Astron. Astrophys.,* **350**, 423–433 (1999).

Rosner R., Magnetic fields of stars: using stars as tools for understanding the origins of cosmic magnetic fields, *Phil. Trans. R. Soc. Lond.*, A **358**, 689–709 (2000).

Rotvig J. & Jones C.A., Rotating convection-driven dynamos at low Ekman number, *Phys. Rev.*, E **66**, 056308:1-15 (2002).

Rotvig J. & Jones C.A., Multiple jets and bursting in the rapidly convecting two dimensional annulus model with nearly plane-parallel boundaries, *J. Fluid Mech.*, **567**, 117-140 (2007).

Rüdiger G., *Differential Rotation and Stellar Convection. Sun and Solar Type Stars*, Gordon & Breach, New York (1989).

Rüdiger G. & Arlt R., Physics of the solar cycle, in *Advances in Nonlinear Dynamos*, A. Ferriz-Mas & M. Núñez Eds, Taylor & Francis, London, 147–194 (2003).

Rüdiger G. & Kitchatinov L.L., The turbulent stresses in the theory of the solar torsional oscillation, *Astron. Astrophys.*, **236**, 503–508 (1990).

Rüdiger G., von Rekowski B., Donahue R.A. & Baliunas S.L., Differential rotation and meridional flow for fast-rotating solar-type stars, *Astrophys. J.*, **494**, 691–699 (1998).

Russell C.T., Magnetic fields of the terrestrial planets, *J. Geophys. Res – Planet*, **98**, 18681–18695 (1993).

Russell C.T., Yu Z.J. & Kivelson M.G., The rotation period of Jupiter, *Geophys. Res. Lett.*, **28**, 1911–1912 (2001).

Ruzmaikin A.A., The Martian dynamo, *Phys. Earth Planet. Inter.*, **67**, 268–274 (1991).

Ruzmaikin A.A. & Feynman J., Strength and phase of the solar dynamo during the last 12 cycles, *J. Geophys. Res.*, **106**, 15783–15789 (2001).

Ruzmaikin A.A. & Shukurov A.M., Magnetic field generation in the Galactic disk, *Sov. Astron.*, **25**, 553–558 (1981).

Ruzmaikin A.A., Shukurov A.M. & Sokoloff D.D., *Magnetic Fields of Galaxies,* Kluwer, Dordrecht (1988).

Ruzmaikin A.A., Shukurov A.M., Sokoloff D.D. & Starchenko S.V., Maximally–Efficient–Generation Approach in the dynamo theory, *Geophys. Astrophys. Fluid Dyn.*, **52**, 125–139 (1990).

Ruzmaikin A.A., Sokoloff D.D. & Shukurov A.M., Magnetic field distribution in spiral galaxies, *Astron. Astrophys.,* **148**, 335–343 (1985).

Ruzmaikin A.A., Sokoloff D.D. & Shukurov A.M., Magnetic fields in spiral galaxies, in *Plasma Astrophysics*, ESA Publ. SP-251, 539–544 (1986).

Ruzmaikin A.A., Sokoloff D.D. & Shukurov A.M., A hydromagnetic screw dynamo, *J. Fluid Mech.*, **197**, 39–56 (1988).

Ruzmaikin A.A. & Starchenko S.V., On the origin of Uranus and Neptune magnetic fields, *Icarus*, **93**, 82–87 (1991).

Saar S.H & Brandenburg A., Time evolution of the magnetic activity period. II. Results for an expanded stellar sample, *Astrophys. J.*, **524**, 295–310 (1999).

Sakuraba A. & Kono M., Effect of the inner core on the numerical solution of the magnetohydrodynamic dynamo, Phys. Earth Planet. Inter., **111**, 105–121 (1999).

Sarson G.R., Reversal models from dynamo calculations, Phil. Trans. R. Soc. Lond., A **358**, 921–942 (2000).

Sarson G.R. & Jones C.A., A convection driven geodynamo reversal model, Phys. Earth Planet. Inter., **111**, 3–20 (1999).

Sarson G.R., Jones C.A. & Longbottom A.W., The influence of boundary region heterogeneities on the geodynamo, Phys. Earth Planet. Inter., **101**, 13–32 (1997).

Sarson G.R., Jones C.A. & Longbottom A.W., Convection driven geodynamo models of varying Ekman number, Geophys. Astrophys. Fluid. Dyn., **88**, 225–259 (1998).

Sarson G.R., Jones C.A. & Zhang K., Dynamo action in a uniform ambient field, *Phys. Earth Planet. Inter.*, **111**, 47–68 (1999).

Sarson G.R., Jones C.A., Zhang K. & Schubert G., Magnetoconvection dynamos and the magnetic fields of Io and Ganymede, *Science,* **276**, 1106–1108 (1997).

Sawa T. & Fujimoto M., Bisymmetric spiral configuration of magnetic fields in spiral galaxies. I. Local theory, *Publ. Astron. Soc. Japan,* **38**, 551–566 (1986).

Schaeffer N. & Cardin P., Quasi-geostrophic model of the instabilities of the Stewartson layer in flat and depth varying containers, *Physics of Fluids*, **17**, 104111 (2005).

Schaeffer N. & Cardin P., Quasi-geostrophic kinematic dynamos at low magnetic Prandtl number, *Earth Planet. Sci. Lett.*, **245**, 595–604 (2006).

Schekochihin A., Cowley S., Hamnett G.W., Maron J.L. & McWilliams J.C., A model of nonlinear evolution and saturation of the turbulent MHD dynamo, *New J. Phys.,* **4**, 84.1–84.22 (2002) [astro-ph/0207503].

Schekochihin A.A., Cowley S.C., Maron J.L. & McWilliams J.C., Self-Similar Turbulent Dynamo, *Phys. Rev. Lett.*, **92**, 054502 (2004) .

Schmalz S. & Stix M., An $\alpha\Omega$ dynamo with order and chaos, *Astron. Astrophys.*, **245**, 654–661 (1991).

Schmitt D., An $\alpha\omega$–dynamo with an α–effect due to magnetostrophic waves, *Astron. Astrophys.*, **174**, 281–287 (1987).

Schmitt D., Schüssler M. & Ferriz-Mas A., Intermittent solar activity by an on-off dynamo, *Astron. Astrophys.*, **311**, L1–L4a (1996).

Schnaubelt M. & Busse F.H., Convection in a rotating cylindrical annulus. Part 3. Vacillating and spatially modulated flow, *J. Fluid Mech.*, **245**, 155–173 (1992).

Schrijver C.J. & Title A.M., On the formation of polar spots in sun-like stars, *Astrophys. J.*, **551**, 1099–1106 (2001).

Schrijver C.J. & Zwaan C., *Solar and Stellar Magnetic Activity*, Cambridge University Press, Cambridge (2000).

Schubert G., Turcotte D.L. & Olson, P. *Mantle Convection in the Earth and Planets*. Cambridge University Press (2001).

Schubert G. & Zhang K., Effects of an electrically conducting inner core on planetary and stellar dynamos, Astrophys. J., **557**, 930–942 (2001).

Schüssler M., The solar torsional oscillation and dynamo models of the solar cycle, *Astron. Astrophys.*, **94**, 755–756 (1981).

Schüssler M., Caligari P., Ferriz-Mas A., Solanki S.K. & Stix M., Distribution of starspots on cool stars. I. Young and main sequence stars of $1M_\odot$, *Astron. Astrophys.*, **314**, 503–512 (1996).

Segalovitz A., Shane W.W., de Bruyn A.G., Polarisation detection at radio wavelengths in three spiral galaxies, *Nature,* **264**, 222–226 (1976).

Shakura N.I. & Sunyaev R.A., Black holes in binary systems. Observational appearance, *Astron. Astrophys.,* **24**, 337–355 (1973).

Shapiro P.R. & Field G.B., Consequences of a new hot component of the interstellar medium, *Astrophys. J.,* **205**, 762–765 (1976).

Shew W.L., Sisan D.R. & Lathrop D., Hunting for dynamos: Eight different liquid sodium flows. in P. Chossat *et al.*Eds., Dynamo and dynamics, a Mathematical challenge, Kluwer Academic Publisher, 83–92 (2001).

Shew W.L., Sisan D.R. & Lathrop D., Mechanically forced and thermally driven flows in liquid sodium, *Magnetohydrodynalics*, **38**, 121–127 (2002).

Showman A.P., Stevenson D.J. & Malhotra R., Coupled orbital and thermal evolution of Ganymede, *Icarus*, **129**, 367–383 (1997).

Shu F., *The Physics of Astrophysics*, University Science Books (1992).

Shukurov A., Magnetic spiral arms in galaxies, *Mon. Not. R. Astron. Soc.,* **299**, L21–L24 (1998).

Shukurov A. & Berkhuijsen E.M., Faraday ghosts: depolarization canals in the Galactic radio emission, *Mon. Not. R. Astron. Soc.,* **342**, 496–500 (2003). Erratum: *Mon. Not. R. Astron. Soc.,* **345**, 1392 (2003) [astro-ph/0303087].

Shukurov A. & Sokoloff D., Galactic spiral arms and dynamo control parameters, *Studia Geoph. et Geod.,* **42**, 391–396 (1998).

Shukurov A., Sarson G.R., Nordlund Å., Gudiksen B. & Brandenburg A., The effects of spiral arms on the multi-phase ISM, Astrophys.Space Sci., **289**, 319–322 (2004) [astro-ph/0212260].

Simard-Normandin M. & Kronberg P.P., Rotation measures and the galactic magnetic field, *Astrophys. J.,* **242**, 74–94 (1980).

Simitev R. & Busse F.H., Parameter dependences of convection driven spherical dynamos, in *High performance computing in science and engineering* (E. Krause & W. Jager Eds.), Springer-Verlag, Heidelberg (2002).

Simitev R. & Busse F.H., Patterns of convection in rotating spherical shells, *New J. Phys.* **5**, 1.1–1.20 (2003a).

Simitev R. & Busse F.H., Low Prandtl number convection in rotating spherical fluid shells and its dynamo states, in *High performance computing in science and engineering* (E. Krause & W. Jager Eds.), Springer-Verlag, Heidelberg (2003b).

Simitev R. & Busse F.H., Prandtl number dependence of convection driven dynamos in rotating spherical fluid shells, *J. Fluid Mech.*, **532**, 365–388 (2005).

Simon P.A & Legrand J.-P., Some solar cycle phenomena related to the geomagnetic activity from 1868 to 1980 II. High velocity wind streams and cyclical behaviour of poloidal field, *Astron. Astrophys.*, **155**, 227–236 (1986).

Sisan D., Shew W. & Lathrop D., Lorentz force effects in magneto-turbulence, *Phys. Earth Planet. Inter.*, **135**, 137–159 (2003).

Siso-Nadal F. & Davidson P.A., Anisotropic evolution of small isolated vortices within the core of the Earth, *Phys. Fluids*, **16**, 1242–1254 (2004).

Skinner P.H. & Soward A.M., Convection in a rotating magnetic system and Taylor's constraint, Geophys. Astrophys. Fluid. Dyn., **44**, 91–116 (1988).

Skinner P.H. & Soward A.M., Convection in a rotating magnetic system and Taylor's constraint II. Geophys. Astrophys. Fluid. Dyn., **60**, 335–356 (1990).

Smagorinsky J., General circulation with the primitive equations. I. The basic experiment, *Mon. Weather Rev.*, **91**, 99 (1983).

Soderblom D.R., A survey of chromospheric emission and rotation among solar-type stars in the solar neighborhood, *Astron. J.*, **90**, 2103–2115 (1985).

Soderblom D.R., Jones B.F. & Fischer D., Rotational studies of late-type stars. VII. M34 (NGC 1039) and the evolution of angular momentum and activity in young solar-type stars, *Astrophys. J.*, **563**, 334–340 (2001).

Sofue Y., Fujimoto M., A bisymmetric spiral magnetic field and the spiral arms in our Galaxy, *Astrophys. J.*, **265**, 722–729 (1983).

Sofue Y. & Rubin V., Rotation curves of spiral galaxies, *Ann. Rev. Astron. Astrophys.*, **39**, 137–174 (2001).

Sofue Y., Fujimoto M. & Wielebinski R., Global structure of magnetic fields in spiral galaxies, *Ann. Rev. Astron. Astrophys.*, **24**, 459–497 (1986).

Sokoloff D.D., Bykov A.A., Shukurov A., Berkhuijsen E.M., Beck R. & Poezd A.D., Depolarisation and Faraday effects in galaxies and other extended radio sources, *Mon. Not. R. Astron. Soc.*, **299**, 189–206 (1998); Erratum: *Mon. Not. R. Astron. Soc.*, **303**, 207–208 (1999).

Sokoloff D.D. & Shukurov A., Regular magnetic fields in coronae of spiral galaxies, *Nature*, **347**, 51–53 (1990).

Soward A.M., A convection-driven dynamo, *Phil. Trans. R. Soc. Lond.*, A **275**, 611–646 (1974).

Soward A.M., On the finite amplitude thermal instability in a rapidly rotating fluid sphere, *Geophys. Astrophys. Fluid Dyn.*, **9**, 19–74 (1977).

Soward A.M., A thin disc model of the Galactic dynamo, *Astron. Nachr.*, **299**, 25–33 (1978).

Soward A.M., Non-linear marginal convection in a rotating magnetic system, Geophys. Astrophys. Fluid. Dyn., **35** 329–371 (1986).

Soward A.M., Fast dynamo action in a steady flow, *J. Fluid Mech.*, **180**, 267–295 (1987).

Soward A.M., Thin disc $\alpha\omega$-dynamo models. I. Long length scale modes, *Geophys. Astrophys. Fluid. Dyn.*, **64**, 163–199 (1992a).

Soward A.M., Thin disc $\alpha\omega$-dynamo models. II. Short length scale modes, *Geophys. Astrophys. Fluid. Dyn.*, **64**, 201–225 (1992b).

Soward A.M., The Earth's dynamo, *Geophys. Astrophys. Fluid Dyn.*, **62**, 219–238 (1992c).

Soward A.M., An asymptotic solution of a fast dynamo in a two-dimensional pulsed flow, *Geophys. Astrophys. Fluid Dyn.*, **73**, 179–215 (1993).

Soward A.M., Fast dynamos, in *Lectures on Solar and planetary dynamos*, M.R.E. Proctor, A.D. Gilbert Eds., 181–217. Cambridge University Press (1994).

Soward A.M., Thin aspect ratio $\alpha\Omega$-dynamos in galactic discs and stellar shells. In *Advances in Nonlinear Dynamos,* A. Ferriz-Mas & M. Núñez Eds., Taylor & Francis, London, 224–268 (2003).

Soward A.M. & Hollerbach R., Non-axisymmetric magnetohydrodynamic shear layers in a rotating spherical shell, J. Fluid Mech., **408**, 239–274 (2000).

Soward A.M. & Jones C.A., α^2-dynamos and Taylor's constraint, Geophys. Astrophys. Fluid. Dyn., **27**, 87–122 (1983a).

Soward A.M. & Jones C.A., The linear stability of the flow in the narrow gap between two concentric rotating spheres, *Q. J. Mech. Appl. Maths.*, **36**, 19–42 (1983b).

Spiegel E.A. & Weiss N.O., Magnetic Buoyancy and the Boussinesq Approximation, *Geophys. Astrophys. Fluid Dyn.*, **22**, 219–234 (1982).

Spitzer L., *Physical Processes in the Interstellar Medium*, Wiley (1978).

Spohn T. & Schubert G., Oceans in the icy Galilean satellites of Jupiter? *Icarus*, **161**, 456–467 (2003).

Sreenivasan B. & Jones C.A., The role of inertia in the evolution of spherical dynamos, Geophys. J. Int., **164** (2), 467–476 (2006).

St. Pierre M.G., The strong field branch of the Childress–Soward dynamo, in *Theory of solar and planetary dynamos*, M.R.E. Proctor *et al.*Eds., 295–302 (1993).

St. Pierre M.G., On the local turbulence in the Earth's core, *Geophys. Astrophys. Fluid Dyn.*, **83**, 293–306 (1996).

Stacey F.D., *Physics of the Earth, 3rd ed,* Brisbane, Brookfield (1992).

Starchenko S. & Jones C.A., Typical velocities and magnetic field strengths in planetary interiors, *Icarus*, **157**, 426–435 (2002).

Starchenko S.V. & Shukurov A., Observable parameters of spiral galaxies and galactic magnetic fields, *Astron. Astrophys.*, **214**, 47–60 (1989).

Steenbeck M., Kirko I.M., Gailitis A., Klyavinya A.P., Krause F., Laumanis I.Ya. & Lielausis O., Der experimentelle Nachweis einer elektromotorischen Kraft längs äuXXeren Magnetfeldes, induziert durch eine Strömung flüssigen Metalls (α–effect), *Sov. Phys. Dok.*, **13**, 443–445 (1968).

Steenbeck M., Krause F. & Rädler K.-H., A Calculation of the Mean Electromotive Force in an Electrically Conducting Fluid in Turbulent Motion Under the Influence of Coriolis Forces, *Z. Naturforsch*, **21a**, 369–376 (1966).

Steenbeck M. & Krause F., Zur Dynamotheorie stellarer und planetarer Magnetfelder I. Berechnung sonnenähnlicher Wechselfeldgeneratoren, *Astron. Nachr.*, **291**, 49–84 (1969).

Stefani F., Gerbeth G. & Gailitis A., Velocity profile optimisation in the Riga dynamo experiment, *Tranfer phenomena in Magnetohydrodynamic and electro-conducting flows*, Alémany *et al.*Eds., Kluwer Academic Press, 31–44 (1999).

Stepinski T.F. & Levy E.H., Generation of dynamo magnetic fields in protoplanetary and other astrophysical accretion disks, *Astrophys. J.*, **331**, 416–434 (1988).

Stepinski T.F. & Levy E.H., Dynamo magnetic field modes in thin astrophysical disks: an adiabatic computational approximation, *Astrophys. J.*, **379**, 343–355 (1991).

Stern D.P., A Millennium of Geomagnetism, *Rev. Geophys.*, **40**(3), 1-1–1-30 (2002).

Stevenson, D.J., Turbulent thermal convection in the presence of rotation and magnetic field: a heuristic theory, *Geophys. Astrophys. Fluid Dyn.*, **12**, 139–169 (1979).

Stevenson, D.J., Interiors of the giant planets, *Ann Rev. Earth Planet. Sci.*, **10**, 257–295,(1982a).

Stevenson, D.J., Reducing the non-axisymmetry of a planetary dynamo and an application to Saturn, *Geophys. Astrophys. Fluid Dyn.*, **21**, 113–127 (1982b).

Stevenson, D.J., Planetary magnetic fields, *Rep. Prog. Phys.*, **46**, 555–620 (1983).

Stevenson, D.J., Mercury's magnetic field: a thermoelectric dynamo?, *Earth Planet. Sci. Lett.*, **82**, 114–120 (1987).

Stevenson, D.J., Mars' core and magnetism, *Nature*, **412**, 214–219 (2001).

Stevenson D.J., Planetary magnetic fields, *Earth Planet. Sci. Lett.*, **208**, 1–11 (2003).

Stevenson D.J., Spohn, T. & Schubert, G., Magnetism and thermal evolution of the terrestrial planets, *Icarus*, **54**, 466–489 (1983).

Stewartson K., On almost rigid rotations, *J. Fluid Mech.*, **3**, 299–303 (1957).

Stewartson K., On almost rigid rotations, Part 2, *J. Fluid Mech.*, **26**, 131–144 (1966).

Stewartson K., *Homogeneous fluids in rotation. Section B: Waves*, in: Rotating Fluids in Geophysics (P.H. Roberts & A.M. Soward Eds.), 67–103, Academic Press, London (1978).

Stieglitz R. & Müller U., Experimental demonstration of a homogeneous two-scale dynamo, *Phys. Fluids*, **13**, 561–564 (2001).

Stieglitz R. & Müller U., Experimental demonstration of a homogeneous two-scale dynamo, *Magnetohydrodynamics*, **38**, 27–33 (2002).

Stix M., Two examples of penetrative convection, *Tellus*, **22**, 517–520 (1970).

Stix M., Differential rotation and the solar dynamo, *Astron. Astrophys.*, **47**, 243–254 (1976).

Stix M., The solar dynamo, *Geophys. Astrophys. Fluid Dyn.*, **62**, 211–228 (1991).

Stix M., *The Sun, an Introduction*, 2nd ed., Springer, Berlin (2002).

Stone J.M., Hawley J.F., Gammie C.F. & Balbus S.A., Three-dimensional magnetohydrodynamical simulations of vertically stratified accretion disks, *Astrophys. J.*, **463**, 656–673 (1996).

Strassmeier K.G., Doppler images of starspots, *Astron. Nachr.*, **323**, 309–316 (2002).

Strong A.W., Moskalenko I.V. & Reimer O., Diffuse continuum gamma rays from the Galaxy, *Astrophys. J.*, **537**, 763–784 (2000); Erratum: *Astrophys. J.*, **541**, 1109 (2000).

Stuiver M. & Braziunas T.F., Sun, ocean, climate and atmospheric $^{14}CO_2$, an evaluation of causal and spectral relationships, *Holocene*, **3**, 289–305 (1993).

Stuiver M., Reimer P.J., Bard E., Beck J.W., Burr G.S., Hughen K.A., Kromer B., McCormac G., Van der Plicht J. & Spurk M., INTCAL98 radiocarbon age calibration, 24,000-0 cal BP, *Radiocarbon*, **40**, 1041–1083 (1998).

Subramanian K., Can the turbulent galactic dynamo generate large-scale magnetic fields? *Mon. Not. R. Astron. Soc.*, **294**, 718–728 (1998).

Subramanian K., Unified treatment of small and large-scale dynamos in helical turbulence, *Phys. Rev. Lett.*, **83**, 2957–2960 (1999).

Subramanian K. & Mestel L., Galactic dynamos and density wave theory – II. An alternative treatment for strong non-axisymmetry, *Mon. Not. R. Astron. Soc.*, **265**, 649–654 (1993).

Sumita I. & Olson P., Laboratory experiments on high Rayleigh number thermal convection in a rapidly rotating spherical shell, *Phys. Earth Planet. Inter.*, **117**, 153–170 (2000).

Sun Z.-P., Schubert G. & Glatzmaier G.A., Transitions to chaotic thermal convection in a rapidly rotating spherical fluid shell, *Geophys. Astrophys. Fluid Dyn.*, **69**, 95–131 (1993).

Takehiro S.-I. & Lister J.R., Penetration of colomnnar convection into an outer stably stratified layer in rapidly rotating spherical fluid shells, *Earth Planet. Sci. Lett.*, **187**, 357–366 (2001).

Tavakol R., Covas E., Moss D. & Tworkowski A., Effects of boundary conditions on the dynamics of the solar convection zone, *Astron. Astrophys.*, **387**, 1100–1106 (2002).

Tayler R.J., *The Sun as a Star*, Cambridge University Press, Cambridge (1997).

Taylor J.B., The magnetohydrodynamics of a rotating fluid and the Earth's dynamo problem, Proc. R. Soc. Lond., A **274**, 274–283 (1963).

Taylor, G.I., Diffusion by continuous movements, *Proc. London Math. Soc.*, A **20**, 196–212 (1921).

Tenorio-Tagle G. & Bodenheimer P., Large-scale expanding superstructures in galaxies, *Ann. Rev. Astron. Astrophys.*, **26**, 145–197 (1988).

Thelen J.-C., A mean electromotive force induced by magnetic buoyancy instabilities, Mon. Not. R. Astron. Soc., **315**, 155–164 (2000a).

Thelen J.-C., Non-linear $\alpha\omega$–dynamos driven by magnetic buoyancy, *Mon. Not. R. Astron. Soc.*, **315**, 165–183 (2000b).

Thompson M.J., Christensen-Dalsgaard J., Miesch M.S. & Toomre J., The internal rotation of the Sun, *Annu. Rev. Astron. Astrophys.*, **41**, 599–643 (2003).

Tilgner A., Predictions on the behavior of the Karlsruhe dynamo, *Acta Atron. et Geophys. Univ. Commenianae*, **XIX**, 51–62 (1997).

Tilgner A. & Busse F.H., Finite amplitude convection in rotating spherical fluid shells, *J. Fluid Mech.*, **332**, 359–376 (1997).

Tilgner A. & Busse F.H., Saturation mechanism of a model of the Karlsruhe dynamo, in *Dynamo and dynamics, a mathematical challenge*, P. Chossat *et al.*Eds., 109–116, Kluwer Academic Publishers (2001).

Tilgner A. & Busse F.H., Simulation of the bifurcation diagram of the Karlsruhe dynamo, *Magnetohydrodynamics*, **38**, 35–40 (2002).

Tobias S.M., Grand minima in stellar dynamos, *Astron. Astrophys.*, **307**, L21–L24 (1996).

Tobias S.M., Properties of nonlinear dynamo waves, *Geophys. Astrophys. Fluid Dyn.*, **86**, 287–343 (1997a).

Tobias S.M., The solar cycle: parity interactions and amplitude modulation, *Astron. Astrophys.*, **322**, 1007–1017 (1997b).

Tobias S.M., Relating stellar magnetic cycle periods to dynamo calculations, *Mon. Not. R. Astron. Soc.*, **296**, 653–661 (1998).

Tobias S.M., The solar dynamo, *Phil. Trans. R. Soc. Lond.*, A **360**, 2741–2756 (2002a).

Tobias S.M., Modulation of solar and stellar dynamos, *Astron. Nachr.*, **323**, 417–423 (2002b).

Tobias S.M., The solar tachocline: formation, stability and its rôle in the solar dynamo, in *Fluid Dynamics and Dynamos in Astrophysics and Geophysics*, ed. A.M. Soward, C.A. Jones, D.W. Hughes & N.O. Weiss, Taylor & Francis, London (2004).

Tobias S.M., Brummell N.H., Clune T.L. & Toomre J., Pumping of magnetic fields by turbulent convection, *Astrophys. J.*, **502**, L177–L180 (1998a).

Tobias S.M., Brummell N.H., Clune T.L. & Toomre J., Transport and storage of magnetic field by overshooting turbulent compressible convection, *Astrophys. J.*, **549**, 1183–1203 (2001).

Tobias S.M. & Hughes D.W., The influence of velocity shear on magnetic buoyancy instability in the solar tachocline, *Astrophys. J.*, **603**, 785–802 (2004).

Tobias S.M., Hughes D.W. & Weiss N.O., Unpredictable Sun leaves researchers in the dark, *Nature*, **442**, 7098, 26 (2006).

Tobias S.M, Proctor M.R.E. & Knobloch E., Convective and absolute instabilities in fluid flows in finite geometries, *Physica*, D **113**, 43–72 (1998b).

Tobias S.M., Weiss N.O. & Kirk V., Chaotically modulated stellar dynamos, *Mon. Not. R. Astron. Soc.*, **273**, 1150–1166 (1995).

Tromp J., Inner-core anisotropy and rotation, *Ann. Rev. Earth Planet. Sci.*, **29**, 47–69 (2001).

Turcotte D.L., *Fractals and Chaos in Geology and Geophysics*, Cambridge University Press (1993).

Tworkowski A., Tavakol R., Brandenburg A., Brooke J., Moss D. & Tuominen I., Intermittent behaviour in axisymmetric mean-field dynamo models in spherical shells, *Mon. Not. R. Astron. Soc.*, **296**, 287–295 (1998).

Ulrich R.K., Boyden J.E., Webster L., Snodgrass H.B., Podilla S.P., Gilman P.A. & Shieber T., Solar rotation measurements at Mount Wilson. V. Reanalysis of 21 years of data, *Solar Phys.*, **117**, 291–328 (1988).

Vainshtein S.I. & Cattaneo F., Nonlinear restrictions on dynamo action, *Astrophys. J.*, **393**, 165–171 (1992).

Vainshtein S.I. & Ruzmaikin A.A., Generation of the large-scale galactic magnetic field, *Astron. Zh.*, **48**, 902–909 (1971) [*Sov. Astron.*, **15**, 714 (1972)].

Vainshtein S.I. & Ruzmaikin A.A., Generation of the large-scale galactic magnetic field. II, *Astron. Zh.*, **49**, 449–452 (1972) [*Sov. Astron.*, **16**, 365 (1972)].

Vainshtein S.I., Sagdeev R.Z., Rosner R. & Kim, E.-J., Fractal properties of the stretch–twist–fold magnetic dynamo, *Phys. Rev.*, E **53**, 4729–4744 (1996).

Vainshtein S.I. & Zeldovich Ya.B., Origin of magnetic fields in astrophysics, *Sov. Phys. Usp.*, **15**, 159–172 (1972).

Valet J.P., Meynadier L. & Guyodo Y., Geomagnetic dipole strength and reversal rate over the past two million years, *Nature*, **435**, 802–805 (2005).

Vaughan A.H. & Preston G.W., A survey of chromospheric Ca II H end K emission in field stars of the solar neighborhood, *Publ. Astron. Soc. Pac.*, **92**, 385–391 (1980).

Vempaty S. & Loper D., Hydromagnetic boundary layers in a rotating cylindrical container, *Phys. Fluids*, **18**, 1678–1686 (1975).

Vempaty S. & Loper D., Hydrodynamic free shear layers in rotating flows, *ZAMP*, **29**, 450–461 (1978).

Verhoogen J., Heat balance in the Earth's core, *Geophys. J.*, **4**, 276–281 (1961).

Verhoogen J., *E*nergetics of the Earth, National Academy Press Washington D.C. (1980).

Vočadlo L., Alfè D., Gillan M.J. & Price G.D., The properties of iron under core conditions from first principles calculations, *Phys. Earth Planet. Inter.*, **140**, 101–125 (2003).

Vorontsov S.V., Christensen-Dalsgaard J., Schou J., Strakhov G.N. & Thompson M.J., Helioseismic measurement of solar torsional oscillations, *Science*, **296**, 101–103 (2002).

Wagner G., Beer J., Masarik J., Kubik P.W., Mende W., Laj C., Raisbeck G.M. & Yiou F., Presence of the solar de Vries cycle ($\sim$205 years) during the last ice age, *Geophys. Res. Lett.*, **28**, 303–306 (2001).

Wakker B.P. & van Woerden H., High-velocity clouds, *Ann. Rev. Astron. Astrophys.*, **35**, 217–266 (1997).

Walker M.R., Barenghi C.F. & Jones C.A., A note on dynamo action at asymptotically small Ekman number, Geophys. Astrophys. Fluid. Dyn., **88**, 261–275 (1998).

Weiss N.O., Solar and stellar dynamos, in *Lectures on solar and planetary dynamos*, M.R.E. Proctor & A.D. Gilbert Eds., 59–95, Cambridge University Press (1994).

Weiss N.O., Cattaneo F. & Jones C.A., Periodic and aperiodic dynamo waves, Geophys. Astrophys. Fluid. Dyn., **30**, 305–341 (1984).

Weiss N.O. & Tobias S.M., Physical causes of solar activity, *Space. Sci. Rev.*, **94**, 99–112 (2000).

Westerburg M. & Busse F.H., Centrifugally driven convection in the rotating cylyndrical annulus with modulated boundaries, *Nonlin. Processes in Geophys.*, **10**, 275–280 (2003).

Wicht K., Inner-core conductivity in numerical dynamo simulations, Phys. Earth Planet. Inter., **132**, 281–302 (2002).

Widrow L.M., Origin of galactic and extragalactic magnetic fields, *Rev. Mod. Phys.*, **74**, 775–823 (2002).

Wielebinski R. & Krause F., Magnetic fields in galaxies, *Astron. Astrophys. Rev.*, **4**, 449–485 (1993).

Williams R.M., Chu Y.-H., Dickel J.R., Beyer R., Petre R., Smith R.C. & Milne D.K., Supernova remnants in the Magellanic Clouds. I. The colliding remnants DEM L316, *Astrophys. J.*, **480**, 618–632 (1997).

Willis A.P., Shukurov A., Soward A.M. & Sokoloff D., Nonlocal effects in the mean-field disc dynamo. II. Numerical and asymptotic solutions, Geophys. Astrophys. Fluid. Dyn., **98**, 537–554 (2004).

Wilmot-Smith A.L., Martens P.C.H., Nandy D., Priest E.R. & Tobias S.M., Low-order stellar dynamo models, *Mon. Not. R. Astron. Soc.*, **363**, 1167–1172 (2005).

Wilson P.R., *Solar and Stellar Magnetic Cycles*. Cambridge University Press (1994).

Wissink J.G., Hughes D.W., Matthews P.C. & Proctor M.R.E., The three-dimensional breakup of a magnetic layer, *Mon. Not. R. Astron. Soc.*, **318**, 501–510 (2000).

Witkowsky M. & Marty P., Effect of a rotating magnetic field of arbitrary frequency on a liquid metal column, *Eur. J. Mech.* B/*Fluids*, **17**, 239–254 (1998).

Woltjer L., Remarks on the Galactic magnetic field. In *Radio Astronomy and the Galactic System,* Proc. IAU Symp. 31, ed. H. van Woerden, Academic Press, London, 1967, 479–485 (1967).

Worledge D., Knobloch E., Tobias S.M. & Proctor M.R.E., Dynamo waves in semi-infinite and finite domains, *Proc. R. Soc. Lond.*, A **453**, 119–143 (1997).

Yano J.-I., Asymptotic theory of thermal convection in rapidly rotating system, *J. Fluid Mech.*, **243**, 103–131 (1992).

Yoder C.F., Konopliv A.S., Yuan D.N., Standish E.M. & Folkner W.M., Fluid core size of mars from detection of the solar tide, *Science*, **300**, 299–303 (2003).

Yoshimura H., Nonlinear astrophysical dynamos: multiple-period dynamo wave oscillations and long-term modulations of the 22 year solar cycle, *Astrophys. J.*, **226**, 706–719 (1978).

Yoshimura H., Solar cycle Lorentz force: waves and the torsional oscillations of the Sun, *Astrophys. J.*, **247**, 1102–1111 (1981).

Yoshizawa A., Itoh S.-I. & Itoh K., *Plasma and Fluid Turbulence, Theory and Modelling*, Bristol: Institute of Physics Publishing (2003).

Zeldovich Ya.B., The magnetic field in the two-dimensional motion of a conducting turbulent fluid, *Sov. Phys. JETP*, **4**, 460–462 (1957).

Zeldovich Ya.B., Ruzmaikin A.A., Sokoloff D.D., *Magnetic Fields in Astrophysics,* Gordon & Breach, New York (1983).

Zeldovich Ya.B., Ruzmaikin A.A. & Sokoloff D.D., *The Almighty Chance,* World Sci., Singapore (1990).

Zhang K., Convection in a rapidly rotating spherical fluid shell at infinite Prandtl number: steadily drifting rolls, *Phys. Earth Planet. Inter.*, **68**, 156–169 (1991).

Zhang K., Spiraling columnar convection in rapidly rotating spherical fluid shells, *J. Fluid Mech.*, **236**, 535–556 (1992a).

Zhang K., Convection in a rapidly rotating spherical shell at infinite Prandtl number: transition to vacillating flows, *Phys. Earth. Planet. Inter.*, **72**, 236–248 (1992b).

Zhang, K., On coupling between the Poincaré equation and the heat equation, *J. Fluid Mech.*, **268**, 211–229 (1994).

Zhang K., On coupling between the Poincaré equation and the heat equation: non-slip boundary conditions, *J. Fluid Mech.*, **284**, 239–256 (1995).

Zhang K. & Busse F.H., On the onset of convection in rotating spherical shells, *Geophys. Astrophys. Fluid Dyn.*, **39**, 119-147 (1987).

Zhang K. & Busse F.H., Convection driven magnetohydrodynamic dynamos in rotating spherical shells, Geophys. Astrophys. Fluid. Dyn., **49**, 97–116 (1989).

Zhang K. & Busse F.H., Generation of magnetic fields by convection in a rotating spherical fluid shell of infinite Prandtl number, Phys. Earth Planet. Inter., **59**, 208–222 (1990).

Zhang K., Earnshaw P., Liao X. & Busse F.H., On inertial waves in a rotating fluid sphere, *J. Fluid Mech.*, **437**, 103–109 (2001).

Zhang K. & Gubbins D., Is the geodynamo process intrinsically unstable?, Geophys. J. Int., **140**, F1–F4 (2000).

Zhang K. & Jones C.A., On small Roberts number magnetoconvection in rapidly rotating systems, Proc. R. Soc. Lond., A **452**, 981–995 (1996).

Zhang K. & Jones C.A., The effect of hyperviscosity on geodynamo models, Geophys. Res. Lett., **24**, 2869–2872 (1997).

Zhang K., Jones C.A. & Sarson G.R., The dynamical effects of hyperviscosity on geodynamo models, *Studia Geophys. et Geod.*, **42**, 247–253 (1998).

Zhang K. & Schubert G., Linear penetrative spherical rotating convection, *J. Atmos. Sci.*, **54**, 2509–2518 (1997).

Zhang K. & Schubert G., From Penetrative Convection to Teleconvection, *Astrophys. J.*, **572**, 461–470 (2002).

Reference Index

Subject Index